Elementary Real and Complex Analysis

Georgi E. Shilov

TRANSLATED AND EDITED BY

Richard A. Silverman

REVISED ENGLISH EDITION

DOVER PUBLICATIONS, INC.
New York

Bibliographical Note

This Dover edition, first published in 1996, is an unabridged, corrected republication of the work first published in English by The MIT Press, Cambridge, Massachusetts, 1973, as Volume 1 of the two-volume course "Mathematical Analysis."

Library of Congress Cataloging-in-Publication Data

Shilov, G. E. (Georgiĭ Evgen'evich)
    [Matematicheskiĭ analiz. Chasti 1–2. English]
    Elementary real and complex analysis / Georgi E. Shilov ; revised English edition translated and edited by Richard A. Silverman.
       p.   cm.
    Originally published in English: Cambridge, Mass. : MIT Press, 1973.
    Includes index.
    ISBN-13: 978-0-486-68922-7 (pbk.)
    ISBN-10: 0-486-68922-0 (pbk.)
    1. Mathematical analysis.   I. Silverman, Richard A. II. Title.
QA300.S4552   1996
515—dc20   95-37030
                                                CIP

Manufactured in the United States by Courier Corporation
68922006
www.doverpublications.com

Contents

4 Limits

5 Continuous Functions

6 Series

7 The Derivative

8 Higher Derivatives

9 The Integral

10 Analytic Functions

Preface

It was with great delight that I learned of the imminent publication of an English-language edition of my introductory course on mathematical analysis under the editorship of Dr. R. A. Silverman. Since the literature already includes many fine books devoted to the same general subject matter, I would like to take this opportunity to point out the special features of my approach.

Mathematical analysis is a large "continent" concerned with the concepts of function, derivative, and integral. At present this continent consists of many "countries" such as differential equations (ordinary and partial), integral equations, functions of a complex variable, differential geometry, calculus of variations, etc. But even though the subject matter of mathematical analysis can be regarded as well-established, notable changes in its structure are still under way. In Goursat's classical "cours d'analyse" of the twenties all of analysis is portrayed on a kind of "great plain," on a single level of abstraction. In the books of our day, however, much attention is paid to the appearance in analysis of various "stages" of abstraction, i.e., to various "structures" (Bourbaki's term) characterizing the mathematico-logical foundations of the original constructions. This emphasis on foundations clarifies the gist of the ideas involved, thereby freeing mathematics from concern with the idiosyncracies of each object under consideration. At the same time, an understanding of the nub of the matter allows one to take account immediately of new objects of a different individual nature but of exactly the same "structural depth."

Consider, for example, Picard's proof of the existence and uniqueness of the solution of a differential equation in which the desired function is successively approximated on a given interval by other functions in accordance with certain rules. This proof had been known for some time when Banach and others formulated the "fixed point method." The latter plainly reveals the nub of Picard's proof, namely the presence of a contraction operator in a certain metric space. In this regard, the specific context of Picard's problem, i.e., numerical functions on an interval, a differential equation, etc., turns out to be quite irrelevant. As a result, the fixed point method not only makes the "geometrical" proof of Picard's theorem more transparent, but, by further developing the key idea of Picard's proof, even leads to the proof of existence theorems involving neither functions on an interval nor a differential equation. Considerations of the same kind apply equally well to the geometry of Hilbert space, the study of differentiable functionals, and many other topics.

Analysis presented from this point of view can be found, for example, in the superb books by J. Dieudonné. However it seems to me that Dieudonné's books, for all their formal perfection, require that the reader's "mathemati-

cal I. Q." be too high. Thus, for my part, I have tried to accomodate the
interests of a larger population of those concerned with mathematics. Therefore in many cases where Dieudonné instantly and almost miraculously
produces deep classical results from general considerations, so that the
reader can only take off his hat in silent admiration, the reader of my course
is invited to climb with me from the foothills of elementary topics to successive levels of abstraction and then look down from above on the various
valleys which now come into his field of view. Perhaps this approach is
thornier, but in any event the mathematical traveler will thereby acquire
the training needed for further exploration on his own.

The present course begins with a systematic study of the real numbers,
understood to be a set of objects satisfying certain definite axioms. There
are other approaches to the theory of real numbers where things I take as
axioms are proved, starting from set theory and the axioms for the natural
numbers (for example, a rigorous treatment in this vein can be found in
Landau's famous course). Both treatments have a key deficiency, namely
the absence of a proof of the compatibility of the axioms. Evidently modern
mathematics lacks a construction of the real numbers which is free of this
shortcoming. The whole question, far from being a mere technicality, involves the very foundations of mathematical thought. In any event, this
being the case, it is really not very important where one starts a general
treatment of analysis, and my choice is governed by the consideration that
the starting point bear as close a resemblance as possible to analytic constructions proper.

The concepts of a mathematical structure and an isomorphism are introduced in Chapter 2, after a brief digression on set theory, and a proof of
the uniqueness of the structure of real numbers (to within an isomorphism) is given as an illustration. Two other structures are then introduced,
namely n-dimensional space and the field of complex numbers. After a
detailed treatment of metric spaces in Chapter 3, a general theory of limits
is developed in Chapter 4. The starting point of this theory is taken to be
on the one hand a set E equipped with a "direction," i.e., a system of subsets
of E with an empty intersection (this notion, closely related to the "filters"
of H. Cartan, is more restricted than that of a filter but is entirely adequate
for the purposes of analysis), and on the other hand a function defined on
E taking values in a metric space. All the limits considered in analysis, from
limits of a numerical sequence to the notions of the derivative and integral,
are comprised in this scheme. Chapter 5 is concerned first with some theorems on continuous numerical functions on the real line and then with the
use of functional equations to introduce the logarithm (from which the
exponential is obtained by inversion) and the trigonometric functions. The

algebra and topology of complex numbers and the fundamental theorem of algebra are presented as applications. Chapter 6 is on infinite series, dealing not only with numerical series but also with series whose terms are vectors and functions (including power series). Chapters 7 and 8 treat differential calculus proper, with Taylor's series leading to a natural extension of real analysis into the complex domain. Chapter 9 presents the general theory of Riemann integration, together with a number of its applications. The further development of analysis requires the technique of analytic functions, which is considered in detail in Chapter 10. Finally Chapter 11 is devoted to improper integrals, and makes full use of the technique of analytic functions now at our disposal.

Each chapter is equipped with a set of problems; hints and answers to most of these problems appear at the end of the book. To a certain extent, the problems help to develop necessary technical skill, but they are primarily intended to illustrate and amplify the material in the text.

G. E. S.

1 Real Numbers

1.1. Set-Theoretic Preliminaries

1.11. Words like "aggregate," "collection," and "set" come up at once when talking about "objects" (or "elements") of any kind. Thus one can talk about the set of students in an auditorium, the set of grains of sand on a beach, the set of vertices or the set of sides of a polygon, and so on. In each of these examples the set in question consists of a definite number of elements, which can be estimated within certain limits, even though it may be difficult in practice to find the number of elements exactly.† Such sets are said to be *finite*.

In mathematics one must often deal with sets consisting of a number of objects which is not finite. The simplest examples of such sets are the set 1,2,3,... of all natural numbers (positive integers) and the set of all the points on a line segment (precise definitions of these objects will be given later). Such sets are said to be *infinite*. To the category of sets we also assign the *empty set*, namely, the set containing no elements of all.

As a rule, sets will be denoted by large letters $A,B,C,...$ and elements of sets will be denoted by small letters. By $a \in A$ (or $A \ni a$) we mean that a is an element of the set A, while by $a \notin A$ (or $a \bar{\in} A$) we mean that a is not an element of the set A. By $A \subset B$ (or $B \supset A$) we mean that every element of the set A is an element of the set B; the set A is then said to be a *subset* of the set B. The largest subset of the set B is obviously the set B itself, while the smallest subset of B is the empty set. Any other subset of the set B, containing some but not all elements of B, is called a *proper subset* of B. The symbols $\in, \ni, \subset, \supset$ are called *inclusion relations*. Suppose both $A \subset B$ and $B \subset A$. Then every element of the set A is an element of the set B, and conversely, every element of the set B is an element of the set A. It follows that the sets A and B consist of precisely the same elements and hence coincide, a fact expressed by writing $A = B$. The analogous formula for elements, namely $a = b$, simply means that a and b are one and the same element.

Sets can be specified in various ways. The simplest way is to write the elements of the set explicitly between curly brackets; e.g., $A = \{1, 2, 3, ...\}$ is the set of all natural numbers (positive integers). Another way is to specify some property of the elements of the set; e.g., $A = \{x : x^2 - 1 < 0\}$ is the set of all x satisfying the inequality $x^2 - 1 < 0$ written after the colon.

1.12. Unions and intersections. Let $A,B,C,...$ be given sets. Then the set

---

† "Some people, oh king Hieron, think that the number of grains of sand in the world is infinitely large. However, I can give you a convincing proof of my ability to name certain numbers larger than the number of grains of sand in a pile as large as the Earth itself." (Archimedes)

of all elements of $A,B,C,\ldots$ belonging to *at least one* of the sets $A,B,C,\ldots$ is called the *union* of the sets $A,B,C,\ldots$, while the set of all elements of $A,B,C,\ldots$ belonging to *every one* of the sets $A,B,C,\ldots$ is called the *intersection* of the sets $A,B,C,\ldots$. For example, let

$$A=\{6,7,8,\ldots\}, \quad B=\{3,6,9,\ldots\},$$

i.e., let A be the set of all natural numbers greater than 5 and B the set of all natural numbers divisible by 3. Then the union of A and B is the set

$$S=\{3,6,7,8,9,10,\ldots\},$$

the set of all natural numbers except 1,2,4, and 5, while the intersection of A and B is the set

$$D=\{6,9,12\ldots\},$$

the set of all natural numbers divisible by 3 except 3 itself. In the case where the sets $A,B,C,\ldots$ have no elements in common, the intersection of $A,B,C,\ldots$ is the empty set, and the sets $A,B,C,\ldots$ are then said to be *nonintersecting*. For example, the three sets

$$A=\{1,2\}, \quad B=\{2,3\}, \quad C=\{1,3\}$$

are nonintersecting, even though any *two* of the sets share a common element.

We can consider the union and intersection of both a finite number of sets and an infinite number of sets. For example, the union of the sets of points on all (infinitely many) lines in the plane passing through a given point O is clearly the set of all points in the plane, while the intersection of all these sets consists of the single point O.

To denote the union S of given sets, say $A_1, A_2, \ldots, A_\nu, \ldots$, we use the symbols

$$\sum \text{ or } \bigcup,$$

writing

$$S = \sum_{\nu=1}^{\infty} A_\nu \quad \text{or} \quad S = \bigcup_{\nu=1}^{\infty} A_\nu,$$

while to denote the intersection D of the sets, we use the symbols

$$\prod \text{ or } \bigcap,$$

writing

$$D = \prod_{\nu=1}^{\infty} A_\nu \quad \text{or} \quad D = \bigcap_{\nu=1}^{\infty} A_\nu.$$

1.2. Axioms for the Real Number System

The following considerations stem from the simplest properties of numbers, known partly from everyday experience and partly from elementary mathematics.† Rather than define the real numbers separately, we will define the whole set of real numbers at once as a set of elements equipped with certain operations and relations which in turn satisfy four groups of axioms. The first group consists of the *addition axioms*, the second of the *multiplication axioms*, the third of the *order axioms*, and the fourth of a single axiom called the *least upper bound axiom*.

Definition. By the *real number system* R is meant the set whose elements $x, y, z, \ldots$, called *real numbers*, satisfy the four groups of axioms given in Secs. 1.21–1.24. The set R is often called the *real line*, with its elements in turn called *points*.

1.21. The addition axioms. To every pair of elements x and y in R there corresponds a (unique) element $x + y$, called the *sum* of x and y, where the rule associating $x + y$ with x and y has the following properties:

(a) $x + y = y + x$ for every x and y in R (*addition is commutative*);
(b) $(x + y) + z = x + (y + z)$ for every x, y, z in R (*addition is associative*);‡
(c) R contains an element 0, called the *zero element*, such that $x + 0 = x$ for every x in R;
(d) For every x in R there exists an element y in R, called the *negative* of x, such that $x + y = 0$.

1.22. The multiplication axioms. To every pair of elements x and y in R there corresponds a (unique) element $x \cdot y$ (or xy), called the *product* of x and

---

† It is stretching things quite a bit to say that the least upper bound axiom (Sec. 1.24) is known from "everyday experience." But the same is true of Euclid's axiom on the existence of a unique line through a given point parallel to a given line (a key axiom of ordinary plane geometry). Experience does not present us with uniquely determined sets of mathematical axioms. Rather, between experience and any scientific theory there is an intervening stage of formulating appropriate axioms, which may vary greatly within the context of one and the same experimental data. Thus besides Euclidean geometry we have non-Euclidean (Lobachevskian) geometry in which there are many lines through a given point parallel to a given line. By the same token, there are other theories of the real numbers besides the one given here, in which a set bounded from above can fail to have a least upper bound. See V. A. Uspensky, *Lectures on Computable Functions* (in Russian), Moscow (1960), Sec. 12.
‡ Thus the expression $x + y + z$ has a unique meaning.

y, where the rule associating xy with x and y has the following properties:
(a) $xy = yx$ for every x and y in R (*multiplication is commutative*);
(b) $(xy)z = x(yz)$ for every x, y, z in R (*multiplication is associative*);†
(c) R contains an element $1 \neq 0$, called the *unit element*, such that $1 \cdot x = x$ for every x in R;
(d) For every $x \neq 0$ in R there exists an element u in R, called the *reciprocal* of x, such that $xu = 1$;
(e) The formula

$$x(y + z) = xy + xz$$

holds for every x, y, z in R (*multiplication is distributive over addition*).

The last axiom connects the operation of multiplication with the operation of addition introduced in Sec. 1.21.

A set of objects $x, y, z, \ldots$ satisfying the axioms of Secs. 1.21 and 1.22 is called a *number field* or simply a *field*.

1.23. The order axioms. For every pair of elements x and y in R one (or both) of the relations $x \leqslant y$ (x *is less than or equal to* y) or $y \leqslant x$ holds, where $\leqslant$ has the following properties:
(a) $x \leqslant x$ for every x in R, and $x \leqslant y$, $y \leqslant x$ together imply $x = y$;
(b) If $x \leqslant y$, $y \leqslant z$, then $x \leqslant z$.
(c) If $x \leqslant y$, then $x + z \leqslant y + z$ for every z in R;
(d) If $0 \leqslant x$, $0 \leqslant y$, then $0 \leqslant xy$.

The relation $x \leqslant y$ can also be written in the form $y \geqslant x$ (y *is greater than or equal to* x). If $x \leqslant y$ and $x \neq y$, we write $x < y$ (x *is less than* y) or $y > x$ (y *is greater than* x).

1.24. A set $E \subset R$ is said to be *bounded from above* if there exists an element $z \in R$ such that $x \leqslant z$ for every $x \in E$, a fact expressed concisely by writing $E \leqslant z$. Every number z with the above property relative to a set E is called an *upper bound* of E. An upper bound z_0 of the set E is called the *least upper bound* of E if every other upper bound z of E is greater than or equal to z_0 (why is z_0 unique?). The least upper bound of E is denoted by sup E (from the Latin "supremum"). We can now state the following

Least upper bound axiom. *Every set $E \subset R$ which is bounded from above has a least upper bound.*

1.3. Consequences of the Addition Axioms

Our next task is to deduce various implications of the above axioms which

† Thus the expression xyz has a unique meaning.

will be needed later. We start with certain consequences of the addition axioms.

1.31. THEOREM. *The system R contains a unique zero element.*

Proof. Suppose R contains two zero elements 0_1 and 0_2. Then it follows from Axioms a and c of Sec. 1.21 that

$$0_2 = 0_2 + 0_1 = 0_1 + 0_2 = 0_1. \blacksquare†$$

1.32. THEOREM. *Every element x in R has a unique negative.*

Proof. Suppose x has two negatives y_1 and y_2, so that $x + y_1 = x + y_2 = 0$. Then it follows from Axioms a–c of Sec. 1.21 that

$$y_2 = y_2 + 0 = y_2 + (x + y_1) = (y_2 + x) + y_1$$
$$= y_1 + (x + y_2) = y_1 + 0 = y_1. \blacksquare$$

The negative of the element x is denoted by $-x$. The sum $x + (-y)$, written more concisely as $x - y$, is called the *difference* of x and y. *The negative* $-(x + y)$ *of the sum* $x + y$ *is the sum of the negatives of x and y*, since

$$x + y - x - y = x - x + y - y = 0 + 0 = 0.$$

1.33. THEOREM. *The equation*

$$a + x = b \tag{1}$$

has a unique solution in R, equal to b − a.

Proof. Adding the number $-a$ to both sides of (1) and using Axioms a–c of Sec. 1.21, we find that

$$a + x - a = x + a - a = x + 0 = x = b - a,$$

so that the solution, if it exists, equals $b - a$. But $b - a$ is a solution, since

$$a + (b - a) = a + b + (-a) = b + a + (-a) = b + 0 = b. \blacksquare$$

1.4. Consequences of the Multiplication Axioms

1.41. a. THEOREM. *The system R contains a unique unit element.*

Proof. Suppose R contains two unit elements 1_1 and 1_2. Then it follows from Axiom a of Sec. 1.22 that

$$1_2 = 1_1 \cdot 1_2 = 1_2 \cdot 1_1 = 1_1. \blacksquare$$

† The symbol $\blacksquare$ means Q.E.D. and indicates the end of a proof.

b. THEOREM. *Every element $x \neq 0$ in R has a unique reciprocal.*

Proof. Suppose x has two reciprocals u_1 and u_2, so that $xu_1 = xu_2 = 1$. Then it follows from Axioms a–c of Sec. 1.22 that

$$u_2 = 1 \cdot u_2 = (xu_1)u_2 = x(u_1 u_2) = x(u_2 u_1)$$
$$= (xu_2)u_1 = 1 \cdot u_1 = u_1. \quad \blacksquare$$

1.42. The reciprocal of the element x is denoted by $1/x$. *The reciprocal $1/xy$ of the product xy is the product of the reciprocals of x and y*, since

$$xy\frac{1}{x}\frac{1}{y} = x\frac{1}{x}y\frac{1}{y} = 1 \cdot 1 = 1.$$

The product $x \cdot 1/z$, written more concisely as x/z, is called the *quotient* (or *ratio*) of x and z.

1.43. Definition. The numbers

$$1, 2 = 1+1, 3 = 2+1, \dots, n = (n-1)+1, \dots$$

are called *natural numbers* (or *positive integers*). Thus the set of natural numbers can be defined as the smallest numerical set containing the number 1 and containing the number $n+1$ whenever it contains the natural number n.

In many problems one must show that some numerical set A (e.g., the set of all natural numbers n for which some property T_n, depending on n, is valid) contains *all* natural numbers. The *method of mathematical induction*, used in such problems, consists in verifying that
(1) A contains 1;
(2) If A contains a natural number n, then A also contains $n+1$.
If these two conditions are satisfied, it is clear from the foregoing that A contains all natural numbers. Thus the method of mathematical induction is a consequence of the very definition of the natural numbers.

1.44. a. By the *integers* we mean the natural numbers together with their negatives and the number zero.

b. Let m be an integer. Then the integer $2m$ is said to be *even*, while the integer $2m+1$ is said to be *odd*.

c. By the *rational numbers* we mean all quotients of the form m/n, where m and n are integers and $n \neq 0$.

d. All other real numbers are said to be *irrational*.

1.45. THEOREM. *The equation*

$$ax = b \qquad (a \neq 0) \qquad (1)$$

has a unique solution in R, equal to b/a.

Proof. Dividing both sides of (1) by a and using Axioms a–c of Sec. 1.31, we find that

$$\frac{1}{a}(ax) = \left(\frac{1}{a}a\right)x = 1 \cdot x = x = \frac{b}{a},$$

so that the solution, if it exists, equals b/a. But b/a is a solution, since

$$a\frac{b}{a} = ab\frac{1}{a} = \left(a\frac{1}{a}\right)b = 1 \cdot b = b. \ \blacksquare$$

1.46. By definition,

$$x^n = \underbrace{x \cdot x \cdots x}_{n \text{ times}} \qquad (n = 1, 2, \ldots),$$

and hence obviously

$$x^m x^n = x^{m+n}, \quad (x^m)^n = x^{mn} \qquad (2)$$

for arbitrary $m, n = 1, 2, \ldots$. The expression x^n is called the "nth power of x." To define x^n for arbitrary integers, we set

$$x^0 = 1, \quad x^{-n} = \frac{1}{x^n}$$

for arbitrary $x \neq 0$. We now verify that the formulas (2) continue to hold for arbitrary integers m and n. Suppose first that $m > 0$, $n = -q < 0$. Then

$$x^m x^n = x^m \frac{1}{x^q} = x^{m-q} x^q \frac{1}{x^q} = x^{m-q} = x^{m+n}$$

if $q \leqslant m$, while

$$x^m x^n = x^m \frac{1}{x^q} = x^m \frac{1}{x^m x^{q-m}} = x^m \frac{1}{x^m} \frac{1}{x^{q-m}} = \frac{1}{x^{q-m}} = x^{m+n}$$

if $q > n$, by Sec. 1.42. On the other hand, if $m = -p < 0$, $n = -q < 0$, then

$$x^m x^n = \frac{1}{x^p} \frac{1}{x^q} = \frac{1}{x^{p+q}} = x^{m+n},$$

again by Sec. 1.42. The second of the formulas (2) is proved similarly.

1.47. a. THEOREM. *The formula*

$$0 \cdot x = 0$$

holds for every x in R.

Proof. Clearly

$$0 \cdot x + 1 \cdot x = (0+1)x = 1 \cdot x = x,$$
$$0 \cdot x + 1 \cdot x = 0 \cdot x + x,$$

and hence

$$x = 0 \cdot x + x.$$

But then

$$0 \cdot x = x - x = 0,$$

by Theorem 1.33. ∎†

It follows that 0 has no inverse, since the equation $0 \cdot x = 1$ is impossible. This justifies the high school rule: "Don't divide by zero."

b. On the other hand, if $xy = 0$ and $x \neq 0$, then

$$y = \left(\frac{1}{x} x\right) y = \frac{1}{x}(xy) = \frac{1}{x} \cdot 0 = 0.$$

Thus, *if a product vanishes, so does at least one of its factors.*

1.48. THEOREM. *If* $u \neq 0$, $v \neq 0$, *then*

$$\frac{x}{u} + \frac{y}{v} = \frac{xv + yu}{uv}.$$

Proof. Merely note that

$$\frac{xv + yu}{uv} = \frac{1}{uv}(xv + yu) = \frac{1}{uv}xv + \frac{1}{uv}yu$$

$$= \left(\frac{1}{u}x\right)\left(\frac{1}{v}v\right) + \left(\frac{1}{v}y\right)\left(\frac{1}{u}u\right)$$

$$= \left(\frac{1}{u}x\right) \cdot 1 + \left(\frac{1}{v}y\right) \cdot 1 = \frac{x}{u} + \frac{y}{v}. \quad \blacksquare$$

1.49. THEOREM. *The formula*

$$-x = (-1)x \tag{3}$$

holds for every x in R.

---

† Theorem 1.33 refers to the (unique) theorem in Sec. 1.33, Lemma 1.51a to the lemma in Sec. 1.51a, etc.

Proof.† By Theorem 1.47a,

$$(-1)x + x = [(-1) + 1]x = 0 \cdot x = 0,$$

which implies (3). ∎

The results proved in Secs. 1.3 and 1.4 guarantee that the real numbers satisfy all the familiar identities of elementary algebra (like the binomial theorem, the formulas for the sums of arithmetic and geometric progressions, and various formulas involving determinants).

1.5. Consequences of the Order Axioms

1.51. Order and addition

a. LEMMA. *If $x \leqslant y$, $y \leqslant z$, and $x = z$, then $x = y = z$.*

Proof. Since $y \leqslant z = x$, we have $y \leqslant x$ and hence $y = x$ by Axiom a of Sec. 1.23. ∎

b. LEMMA.‡ *If $x < y$, $y \leqslant z$, then $x < z$. Similarly, if $x \leqslant y$, $y < z$, then $x < z$.*

Proof. An immediate consequence of Lemma 1.51a. ∎

c. THEOREM. *The following inequalities are equivalent:*

$$x \leqslant y, \quad 0 \leqslant y - x, \quad -y \leqslant -x, \quad x - y \leqslant 0.$$

Proof. Adding $-x$ to both sides of the first inequality and invoking Axiom c of Sec. 1.23, we get the second inequality. Similarly, adding $-y$ to both sides of the second inequality, we get the third, adding x to both sides of the third inequality, we get the fourth, and, finally, adding y to both sides of the fourth inequality, we get back the first. ∎

d. LEMMA. *If $x < y$, then $x + z < y + z$ for every z in R.*

Proof. Clearly $x < y$ implies $x \leqslant y$ and hence $x + z \leqslant y + z$. But if $x + z = y + z$, then, adding $-z$ to both sides, we get $x = y$ which contradicts $x < y$. It follows that $x + z < y + z$. ∎

e. THEOREM. *If $x_1 \leqslant y_1, \ldots, x_n \leqslant y_n$, then*

$$x_1 + \cdots + x_n \leqslant y_1 + \cdots + y_n,$$

where

$$x_1 + \cdots + x_n < y_1 + \cdots + y_n$$

if $x_j < y_j$ for at least one pair x_j, y_j.

---

† A proof is necessary, since each side of (3) is defined independently.
‡ Note the sense in which Lemmas 1.51b and 1.51d strengthen Axioms b and c of Sec. 1.23.

Proof. By Axiom c of Sec. 1.23,

$$x_1 + \cdots + x_n \leqslant y_1 + x_2 + \cdots + x_n$$
$$\leqslant y_1 + y_2 + \cdots + x_n \leqslant \cdots \leqslant y_1 + \cdots + y_n,$$

where, by the Lemma 1.51d, if $x_j < y_j$ for at least one pair x_j, y_j, then the sign $<$ appears instead of $\leqslant$ in the appropriate place and in all subsequent places as well, by the same lemma. ∎

Thus inequalities "going in the same direction" can be added. In particular, $x_1 \leqslant 0, \ldots, x_n \leqslant 0$ implies $s = x_1 + \cdots + x_n \leqslant 0$, where $s < 0$ if $x_j < 0$ for at least one j. A similar result holds if $\leqslant$ is replaced by $\geqslant$ and $<$ by $>$.

f. THEOREM. *The following inequalities are equivalent:*

$$x < y, \quad 0 < y - x, \quad -y < -x, \quad x - y < 0.$$

Proof. This follows from Lemma 1.51d, in the same way that Theorem 1.51c follows from Axiom c of Sec. 1.23. ∎

1.52. Definition. A real number x is said to be *nonnegative* if $x \geqslant 0$, *positive* if $x > 0$, *nonpositive* if $x \leqslant 0$, and *negative* if $x < 0$.

Thus the number 0 is simultaneously nonnegative and nonpositive.

1.53. Definition. Given two real numbers x and y, suppose $x \leqslant y$, say. Then x is called the *minimum* of the numbers x and y, and we write $x = \min\{x, y\}$. By the same token, y is called the *maximum* of the numbers x and y, and we write $y = \max\{x, y\}$. Using induction, we can define $\min\{x_1, \ldots, x_n\}$ and $\max\{x_1, \ldots, x_n\}$ for any finite set of numbers $x_1, \ldots, x_n$, for example, by setting

$$\max\{x_1, \ldots, x_n\} = \max\{\max\{x_1, \ldots, x_{n-1}\}, x_n\}.$$

The number

$$|x| = \max\{x, -x\}$$

is called the *absolute value* or *modulus* of the number x. Thus $|x| = x$ if $x \geqslant 0$, while $|x| = -x$ if $x \leqslant 0$. The number $|x|$ is nonnegative for every x, and $|-x| = |x|$.

1.54. a. THEOREM. *If $a > 0$, the inequality $|x| \leqslant a$ is equivalent to the inequalities*

$$x \leqslant a, \quad -x \leqslant a. \tag{1}$$

Proof. Obviously $x \leqslant a$ if $x \leqslant 0$, while $-x \leqslant a$ if $x \geqslant 0$. Moreover, $x = |x| \leqslant a$ if $x \geqslant 0$, while $-x = |x| \leqslant a$ if $x \leqslant 0$. ∎

Noting that $-x \leqslant a$ is equivalent to $-a \leqslant x$ (by Theorem 1.51c), we can write (1) in the form

$-a \leqslant x \leqslant a.$ $\hfill (1')$

b. THEOREM. *The inequality*

$$|x+y| \leqslant |x| + |y| \hfill (2)$$

holds for arbitrary real numbers x and y.

Proof. If x and y are both nonnegative or both nonpositive, the inequality follows at once from the definition of the absolute value. If, say, $x \geqslant 0, y \leqslant 0$, then

$$x+y \leqslant x \leqslant x+|y| = |x|+|y|,$$
$$-x-y \leqslant -y = |y| \leqslant |x|+|y|,$$

so that

$$|x+y| = \max\{x+y, -x-y\} \leqslant |x|+|y|. \hspace{0.5em} \blacksquare$$

c. Applying induction to (2), we get

$$|x_1 + \cdots + x_n| \leqslant |x_1| + \cdots + |x_n|.$$

1.55. Order and multiplication

a. LEMMA. *If $x > 0, y > 0$, then $xy > 0$.*

Proof. Use Axiom d of Sec. 1.23, together with Sec. 1.47b. $\hspace{0.5em} \blacksquare$

b. THEOREM. *If $x \leqslant y, z > 0$, then $xz \leqslant yz$.*

Proof. We need only note that

$$yz - xz = (y-x)z \geqslant 0,$$

by Axiom d of Sec. 1.23. $\hspace{0.5em} \blacksquare$

c. It follows from Sec. 1.47b that $\leqslant$ can be changed to $<$ in both places in Theorem 1.55b.

d. In particular, $x^2 < x$ if $0 < x < 1$, while $x^2 > x$ if $x > 1$.

e. If $0 < x \leqslant y, 0 < z \leqslant u$, then $xz \leqslant yz \leqslant yu$, so that inequalities can be multiplied under these conditions.

f. In particular, *if $0 < x < y$, then $x^2 < y^2, \ldots, x^n < y^n$.*

g. THEOREM. *If $x \leqslant 0, y \geqslant 0$, then $xy \leqslant 0$, while if $x \leqslant 0, y \leqslant 0$, then $xy \geqslant 0$.*

Proof. In the first case $-x \geqslant 0$. Hence, by Axiom d of Sec. 1.23 and Theorem 1.49,

$$(-x)y = (-1)xy = -(xy) \geqslant 0,$$

which implies $xy \leqslant 0$. In the second case $-y \geqslant 0$, and hence $-xy \leqslant 0$, $xy \geqslant 0$, by the first result. ∎

Theorem 1.55g remains true if the signs $\leqslant$ and $\geqslant$ are replaced everywhere by $<$ and $>$.

h. In particular, $x^2 = x \cdot x > 0$ for all $x \neq 0$. It follows that $1 = 1 \cdot 1 > 0$ and then from Lemma 1.51d that $2 = 1 + 1 > 1 + 0 = 1$, $3 = 2 + 1 > 2$, etc.

i. THEOREM. *The formula* $|xy| = |x||y|$ *holds for all x and y.*

Proof. If $x \geqslant 0$, $y \geqslant 0$, then $|x| = x$, $|y| = y$ and $xy \geqslant 0$, by Axiom d of Sec. 1.23, so that $|xy| = xy = |x||y|$. Similarly, if $x \leqslant 0$, $y \leqslant 0$, then $|x| = -x$, $|y| = -y$ and $xy \leqslant 0$, by Theorem 1.55g, so that $|xy| = xy = (-x)(-y) = |x||y|$. If $x \geqslant 0$, $y \leqslant 0$, then $|x| = x$, $|y| = -y$ and $xy \leqslant 0$, by Theorem 1.55g again, so that $|xy| = -xy = x(-y) = |x||y|$, and similarly for the case $x \leqslant 0$, $y \geqslant 0$. ∎

1.56. THEOREM. *If $x > 0$, then $1/x > 0$. Moreover $0 < x < y$ implies*

$$0 < \frac{1}{y} < \frac{1}{x}.$$

Proof. The first assertion follows from

$$x\frac{1}{x} = 1 > 0$$

and Theorem 1.55g (in its stronger form). To prove the second assertion, multiply $0 < x < y$ by $1/xy$. ∎

In particular, all rational numbers of the form p/q, where p and q are natural numbers, are positive.

1.57. The following principle is often used in proofs:

THEOREM. *If a number z is nonnegative and less than every positive number, then $z = 0$.*

Proof. If $z > 0$, then, by hypothesis, $z < z$, which is impossible. ∎

1.6. Consequences of the Least Upper Bound Axiom

1.61.† A set $E \subset R$ is said to be *bounded from below* if there exists an element $z \in R$ such that $z \leqslant x$ for every $x \in E$, a fact expressed concisely by writing $z \leqslant E$. Every number z with the above property relative to a set E is called a *lower bound* of E. If E is bounded from above, i.e., if there exists a number z such

† The definitions in this section are obvious analogues of those given in Sec. 1.24.

that $E \leqslant z$, then $-E$ (the set of all numbers $-x$ with $x \in E$) is bounded from below, since $x \leqslant z$ implies $-z \leqslant -x$. In particular, $-z$ is a lower bound of the set $-E$. Conversely, if E is bounded from below by a number z, the same argument shows that $-E$ is bounded from above by the number $-z$.

Suppose the set E is bounded from below. Then a lower bound z_0 of E is called the *greatest lower bound* of E if every other lower bound z of E is less than or equal to z_0 (why is z_0 unique?). The greatest lower bound of E is denoted by inf E (from the Latin "infimum").

THEOREM. *Every set $E \subset R$ which is bounded from below has a greatest lower bound, equal to $-\sup(-E)$.*

Proof. The set $-E$ is bounded from above and hence has a least upper bound $\xi = \sup(-E)$, by the axiom of Sec. 1.24. If $x \in E$, then $-x \leqslant \xi$ and hence $-\xi \leqslant x$, i.e., $-\xi$ is a lower bound of the set E. Let η be any other lower bound of E. Then $-\eta$ is an upper bound of $-E$, and hence, by the definition of the least upper bound, $-\eta \geqslant \sup(-E) = \xi$ or equivalently $\eta \leqslant -\xi$. In other words, inf E exists and equals $-\xi = -\sup(-E)$. ∎

1.62. a. THEOREM. *If the sets E and F are bounded from above and $E \subset F$, then $\sup E \leqslant \sup F$, while if E and F are bounded from below and $E \subset F$, then inf $F \leqslant$ inf E.*

Proof. In the first case, sup F is an upper bound for F and hence all the more so for $E \subset F$, so that $\sup E \leqslant \sup F$. In the second case, inf F is a lower bound for F and hence all the more so for $E \subset F$, so that inf $F \leqslant$ inf E. ∎

b. THEOREM. *If $x \leqslant y$ for arbitrary $x \in E, y \in F$, then E is bounded from above, F is bounded from below, and sup $E \leqslant$ inf F.*

Proof. The set E is bounded from above by any $y \in F$. Hence sup E exists, and sup $E \leqslant y$ for any $y \in F$. It follows that F is bounded from below by the number sup E, and hence that sup $E \leqslant$ inf F. ∎

1.63. Next we prove the existence and uniqueness of the "nth root" of any positive number.

THEOREM. *Given any real $x > 0$ and integer $n > 0$, there exists a unique nth root of x, i.e., a number $y > 0$ such that $y^n = x$.*

Proof (after W. Rudin). Let A be the set of all positive z such that $z^n \leqslant x$. Then A is bounded from above (by 1 if $x \leqslant 1$ and by x if $x \geqslant 1$). Let

$$y = \sup A. \tag{1}$$

We now show that $y^n = x$, thereby proving the theorem.

Suppose $y^n < x$ and let $x - y^n = \varepsilon$. Then, by the binomial theorem,

$$(y+h)^n = y^n + ny^{n-1}h + \frac{n(n-1)}{1 \cdot 2}y^{n-2}h^2 + \cdots$$

$$= y^n + h\left[ny^{n-1} + \frac{n(n-1)}{1 \cdot 2}hy^{n-2} + \cdots\right]$$

$$\leqslant y^n + h\left[ny^{n-1} + \frac{n(n-1)}{1 \cdot 2}y^{n-2} + \cdots\right] = y^n + h[(1+y)^n - y^n]$$

for any positive $h \leqslant 1$. Choosing

$$h < \frac{\varepsilon}{(1+y)^n - y^n},$$

we get $(y+h)^n \leqslant y^n + \varepsilon = x$, which contradicts the definition (1). Therefore $y^n \geqslant x$. Suppose $y^n > x$ and let $y^n - x = \varepsilon$. Then

$$(y-h)^n = y^n - ny^{n-1}h + \frac{n(n-1)}{1 \cdot 2}y^{n-2}h^2 - \cdots$$

$$= y^n - h\left[ny^{n-1} - \frac{n(n-1)}{1 \cdot 2}y^{n-2}h + \cdots\right]$$

$$\geqslant y^n - h\left[ny^{n-1} + \frac{n(n-1)}{1 \cdot 2}y^{n-2}h + \cdots\right]$$

$$\geqslant y^n - h\left[ny^{n-1} + \frac{n(n-1)}{1 \cdot 2}y^{n-2} + \cdots\right] = y^n - h[(1+y)^n - y^n]$$

for any positive $h \leqslant 1$. Again choosing

$$h < \frac{\varepsilon}{(1+y)^n - y^n},$$

we get $(y - h)^n \geqslant y^n - \varepsilon = x$, which again contradicts (1). It follows that $y^n = x$, as asserted. The uniqueness of the nth root follows from the inequality $y_1^n < y_2^n$ implied by $y_1 < y_2$ (cf. Sec. 1.55f). ∎

The nth root of x will henceforth be denoted by $\sqrt[n]{x}$.

1.64. THEOREM. *The formula*

$$\sqrt[n]{xy} = \sqrt[n]{x}\,\sqrt[n]{y} \tag{2}$$

holds for arbitrary positive x and y.

Proof. Let $\xi = \sqrt[n]{x}, \eta = \sqrt[n]{y}, \tau = \sqrt[n]{xy}$. Since $\xi^n = x, \eta^n = y$, we have $(\xi\eta)^n = \xi^n\eta^n$

$= xy = \tau^n$. But then

$$\tau = \sqrt[n]{xy} = \xi\eta,$$

by the uniqueness of the nth root. ∎

Similarly, it can be shown that

$$\sqrt[m]{\sqrt[n]{x}} = \sqrt[mn]{x} \tag{3}$$

for arbitrary $x > 0$ and integers $m, n > 1$.

1.65. Suppose n is an even integer. Then $(-x)^n = (-1)^n x^n = x^n > 0$ for all $x \neq 0$, so that the equation $y^n = x > 0$ has both a positive solution $y_1 = \sqrt[n]{x}$ and a negative solution $y_2 = -\sqrt[n]{x}$, while the equation $y^n = x < 0$ has no real solutions at all. On the other hand, if n is an odd integer, the equation $y^n = x > 0$ has a unique real solution $y = \sqrt[n]{x}$ and the equation $y^n = x < 0$ also has a unique solution, namely $y = -\sqrt[n]{|x|}$.

Formulas (2) and (3) guarantee the validity for the real numbers of all the usual elementary algebraic results involving radicals (like the formulas for the solution of quadratic and cubic equations).

1.66. A set $E \subset R$ is said to be *bounded from both sides*, or simply *bounded*, if it is bounded both from above and from below. Every bounded set E has both a least upper bound sup E and a greatest lower bound inf E. The following are particularly important examples of bounded sets:

1.67. Intervals. The set of all real numbers x satisfying the inequality $a \leqslant x \leqslant b$ is denoted by $[a,b]$ and called a *closed interval*, with left-hand *end point* a and right-hand *end point* b (it is assumed that $a < b$). The set of all real numbers x satisfying the inequality $a < x < b$ is denoted by (a,b) and called an *open interval*, again with left-hand end point a and right-hand end point b. Thus the end points of a closed interval belong to the interval itself, while the end points of an open interval do not. In any event,

$$\sup[a,b] = \sup(a,b) = b, \qquad \inf[a,b] = \inf(a,b) = a.$$

It is also convenient to introduce "half-closed" and "half-open" intervals. Thus the sets $\{x: a < x \leqslant b\} = (a,b]$ and $\{x: a \leqslant x < b\} = [a,b)$ are both called *intervals*, the former half-open on the left and half-closed on the right, the latter half-closed on the left and half-open on the right. For completeness, we will sometimes regard a single point a as a closed interval, writing

$$\{a\} = [a,a] = \{x: a \leqslant x \leqslant a\}.$$

1.7. The Principle of Archimedes and Its Consequences

1.71. THEOREM (**Principle of Archimedes**).† *Given arbitrary real numbers $x > 0$ and y, there exists an integer n such that $(n-1)x \leqslant y < nx$.*

Proof. Suppose $px \leqslant y$ for every integer p. Then the set A of all numbers px is bounded from above, with y as an upper bound. It follows from the axiom of Sec. 1.24 that A has a least upper bound $\xi = \sup A$. Since the number $\xi - x < \xi$ is no longer an upper bound of A, there exists an integer p such that $px > \xi - x$. But then $(p+1)x > \xi$, so that ξ cannot be an upper bound of A. This contradiction proves the existence of an integer p such that $px > y$. An analogous argument involving lower bounds shows that there exists an integer q such that $qx < y$, where clearly $q \leqslant p$. Examining all the pairs $(q, q+1)$, $(q+1, q+2), \ldots, (p-1, p)$, we find one among them, say $(n-1, n)$, such that $(n-1)x \leqslant y$ while $y < nx$. ∎

In particular, choosing $x = 1$, we find that there exists an integer n such that $n - 1 \leqslant y < n$ for any given $y \in R$. The number $n - 1$ is called the *integral part* of y and is denoted by $[y]$, while the number $y - [y]$ is called the *fractional part* of y and is denoted by (y). Thus every number y is the sum

$$y = [y] + (y)$$

of its integral and fractional parts.

1.72. THEOREM. *Given arbitrary real numbers $x > 1$, $y > 0$, there exists an integer n such that $x^{n-1} \leqslant y < x^n$.*

Proof. This is the "multiplicative version" of the principle of Archimedes, and is proved by going over from integral multiples‡ of x to integral powers of x (see Sec. 1.46) in the proof of Theorem 1.71 (give the details). ∎

1.73. THEOREM. *Given arbitrary real numbers $x > 0$ and $y > 0$, there exists an integer $n > 0$ such that*

$$\frac{y}{n} < x. \tag{1}$$

Proof. As in Theorem 1.71, $y < nx$, where now y and

$$n > \frac{y}{x} \tag{2}$$

---

† There are other modern axiomatic treatments of the real numbers in which the principle of Archimedes and the principle of nested intervals (see Sec. 1.8) appear as *axioms*, with the least upper bound axiom of Sec. 1.24 then becoming a *theorem*.
‡ The numbers $(n-1)x$ and nx are called *integral multiples* of x, for an obvious reason.

are positive. Multiplying both sides of (2) by x/n, we get (1). ∎

In particular, it follows that

$$\inf\left\{\frac{y}{n} : n = 1,2,\ldots\right\} = 0$$

for any $y > 0$. In fact, the set in curly brackets consists of positive numbers only and hence has a nonnegative greatest lower bound. But, as just shown, this greatest lower bound cannot be positive, and hence must equal zero.

1.74. COROLLARY. *Each of the following systems of half-open intervals has an empty intersection*:

$$(0,y] \supset \left(0,\frac{y}{2}\right] \supset \cdots \supset \left(0,\frac{y}{n}\right] \supset \cdots \qquad (y > 0), \tag{3}$$

$$(a,a+y] \supset \left(a,a+\frac{y}{2}\right] \supset \cdots \supset \left(a,a+\frac{y}{n}\right] \supset \cdots, \tag{4}$$

$$[a-y,a) \supset \left[a-\frac{y}{2},a\right) \supset \cdots \supset \left[a-\frac{y}{n},a\right) \supset \cdots. \tag{5}$$

Proof. If the intervals of the system (4) had a common point ξ, then $\xi - a$ would be a common point of the system (3), while if the intervals of the system (5) had a common point η, then $a - \eta$ would be a common point of the system (3). But the intervals of the system (3) cannot have any common points at all, because of Theorem 1.73. ∎

The corollary clearly remains true if y/n $(n = 1,2,\ldots)$ is replaced by y/nx $(x > 0$ arbitrary) or by y/x^{n-1} $(x > 1$ arbitrary), in particular, by $y/10^{n-1}$.

1.75. THEOREM. *Every open interval (a,b) contains a rational point.*

Proof. Let $h = b - a > 0$, and let n be an integer greater than $1/h$ (the existence of n follows from Theorem 1.71), so that $1/n < h$. By Theorem 1.71 again, there exists an integer m such that

$$\frac{m}{n} \leqslant a < \frac{m+1}{n},$$

where clearly

$$\frac{m+1}{n} - a \leqslant \frac{1}{n} < b - a$$

and hence

$$\frac{m+1}{n} < b.$$

It follows that

$$a < \frac{m+1}{n} < b,$$

i.e., the rational number

$$\frac{m+1}{n}$$

belongs to the interval (a,b). ∎

There are actually an infinite number of rational points in (a,b). In fact, applying the above theorem to the interval

$$\left(\frac{m+1}{n}, b\right)$$

gives a new rational number p/q such that

$$\frac{m+1}{n} < \frac{p}{q} < b,$$

and this process can clearly be continued indefinitely.

1.76. THEOREM. *Given any real number ξ, let N_ξ be the set of all rational numbers $s \leqslant \xi$, and let P_ξ be the set of all rational numbers $r \geqslant \xi$. Then*

$$\sup N_\xi = \xi = \inf P_\xi.†$$

Proof. Let $\alpha = \sup N_\xi$. Then, since $s \leqslant \xi$ for every $s \in N_\xi$, we have $\alpha \leqslant \xi$ by the very definition of the least upper bound. Suppose $\alpha < \xi$. By Theorem 1.75, there is a rational point p in (α, ξ). Since $p < \xi$, we have $p \in N_\xi$, which implies $p \leqslant \sup N_\xi = \alpha$, contrary to the condition $p \in (\alpha, \xi)$. Therefore the inequality $\alpha < \xi$ is impossible, and hence $\alpha = \sup N_\xi = \xi$. The fact that $\inf P_\xi = \xi$ is proved similarly. ∎

1.77. Decimal representation of real numbers. Next we show how an arbitrary real number ξ can be represented by a suitable sequence of the symbols ("digits") $0,1,2,\ldots,9$.

a. Suppose $\xi > 0$. By Theorem 1.72, there exists a (unique) integer p such

† Note that N_ξ is bounded from above (by ξ), while P_ξ is bounded from below (by ξ).

that

$$10^p \leqslant \xi < 10^{p+1}.$$

Having found the "exponent" p, we next find a number θ_0 (from the set $1,2,\ldots,9$) such that

$$\theta_0 \cdot 10^p \leqslant \xi < (\theta_0 + 1) \cdot 10^p$$

(where $9 + 1 = 10$). The number θ_0 is also uniquely determined, since the intervals

$$\theta \cdot 10^p \leqslant x < (\theta + 1) \cdot 10^p \qquad (\theta = 0,1,\ldots,9)$$

are nonintersecting. Next, having found θ_0, we find a number θ_1 (from the set $0,1,\ldots,9$) such that

$$\theta_0 \cdot 10^p + \theta_1 \cdot 10^{p-1} \leqslant \xi < \theta_0 \cdot 10^p + (\theta_1 + 1) \cdot 10^{p-1}.$$

Continuing this process indefinitely, we get a sequence of symbols (digits from 0 to 9)

$$\theta_0 \theta_1 \theta_2 \cdots \qquad (\theta_0 \neq 0). \tag{6}$$

The underlying presence of the number p is indicated as follows: If $p \geqslant 0$, we put a decimal point between the symbols θ_p and θ_{p+1}, while if $p < 0$, i.e., if $p = -q$, $q > 0$, we write q additional zeros in front of the sequence (6) and then put a decimal point after the first zero. In this way, the influence of the number p is reflected in the expression (6).

Thus to every real number $\xi > 0$ there corresponds, in accordance with this rule, an expression of the form (6) (possibly preceded by a "string of zeros"), with a decimal point in some position. The expression (6) is called the *decimal representation* or *expansion* of ξ, with (decimal) *digits* $\theta_0, \theta_1, \theta_2, \ldots$ (in that order). The decimal representation of the number 1 is just $1.000\ldots$, and similarly for the numbers $2,3,\ldots,9$. Similarly, the decimal representation of the number 10 is $10.000\ldots$, while for numbers of the form

$$\frac{s}{10^t} \qquad (s,t = 0,1,2,\ldots)$$

("rational decimals"), and only for such numbers, there are no more than t nonzero digits after the decimal point.

b. THEOREM. *It is impossible for all the digits in an expression of the form* (6) *to be nines starting from some position, i.e.,* (6) *cannot have an "infinite run of nines."*

Proof. The presence of an infinite run of nines beginning with some number n after the decimal point (the "nth decimal place") would mean that the number ξ lies in all the intervals

$$r + \frac{9}{10^n} \leqslant \xi < r + \frac{10}{10^n} = r + \frac{1}{10^{n-1}},$$

$$r + \frac{9}{10^9} + \frac{9}{10^{n+1}} \leqslant \xi < r + \frac{9}{10^n} + \frac{10}{10^{n+1}} = r + \frac{1}{10^{n-1}},$$

$$\cdots \tag{7}$$

$$r + \frac{9}{10^n} + \frac{9}{10^{n+1}} + \cdots + \frac{9}{10^{n+k}} \leqslant \xi < r + \frac{1}{10^{n-1}},$$

$$\cdots$$

But this is impossible, since the system of intervals (7) has an empty intersection, by the remark following Corollary 1.74. ∎

c. THEOREM. *Let*

$$\tau_1 \tau_2 \cdots \tag{8}$$

be an arbitrary sequence of digits from 0 to 9, with a decimal point in some position, where not all the τ_i are zero and there are digits other than 9 arbitrarily far from the decimal point. Then there exists a real number $\xi > 0$ with (8) as its decimal expansion.

Proof. Let τ_m be the first nonzero digit in (8). The decimal point is either to the right of τ_m by $q \geqslant 0$ digits (not including τ_m), or to the left by $t \geqslant 1$ digits (including τ_m); in the second case we set $q = -t$. We now show that the decimal expansion of the number ξ, defined as

$$\xi = \sup_k \{10^q \cdot \tau_m + 10^{q-1} \cdot \tau_{m+1} + 10^{q-2} \cdot \tau_{m+2} + \cdots + 10^{q-k} \cdot \tau_{m+k}\},$$

coincides with (8). Let s be a fixed positive integer, choose $r > s$ such that $\tau_{m+r} \leqslant 8$, and let $k > r$. Then, summing a geometric progression, we get

$$10^{q-(s+1)} \cdot \tau_{m+s+1} + \cdots + 10^{q-r} \cdot \tau_{m+r} + \cdots + 10^{q-k} \cdot \tau_{m+k}$$
$$\leqslant 9 \cdot 10^{q-(s+1)} + \cdots + 9 \cdot 10^{q-r} + \cdots + 9 \cdot 10^{q-k} - 10^{q-r}$$
$$= 9 \cdot \frac{10^{q-(s+1)} - 10^{q-(k+1)}}{1 - 10^{-1}} - 10^{q-r} < 10^{q-s} - 10^{q-r}.$$

It follows that

$$\xi = \sup_k \{10^q \cdot \tau_m + \cdots + 10^{q-s} \cdot \tau_{m+s} + \cdots + 10^{q-k} \cdot \tau_{m+k}\}$$
$$\leqslant 10^q \cdot \tau_m + \cdots + 10^{q-s} \cdot \tau_{m+s} + 10^{q-s} - 10^{q-r}$$
$$< 10^q \cdot \tau_m + 10^{q-s}(\tau_{m+s} + 1),$$

and hence

$$10^q \cdot \tau_m + \cdots + 10^{q-s} \cdot \tau_{m+s} \leqslant \xi < 10^q \cdot \tau_m + \cdots + 10^{q-s}(\tau_{m+s}+1)$$

for every $s = 0,1,2,\ldots$ Setting $s = 0,1,2,\ldots$ and recalling the definition of the number p and the digits $\theta_0, \theta_1, \ldots$ of the number ξ, we find that $p = q$, $\theta_0 = \tau_m$, $\theta_1 = \tau_{m+1}, \ldots$, i.e., the decimal expansion of ξ is just (8). ∎

d. If $\xi < 0$, then $-\xi > 0$, and hence

$$-\xi = \tau_1 \tau_2 \cdots,$$

as shown in Sec. 1.77c. We now set

$$\xi = -\tau_1 \tau_2 \cdots,$$

by definition. Finally, if $\xi = 0$, we set

$$0 = 0.0000\ldots$$

1.78. Suppose we repeat the considerations of Sec. 1.77, replacing the number 10 figuring in the powers 10^p, $10^{p+1}, \ldots$ by some other integer $P > 1$. This gives a so-called "P-ary system" for representing the real numbers. The most commonly encountered P-ary systems are the *binary* and *ternary* systems, where $P = 2$ and $P = 3$, respectively. In the binary system the representation of an arbitrary real number involves only the digits 0 and 1, while in the ternary system it involves only the digits 0, 1, and 2.

1.8. The Principle of Nested Intervals

1.81. A set Q of intervals on the real line R with the property that given any two intervals $I, J \in Q$, either $I \subset J$ or $J \subset I$, is called a *system of nested intervals*. According to Corollary 1.74, a system of nested half-open intervals may well have an empty intersection, and the same is all the more true of a system of nested open intervals. However, a system of nested *closed* intervals always has a nonempty intersection, as shown by the following proposition due to Cantor:

THEOREM (**Principle of nested intervals**). *Let Q be any system of nested closed intervals $[a,b]$. Then there exists at least one point of R belonging to every interval of Q. More exactly, the numbers*

$$\xi = \sup\{a : [a,b] \in Q\}, \qquad \eta = \inf\{b : [a,b] \in Q\} \tag{1}$$

exist ($\xi \leqslant \eta$) and the interval $[\xi, \eta]$ is the intersection of all the intervals of Q.†

Proof. Let $E = \{a : [a,b] \in Q\}$ be the set of all left-hand end points of the inter-

† If $\xi = \eta$, then by the interval $[\xi, \eta]$ we mean just the point $\xi = \eta$.

vals of Q, and let $F=\{b: [a,b]\in Q\}$ be the set of all right-hand end points of the intervals of Q. Given any two intervals $[a_1, b_1]$, $[a_2, b_2] \in Q$, one must be contained in the other, say $[a_1,b_1] \subset [a_2,b_2]$, so that $a_2 \leqslant a_1 \leqslant b_1 \leqslant b_2$. Hence no $a\in E$ exceeds any $b\in F$. It follows from Theorem 1.62b that the numbers (1) exist and satisfy the inequality $\xi \leqslant \eta$. For any $[a,b]\in Q$ we have $a \leqslant \xi \leqslant \eta \leqslant b$, so that $[a,b] \supset [\xi,\eta]$ and hence $\prod[a,b] \supset [\xi,\eta]$. Moreover, $\prod[a,b]$ consists only of points of the interval $[\xi,\eta]$. In fact, for any point $x \notin [\xi,\eta]$, e.g., for any $x < \xi$, there is a left-hand end point a such that $x < a < \xi = \sup\{a\}$, and then x does not belong to the corresponding interval $[a,b]$. ∎

1.82. The following theorem gives conditions under which the intersection of all the intervals of Q reduces to a single point:

THEOREM. *The intersection of all the intervals of a system Q of nested closed intervals consists of a single point if and only if, given any $\varepsilon > 0$, there is an interval $[a,b]\in Q$ of length $b - a < \varepsilon$.*

Proof. By Theorem 1.81, the intersection of all the intervals of Q is an interval $[\xi,\eta]$, which reduces to a single point if and only if $\xi = \eta$. If $\xi \neq \eta$, then the length of every interval $[a,b] \supset [\xi,\eta]$ of the system Q is no less than $\eta - \xi$, and hence if Q contains intervals of arbitrarily small length, the intersection of all these intervals must reduce to a single point. Conversely, because of (1), given any $\varepsilon > 0$, there is an interval $[a_1,b_1] \in Q$ such that

$$a_1 > \xi - \frac{\varepsilon}{2}$$

and an interval $[a_2,b_2] \in Q$ such that

$$b_2 < \eta + \frac{\varepsilon}{2}.$$

If $[a_1,b_1] \supset [a_2,b_2]$, say, then

$$a_2 \geqslant a_1 > \xi - \frac{\varepsilon}{2}, \qquad b_2 < \eta + \frac{\varepsilon}{2}.$$

Hence, if $\xi = \eta$, then

$$b_2 - a_2 < \varepsilon,$$

so that Q contains an interval of length less than ε. ∎

1.9. The Extended Real Number System

1.91. Definition. By the *extended real number system* $\overline{R}$ is meant the set consisting of the elements of the real number system R and two symbols or "points" $-\infty$ and ∞ (more exactly, $-\infty$ and $+\infty$), called "minus infinity" and "plus infinity."† The order relations are extended to these symbols by the following rules:

(a) $-\infty < x$ for every $x \in R$;
(b) $x < \infty$ for every $x \in R$;
(c) $-\infty < \infty$.

The axioms of order (Sec. 1.23) continue to hold in the extended system $\bar{R}$. The ordinary real numbers (elements of R) are said to be *finite*, as opposed to the symbols $-\infty$ and ∞.

1.92. Given any nonempty set $E \subset \bar{R}$, the expressions sup E and inf E are defined as follows: If E does not contain the point ∞ and is bounded from above, then sup E has the same meaning as in Sec. 1.24; otherwise‡ we set sup $E = \infty$. Similarly, if E does not contain the point $-\infty$ and is bounded from below, then inf E has the same meaning as in Sec. 1.61; otherwise we set inf $E = -\infty$. Thus *every nonempty subset of the extended real number system $\bar{R}$ has a least upper bound and a greatest lower bound.*

1.93. If a and b $(a < b)$ are any two points of $\bar{R}$, then, just as in the case of R (see Sec. 1.67), the set

$$[a,b] = \{x \in \bar{R}: a \leqslant x \leqslant b\}$$

is called a *closed interval* with end points a and b, while the set

$$(a,b) = \{x \in \bar{R}: a < x < b\}$$

is called an *open interval* with end points a and b.

1.94. The principle of nested intervals (Theorem 1.81) is readily generalized to the case of the extended real line $\bar{R}$. Let Q be any system of closed intervals on $\bar{R}$ which are nested in the same sense as in Sec. 1.81. Then there exists at least one point of $\bar{R}$ belonging to every interval of Q, i.e., the intersection of all the intervals of Q is nonempty. In fact, this intersection is just the interval $[\xi,\eta]$, where

$$\xi = \sup\{a: [a,b] \in Q\}, \qquad \eta = \inf\{b: [a,b] \in Q\} \qquad (\xi \leqslant \eta),$$

as in Sec. 1.81. The proof is virtually the same as for the ordinary principle of nested intervals (give the details).

---

† The set $\bar{R}$ is often called the *extended real line*.
‡ That is, if E contains ∞, or if E fails to contain ∞ but is not bounded from above.

Problems

1. Let addition and multiplication of the elements of the set $\{0,1\}$ be defined by the rules

$$0+0=0, \qquad 0+1=1, \qquad 1+0=1, \qquad 1+1=0,$$
$$0\cdot0=0, \qquad 0\cdot1=0, \qquad 1\cdot0=0, \qquad 1\cdot1=1.$$

Verify that $\{0,1\}$ is a field.

2. Prove that

(a) $|x-y| \geqslant ||x|-|y||$ for all real x,y;

(b) $|x-y_1-\cdots-y_n| \geqslant ||x|-|y_1|-\cdots-|y_n||$ for all real $x,y_1,\ldots,y_n$.

3. Prove that the sum (or product) of a rational number and an irrational number is irrational. Can the sum (or product) of two irrational numbers be rational?

4. Prove that $\sqrt{3}$ is irrational.

5. Which is larger, $\sqrt{3}+\sqrt{5}$ or $\sqrt{2}+\sqrt{6}$?

6. Prove that

$$a+\frac{1}{a} \geqslant 2$$

if a is a positive real number.

7. Let $x_1,x_2,\ldots,x_n$ be positive real numbers. Prove that if $x_1 x_2 \ldots x_n = 1$, then

$$x_1+x_2+\cdots+x_n \geqslant n,$$

where equality occurs if and only if $x_1 = x_2 = \cdots = x_n = 1$.

8. The *geometric mean* of the positive real numbers $x_1,x_2,\ldots,x_n$ is defined by

$$g=\sqrt[n]{x_1 x_2 \cdots x_n}$$

and the *arithmetic mean* by

$$a=\frac{x_1+x_2+\cdots+x_n}{n}.$$

Prove that $g<a$ unless $x_1=x_2=\cdots=x_n$, in which case $g=a$.

9. Prove that

$$n!=1\cdot2\cdots n<\left(\frac{n+1}{2}\right)^n \qquad (n\geqslant2).$$

10. Solve the following equations:

(a) $|x-1|=2$;   (b) $|x-1|=|3-x|$;   (c) $|x+1|=|3+x|$;   (d) $|2x|=|x-2|$.

11. Find the set of all real x such that

$$\frac{x-4}{2x+3} < \frac{1}{3}.$$

12. Find the set of all real x such that $|x+1|+|x-1|>2$ and $|x-2|\leqslant 6$.

13. Let

$$A=\{a,a^2,a^3,\dots\},$$

where $0\leqslant a\leqslant 1$. Find $\max A$, $\min A$, $\sup A$, and $\inf A$. What happens if $a>1$? Discuss the case of negative a.

14. Give an example where $[|x|]\neq|[x]|$.

15. Prove that

$$\alpha=\tfrac{1}{2}\ \tfrac{3}{4}\ \tfrac{5}{6}\ \cdots\ \tfrac{99}{100}<\tfrac{1}{10}.$$

16. By a "repeating decimal" is meant a decimal like

$$\tfrac{1}{3}=0.333333\cdots,$$

where a single digit (possibly 0) "repeats itself forever" or a decimal like

$$\tfrac{1}{7}=0.142857142857\dots,$$

where a whole block of digits repeats itself forever. Prove that every repeating decimal represents a rational number.

17. What are the rational numbers represented by the following repeating decimals:

(a) $0.919999\dots$;   (b) $0.417417\dots$;   (c) $2.331331\dots$;   (d) $-8.191919\dots$?

18. By the *arithmetic sum* $A+B$ of two numerical sets A and B is meant the set of all sums $x+y$ where $x\in A, y\in B$. Prove that if A and B are bounded from above, then $A+B$ is bounded from above and

$$\sup\ (A+B)=\sup\ A+\sup\ B.$$

19. By the *arithmetic product* AB of two numerical sets A and B is meant the set of all products xy where $x\in A, y\in B$. Prove that if A and B are bounded from above and consist of positive numbers only, then AB is bounded from above and

$$\sup AB=\sup A\cdot\sup B.$$

20. By the *arithmetic nth power* A^n of a numerical set A is meant the set of all numbers x^n where $x \in A$. (Note that in general $A^2 \neq AA$.) Prove that if A is bounded from above and consists of positive numbers only, then A^n is bounded from above and

$$\sup A^n = (\sup A)^n.$$

21. Prove that if the set of all real numbers is partitioned into two nonempty nonintersecting subsets A and B such that $a < b$ for all $a \in A$, $b \in B$ (a "Dedekind cut"), then there exists one and only one number γ such that $a \leqslant \gamma \leqslant b$ for all $a \in A$, $b \in B$.

2 Sets

2.1. Operations on Sets

2.11. The operations on sets introduced in Sec. 1.12 will now be examined in more detail. It will be recalled that the *union* of given sets $A,B,C,\ldots$ is defined as the set of all elements belonging to *at least one* of the sets $A,B,C,\ldots$, while the *intersection* of the given sets is defined as the set of all elements belonging to *every one* of the sets $A,B,C,\ldots$

The union S and intersection D of the sets $A,B,C,\ldots$ are sometimes called the *sum* and *product*, respectively,† and correspondingly we write

$$S = A + B + C + \cdots, \quad D = ABC \cdots.$$

There is some justification for this "arithmetic language," as shown by the following

THEOREM. *The formula*

$$(A+B)C = AC+BC \tag{1}$$

holds for any three sets A, B, and C.

Proof. Recalling that equality of two sets means that every element of one set is also an element of the other set, we see that to prove (1) it must be shown that every element x belonging to the left-hand side $(A+B)C$ also belongs to the right-hand side $AB+AC$, and conversely that every element y belonging to the right-hand side $AB+AC$ also belongs to the left-hand side $(A+B)C$.

Thus suppose x belongs to $(A+B)C$. Then, being an element of the intersection of $A+B$ and C, x must belong to both $A+B$ and C, so that $x \in A+B$, $x \in C$. Since x belongs to the union of A and B, x must belong to at least one of the sets A and B, say to A. But $x \in A$, $x \in C$ together imply $x \in AC$ and hence $x \in AC+BC$. If, on the other hand, x belongs to B rather than to A, we deduce in the same way that $x \in BC$ and hence $x \in AC+BC$.

Conversely, suppose y belongs to the sum $AC+BC$. Then y belongs to (at least) one of the sets AC and BC, say to BC, so that $y \in B$, $y \in C$. But $y \in B$ implies $y \in A+B$, and hence finally $y \in (A+B)C$. The case $y \in AC$ is analyzed similarly. ∎

It must not be thought that all the rules of arithmetic carry over to operations on sets. For example, given arbitrary sets A,B, and C, we have the

---

† Not to be confused with the "arithmetic sum" and "arithmetic product" introduced in Problems 18 and 19 of Chapter 1.

formulas†

$$A + A = A,$$
$$AA = A,$$
$$A + BC = (A + B)(A + C),$$

none of which makes any sense as a piece of ordinary arithmetic.

2.12. Next we introduce a new operation, known as *complementation*. Let B be a subset of the set A. Then the set of all elements of A not belonging to B, denoted by $\mathbf{C}B$ or $A - B$, is called the *complement* of B (relative to A). Obviously, we have

$$\mathbf{C}(\mathbf{C}B) = A - (A - B) = B,$$
$$(A - B) + B = A.$$

Note, however, that the formula

$$(A + B) - B = A$$

is in general false, and in fact holds only if A and B have no common elements.

THEOREM. *The formula*

$$\mathbf{C} \sum_{\nu} B_{\nu} = \prod_{\nu} \mathbf{C}B_{\nu} \tag{2}$$

holds for arbitrary sets B_{ν},‡ i.e., the complement of a union of sets equals the intersection of the complements of the sets.

Proof. If

$$x \in \mathbf{C} \sum_{\nu} B_{\nu}, \tag{3}$$

then

$$x \notin \sum_{\nu} B_{\nu}, \tag{4}$$

i.e., for every ν, $x \notin B_{\nu}$ or equivalently $x \in \mathbf{C}B_{\nu}$. But then

$$x \in \prod_{\nu} \mathbf{C}B_{\nu}. \tag{5}$$

Conversely, suppose (5) holds. Then for every ν, $x \in \mathbf{C}B_{\nu}$ or equivalently $x \notin B_{\nu}$. But then (4) holds, which implies (3). ∎

---

† The reader should prove these formulas as an exercise.
‡ Here the sets B_{ν} are indexed by a parameter ν (think of ν as ranging from 1 to n or over all positive integers or over some more general "index set").

Applying the operation $\mathbf{C}$ to both sides of (2) and setting $A_v = \mathbf{C}B_v$, we get

$$\sum_v \mathbf{C}A_v = \mathbf{C}\prod_v A_v, \tag{2'}$$

i.e., *the complement of an intersection of sets equals the union of the complements of the sets.*

Formulas (2) and (2') can be combined into the following simple rule: *In interchanging the symbol $\mathbf{C}$ with the symbols $\sum$ and $\prod$, we must change $\sum$ to $\prod$ and $\prod$ to $\sum$.*

2.2. Equivalence of Sets

We now establish rules allowing us to compare different sets with regard to the "number of elements" in each. In the case of finite sets, there is no problem at all. In fact, given two finite sets A and B, we need only count the number of elements in each set to ascertain which of the sets (if either) has more elements than the other, regarding A and B as equivalent if both sets turn out to have the same number of elements. As it stands, this definition of equivalence is not immediately applicable to the case of infinite sets, and we now cast it in another form making it suitable for infinite sets. To this end, we note that in establishing the equivalence or nonequivalence of two finite sets A and B, there is actually no need to count the number of elements in each set. For example, if A is the set of people in an auditorium and B the set of seats in the auditorium, then, instead of counting people and seats separately, we can immediately ascertain, without calculations, whether the two sets A and B are equivalent or not (equivalent if there are no empty seats or standees, nonequivalent otherwise). This leads to the key idea of *establishing a one-to-one correspondence between two sets A and B.*

2.21. Definition. Given two sets A and B, suppose a unique element $b \in B$ is associated with each element $a \in A$, in accordance with some rule (or "mapping"), a fact indicated by writing $a \mapsto b$. Suppose further that each $b \in B$ is associated with one and only one element $a \in A$. Then A and B are said to be "in one-to-one correspondence," a fact indicated by writing $a \leftrightarrow b$, and the rule in question is called a *one-to-one correspondence* (or "one-to-one mapping"). Two sets A and B are said to be *equivalent* if they are in one-to-one correspondence.

The new definition of equivalence works for arbitrary sets, not just for finite sets. For example, the infinite set A of all positive integers $1, 2, \ldots$ is equivalent to the set B of all negative integers $-1, -2, \ldots$ In fact, here the appropriate one-to-one correspondence is just $n \leftrightarrow -n$. In just the same way, the set of all positive integers $1, 2, \ldots$ is equivalent to the set of all even integers 2,4,..., with $n \leftrightarrow 2n$ as the appropriate one-to-one correspondence. The last example shows that a set can be equivalent to one of its proper subsets. Naturally, this can only happen if the set is infinite.

If two sets A and B are equivalent, we write $A \sim B$. It is easy to see that the relation $\sim$ is *symmetric* ($A \sim A$), *reflexive* (if $A \sim B$, then $B \sim A$), and *transitive* (if $A \sim B$ and $B \sim C$, then $A \sim C$). Equivalent sets are also said to *have the same power*.

2.22. Examples. The interval $[0,a]$, $a > 0$ is equivalent to the unit interval $[0,1]$, with $x \in [0,1] \leftrightarrow y = ax \in [0,a]$ as the appropriate correspondence. Moreover, any two intervals $[a,b]$ and $[a+h,b+h]$ of equal length $b-a$ are equivalent, with $x \in [a,b] \leftrightarrow y = x+h \in [a+h,b+h]$ as the correspondence. It follows that any two closed intervals on the real line R are equivalent, and the same is true of any two open intervals on R.

The correspondence

$$x \leftrightarrow y = \frac{1}{x}$$

establishes the equivalence of the interval $0 < x < 1$ and the half-line $y > 1$, while the correspondence

$$x \leftrightarrow y = \frac{x}{1+|x|}$$

establishes the equivalence of the whole real line $-\infty < x < \infty$ and the interval $-1 < y < 1$.

THEOREM. *The closed interval $[a,b]$ and the open interval (a,b) are equivalent.*

Proof. Let A be any sequence (see Sec. 2.81) of distinct points

$$x_1 = a, \; x_2 = b, \; x_3, \; x_4, \ldots$$

of the closed interval $[a,b]$, with the end points a and b as the first two points. Clearly, the points $x_3, x_4, \ldots$ and any point $y \notin A$ all belong to the open interval (a,b). The rule

$$x_1 \leftrightarrow x_3,$$
$$x_2 \leftrightarrow x_4,$$
$$\cdots$$
$$x_n \leftrightarrow x_{n+2},$$
$$\cdots$$
$$y \notin A \leftrightarrow y$$

then establishes a one-to-one correspondence between $[a,b]$ and (a,b). ∎

It follows from the above considerations that any open interval is equivalent to any closed interval, to any half-line, or even to the whole real line R.

2.3. Countable Sets

2.31. Definition. A set equivalent to the set of all natural numbers 1,2,... is said to be *countable*. In other words, a set is called countable if all its elements can be "numbered" or "counted," i.e., labeled with the positive integers. We now prove some simple theorems involving countable sets.

2.32. THEOREM. *Every infinite subset B of a countable set A is itself countable.*

Proof. We need only renumber the elements of B in the order of their appearance in the original numbering of the elements of A (since B is infinite, this will require all the natural numbers). ∎

2.33. THEOREM. *The union of a finite or countable collection of countable sets is itself a countable set.*

Proof. First we consider the case of two countable sets $A = \{a_1, a_2, \ldots\}$ and $B = \{b_1, b_2, \ldots\}$. Writing all the elements of both sets in a single row

$$a_1, b_1, a_2, b_2, a_3, b_3, \ldots, \tag{1}$$

we can then renumber the elements in the order of their appearance in (1), with the proviso that any element appearing twice (any element belonging to both A and B) is numbered when it appears for the first time and omitted when it appears for the second time. This gives every element in the union of A and B a number, as required.

The theorem is proved in a completely analogous way in the case of three, four, or, more generally, any finite number of countable sets. In the case of a countable collection of countable sets

$$A_1 = \{a_{11}, a_{12}, \ldots, a_{1n}, \ldots\},$$
$$A_2 = \{a_{21}, a_{22}, \ldots, a_{2n}, \ldots\},$$
$$\cdots$$
$$A_k = \{a_{k1}, a_{k2}, \ldots, a_{kn}, \ldots\},$$
$$\cdots$$

we need only use a somewhat different rule for writing all the elements of $A_1, A_2, \ldots, A_k, \ldots$ in a single row, with the rest of the proof remaining the same. For example, we can group elements with the same sum of indices, writing

$$a_{11}; \; a_{21}, a_{12}; \; a_{31}, a_{22}, a_{13}; \; a_{41}, a_{32}, a_{23}, a_{14}; \cdots \quad ∎$$

2.34. THEOREM. *The set of all rational numbers (i.e., numbers of the form p/q where p and $q \neq 0$ are integers) is countable.*

Proof. The set Q of all rational numbers is clearly the union of the following countable collection of countable sets:†

$A_1 = \{n: n = 0, \pm 1, \pm 2, \ldots\}$,
$A_2 = \{n/2: n = 0, \pm 1, \pm 2, \ldots\}$,
$\cdots$
$A_k = \{n/k: n = 0, \pm 1, \pm 2, \ldots\}$,
$\cdots$

Hence Q is itself countable, by Theorem 2.33. ∎

2.35. THEOREM. *Given two countable sets*

$A = \{a_1, a_2, \ldots, a_k, \ldots\}$

and

$B = \{b_1, b_2, \ldots, b_n, \ldots\}$,

the set of all ordered pairs

$$(a_k, b_n) \qquad (k, n = 1, 2, \ldots) \tag{2}$$

is itself countable.

Proof. The set C of all pairs (2) is clearly the union of the following countable collection of countable sets:

$A_1 = \{(a_1, b_1), (a_1, b_2), \ldots, (a_1, b_n), \ldots\}$,
$A_2 = \{(a_2, b_1), (a_2, b_2), \ldots, (a_2, b_n), \ldots\}$,
$\cdots$
$A_k = \{(a_k, b_1), (a_k, b_2), \ldots, (a_k, b_n), \ldots\}$
$\cdots$

Hence C is itself countable, again by Theorem 2.33. ∎

Geometrically, the pair (a_k, b_n) corresponds to the point in the plane with coordinates a_k, b_n. Thus Theorem 2.35 implies in particular that the set of all points in the plane with rational coordinates is countable.

2.36. THEOREM. *The set of all polynomials*

† Thus A_k $(k = 1, 2, \ldots)$ is the set of all fractions of the form n/k, where $n = 0, \pm 1, \pm 2, \ldots$

$$P(x) = a_0 + a_1 x + \cdots + a_n x^n$$

(*of arbitrary degree*) *with rational coefficients* $a_0, a_1, \ldots, a_n$ *is countable.*

Proof. The set of all polynomials with rational coefficients is the union of the countable collection of sets $A_0, A_1, \ldots, A_n, \ldots$, where A_n denotes the set of all polynomials of degree $\leqslant n$ with rational coefficients. Hence, by Theorem 2.33, we need only show that each set A_n is countable. For $n = 0$, A_n reduces to the set of all rational numbers, which is countable by Theorem 2.34. We now invoke mathematical induction, showing that countability of the set A_n implies that of the set A_{n+1}. Every element of A_{n+1} can be written in the form

$$Q(x) + a_{n+1} x^{n+1},$$

where a_{n+1} is a rational number and $Q(x)$ is a polynomial of degree $\leqslant n$ with rational coefficients (i.e., an element of A_n). The set of all polynomials $Q(x)$ is countable, by the induction hypothesis, and the set of all numbers a_{n+1} is also countable. Since every element of A_{n+1} can be assigned a pair $(Q(x), a_{n+1})$, where $Q(x)$ and a_{n+1} both range over countable sets, it follows from Theorem 2.35 that A_{n+1} is countable, and the induction is complete. ∎

2.37. THEOREM. *The set of all algebraic numbers (i.e., roots of polynomials with rational coefficients) is countable.*

Proof. According to Theorem 2.36, the set of all polynomials with rational coefficients can be "numbered" with positive integers, thereby forming a sequence

$$P_1(x), P_2(x), \ldots, P_n(x), \ldots$$

But each of these polynomials has only a finite number of roots (see Sec. 5.87). Writing down all the roots of $P_1(x)$, then all the roots of $P_2(x)$ in the same row, and so on, we succeed in "counting" all the algebraic numbers. ∎

2.4. Uncountable Sets

An infinite set which is not countable, i.e., whose elements cannot be put into a one-to-one correspondence with the positive integers, is said to be *uncountable*. A typical uncountable set is the *continuum*, namely the set of all points in a closed interval. This is shown by the following theorem due to Cantor (1874):

2.41. THEOREM. *The set of all points in the unit interval* [0,1] *is uncountable.*

Proof. Suppose to the contrary that the set of all points in the interval $[0,1]$ is countable, so that these points can be arranged in a sequence

$$x_1, x_2, \ldots, x_n, \ldots \tag{1}$$

Starting from (1), we construct a sequence of nested closed intervals as follows. First we divide the interval $[0,1]$ into three equal parts. Regardless of the location of x_1, it cannot belong simultaneously to all three intervals $[0, \frac{1}{3}]$, $[\frac{1}{3}, \frac{2}{3}]$, $[\frac{2}{3}, 1]$, and hence one of these intervals must fail to contain x_1 (either inside the interval or on its boundary). Let Δ_1 denote this interval. We now divide Δ_1 in turn into three equal parts, choosing a part Δ_2 which does not contain x_2. More generally, once having used this procedure to construct the intervals $\Delta_1 \supset \Delta_2 \supset \cdots \supset \Delta_n$ where $x_n \notin \Delta_n$, we divide Δ_n into three equal parts, choosing a part Δ_{n+1} such that $x_{n+1} \notin \Delta_{n+1}$, and so on indefinitely. By Theorem 1.81, the resulting infinite sequence of nested intervals $\Delta_1 \supset \Delta_2 \supset \cdots \supset \Delta_n \supset \cdots$ has a nonempty intersection, containing some point ζ. Since ζ belongs to all the intervals Δ_n, it cannot coincide with any of the points x_n. But then, contrary to our assumption, the sequence (1) does not include every point in $[0,1]$, and the theorem is proved by contradiction. ∎

2.42. It has already been shown (see Theorem 2.34) that the set of all rational numbers in $[0,1]$ is countable. Calling the remaining numbers *irrational*, we now see that $[0,1]$ contains "many more" irrational numbers than rational numbers. More exactly, the irrational numbers in $[0,1]$ form an uncountable set, since if the set of all irrational numbers in $[0,1]$ were countable, the set of all numbers in $[0,1]$ would also be countable (as the union of two countable sets). Moreover, the set of all irrational algebraic numbers, i.e., irrational roots of polynomials with rational coefficients, is countable, by Theorem 2.37. Hence the set of all *transcendental numbers*, i.e., numbers which are not roots of polynomials with rational coefficients, must be uncountable. Incidentally, this proves the very existence of transcendental numbers, a fact which is hardly obvious in advance.

2.43. Definition. Every set equivalent to the set of all points in the interval $[0,1]$ is called a *set with the power of the continuum.*

According to Sec. 2.22, the set of points in every closed interval, in every open interval, or even in the whole real line $-\infty < x < \infty$, is equivalent to the set of points in $[0,1]$. Hence these sets all have the power of the continuum.

2.5. Mathematical Structures

2.51. From a very general point of view, mathematics deals only with sets. But the richness of one or another mathematical theory depends on further connections between the elements (or subsets) of the sets figuring in the given theory. These connections are formulated abstractly with the help of axioms. For example, in Chapter 1 the real number system R is defined as a set whose elements satisfy a certain rather complicated system of axioms. A set equipped with extra conditions (on its elements or subsets) is called a *mathematical structure*. An exact definition of a mathematical structure should contain a general definition of such conditions, and can be given.†
Rather than go into this matter here, we confine ourselves to the above remarks, to be illustrated by a number of examples.

2.52. Isomorphic structures. Automorphisms. Two structures equipped with conditions of the same kind are said to be *isomorphic* if there is a one-to-one correspondence between the elements of the structures which preserves the validity (or nonvalidity) of the conditions defining the given structure. Every structure is isomorphic to itself, since the identity mapping (carrying every element into itself) is obviously one-to-one and preserves all conditions satisfied by the elements and subsets of the structure. But there may also be nonidentical one-to-one mappings of a structure onto itself (cf. Sec. 2.81) which preserve the conditions defining the structure. Such mappings are called *automorphisms* of the structure. In other words, an automorphism of a structure is an isomorphism between the structure and itself.

For example, consider the structure known as a *linearly ordered set*, by which we mean a set E of elements $x, y, \ldots$ with the extra condition that given any two distinct elements x and y, either $x < y$ ("x is less than y") or $y < x$, where moreover $x < y$, $y < z$ implies $x < y$ ("transitivity of the relation $<$"). Thus the (extended) real line is a linearly ordered set, and so is any subset of the real line (with the usual order relation $<$). Two linearly ordered sets E_1 and E_2 are said to be *isomorphic* if there is a one-to-one correspondence between E_1 and E_1 such that $x_1 \in E_1, y_1 \in E_1, x_1 < y_1$ implies $x_2 < y_2$ for the corresponding elements $x_2 \in E_2, y_2 \in E_2$. In this sense, the whole real line is isomorphic to any open interval, but not to a closed interval (unlike a closed interval, the real line has no least element). An automorphism of a linearly

† See N. Bourbaki, *Elements of Mathematics, Theory of Sets*, Addison-Wesley Publishing Co., Reading, Mass. (1968), Chapter 4.

ordered set is given by any one-to-one mapping$\leftrightarrow$of the set onto itself which preserves order (i.e., if $x_1 < y_1$, $x_1 \leftrightarrow x_2$, $y_1 \leftrightarrow y_2$, then $y_1 < y_2$). Thus the formula $x^n \leftrightarrow y$ represents an automorphism of the interval $(0,1)$, regarded as a linearly ordered set.

There are structures whose systems of axioms define them to within an isomorphism, i.e., such that any two structures satisfying the given system of axioms are isomorphic. The system of axioms defining the given structure is then called *complete*, and the structure itself is said to be *unique*.

2.53. To illustrate the above remarks, we now investigate the uniqueness of the real number system. It will be recalled from Chapter 1 that the real number system R is defined as any set of objects satisfying the axioms of Secs. 1.21–1.24. This suggests the following question: Is any system of objects satisfying all these axioms unique, i.e., are any two structures of real numbers isomorphic? More exactly, given any two "specimens" of the real number system, can we establish a one-to-one correspondence between the elements of the specimens such that the results of adding and multiplying numbers of the first specimen correspond to the results of adding and multiplying corresponding numbers of the second system, and such that order relations between both corresponding elements and least upper bounds of corresponding subsets are preserved? The answer to this question is in the affirmative, as shown by the following

THEOREM. *Let R_1 and R_2 be two specimens of the real number system, where elements of R_1 are equipped with the subscript 1 and those of R_2 with the subscript 2. Then there exists a one-to-one correspondence $x_1 \leftrightarrow x_2$ between the elements $x_1 \in R_1$ and $x_2 \in R_2$ such that*

(1) *If $x_1 \leftrightarrow x_2$, $y_1 \leftrightarrow y_2$, then $x_1 + y_1 \leftrightarrow x_2 + y_2$;*
(2) *If $x_1 \leftrightarrow x_2$, $y_1 \leftrightarrow y_2$, then $x_1 y_1 \leftrightarrow x_2 y_2$;*
(3) *If $x_1 \leftrightarrow x_2$, $y_1 \leftrightarrow y_2$, $x_1 < y_1$, then $x_2 < y_2$;*
(4) *If the set $A_1 \subset R_1$ is bounded from above, then so is the set A_2 consisting of the corresponding elements of R_2, and moreover*

$$\sup A_1 \leftrightarrow \sup A_2.$$

Proof. The proof will be given in several steps (Secs. 2.54–2.58). First we note in passing that the isomorphism of R_1 and R_2 as structures is already implicit in the possibility of representing the elements of R_1 and R_2 as decimals, as in Sec. 1.77. However, here we use a different proof, which is particularly suited to the direct verification of Properties 1–4 above.

2.54. *Step 1.* Let Q_1 be the set of all rational numbers in R_1 and Q_2 the set of all rational numbers in R_2. Then there is a natural one-to-one correspondence between Q_1 and Q_2, established by the very way of writing rational numbers in the form m/n. Properties 1–3 obviously hold for the elements of Q_1 and Q_2, since the operations and relations in question can be expressed directly in terms of the symbols m and n.

Now let $\xi_1 \in R_1$ be arbitrary. Then, by Theorem 1.76,

$$\xi_1 = \sup N_1(\xi_1) = \inf P_1(\xi_1),$$

where $N_1(\xi_1)$ is the set of all rational numbers $r_1 \leqslant \xi_1$ and $P_1(\xi_1)$ is the set of all rational numbers $s_1 \geqslant \xi_1$ (note that $r_1 \leqslant s_1$ always holds). Let N_2 be the set of all rational numbers in R_2 corresponding to the rational numbers in $N_1(\xi_1)$, i.e., with the same representations as fractions of the form m/n, and let P_2 be the set of all rational numbers in R_2 corresponding to the rational numbers in $P_1(\xi_1)$. Then clearly $r_2 \leqslant s_2$ for all $r_2 \in N_2$, $s_2 \in P_2$, and hence

$$\sup N_2 \leqslant \inf P_2,$$

by Theorem 1.62b. Since all the rational numbers in R_2 fall either in N_2 or in P_2, there are no rational numbers between $\alpha_2 = \sup N_2$ and $\beta_2 = \inf P_2$. But then $\alpha_2 = \beta_2$, since otherwise there would be a rational point between α_2 and β_2, by Theorem 1.75. We now associate the number $\xi_2 = \alpha_2 = \beta_2 \in R_2$ with the given number $\xi_1 \in R_1$. The fact that this correspondence is one-to-one follows by "symmetry" (give the details).

2.55. *Step 2.* Next we prove that the correspondence $\leftrightarrow$ just constructed has Properties 3 and 4. If $x_1 < y_1$, then, by Theorem 1.75, there is a rational number r_1 such that $x_1 < r_1 < y_1$. But $x_2 < r_2 < y_2$ for the corresponding numbers $x_2, r_2, y_2 \in R_2$, by the very definition of the correspondence, and hence Property 3 holds. Next, given any set $A_1 \subset R_1$ which is bounded from above, let $\xi_1 = \sup A_1$, let $A_2 \subset R_2$ be the set corresponding to A_1, and let $\xi_2 \in R_2$ be the number corresponding to ξ_1. Then $\xi_1 \geqslant x_1$ for every $x_1 \in A_1$, and hence $\xi_2 \geqslant x_2$ for every $x_2 \in A_2$, so that $\sup A_2 \leqslant \xi_2$ (in particular, this shows that A_2 is bounded from above). If $\sup A_2 < \xi_2$, then there is a number (in fact, a rational number) y_2 such that $x_2 < y_2 < \xi_2$ for every $x_2 \in A_2$, so that $x_1 < y_1 < \xi_1$ for every $x_1 \in A_1$, where y_1 corresponds to y_2. But then

$$\xi_1 = \sup A_1 \leqslant y_1 < \xi_1,$$

which is impossible. Therefore $\sup A_2 = \xi_2$, which proves Property 4.

2.56. *Step 3.* To prove Properties 1 and 2, we will need the following two lemmas:

a. LEMMA. *Given any real numbers y, z, and any rational number $r > y + z$, there are rational numbers $u > y$, $v > z$ such that $u + v = r$.*

Proof. Let $r - (y + z) = h$. Choosing u to be any rational number in the interval $(y, y+h)$, let $v = r - u$. Then obviously v is rational and

$$v > r - (y + h) = z. \quad \blacksquare$$

b. LEMMA. *Given any positive real numbers y, z and any rational number $r > yz$, there are rational numbers $u > y$, $v > z$ such that $uv = r$.*

Proof. Let $r/yz = h > 1$. Choosing u to be any rational number in the interval (y, yh), let $v = r/u$. Then obviously v is rational and

$$v > r/yh = z. \quad \blacksquare$$

2.57. *Step 4.* We are now able to verify Property 1. Suppose $x_1 \leftrightarrow x_2, y_1 \leftrightarrow y_2$, $x_1 + y_1 \leftrightarrow z_2 \neq x_2 + y_2$, where $z_2 > x_2 + y_2$, say, and let t_2 be a rational number in the interval $(x_2 + y_2, z_2)$. Then, by Lemma 2.56a, there are rational numbers $u_2 > x_2$, $v_2 > y_2$ such that $u_2 + v_2 = t_2$. For the corresponding numbers z_1, t_1, u_1, v_1, we have $x_1 < u_1, y_1 < v_1, x_1 + y_1 < u_1 + v_1 = t_1$. But $t_2 < z_2 \leftrightarrow x_1 + y_1$ implies $t_1 < x_1 + y_1$, which is incompatible with $t_2 \in (x_2 + y_2, z_2)$. An analogous argument shows that $z_2 < x_2 + y_2$ is also impossible, and hence

$$z_2 = x_2 + y_2.$$

2.58. *Final step.* Turning to Property 2, we first consider the case of positive x_1 and y_1, which by Property 3 corresponds to positive x_2 and y_2. Suppose $x_1 y_1 \leftrightarrow z_2 \neq x_2 y_2$, where $z_2 > x_2 y_2$, say, and let t_2 be a rational number in the interval $(x_2 y_2, z_2)$. Then, by Lemma 2.56b, there are rational numbers $u_2 > x_2$, $v_2 > y_2$ such that $u_2 v_2 = t_2$. For the corresponding numbers z_1, t_1, u_1, v_1, we have $x_1 < u_1$, $y_1 < v_1$, $x_1 y_1 < u_1 v_1 = t_1$. But $t_2 < z_2 \leftrightarrow x_1 y_1$ implies $t_1 < x_1 y_1$, which is incompatible with $t_2 \in (x_2 y_2, z_2)$. An analogous argument shows that $z_2 < x_2 y_2$ is also impossible, and hence $z_2 = x_2 y_2$.

To prove the general case, we need only verify that $-1 \cdot x_1 \leftrightarrow -1 \cdot x_2$ (because of the usual sign rule for multiplication). Suppose $-1 \cdot x_1 \leftrightarrow z_2$. It follows from $-1 \cdot x_1 + x_1 = 0$ that $z_2 + x_2 = 0$ and hence $z_2 = -1 \cdot x_2$. The proof of our theorem is now complete. $\quad \blacksquare$

2.59. Having just proved the uniqueness of the real number system R to within an isomorphism, we now look for automorphisms of R, by which we mean one-to-one mappings of R onto itself preserving sums, products, inequalities, and least upper bounds (cf. Secs. 2.52–2.53). It turns out that the only automorphism of R is the identity mapping, carrying every element of R into itself. To see this, consider an automorphism of R carrying the element $x \in R$ into the element $x' \in R$. Then the elements $0'$ and $1'$ are the zero and unit elements in the field R, since $0' + x' = x'$ and $1' \cdot x' = x'$ for every x'. Therefore $0' = 0$, $1' = 1$, by the uniqueness of the zero and unit elements

(Theorems 1.31 and 1.41a). It follows that

$$2' = 1' + 1' = 2, \ldots, n' = (n-1)' + 1' = n$$

and

$$\left(\frac{m}{n}\right)' = \frac{m'}{n'} = \frac{m}{n} \qquad (m = 0, 1, 2, \ldots; \ n = 1, 2, \ldots).$$

Moreover

$$-\left(\frac{m}{n}\right)' = -\frac{m}{n},$$

by the uniqueness of the negative element (Theorem 1.32). Thus our automorphism leaves the rational numbers unchanged. But then it also leaves every real number ξ unchanged, since, by Theorem 1.76,

$$\xi = \sup N_\xi,$$

where N_ξ is the set of all rational numbers $\leqslant \xi$, and any automorphism of R must preserve inequalities and least upper bounds. Hence our automorphism is just the identity mapping $\xi \leftrightarrow \xi$.

2.6. n-Dimensional Space

2.61. Definition. Given a positive integer n, let R_n be the set of all ordered n-tuples

$$x = (x_1, \ldots, x_n)$$

consisting of n real numbers $x_1, \ldots, x_n$. Then x is called a *vector*, with *components* $x_1, \ldots, x_n$.† Two vectors $x = (x_1, \ldots, x_n)$ and $y = (y_1, \ldots, y_n)$ are said to be *equal* if

$$x_1 = y_1, \ldots, x_n = y_n.$$

The set R_n itself is called *n-dimensional (real) space*.

2.62. The operation of addition of elements of R_n is defined by the formula

$$(x_1, \ldots, x_n) + (y_1, \ldots, y_n) = (x_1 + y_1, \ldots, x_n + y_n). \tag{1}$$

Addition in R_n satisfies all the axioms of Sec. 1.21, with the symbols $x, y, \ldots$ regarded as elements of R_n. In fact, if $x = (x_1, \ldots, x_n)$, $y = (y_1, \ldots, y_n)$, $z = (z_1, \ldots, z_n)$, then clearly

† Alternatively, we sometimes call x a *point* of R_n, with *coordinates* $x_1, \ldots, x_n$.

(a) $x+y=(x_1+y_1,...,x_n+y_n)=y+x$;

(b) $(x+y)+z=(x_1+y_1+z_1,...,x_n+y_n+z_n)=x+(y+z)$;

(c) R_n contains an element $0=(0,...,0)$, called the *zero vector*, such that $x+0$ $=(x_1,...,x_n)+(0,...,0)=x$ for every $x \in R_n$;

(d) For every $x \in R_n$ there exists an element $y \in R_n$, called the *negative* (or *opposite*) of the vector x, such that $x+y=0$, namely the element $y=$ $(-x_1,...,-x_n)$.

Naturally, addition in R_n satisfies not only Axioms a–d, but also all the consequences of these axioms (given in Sec. 1.3).

2.63. The operation of multiplication of elements of R_n by real numbers $\alpha \in R$ is defined by the formula

$$\alpha(x_1,...,x_n)=(\alpha x_1,...,\alpha x_n). \tag{2}$$

Then obviously

$$\alpha(x+y)=\alpha x+\alpha y,$$
$$(\alpha+\beta)x=\alpha x+\beta x,$$
$$1 \cdot x=x,$$
$$\alpha(\beta x)=(\alpha\beta)x.$$

2.64. Definition. We say that m vectors

$$x^{(1)}=(x_1^{(1)},...,x_n^{(1)}),...,x^{(m)}=(x_1^{(m)},...,x_n^{(m)})$$

in the space R_n are *linearly dependent* if there exist constants $c_1,...,c_m$, *not all zero*, such that

$$c_1 x^{(1)}+\cdots+c_m x^{(m)}=0. \tag{3}$$

If (3) holds only if $c_1=\cdots=c_m=0$, the vectors $x^{(1)},...,x^{(m)}$ are called *linearly independent*.

According to (1) and (2), the "vector equation" (3) is equivalent to the following system of "numerical equations":

$$c_1 x_1^{(1)}+\cdots+c_m x_1^{(m)}=0,$$
$$c_1 x_2^{(1)}+\cdots+c_m x_2^{(m)}=0,$$
$$\cdots \tag{3'}$$
$$c_1 x_n^{(1)}+\cdots+c_m x_n^{(m)}=0.$$

The reader familiar with linear algebra will infer from (3') that the vectors $x^{(1)},...,x^{(n)}$ are linearly dependent if and only if the rank of the matrix $\|x_j^{(k)}\|$ is less than m.† It should also be noted that any nonsingular square matrix

† See e.g., G. E. Shilov, *Linear Algebra* (translated by R. A. Silverman), Dover Publications, Inc., N.Y. (1977), Theorem 3.12a.

$\|x_j^{(k)}\|$ of order n specifies a system of n linearly independent vectors in R_n, while any $n + 1$ vectors in R_n are linearly dependent.†

2.65. Let $f_1,\ldots,f_n$ be any system of n linearly independent vectors in R_n. Then, given any vector $g \in R_n$, it follows from the linear dependence of the vectors $f_1,\ldots,f_n,g$ that

$$c_0 g + c_1 f_1 + \cdots + c_n f_n = 0, \qquad (4)$$

where certainly $c_0 \neq 0$, since otherwise the vectors $f_1,\ldots,f_n$ would be linearly dependent. Hence we can solve (4) for g, obtaining a relation of the form

$$g = \alpha_1 f_1 + \cdots + \alpha_n f_n.$$

The numbers $\alpha_1,\ldots,\alpha_n$ are called the *components* of the vector g with respect to the *basis* $f_1,\ldots,f_n$.

For example, choosing $f_1,\ldots,f_n$ to be the vectors

$$e_1 = (1,0,\ldots,0),\ldots, e_n = (0, 0,\ldots,1), \qquad (5)$$

we find that any vector $x = (x_1,\ldots,x_n) \in R_n$ can be represented in the form

$$x = x_1 e_1 + \cdots + x_n e_n,$$

so that the numbers $x_1,\ldots,x_n$ are just the components of the vector x with respect to the special basis (5).

2.66. Given a fixed basis $f_1,\ldots,f_n$ in the space R_n, let

$$g = \alpha_1 f_1 + \cdots + \alpha_n f_n, \qquad h = \beta_1 f_1 + \cdots + \beta_n f_n.$$

Then, by Sec. 2.63,

$$g + h = (\alpha_1 + \beta_1) f_1 + \cdots + (\alpha_n + \beta_n) f_n,$$
$$cg = c\alpha_1 f_1 + \cdots + c\alpha_n f_n \text{ for every } c \in R,$$

so that addition of vectors and multiplication of a vector by a number c leads (in any basis) to addition of the corresponding components and multiplication of these components by c. Thus the mapping which associates the vector $(\alpha_1,\ldots,\alpha_n)$ with every vector $x = (x_1,\ldots,x_n)$, where $\alpha_1,\ldots,\alpha_n$ are the components of x with respect to a fixed basis $f_1,\ldots,f_n$, is an automorphism (Sec. 2.52) of the n-dimensional space R_n.

2.67.‡ More generally, let A be any automorphism of the n-dimensional space R_n. Suppose A carries the element $x \in R_n$ into the element $x' \in R_n$, a fact indicated by writing $A: x \mapsto x'$ or $x' = A(x)$. Being an automorphism of

---

† Shilov, *Linear Algebra*, Theorems 3.12a and 3.12b.
‡ This section can be omitted by readers unfamiliar with elementary linear algebra.

R_n, A preserves sums of vectors and products of vectors with real numbers, so that

(a) $A(x+y) = A(x) + A(y)$ for every $x, y \in R_n$;

(b) $A(\alpha x) = \alpha A(x)$ for every $x \in R_n$ and $\alpha \in R$.

Properties a and b immediately imply the following more general formula:

(c) $A(\sum\limits_{k=1}^{p} \alpha_k x_k) = \sum\limits_{k=1}^{p} \alpha_k A(x_k)$ for every $x_1,\ldots,x_p \in R_n$ and $\alpha_1,\ldots,\alpha_p \in R$.

The effect of the automorphism A on the components of the vector x can be expressed in terms of its effect on the vectors of the basis (5). Let

$$x = (x_1,\ldots,x_n), \qquad x' = A(x) = (x'_1,\ldots,x'_n),$$

and suppose

$$e'_k = A(e_k) = (a_{1k},\ldots,a_{nk}),$$

where e_k is the kth vector of the basis (5), with 1 as its kth component and 0 for all its other components. Then clearly

$$x' = \sum_{j=1}^{n} x'_j e_j = A(x) = A\left(\sum_{k=1}^{n} x_k e_k\right) = \sum_{k=1}^{n} x_k A(e_k)$$

$$= \sum_{k=1}^{n} x_k \left(\sum_{j=1}^{n} a_{jk} e_k\right) = \sum_{k=1}^{n} \left(\sum_{j=1}^{n} a_{jk} x_k\right) e_k,$$

which implies

$$x'_j = \sum_{k=1}^{n} a_{jk} x_k \qquad (j=1,\ldots,n). \tag{6}$$

These linear relations, describing the effect of the automorphism A on the components of the vector x, must give a one-to-one mapping (or "transformation") of the space R_n onto itself. It follows that the square matrix $\|a_{jk}\|$ is nonsingular, i.e., has a nonvanishing determinant. Conversely, every nonsingular matrix $\|a_{jk}\|$ specifies a one-to-one mapping of the space R_n onto itself, in accordance with the formulas (6), satisfying Properties a and b.† Thus, finally, *every automorphism of the space R_n is specified by a nonsingular matrix $\|a_{jk}\|$ via the formulas* (6).

We now give some examples illustrating the above considerations. The *identity transformation I* specified by the matrix

$$\begin{Vmatrix} 1 & 0 & \cdots & 0 \\ 0 & 1 & \cdots & 0 \\ \cdot & \cdot & \cdots & \cdot \\ 0 & 0 & \cdots & 1 \end{Vmatrix}$$

† Shilov, *Linear Algebra*, Sec. 4.76.

carries every vector into itself. The automorphism Λ_j specified by the matrix

$$
(\text{row } j) \quad \begin{Vmatrix} 1 & & & & & & 0 \\ & \ddots & & & & & \\ & & 1 & & & & \\ & & & \lambda_j & & & \\ & & & & 1 & & \\ & & & & & \ddots & \\ 0 & & & & & & 1 \end{Vmatrix}
$$

(with all unwritten elements equal to 0) magnifies the jth component of every vector λ_j times, and is accordingly called a λ_j-*fold expansion* along the x_j-axis. The automorphism specified by the matrix

$$
\begin{Vmatrix} \lambda_1 & & & 0 \\ & \lambda_2 & & \\ & & \ddots & \\ 0 & & & \lambda_n \end{Vmatrix} \qquad (\lambda_1 \lambda_2 \cdots \lambda_n \neq 0)
$$

produces a λ_1-fold expansion along the x_1-axis, a λ_2-fold expansion along the x_2-axis, and so on. If the numbers $\lambda_1, \lambda_2, \ldots, \lambda_n$ are all equal, the corresponding automorphism, specified by the matrix

$$
\begin{Vmatrix} \lambda & & & 0 \\ & \lambda & & \\ & & \ddots & \\ 0 & & & \lambda \end{Vmatrix} \qquad (\lambda \neq 0)
$$

carries every vector into the vector λx. Such an automorphism is called a *similarity transformation*, with *ratio of similitude* λ.

2.7. Complex Numbers

2.71. The following question now arises quite naturally: Can an operation of multiplication of vectors be introduced in the space R_n, associating a vector xy with every pair of vectors $x, y \in R_n$, such that the multiplication axioms a–e of Sec. 1.22 are satisfied? It turns out that the answer to this question is in the affirmative if $n = 2$ (cf. Sec. 2.76). One way of introducing multiplication in R_2 goes as follows.

Let

$$
e_1 = (1,0), \qquad e_2 = (0,1).
$$

Then every vector $z \in R_2$ can be written in the form

$$z = (x,y) = xe_1 + ye_2,$$

where x and y are the components of z with respect to the basis e_1, e_2. Now let

$$e_1^2 = e_1, \qquad e_2^2 = -e_1, \qquad e_1 e_2 = e_2 e_1 = e_2, \tag{1}$$

and use "linearity" to extend this operation of multiplication to all vectors $z = xe_1 + ye_2 \in R_2$. To be more explicit, if $w = ue_1 + ve_2$, then

$$zw = (xe_1 + ye_2)(ue_1 + ve_2)$$
$$= xue_1^2 + xve_1 e_2 + yue_2 e_1 + yve_2^2 = (xu - yv)e_1 + (xv + yu)e_2.$$

Thus the definition of multiplication in R_2 is just

$$zw = (x,y)(u,v) = (xu - yv, xv + yu). \tag{2}$$

The product (2) obviously has the following properties:
(a) $wz = (ux - vy, vx + uy) = zw$;
(b) If $\gamma = (\alpha, \beta)$, then

$$(zw)\gamma = (xu\alpha - yv\alpha - xv\beta - uy\beta, \ xu\beta - yv\beta + xv\alpha + yu\alpha)$$
$$= (x(u\alpha - yv) - y(v\alpha + u\beta), \ x(v\alpha + u\beta) + y(u\alpha - v\beta)) = z(w\gamma);$$

(c) R_2 contains an element $e = (1,0)$, called the *unit element*, such that $(1,0)(u,v) = (u,v)$ for every $(u,v) \in R_2$;
(d) For every $(x,y) \neq (0,0)$ in R_2 there exists an element $(u,v) \in R_2$, called the *reciprocal* of x, such that $(x,y)(u,v) = 1$, namely the element

$$(u,v) = \left(\frac{x}{x^2 + y^2}, \ -\frac{y}{x^2 + y^2} \right),$$

since clearly

$$(x,y)(u,v) = \left(\frac{x^2}{x^2 + y^2} + \frac{y^2}{x^2 + y^2}, \ -\frac{xy}{x^2 + y^2} + \frac{xy}{x^2 + y^2} \right) = (1,0);$$

(e) The formula

$$\gamma(z + w) = \gamma z + \gamma w$$

holds for every $z = (x,y)$, $w = (u,v)$, $\gamma = (\alpha, \beta)$ in R_2, since

$$\gamma(z + w) = (\alpha, \beta)(x + u, y + v)$$
$$= (\alpha(x + u) - \beta(y + v), \ \beta(x + u) + \alpha(y + v))$$
$$= ((\alpha x - \beta y) + (\alpha u - \beta v), \ (\beta x + \alpha y) + (\beta u + \alpha v))$$
$$= (\alpha x - \beta y, \ \beta x + \alpha y) + (\alpha u - \beta v, \ \beta u + \alpha v)$$
$$= (\alpha, \beta)(x,y) + (\alpha, \beta)(u,v) = \gamma z + \gamma w.$$

Thus multiplication in R_2, as just defined, satisfies not only Axioms a–e of Sec. 1.22, but also all the consequences of these axioms (given in Sec. 1.4).

2.72. Definition. The space R_2 equipped with the addition operation

$$(x,y) + (u,v) = (x+u, y+v)$$

and the multiplication operation (2) is called the *field of complex numbers* (synonmously, the *complex number system* or *complex plane*), denoted by C. The one-to-one correspondence

$$x \leftrightarrow (x,0) \tag{3}$$

between the real number system R and the complex number system C preserves sums and products, since

$$(x,0) + (y,0) = (x+y,0), \qquad (x,0)(y,0) = (xy,0).$$

In this sense, the correspondence (3) "embeds" the system R in the system C. Hence the complex number $(x,0)$ can be identified with the real number x, a fact expressed by writing

$$(x,0) = x.$$

In particular, the basis vector $e_1 = (1,0)$ can be denoted by 1. Denoting the second basis vector $e_2 = (0,1)$ by i, we can now write the complex number (x,y) in the form $x+iy$. Using (1), we have

$$i^2 = e_2^2 = -e_1 = (-1,0) = -1,$$

and hence $i = \sqrt{-1}$. Moreover, since

$$(-i)^2 = (-1)^2 i^2 = i^2 = -1,$$

the number $-i$ is another complex square root of -1. There are no other complex square roots of -1, since if $z^2 = -1$ for some $z \in C$, then, by Sec. 1.47b (together with the concluding remark of Sec. 2.71), either $z+i=0$ or $z-i=0$, i.e., either $z=i$ or $z=-i$.

In terms of the "imaginary unit" $i = \sqrt{-1}$, we can now write the multiplication rule (2) in the form

$$(x+iy)(u+iv) = (xu-yv) + i(xv+yu),$$

in keeping with the usual rule for multiplying binomials (together with the formula $i^2 = -1$).

Given a complex number $z = x+iy$, the number x is called the *real part* of z, denoted by

$$x = \operatorname{Re} z,$$

while the number y is called the *imaginary part* of z, denoted by

$$y = \operatorname{Im} z.$$

If Re $z = 0$, Im $z \neq 0$, the number z is said to be *purely imaginary*. The set of all complex numbers z with Im $z = 0$ is called the *real axis*, while the set of all z with Re $z = 0$ is called the *imaginary axis*. Because of the correspondence $x \leftrightarrow (x,0) \leftrightarrow x + i0$, the real axis can be identified with the set of all real numbers (i.e., the "real line").

2.73. The complex numbers $x + iy$ and $x - iy$ are said to be *conjugate complex numbers* or *complex conjugates* (of each other). If one of these numbers is denoted by z, the other is denoted by $\bar{z}$.

a. THEOREM. *The formula*

$$z = \bar{z} \tag{4}$$

holds if and only if z is real.

Proof. If $z = x + iy$, then (4) holds if and only if

$$x + iy = x - iy,$$

i.e., if and only if $y = 0$, in which case $z = x$ is real. ∎

b. THEOREM. *The formulas*

$$\overline{z_1 + z_2} = \bar{z}_1 + \bar{z}_2, \tag{5}$$

$$\overline{z_1 z_2} = \bar{z}_1 \bar{z}_2 \tag{6}$$

hold for arbitrary $z_1, z_2 \in C$.

Proof. If $z_1 = x_1 + iy_1$, $z_2 = x_2 + iy_2$, then

$$\overline{z_1 + z_2} = (x_1 + x_2) - i(y_1 + y_2) = (x_1 - iy_1) + (x_2 - iy_2) = \bar{z}_1 + \bar{z}_2,$$

which proves (5), while

$$\overline{z_1 z_2} = (x_1 x_2 - y_1 y_2) - i(x_1 y_2 + x_2 y_1)$$
$$= (x_1 - iy_1)(x_2 - iy_2) = \bar{z}_1 \bar{z}_2,$$

which proves (6). ∎

In particular, (6) implies $\overline{\alpha z} = \alpha \bar{z}$ for all real α.

2.74. Uniqueness of the complex number system

THEOREM. *Let R_2 be a two-dimensional space equipped with a multiplication operation satisfying the axioms of Sec. 1.22. Then R_2 contains two linearly independent* *vectors g_1 and g_2 such that*

$$g_1^2 = g_1, \qquad g_2^2 = -g_1, \qquad g_1 g_2 = g_2 g_1 = g_2.$$

Proof. By Axiom c of Sec. 1.22, R_2 contains a nonzero unit element 1. Let $g_1 = 1$, and let f_1 be any vector linearly independent of g_1. Then

$$f_1^2 = \alpha g_1 + \beta f_1$$

for suitable real α and β. For the vector $f_2 = f_1 + \gamma g_1 \neq 0$ (γ real) we have

$$(f_1 + \gamma g_1)^2 = f_1^2 + 2\gamma f_1 + \gamma^2 g_1 = \alpha g_1 + \beta f_1 + 2\gamma f_1 + \gamma^2 g_1,$$

and hence, choosing $\gamma = -\beta/2$, we get

$$f_2^2 = (\alpha + \gamma^2) g_1.$$

By Theorem 1.47a, the real number $\alpha + \gamma^2$ cannot equal 0. Moreover, $\alpha + \gamma^2$ cannot be positive, since then we would have

$$(f_2 - \sqrt{\alpha + \gamma^2} g_1)(f_2 + \sqrt{\alpha + \gamma^2} g_1) = 0,$$

contrary to Sec. 1.47b. It follows that

$$\alpha + \gamma^2 = -\delta \qquad (\delta > 0).$$

Replacing f_2 by

$$g_2 = \frac{1}{\sqrt{\delta}} f_2,$$

we get

$$g_2^2 = \frac{(\alpha + \gamma^2) g_1}{\delta} = -g_1.$$

Finally, the fact that g_1 is the unit element implies $g_1 g_2 = g_2 g_1 = g_2$. ∎

Let R_2 and R_2' both be two-dimensional spaces, each equipped with a multiplication operation satisfying the axioms of Sec. 1.22. Then, according to Theorem 2.74, we can establish a one-to-one correspondence between R_2 and R_2' preserving multiplication, i.e., the corresponding structures (two-dimensional spaces equipped with multiplication) are isomorphic.

2.75. Automorphisms of the complex number system. As shown in Sec 2.59, the only automorphism of the real number system is the identity automorphism $\xi \leftrightarrow \xi$. We now turn to automorphisms of the complex number system C. By an *automorphism* of C is meant a one-to-one mapping $z \leftrightarrow z'$ of C onto itself such that, given arbitrary $z_1, z_2 \in C$ and real α,

(1) $(z_1 + z_2)' = z_1' + z_2';$

(2) $(z_1 z_2)' = z_1' z_2'$;
(3) $(\alpha z_1)' = \alpha z_1'$.

Properties 1 and 3 show that an automorphism of C as defined here is an automorphism of C regarded as a two-dimensional space (see Sec. 2.67), while Property 3 says that our automorphism must also preserve products.

THEOREM. *The automorphism $x + iy \leftrightarrow x - iy$ is the only automorphism of C other than the identity automorphism.*

Proof. Let A be an automorphism of C carrying the element $x + iy$ into the element $(x + iy)'$. Then Property 3 implies

$$1' \cdot z' = (1 \cdot z)' = z',$$

so that $1'$ is the unit element of C. Hence $1' = 1$, by the uniqueness of the unit element. Moreover

$$x' = (1 \cdot x)' = 1' \cdot x = 1 \cdot x = x$$

by Property 3, so that the automorphism A leaves the whole set of real numbers unchanged. Finally, by Property 2,

$$(i')^2 = i^2 = -1,$$

so that i' equals either i or $-i$ (cf. Sec. 2.72). In the first case,

$$(x + iy)' = x' + iy' = x + iy,$$

while, in the second case,

$$(x + iy)' = x' + i'y' = x - iy. \quad \blacksquare$$

The study of complex numbers will be pursued in Chapter 5 (Sec. 5.72 ff).

2.76. It is natural to ask whether a multiplication operation satisfying Axioms a–e of Sec. 1.22 can be defined for vectors of a general n-dimensional space R_n. According to a theorem of Frobenius which will not be proved here,[†] this cannot be done if $n > 2$.

2.8. Functions and Graphs

2.81. Given two sets X and Y, suppose a unique element $y \in Y$ is associated with each element $x \in X$, in accordance with some rule or "mapping" f, a fact indicated by writing $f: x \mapsto y$ or $y = f(x)$.[‡] Then f is called a *function* with

---

† For the proof, see A. G. Kurosh, *Lectures on General Algebra* (translated by K. A. Hirsch), Chelsea Publishing Co., N. Y. (1963), Sec. 38.
‡ Cf. Secs. 2.21 and 2.67.

domain (of definition) X,† while x is called the *independent variable* (or *argument*) of the function f and y is called the *dependent variable* (or *value*) of f. Following the customary slight abuse of notation, we will often refer to the "function $f(x)$" as well as to the "function f," although, strictly speaking, $f(x)$ is the value of f corresponding to the argument x.

It should be emphasized that there is no need for *every* $y \in Y$ to be a value of f for some $x \in X$, nor is there any need for distinct values of f to correspond to distinct values of x. However, if both of these extra conditions are satisfied, then f becomes a *one-to-one function*, establishing a one-to-one correspondence (Sec. 2.21) between the sets X and Y, or synonymously, "mapping X onto Y in a one-to-one fashion."

If the set Y in which $f(x)$ takes its values is the real line R, we call $f(x)$ a *numerical* (or *real*) *function*, while if Y is the "vector space" R_n of Sec. 2.6, we call $f(x)$ a *vector function*. If the domain X of the function $f(x)$ is the real line R, the extended real line $\overline{R}$ (Sec. 1.91) or some set $E \subset \overline{R}$, we call $f(x)$ a *function of a real variable*.

If the domain of $f(x)$ is the set of positive integers $1,2,3,\ldots$, we call $f(x)$ a *sequence* of points in Y. In this case, we usually denote $f(x)$ by $f(n)$ or f_n. Note that the concept of a sequence of points in a set Y does not reduce to that of a (at most countable) subset of Y, since points can "repeat themselves" in a sequence but not in a subset. For example, the sequence

$$f_1 = a, \quad f_2 = a, \quad f_3 = a,\ldots$$

(where a is a fixed element of Y) is certainly not the same as the set consisting of the single element a, while the sequence

$$f_1 = a, \quad f_2 = b_1, \quad f_3 = a, \quad f_4 = b_2, \quad f_5 = a, \quad f_6 = b_3,\ldots$$

is certainly not the same as the countable subset $\{a, b_1, b_2, b_3, \ldots\} \subset Y$.

2.82. Given any two sets X and Y, by the *direct product* of X and Y, written $X \times Y$, we mean the set of all ordered pairs (x,y) where the first element x belongs to X and the second element y belongs to Y. For example, the two-dimensional space R_2 (the "xy-plane") is the direct product of two real lines $R_1 = R$. Two elements of $X \times Y$, i.e., two ordered pairs (x,y) and (x',y'), are said to be *equal* if $x = x', y = y'$.

Suppose we fix an element $y_0 \in Y$ and consider the subset of all points $(x,y_0) \in X \times Y$. This subset, which is obviously equivalent to the set X itself, is called the *section* of $X \times Y$ corresponding to the element y_0. The whole direct product $X \times Y$ is clearly the union of all the different sections of $X \times Y$, each equivalent to X (and to every other section as well).

† We also call f a function (*defined*) *on* X. The set of all values actually taken by f, i.e., the set $\{y : y = f(x), x \in X\}$, is called the *range* of f.

2.83. By the *graph* of a given function $y = f(x)$ with domain X and values in Y we mean the subset of the direct product $X \times Y$ consisting of all pairs (x, y) for which $y = f(x)$. If $X = R_1, Y = R_1$, this coincides with the usual definition of the graph of a numerical function of a real variable; in most cases of practical importance, the graph of such a function is represented by some curve in the xy-plane. If the domain X of the function $f(x)$ is the plane of points (x_1, x_2) while the set Y is the real line, then the graph of $f(x)$ is some set in R_3 which can often be thought of as a surface. If the domain X of the function $f(x)$ is the real line while the set Y is the plane of points (y_1, y_2), then the graph of $f(x)$ is again a set in R_3, but this time the graph is best thought of as a curve (since it intersects every plane $x = \text{const}$ in just one point). The above examples constitute some of the most important objects studied in mathematical analysis.

2.84. "Single-valued" versus "multiple-valued" functions. According to the definition of Sec. 2.81, every function f is *single-valued* in the sense that it associates a *unique* element $y \in Y$ with each element $x \in X$. However, the expression *multiple-valued function* is often encountered in the literature, by which is meant a "function" which associates not just one but several elements $y \in Y$ with each $x \in X$. We could extend the definition of a "function" in this direction, but we prefer not to do so, since it would lead to difficulties in defining operations on such "functions." Nevertheless, such functions will be found useful now and then. For example, several single-valued functions are often combined in a single formula for the sake of brevity, and such a combination can be regarded as a "multiple-valued" function. Thus the "double-valued" function

$$y = x \pm \sqrt{1 - x^2} \qquad (-1 \leqslant x \leqslant 1)$$

is simply a combination of the two single-valued functions

$$y_1 = x + \sqrt{1 - x^2}, \qquad y_2 = x - \sqrt{1 - x^2} \qquad (-1 \leqslant x \leqslant 1).$$

Problems

1. Prove that the following sets are countable:
(a) The set of all intervals $a < x < b$ with rational end points;
(b) The set of all polygonal lines in the plane with finitely many segments and vertices at rational points.
2. Prove that the following sets are either finite or countable:
(a) The set of all pairwise nonintersecting intervals on the line;
(b) The set of all closed self-intersecting "figure eights" in the plane with no points in common;

(c) The set M of all positive real numbers such that all finite sums

$$\sum x_j \qquad (x_j \in M)$$

are bounded by a fixed number A.

Comment. Analogous results have been proved for the set of all nonintersecting figures in the plane with triple points (like the letter T), as well as for the set of all nonintersecting figures in space with singular points of the "button" type or pieces of the "Möbius strip" type.†

3. Represent the set of natural numbers 1,2,... as the union of a countable collection of nonintersecting countable sets.

4 (*A riddle*). I. As related by the mathematician X, he was once visited by the brothers N, who, upon entering, took off their hats and hung them up on a rack in the hall. Later, when the guests were leaving and getting ready to put on their hats, it turned out to the host's chagrin that one hat was missing, although nobody had entered the hall during the time of the visit.

II. When the brothers N visited X on another occasion, they again hung up their hats on the rack in the hall. Later, when the guests were leaving and getting ready to put on their hats, it turned out that there was an extra hat, although both the host and the guests were certain that there had been no hat on the rack when the guests arrived.

III. On the next visit, the guests put on their hats and left, and the host accompanied them to the street. Upon returning, he discovered the same number of hats on the rack as before the guests had left.

IV. Finally, on still another visit, the guests arrived without hats, and, upon leaving, put on the hats left over from the last visit. After accompanying the guests to the street, the host returned to discover once again the same number of hats on the rack as before the guests had left.

Explain all these seemingly paradoxical events!

5. Suppose we add the elements of a finite or countable set B to an infinite set A. Prove that the resulting set is equivalent to the original set A.

6. Prove that the set I of all irrational numbers and the set T of all transcendental numbers both have the power of the continuum.

7. Prove that the set A of all sequences consisting only of zeros and ones has the power of the continuum.

8. Prove that the set of all increasing sequences of natural numbers $n_1, n_2, \ldots$ $(0 < n_1 < n_2 < \ldots)$ has the power of the continuum.

9. Prove that the set of all sequences of natural numbers $n_1, n_2, \ldots$ (not necessarily increasing) has the power of the continuum.

† See V. V. Grushin and V. P. Palamodov, Uspekhi Mat. Nauk, vol. 17, no. 3 (1962), pp. 163–168.

10. Prove that the set of all sequences of real numbers ξ_1, ξ_2,... has the power of the continuum.

11. Prove that the set of all points of the n-dimensional space R_n $(n=1,2,...)$ has the power of the continuum.

12. Prove that the set E of all functions $y=f(x)$ on the interval $[0,1]$ whose range consists of two distinct points does not have the power of the continuum.

13. Prove that the set of all subsets of a given set A is not equivalent to A itself.

3 Metric Spaces

Metric spaces are among the most important mathematical structures. In particular, the general theory of limits to be given in Chapter 4 will be developed in a metric space context.

3.1. Definitions and Examples

3.11. By a *metric space* M is meant any set of elements $x, y, z, \ldots$, called "points," equipped with a rule associating a number $\rho(x, y)$, called the "distance from x to y,"† with every pair of points x and y in M, where the rule satisfies the following requirements (axioms):

(a) $\rho(x, y) > 0$ if $x \neq y$, $\rho(x, x) = 0$ for every x;

(b) $\rho(y, x) = \rho(x, y)$ for every x and y *(symmetry of the distance)*;

(c) $\rho(x, z) \leqslant \rho(x, y) + \rho(y, z)$ for every x, y, z *(triangle inequality)*.

The triangle inequality generalizes a fact familiar from elementary geometry, namely that the length of the side xz of a triangle xyz does not exceed the sum of the lengths of the sides xy and yz.

The rule associating the number $\rho(x, y)$ with the pair of points $x, y \in M$ is called the *metric* (or *distance*) of the space M. Note that every subset $E \subset M$ of a metric space M is itself a metric space, with the same metric as specified in the whole space M.

We now prove a number of simple consequences of Axioms a–c:

a. THEOREM. *The inequality*

$$\rho(x_1, x_n) \leqslant \rho(x_1, x_2) + \rho(x_2, x_3) + \cdots + \rho(x_{n-1}, x_n)$$

holds for arbitrary points $x_1, \ldots, x_n$ of a metric space M.‡

Proof. It follows from Axiom c that

$$\begin{aligned}
\rho(x_1, x_n) &\leqslant \rho(x_1, x_2) + \rho(x_2, x_n) \\
&\leqslant \rho(x_1, x_2) + \rho(x_2, x_3) + \rho(x_3, x_n) \\
&\leqslant \cdots \leqslant \rho(x_1, x_2) + \rho(x_2, x_3) + \cdots + \rho(x_{n-1}, x_n). \quad \blacksquare
\end{aligned}$$

b. THEOREM. *The "quadrilateral inequality"*

$$|\rho(x, y) - \rho(z, v)| \leqslant \rho(x, z) + \rho(y, v) \tag{1}$$

holds for arbitrary points x, y, z, v of a metric space M.

---

† Because of the symmetry of the distance, we can also call $\rho(x, y)$ the "distance between x and y."

‡ This theorem generalizes a fact familiar from elementary geometry, namely that the length of a polygonal line does not exceed the sum of the lengths of the segments making up the line.

Proof. By Theorem 3.11a, we have

$$\rho(x,y) \leqslant \rho(x,z) + \rho(z,v) + \rho(v,y),$$
$$\rho(z,v) \leqslant \rho(z,x) + \rho(x,y) + \rho(y,v),$$

or

$$\rho(x,y) - \rho(z,v) \leqslant \rho(x,z) + \rho(v,y),$$
$$\rho(z,v) - \rho(x,y) \leqslant \rho(z,x) + \rho(y,v).$$

The right-hand sides of the last two inequalities coincide, by Axiom b, while the left-hand sides differ only in sign. Thus the two inequalities together are equivalent to (1). ∎

c. Setting $v = y$ in the inequality (1), we get

$$|\rho(x,y) - \rho(z,y)| \leqslant \rho(x,z) \tag{1'}$$

(in elementary geometry, this means that the difference between the lengths of two sides of a triangle cannot exceed the length of the third side).

3.12. a. A set E in a metric space M is said to be *bounded* if the distance $\rho(a,x)$ from some fixed point $a \in M$ to a variable point $x \in E$ is bounded by a fixed constant. In this case, the distance from any other fixed point $b \in M$ to a variable point $x \in E$ is also bounded by a fixed constant, and the same is true of the distance $\rho(x,y)$ between two variable points $x, y \in E$, as follows at once from the triangle inequalities

$$\rho(b,x) \leqslant \rho(b,a) + \rho(a,x),$$
$$\rho(x,y) \leqslant \rho(x,a) + \rho(a,y).$$

The quantity

$$\text{diam } E = \sup_{x,y \in E} \rho(x,y)$$

is called the *diameter* of the (bounded) set E.

b. For $M = R_1$ (the real line) boundedness of the set E in the sense of Sec. 3.12a is equivalent to boundedness of E in the ordinary sense (see Sec. 1.66), since in this case both definitions simply mean that E lies in some (finite) closed interval.

c. A set E in a metric space M is said to be *unbounded* if given any $C > 0$, there exists a pair of points $x, y \in E$ such that $\rho(x,y) > C$. If E is unbounded, then, given any $C > 0$ and any fixed point $a \in M$, we can find a point $x \in E$ such that $\rho(a,x) > C$, since otherwise E would be bounded, and hence, as shown above, so would the distance between any pair of points $x, y \in E$.

To get the definition of a *bounded* or an *unbounded metric space*, we need only set $E = M$ in the above definitions.

d. The set of all points x of a metric space M whose distances from a fixed point x_0 is less than a given number $r > 0$, i.e., such that

$$\rho(x,x_0) < r,$$

is called a *ball* (more exactly, an *open ball*) *of radius r centered at x_0*. The set of all points x such that

$$\rho(x,x_0) \leqslant r$$

is called a *closed ball* of radius r centered at x_0. Finally, the set of all points x whose distances from x_0 are exactly equal to r, i.e., such that

$$\rho(x,x_0) = r,$$

is called a *sphere* of radius r centered at x_0.

e. Any (open) ball centered at x_0 is called a *neighborhood* of the point x_0. A point x_0 is called an *interior point* of a set $E \subset M$ if E contains both x_0 and some neighborhood of x_0.

3.13. Examples

a. Any set M on the real line R_1 is a metric space when equipped with the distance

$$\rho(x,y) = |x - y|.$$

In fact, the validity of Axioms a–c of Sec. 3.11 follows at once from the familiar properties of the absolute value (in particular, see Theorem 1.54b). If M is the whole real line, then the open ball of radius r centered at x_0 is the open interval

$$|x - x_0| < r,$$

the closed ball of radius r centered at x_0 is the closed interval

$$|x - x_0| \leqslant r,$$

while the sphere of radius r centered at x_0 is just the pair of points

$$x = x_0 \pm r.$$

b. In just the same way, a set M in the plane R_2 or in the three-dimensional space R_3 is a metric space if the distance between two points

$$x = (\xi_1, \xi_2, \xi_3), \quad y = (\eta_1, \eta_2, \eta_3)$$

(where, to be explicit, we consider the case of R_3) is defined as the usual geometric distance†

$$\rho(x,y) = \sqrt{(\xi_1 - \eta_1)^2 + (\xi_2 - \eta_2)^2 + (\xi_3 - \eta_3)^2}.$$

Here the triangle inequality (Axiom c) becomes an ordinary geometric inequality, i.e., the length of one side of a triangle does not exceed the sum of the lengths of the other two sides. A more general example is considered in the next section.

3.14. a. Consider the n-dimensional space R_n (Sec. 2.61). Given any two vectors

$$x = (\xi_1,...,\xi_n), \quad y = (\eta_1,...,\eta_n) \tag{2}$$

in R_n, the number

$$(x,y) = \sum_{k=1}^{n} \xi_k \eta_k \tag{3}$$

is called the *scalar product* of x and y. The number

$$|x| = \sqrt{\sum_{k=1}^{n} \xi_k^2} = \sqrt{(x,x)} \geqslant 0$$

is called the *length* or *norm* of the vector x, and a vector x such that $|x| = 1$ is called a *unit vector* (or *normalized vector*). As we will see later, the presence of a scalar product allows us to construct a geometry in R_n with suitable lengths and angles. The space R_n equipped with the scalar product (3) is called *n-dimensional Euclidean space*. It should be noted that the scalar product (3) has the following four easily verified properties (for arbitrary $x,y,z \in R_n$):
(1) $(x,x) \geqslant 0$;
(2) $(x,y) = (y,x)$;
(3) $(\alpha x,y) = \alpha(x,y)$ for every real α;
(4) $(x+y,z) = (x,z) + (y,z)$.

Next we define the distance between the vectors (2) as the length of the difference vector $x - y$, i.e., by the formula

$$\rho(x,y) = \sqrt{\sum_{k=1}^{n} (\xi_k - \eta_k)^2}. \tag{4}$$

The validity of the first two axioms of Sec. 3.11 is obvious in this case. To prove the validity of Axiom c, i.e., of the triangle inequality‡

---

† As always, the radical denotes the *positive* square root.
‡ Formula (5) is obviously an alternative version of the triangle inequality of Sec. 3.11.

$$\rho(x,y) \leqslant \rho(x,z) + \rho(y,z), \tag{5}$$

we argue as follows. The inequality (5) is equivalent to the inequality

$$|x+y| \leqslant |x| + |y|, \tag{6}$$

since (6) can be obtained from (5) by replacing y by $-y$ and z by 0, while conversely, (5) can be obtained from (6) by replacing x by $x-z$ and y by $z-y$. But Properties 1–4 above imply (6) quite generally, as shown by the following

THEOREM. *Let M be any set equipped with a scalar product satisfying Properties* 1–4. *Then the inequality* (6) *holds for arbitrary $x, y, z \in M$, where $|x| = \sqrt{(x,x)}$.*

Proof. Consider the expression

$$\varphi(\lambda) = (x - \lambda y, x - \lambda y),$$

where λ is an arbitrary real number. By Properties 2–4,

$$\varphi(\lambda) = (x,x) - 2\lambda(x,y) + \lambda^2(y,y) = A - 2B\lambda + C\lambda^2,$$

where

$$A = (x,x), \qquad B = (x,y), \qquad C = (y,y).$$

Moreover

$$\varphi(\lambda) \geqslant 0,$$

by Property 1. Hence the polynomial $\varphi(\lambda)$ cannot have distinct real roots, since otherwise it would take both positive and negative values. It follows that the quantity

$$B^2 - AC = (x,y)^2 - (x,x)(y,y)$$

(appearing under the radical in the formula for the solution of the corresponding quadratic equation) cannot be positive, and hence

$$(x,y)^2 \leqslant (x,x)(y,y) = |x|^2|y|^2.$$

This implies the *Schwarz inequality*

$$|(x,y)| \leqslant |x||y|. \tag{7}$$

Using (7), we find that

$$|x+y|^2 = (x+y, x+y) = (x,x) + 2(x,y) + (y,y)$$
$$\leqslant |x|^2 + 2|x||y| + |y|^2 = (|x| + |y|)^2,$$

which is equivalent to (6). ∎

b. It follows from the above considerations that the space R_n equipped with the distance (4) is a metric space, and so is any subset $E \subset R_n$ equipped with the same distance. In terms of the components of the vectors (2), the inequality (7) takes the form

$$\sqrt{\sum_{k=1}^{n} \xi_k \eta_k} \leqslant \sqrt{\sum_{k=1}^{n} \xi_k^2} \sqrt{\sum_{k=1}^{n} \eta_k^2}, \tag{8}$$

known as *Cauchy's inequality*. Similarly, (6) becomes

$$\sqrt{\sum_{k=1}^{n} (\xi_k + \eta_k)^2} \leqslant \sqrt{\sum_{k=1}^{n} \xi_k^2} + \sqrt{\sum_{k=1}^{n} \eta_k^2} \tag{9}$$

in component form. Moreover, setting $y = 0$, $z = (\zeta_1, \ldots, \zeta_n)$ in the inequality (1'), we get

$$\left| \sqrt{\sum_{k=1}^{n} \xi_k^2} - \sqrt{\sum_{k=1}^{n} \zeta_k^2} \right| \leqslant \sqrt{\sum_{k=1}^{n} (\xi_k - \zeta_k)^2}. \tag{10}$$

The inequalities (8)–(10), valid for all real numbers $\xi_1, \ldots, \xi_n, \eta_1, \ldots, \eta_n$, $\zeta_1, \ldots, \zeta_n$, are very often used in analysis to make all kinds of estimates, even quite independently of their geometric origin in the theory of metric spaces.

THEOREM. *The inequality*

$$\max_{1 \leqslant k \leqslant n} |\xi_k - \eta_k| \leqslant \rho(x, y) = \sqrt{\sum_{k=1}^{n} (\xi_k - \eta_k)^2} \leqslant \sqrt{n} \max_{1 \leqslant k \leqslant n} |\xi_k - \eta_k| \tag{11}$$

holds for arbitrary real $\xi_1, \ldots, \xi_n, \eta_1, \ldots, \eta_n$.

Proof. Since obviously

$$(\xi_k - \eta_k)^2 \leqslant \sum_{k=1}^{n} (\xi_k - \eta_k)^2,$$

we have

$$|\xi_k - \eta_k| \leqslant \sqrt{\sum_{k=1}^{n} (\xi_k - \eta_k)^2}$$

(cf. Sec. 1.55f). Taking the maximum of the left-hand side with respect to k, we get part of (11). On the other hand, clearly

$$\sum_{k=1}^{n} (\xi_k - \eta_k)^2 \leqslant \sum_{k=1}^{n} \max_{1 \leqslant k \leqslant n} (\xi - _k \eta_k)^2$$
$$= n \max_{1 \leqslant k \leqslant n} (\xi_k - \eta_k)^2 = n \{ \max_{1 \leqslant k \leqslant n} |\xi_k - \eta_k| \}^2,$$

which implies the other part of (11). ∎

c. To say that a set $E \subset R_n$ is bounded means that the distance between a variable point $x = (\xi_1, \ldots, \xi_n) \in E$ and a fixed point of R_n, say the point $0 = (0, \ldots, 0)$, is bounded by a fixed constant (see Sec. 3.12a). Thus if E is bounded, there exists a constant $C > 0$ such that

$$|x| = \sqrt{\sum_{k=1}^{n} \xi_k^2} \leqslant C$$

for all $x \in E$. But then obviously

$$|\xi_k| \leqslant C \qquad (k = 1, \ldots, n) \tag{12}$$

for all $x \in E$, i.e., every coordinate of the point $x \in E$ is bounded (on the real line). Conversely, let $E \subset R_n$ be a set of points all of whose coordinates are bounded, i.e., such that (12) holds for some constant $C > 0$. Then

$$|x| = \sqrt{\sum_{k=1}^{n} \xi_k^2} \leqslant \sqrt{\sum_{k=1}^{n} C^2} = \sqrt{n}\, C,$$

so that E is bounded in the space R_n. Thus, finally, *a set $E \subset R_n$ is bounded if and only if the set of values of each coordinate of the points $x \in E$ is bounded (on the real line).*

3.15. a. Regarding metric spaces as mathematical structures and recalling the considerations of Sec. 2.52, we now introduce the notion of an isomorphism of metric spaces, called an *isometry* in the present context. Two metric spaces M and M' equipped with metrics ρ and ρ', respectively, are said to be *isometric* if there is a one-to-one correspondence between elements $x, y \in M$ and the corresponding elements $x', y' \in M'$ which "preserves distances," in the sense that

$$\rho(x, y) = \rho'(x', y').$$

For example, two closed intervals of equal length (on the real line) are isometric with the natural "metric" (distance function), but not two intervals of unequal length. Every plane figure is isometric to its reflection in any line. It should be noted that any *linear* isometry in n-dimensional Euclidean space, i.e., any isometry $x \leftrightarrow x'$ such that $(x + y)' = x' + y'$, also preserves scalar products. In fact, we have

$$(x + y, x + y) = (x, x) + 2(x, y) + (y, y),$$
$$((x + y)', (x + y)') = (x' + y', x' + y') = (x', x') + 2(x', y') + (y', y'),$$

and hence

$$(x,y) = (x',y'),$$

since

$$(x,x) = |x|^2 = |x'|^2 = (x',x'),$$
$$(y,y) = |y|^2 = |y'|^2 = (y',y'),$$

by the very definition of an isometry. A simple example of an isometric mapping of the n-dimensional Euclidean space R_n into itself (an automorphism of R_n) is given by reflection in the plane $x_n = 0$, which carries every vector $x = (\xi_1,...,\xi_n)$ into the vector $x' = (\xi_1,...,\xi_{n-1}, -\xi_n)$. The mapping which shifts every vector in R_n by the vector $\beta = (\beta_1,...,\beta_n)$, i.e., which carries every vector $x = (\xi_1,...,\xi_n)$ into the vector

$$x + \beta = (\xi_1 + \beta_1,...,\xi_n + \beta_n),$$

is not an automorphism of R_n (Sec. 2.67) if $\beta \neq 0$, since it does not leave the zero vector unchanged.

b. Suppose there exists an isometric mapping of R_n into itself carrying the set $E \subset R_n$ into the set $F \subset R_n$. Then the sets E and F are said to be *congruent*, in keeping with the terminology of elementary geometry.

3.16. Metrization of the direct product of two metric spaces. Let M_1 be a metric space with metric ρ_1 and points $x_1, y_1, \ldots$, and let M_2 be another metric space with metric ρ_2 and points $x_2, y_2, \ldots$ Let $M = M_1 \times M_2$ be the direct product of the sets M_1 and M_2 (see Sec. 2.82), i.e., the set of all pairs $x = (x_1, x_2)$ with $x_1 \in M_1$, $x_2 \in M_2$. Suppose we define a metric ρ on the set M by the formula

$$\rho(x,y) \equiv \rho((x_1,x_2),(y_1,y_2)) = \max \{\rho_1(x_1,y_1),\rho_2(x_2,y_2)\}. \tag{13}$$

THEOREM. *The set M equipped with the distance function* (13) *is a metric space.*

Proof. We must verify that (13) satisfies Axioms a–c of Sec. 3.11. If $x \neq y$, then either $x_1 \neq y_1$ or $x_2 \neq y_2$, i.e., either $\rho_1(x_1,y_1) > 0$ or $\rho_2(x_2,y_2) > 0$. Hence $\rho(x,y) > 0$ in both cases, by the definition (13). Moreover, if $x = y$, then $x_1 = y_1, x_2 = y_2$. Therefore $\rho_1(x_1,y_1) = 0, \rho_2(x_2,y_2) = 0$, and hence $\rho(x,y) = 0$. Thus Axiom a holds. Axiom b follows at once from the observation that

$$\rho(y,x) = \max\{\rho_1(y_1,x_1),\rho_2(y_2,x_2)\} = \max\{\rho_1(x_1,y_1),\rho_2(x_2,y_2)\} = \rho(x,y).$$

Finally, let $z = (z_1,z_2)$, and consider the quantity

$$\rho(x,z) = \max\{\rho_1(x_1,z_1),\rho_2(x_2,z_2)\}.$$

Suppose, to be explicit, that $\rho(x,z) = \rho_1(x_1,z_1)$. Then

$$\rho(x,z) = \rho_1(x_1,z_1) \leqslant \rho_1(x_1,y_1) + \rho_1(y_1,z_1) \leqslant \rho(x,y) + \rho(y,z),$$

and similarly if $\rho(x,y) = \rho_2(x_2,y_2)$. This establishes Axiom c. ∎

Note that the metric (13) coincides with the metric of the space M_1 on every section (x,z) with fixed $z \in M_2$ (see Sec. 2.82), so that every section (x,z) is not only equivalent to M_1 but isometric to M_1 as well. Note also that (13) is not the only way of defining a metric on the direct product of two metric spaces M_1 and M_2. For example, we can also use the formula

$$\rho(x,y) = \rho_1(x_1,y_1) + \rho_2(x_2,y_2)$$

or the formula

$$\rho(x,y) = \sqrt{\rho_1^2(x_1,y_1) + \rho_2^2(x_2,y_2)}$$

(give a detailed verification of Axioms a–c in both cases).

3.2. Open Sets

3.21. A set G in a metric space M is said to be *open* if every point $x_0 \in G$ is an *interior point* of G, i.e., if whenever G contains x_0, G also contains some open ball centered at x_0 (the radius of the ball depends in general on x_0).

THEOREM. *The open ball*

$$U = \{x \in M : \rho(x,x_1) < r\}$$

is an open set.†

Proof. Given any $x_0 \in U$, let $\rho(x_0,x_1) = \theta < r$, and consider any ball U_0 of radius $r_0 < r - \theta$ centered at x_0. Then U_0 is (entirely) contained in U, since

$$\rho(x,x_1) \leqslant \rho(x,x_0) + \rho(x_0,x_1) < r_0 + \theta < r$$

for all $x \in U_0$, by the triangle inequality. ∎

3.22. THEOREM. *The union of any collection of open sets and the intersection of any* **finite** *collection of open sets are themselves open sets.*

Proof. The first assertion is an immediate consequence of the definition of an open set. As for the second assertion, suppose x_0 belongs to all the open sets $G_1, G_2,...,G_m$. Then G_1 contains not only x_0 but also some ball of radius r_1 (centered at x_0), G_2 contains not only x_0 but also some ball of radius r_2, and so on. Then the ball of radius $\min \{r_1, r_2,...,r_m\}$ centered at x_0 is contained in all the sets $G_1, G_2,...,G_m$, and hence is contained in their intersection. ∎

The proof just given does not work for the intersection of an infinite col-

---

† This, of course, explains why U is called an *open* ball in the first place.

lection of open sets, since the minimum (more exactly, the greatest lower bound) of an infinite set of positive numbers may equal zero. Thus, for example, the intersection of the infinitely many open sets

$$G_n = \{x: \rho(x,x_0) < 1/n\} \qquad (n = 1,2,\ldots)$$

contains only those points x such that $\rho(x,x_0) = 0$, and hence, by Axiom a of Sec. 3.11, only the single point x_0. But this intersection is in general not open.

3.23. Every open interval (α,β), bounded or unbounded, is obviously an open set on the real line $-\infty < x < \infty$. Moreover, every finite or countable union of open intervals†

$$(\alpha_\nu,\beta_\nu) \qquad (\nu = 1,2,\ldots)$$

is also an open set.

THEOREM. *Every open set G on the real line is a finite or countable union of nonintersecting open intervals.*

Proof. Let x be any point in G. Then, by definition, G contains both x and some open ball (i.e., some open interval) centered at x. We now construct the largest open interval containing x and contained in G. To this end, let E denote the set of points which lie to the right of x and do not belong to G. If E is empty, the whole half-line (x,∞) is contained in G, while if E is nonempty, then E has a greatest lower bound η. The point η certainly does not belong to G, since every point of G has a neighborhood entirely contained in G and hence not containing any points of E, while, on the other hand, every neighborhood of η must contain points of E, since η is the greatest lower bound of E. Thus, in particular, $\eta \neq x$. It is also obvious that the whole interval (x,η) is contained in G.

Next we carry out a similar construction to the left of the point x. This leads either to the half line $(-\infty,x)$ or to an interval (ξ,x) contained in G whose left-hand end point ξ does not belong to G. Thus, starting from a given point $x \in G$, we have constructed an open interval (ξ,η) contained in G, neither end point of which belongs to G (unless it is infinite). An interval of this type is called a *component* of the open set G.

Two components (ξ_1,η_1) and (ξ_2,η_2) sharing a common point x_0 must coincide. In fact, the inequality $\eta_1 < \eta_2$ (say) is impossible, since the point η_1 must belong to G, being an interior point of the interval (x_0,η_2), while at the same time η_1 cannot belong to G, being an end point of the interval (ξ_1,η_1). Hence every set G is a union of nonintersecting components. There can be no more than a countable number of such components, since we can choose

† That is, every union of a finite or countable collection of open intervals.

a rational point in each component (by Theorem 1.75) and the set of all rational points is countable (by Theorem 2.34). ∎

3.3. Convergent Sequences and Homeomorphisms

3.31. Convergent sequences. A sequence

$$x_1, x_2, \ldots, x_n, \ldots \tag{1}$$

of points in a metric space M (equipped with a metric ρ) is said to *converge* to a point $x \in M$ if, given any $\varepsilon > 0$, there exists an integer $N > 0$ such that

$$\rho(x, x_n) < \varepsilon$$

for all $n > N$. In other words, a sequence x_n† is said to converge to a point x if every open ball centered at x contains all points of the sequence starting from some value of n (so that only a finite number of points lie outside the ball). The point x is called the *limit* of the sequence,‡ and the fact that x_n converges to x is expressed by writing $x_n \to x$ or

$$x = \lim_{n \to \infty} x_n.$$

The symbol $x_n \to x$ is read "x_n converges to x" or "x_n approaches x" (as $n \to \infty$, i.e., as "n approaches infinity"). It should be borne in mind that the convergence is always with respect to some underlying metric ρ, and that the integer N will in general depend on ε. A sequence x_n is said to be *convergent* if it approaches some limit as $n \to \infty$. Otherwise the sequence is said to be *divergent* (or to *diverge*).

3.32. a. Let M be the real line R, with metric

$$\rho(x, y) = |x - y|. \tag{2}$$

Then, in keeping with the above definition, we say that a sequence of real numbers $x_1, x_2, \ldots, x_n, \ldots$ converges to a (real) limit ξ if, given any $\varepsilon > 0$, there exists an integer $N > 0$ such that $|\xi - x_n| < \varepsilon$ for all $n > N$.

Thus, for example, the sequence of points

$$x_n = \frac{1}{n} \qquad (n = 1, 2, \ldots)$$

on the real line, with the metric (2), converges to the point $x = 0$. In fact,

---

† The sequence (1), with "general term" x_n, is often simply called "the sequence x_n."
‡ The uniqueness of the limit x is proved in Theorem 3.33a.

given any $\varepsilon > 0$, choose an integer $N > 1/\varepsilon$. Then

$$0 < x_n = \frac{1}{n} < \frac{1}{N} < \varepsilon$$

for all $n > N$, i.e., the points x_n with $n > N$ all fall in a ball of radius ε centered at the point $x = 0$.

b. THEOREM.† *If*

$$\lim_{n \to \infty} x_n = x, \qquad \lim_{n \to \infty} y_n = y,$$

then

$$\lim_{n \to \infty} \rho(x_n, y_n) = \rho(x, y).$$

Proof. Given any $\varepsilon > 0$, there is an $N > 0$ such that

$$\rho(x, x_n) < \frac{\varepsilon}{2}, \qquad \rho(y, y_n) < \frac{\varepsilon}{2}$$

for all $n > N$. Hence, by the quadrilateral inequality (1), p. 53,

$$|\rho(x, y) - \rho(x_n, y_n)| \leqslant \rho(x, x_n) + \rho(y, y_n) < \varepsilon$$

for all $n > N$. ∎

c. THEOREM.‡ *If x_n and y_n are two numerical sequences such that*

$$\lim_{n \to \infty} x_n = x, \qquad \lim_{n \to \infty} y_n = y$$

and

$$x_n \leqslant y_n \qquad (n = 1, 2, \ldots), \tag{3}$$

then

$$x \leqslant y.$$

Proof. If $x > y$, let $\varepsilon = x - y > 0$ and choose N such that

$$|x - x_n| < \frac{\varepsilon}{2}, \qquad |y - y_n| < \frac{\varepsilon}{2}$$

---

† The content of this theorem might be summarized by saying that "distance is continuous" (cf. Sec. 5.12b).
‡ A much more systematic treatment of limits will be given below (Chapter 4). For the time being, we confine ourselves to proving some simple properties of convergent numerical sequences, i.e., convergent sequences of real numbers (such sequences will be studied further in Sec. 4.6).

for all $n > N$. Then

$$x_n > x - \frac{\varepsilon}{2} = (y + \varepsilon) - \frac{\varepsilon}{2} > \left(y_n - \frac{\varepsilon}{2}\right) + \varepsilon - \frac{\varepsilon}{2} = y_n$$

for all $n > N$, contrary to (3), and the proof follows by contradiction. ∎

d. THEOREM. *If x_n and y_n are two numerical sequences such that*

$$\lim_{n \to \infty} x_n = x, \qquad \lim_{n \to \infty} y_n = y,$$

then

$$\lim_{n \to \infty} (x_n + y_n) = x + y.$$

Proof. Given any $\varepsilon > 0$, choose N such that

$$|x - x_n| < \frac{\varepsilon}{2}, \qquad |y - y_n| < \frac{\varepsilon}{2}$$

for all $n > N$. Then

$$\begin{aligned}
|(x + y) - (x_n + y_n)| &= |(x - x_n) + (y - y_n)| \\
&\leqslant |x - x_n| + |y - y_n| < \varepsilon
\end{aligned}$$

for all $n > N$. ∎

e. COROLLARY. *If x_n, y_n, z_n are numerical sequences such that*

$$\lim_{n \to \infty} x_n = x, \qquad \lim_{n \to \infty} y_n = y, \qquad \lim_{n \to \infty} z_n = z$$

and

$$x_n + y_n \leqslant z_n \qquad (n = 1, 2, \ldots),$$

then

$$x + y \leqslant z.$$

Proof. An immediate consequence of the preceding two theorems. ∎

f. THEOREM. *Let M be the m-dimensional real space R_m of points $x = (\xi_1, \ldots, \xi_m)$, $y = (\eta_1, \ldots, \eta_m), \ldots$, equipped with the metric*

$$\rho(x, y) = \sqrt{\sum_{k=1}^{m} (\xi_k - \eta_k)^2}$$

(see Sec. 3.14a). Then a sequence of points $x^{(n)} = (\xi_1^{(n)}, \ldots, \xi_m^{(n)}) \in R_m$ converges to a point $x = (\xi_1, \ldots, \xi_m) \in R_m$ if and only if

$$\lim_{n \to \infty} \xi_1^{(n)} = \xi_1, \ldots, \lim_{n \to \infty} \xi_m^{(n)} = \xi_m, \tag{4}$$

i.e., if and only if each component of $x^{(n)}$ converges to the corresponding component of x.

Proof. Suppose (4) holds. Then, given any $\varepsilon > 0$, we can find an N such that all m inequalities

$$|\xi_1 - \xi_1^{(n)}| < \frac{\varepsilon}{\sqrt{n}}, \ldots, |\xi_m - \xi_m^{(n)}| < \frac{\varepsilon}{\sqrt{n}}$$

hold for all $n > N$. It follows that

$$\max_{1 \leq k \leq m} |\xi_k - \xi_k^{(n)}| < \frac{\varepsilon}{\sqrt{m}}$$

if $n > N$. Hence, by the second inequality in formula (11), p. 58,

$$\rho(x, x_n) = \sqrt{\sum_{k=1}^{m} (\xi_k - \xi_k^{(n)})^2} \leq \sqrt{m} \max_{1 \leq k \leq m} |\xi_k - \xi_k^{(n)}| < \varepsilon$$

for all $n > N$, i.e., $x_n \to x$ in the metric of the space R_m.

Conversely, suppose $x_n \to x$ in the metric of R_m. Then, given any $\varepsilon > 0$, there is an N such that

$$\rho(x, x_n) = \sqrt{\sum_{k=1}^{m} (\xi_k - \xi_k^{(n)})^2} < \varepsilon$$

for all $n > N$. But then, by the first inequality in the formula just cited,

$$\max_{1 \leq k \leq m} |\xi_k - \xi_k^{(n)}| \leq \rho(x, x_n) < \varepsilon,$$

which immediately implies (4). ∎

3.33. Returning to the case of a general metric space M with metric ρ, we now prove two further results.

a. THEOREM. *The limit of a convergent sequence x_n is unique.*

Proof. Suppose both $x_n \to x$ and $x_n \to y$. Then, given any $\varepsilon > 0$, there is an N such that

$$\rho(x, x_n) < \varepsilon, \qquad \rho(y, x_n) < \varepsilon$$

for all $n > N$, and hence, by the triangle inequality,

$$\rho(x, y) \leq \rho(x, x_n) + \rho(y, x_n) < 2\varepsilon$$

for all $n > N$. Since $\varepsilon > 0$ is arbitrarily small, it follows from Theorem 1.57 that $\rho(x, y) = 0$ and hence from Axiom b of Sec. 3.11 that $x = y$. ∎

b. THEOREM. *Every convergent sequence x_n is bounded.*†

---

† That is, the numbers $\rho(a, x_n)$, where a is some fixed point of M, form a bounded set on the real line (cf. Sec. 3.12a).

Proof. Let

$$a = \lim_{n \to \infty} x_n.$$

Then, for some fixed $\varepsilon > 0$, say $\varepsilon = 1$, choose N such that $\rho(a, x_n) < \varepsilon = 1$ for all $n > N$. Now let

$$D = \max\{\rho(a, x_1), \dots, \rho(a, x_N)\}.$$

Then

$$\rho(a, x_n) \leqslant \max\{D, 1\} \qquad (n = 1, 2, \dots). \quad \blacksquare$$

3.34. Homeomorphic metric spaces. In many problems involving metric spaces we are interested not so much in the special form of the metric as in knowing which sequences are convergent and which are divergent. This leads to the following definition:

a. Two metric spaces M and M' are said to be *homeomorphic* if there is a one-to-one correspondence $x \sim x'$ between the elements $x \in M$ and $x' \in M'$ which carries convergent sequences into convergent sequences, i.e., which is such that if $x_n \to x$ in M, then $x'_n \to x'$ in M', while if $x'_n \to x'$ in M', then $x_n \to x$ in M (where $x_n \sim x'_n$, $x \sim x'$).† The mapping $x \sim x'$ is then called a *homeomorphism*.

b. Next we give a criterion for a one-to-one correspondence between two metric spaces to be a homeomorphism:

THEOREM. *Let $\sim$ be a one-to-one correspondence between a metric space M with metric $\rho(x, y)$ and a metric space M' with metric $\rho'(x', y')$. Then $\sim$ is a homeomorphism if and only if, given any $x \in M$ and $\varepsilon > 0$, there exists an $\varepsilon' > 0$ such that the mapping $\sim$ carries the ball $\{y' \in M' : \rho'(x', y') < \varepsilon'\}$ into a subset of the ball $\{y \in M : \rho(x, y) < \varepsilon\}$, and conversely, given any $x' \in M'$ and $\varepsilon' > 0$, there exists an $\varepsilon > 0$ such that $\sim$ carries the ball $\{y \in M : \rho(x, y) < \varepsilon\}$ into a subset of the ball $\{y' \in M' : \rho'(x', y') < \varepsilon'\}$.*

Proof. Suppose the mapping $\sim$ is a homeomorphism, so that $x_n \to x$ implies $x'_n \to x'$ and vice versa. Let the point $x \in M$ and the number $\varepsilon > 0$ be fixed, and suppose there does not exist a number ε' with the property figuring in the statement of the theorem. Then the mapping $\sim$ carries every ball

$$\{y' \in M' : \rho'(x', y') < 1/n\} \qquad (n = 1, 2, \dots)$$

into a set with points lying outside the ball $\{y \in M : \rho(x, y) < \varepsilon\}$, so that, in particular, for every $n = 1, 2, \dots$ we can find a point $y'_n \in M'$ such that

---

† To avoid confusion, we now use the symbol $\sim$ to denote a one-to-one correspondence, rather than the symbol $\leftrightarrow$ favored heretofore.

$$\rho'(x',y_n') < 1/n, \qquad \rho(x,y_n) \geqslant \varepsilon.$$

But then y_n does not approach x although $y_n' \to x'$, thereby contradicting the assumption that $\sim$ is a homeomorphism. Hence if $\sim$ is a homeomorphism, it must be possible to find a number ε' with the property figuring in the statement of the theorem. Interchanging the roles of M and M' in this argument, we can prove in the same way that the required number ε exists for every given $x' \in M'$ and $\varepsilon' > 0$.

Conversely, suppose we can find a suitable number ε' for every given $x \in M$ and $\varepsilon > 0$. Then $x_n' \to x'$ implies $x_n \to x$. In fact, having found ε', let N be such that $\rho'(x',x_n') < \varepsilon'$ for all $n > N$. Applying the mapping $\sim$ and using the hypothesis, we get $\rho(x,x_n) < \varepsilon$ for all $n > N$. Therefore $x_n \to x$, since $\varepsilon > 0$ is arbitrary. Interchanging the roles of M and M' in this argument, we find in the same way that $x_n \to x$ implies $x_n' \to x'$. ∎

c. We can also define different metrics on one and the same set E, thereby converting it into different metric spaces. Two metrics $\rho(x,y)$ and $\rho'(x',y')$ defined on the same set E are said to be *homeomorphic* (on E) if the identity mapping $x \sim x$ is a homeomorphism of the resulting metric spaces M and M'. As applied to this situation, Theorem 3.32b takes the following form:

THEOREM. *Two metrics ρ and ρ' defined on the same set E are homeomorphic if and only if, given any $x \in E$ and $\varepsilon > 0$, there exists an $\varepsilon' > 0$ such that the ball $\{y \in E: \rho'(x,y) < \varepsilon'\}$ is contained in the ball $\{y \in E: \rho(x,y) < \varepsilon\}$, and conversely, given any $x \in E$ and $\varepsilon' > 0$, there exists an $\varepsilon > 0$ such that the ball $\{y \in E: \rho(x,y) < \varepsilon\}$ is contained in the ball $\{y \in E: \rho'(x,y) < \varepsilon'\}$.*

d. To illustrate the above considerations, consider the following three metrics defined on the direct product $M = M_1 \times M_2$ of two metric spaces M_1 and M_2 (as in Sec. 3.16):

$$\rho^{(1)}(x,y) = \max \{\rho_1(x_1,y_1), \rho_2(x_2,y_2)\},$$
$$\rho^{(2)}(x,y) = \rho_1(x_1,y_1) + \rho_2(x_2,y_2),$$
$$\rho^{(3)}(x,y) = \sqrt{\rho_1^2(x_1,y_1) + \rho_2^2(x_2,y_2)}.$$

The fact that all three metrics are homeomorphic on M follows from the chain of inequalities

$$\max \{a,b\} \leqslant \sqrt{a^2 + b^2} \leqslant a + b \leqslant 2 \max \{a,b\}, \tag{5}$$

valid for arbitrary numbers $a \geqslant 0$, $b \geqslant 0$. To prove (5), we note that if $a \leqslant b$, say, then

$$\max \{a,b\} = b = \sqrt{b^2} \leqslant \sqrt{a^2 + b^2} \leqslant \sqrt{a^2 + 2ab + b^2} = a + b \leqslant 2b = 2 \max \{a,b\}.$$

3.35. a. We begin by studying a certain special function

$$f(x) = \frac{x}{1+|x|} \qquad (-\infty < x < \infty), \tag{6}$$

which plays an important role in the main result of this section (Theorem 3.35d). To complete the definition of $f(x)$, we set

$$f(-\infty) = -1, \qquad f(+\infty) = +1,$$

thereby defining $f(x)$ on the extended real line $\bar{R}$. Obviously, $f(x) > 0$ if $x > 0$ and $f(x) < 0$ if $x < 0$, while $f(0) = 0$. Moreover, $f(-x) = -f(x)$ for every finite x, and

$$|f(x)| = \frac{|x|}{1+|x|} < \frac{1+|x|}{1+|x|} = 1.$$

The graph of $f(x)$ is shown in Figure 1, from which it is apparent that $f(x)$ is a one-to-one function.

b. LEMMA. *The inequality*

$$|f(x) - f(y)| \leqslant |x - y| \tag{7}$$

holds for arbitrary real numbers x and y.

Proof. For $x \geqslant 0, y \geqslant 0$ we have

$$|f(x) - f(y)| = \left| \frac{x}{1+x} - \frac{y}{1+y} \right| = \frac{|x-y|}{(1+x)(1+y)} \leqslant |x-y|.$$

Similarly,

$$|f(x) - f(y)| = \left| \frac{x}{1-x} - \frac{y}{1-y} \right| = \frac{|x-y|}{(1-x)(1-y)} \leqslant |x-y|$$

if $x \leqslant 0, y \leqslant 0$, while

$$|f(x) - f(y)| \leqslant |f(x)| + f(y) = \frac{|x|}{1+|x|} + \frac{y}{1+y} \leqslant |x| + y = -x + y = |x - y|$$

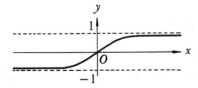

Figure 1

if $x \leqslant 0, y \geqslant 0$, which in turn implies

$$|f(x) - f(y)| = |f(y) - f(x)| \leqslant |y - x| = |x - y|$$

if $x \geqslant 0, y \leqslant 0$. ∎

c. LEMMA. *If x and y are such that*

$$|f(x)| < 1 - \delta, \qquad |f(y)| < 1 - \delta \qquad (0 < \delta < 1), \tag{8}$$

then

$$|x - y| \leqslant \frac{1}{\delta^2} |f(x) - f(y)|. \tag{9}$$

Proof. We begin by solving (6) for x. If $x \geqslant 0$, then $|x| = x$ and

$$x = \frac{f(x)}{1 - f(x)},$$

while if $x \leqslant 0$, then $|x| = -x$ and

$$x = \frac{f(x)}{1 + f(x)}.$$

Thus

$$|x - y| = \left| \frac{f(x)}{1 - f(x)} - \frac{f(y)}{1 - f(y)} \right| = \frac{|f(x) - f(y)|}{[1 - f(x)][1 - f(y)]} \leqslant \frac{1}{\delta^2} |f(x) - f(y)|$$

if $x \geqslant 0, y \geqslant 0$, because of (8). This in turn implies

$$|x - y| = |-y - (-x)| \leqslant \frac{1}{\delta^2} |f(-y) - f(-x)| = \frac{1}{\delta^2} |f(x) - f(y)|$$

if $x \leqslant 0, y \leqslant 0$, since then $-x \geqslant 0$, $-y \geqslant 0$. Moreover

$$|x - y| = |x| + y = \frac{|f(x)|}{1 + f(x)} + \frac{f(y)}{1 - f(y)}$$

$$\leqslant \frac{1}{\delta^2} [|f(x)| + f(y)] = \frac{1}{\delta^2} |f(x) - f(y)|$$

if $x \leqslant 0, y \geqslant 0$, and hence

$$|x - y| = |y - x| \leqslant \frac{1}{\delta^2} |f(y) - f(x)| = \frac{1}{\delta^2} |f(x) - f(y)|$$

if $x \geqslant 0, y \leqslant 0$. ∎

d. We now use the function $f(x)$ to define a new metric on the extended real line $\bar{R}$ homeomorphic to the usual metric on the set R (the ordinary real line):

THEOREM. *Given any two points x and y of the extended real line $\overline{R}$, let*

$$r(x,y) = |f(x) - f(y)|, \tag{10}$$

where $f(x)$ is the function defined in Sec. 3.35a. Then $r(x,y)$ is a metric on $\overline{R}$ which is homeomorphic to the usual metric

$$\rho(x,y) = |x - y| \tag{11}$$

on the set R of all finite real numbers $(-\infty < x < \infty)$.

Proof. The geometric meaning of the metric (10) is clear from Figure 1: The distance $r(x,y)$ between two points x and y on the horizontal axis is defined as the length of the corresponding segment $[f(x), f(y)]$ of the vertical axis. We begin by verifying that (10) satisfies Axioms a–c of Sec. 3.11. The fact that $r(x,y) = r(y,x)$ is immediately apparent from (10), so that Axiom b holds. Moreover, it is also clear from (10), together with the inequality (7), that $r(x,y)$ is positive if $x \neq y$ and zero if $x = y$, so that Axiom a holds. To prove Axiom c, the triangle inequality, we merely note that

$$r(x,z) = |f(x) - f(z)| \leqslant |f(x) - f(y)| + |f(y) - f(z)| = r(x,y) + r(y,z).$$

Next we verify that the metrics $r(x,y)$ and $\rho(x,y)$ are homeomorphic on the set R. Suppose $x_n \to x$ in the metric (11). Then, given any $\varepsilon > 0$, there is an N such that $|x - x_n| < \varepsilon$ for all $n > N$. But then, by (7),

$$r(x,x_n) = |f(x) - f(x_n)| \leqslant |x - x_n| < \varepsilon$$

for all $n > N$, so that $x_n \to x$ in the metric (10). On the other hand, suppose $x_n \to x$ in the metric (11). Noting that $|f(x)| < 1$, we set

$$f(x) = 1 - 2\delta \qquad (0 < \delta < \tfrac{1}{2}).$$

Next, given any $\varepsilon > 0$, we find an N such that

$$r(x,x_n) = |f(x) - f(x_n)| < \delta^2 \min\{\delta, \varepsilon\}$$

for all $n > N$. Then

$$|f(x_n)| \leqslant |f(x)| + \delta^2 \min\{\delta, \varepsilon\} \leqslant |f(x)| + \delta = 1 - \delta$$

for all $n > N$, and since $|f(x)| = 1 - 2\delta < 1 - \delta$, we apply the inequality (9), obtaining

$$\rho(x,x_n) = |x - x_n| \leqslant \frac{1}{\delta^2} |f(x) - f(x_n)| \leqslant \min\{\delta, \varepsilon\} \leqslant \varepsilon$$

for all $n > N$, so that $x_n \to x$ in the metric (11). ∎

e. Consider the set $\{x \in \overline{R}: r(x,\infty) \leqslant c\}$, i.e., the closed ball centered at the point ∞ in the metric space $\bar{R}$ equipped with the metric (10). By definition, this ball is the set of all $x \in \bar{R}$ satisfying the inequality

$$|f(\infty) - f(x)| = \left| 1 - \frac{x}{1 + |x|} \right| \leqslant c. \tag{12}$$

Confining ourselves to the most important case, where c is small and $c \leqslant 1$, we find that only nonnegative values of x can satisfy (12), since $f(x) < 0$ if $x < 0$ which implies $f(\infty) - f(x) > 1$. The inequality (12) thus becomes

$$1 - \frac{x}{1 + x} = \frac{1}{1 + x} \leqslant c,$$

or equivalently,

$$x \geqslant \frac{1}{c} - 1.$$

Thus the ball $\{x \in \bar{R} : r(x, \infty) \leqslant c\}$ is the closed interval

$$\frac{1}{c} - 1 \leqslant x \leqslant \infty.$$

Similarly, the closed ball of radius $c \leqslant 1$ (c small) centered at the point $-\infty$ is just the closed interval

$$-\infty \leqslant x \leqslant -\frac{1}{c} + 1.$$

In particular, the sequence of points

$$1, 2, \ldots, n, \ldots \tag{13}$$

in the space $\bar{R}$ all lie, starting from some value of n, in an arbitrary ball centered at the point ∞, and hence

$$\lim_{n \to \infty} n = \infty$$

in the space $\bar{R}$. At the same time, the sequence (13) obviously has no limit in the space R.

f. It should be noted that the space $\bar{R}$ with the metric (10) is isometric to the closed interval $[-1, 1]$ with the usual metric. In fact, the isometry between $\bar{R}$ and $[-1, 1]$ is established by the one-to-one correspondence $x \sim f(x)$, since the distance between the points $x, y \in \bar{R}$ with the metric (10) equals the ordinary distance between the corresponding points $f(x)$ and $f(y)$. Dropping the end points ($-\infty, \infty$ for the space $\bar{R}$ and $-1, 1$ for the interval

[−1,1]), we get an isometry between the real line R with the metric (10) and the open interval $(−1,1)$ with the usual metric. Using Theorem 3.35d, we find that *the metric space R with the usual metric is homeomorphic to the interval* $(−1,1)$.†

3.4. Limit Points

3.41. Let M be a metric space, equipped with a metric ρ, and let x_n be a sequence of points in M. Then a point $x \in M$ is said to be a *limit point* of the sequence x_n if, given any $\varepsilon > 0$ and any integer $N > 0$, there is an integer $n > N$ such that

$$\rho(x, x_n) < \varepsilon.$$

In other words, we say that a point x is a limit point of the sequence x_n if every ball centered at x contains points of the sequence with *arbitrarily large* values of n, but not necessarily *all* values of n (as in the case of a convergent sequence). A convergent sequence x_n with limit x obviously has x as a limit point, and moreover has no other limit points (the proof is similar to that of Theorem 3.33a). On the other hand, a divergent sequence may have any number of limit points or no limit points at all. For example, the sequence

$$x_n = (−1)^n \left(1 + \frac{1}{n}\right) \qquad (n = 1, 2, \ldots)$$

of points on the real line (with the usual metric) has two limit points $−1$ and 1 (and approaches neither), while the sequence

$$x_n = n^{(−1)^n} \qquad (n = 1, 2, \ldots)$$

has the single limit point 0 (which it also does not approach). As another example, suppose all the rational numbers are written in a single sequence (cf. Theorem 2.34). Then every point of the real line is a limit point of this sequence!

3.42. Given a sequence

$$x_1, x_2, \ldots, x_n, \ldots, \tag{1}$$

let

$$x_{n_1}, x_{n_2}, \ldots, x_{n_m}, \ldots \tag{2}$$

be any sequence of points belonging to (1), where the (distinct positive)

† Of course, this fact can easily be verified directly.

integers $n_1, n_2, \ldots, n_m, \ldots$ are arranged in increasing order. Then (2) is called a *subsequence* of the original sequence (1).

THEOREM. *A point x is a limit point of a sequence x_n if and only if x_n has a subsequence converging to x.*

Proof. Suppose x is a limit point of x_n. Then, given any $m = 1, 2, \ldots$, we can find a point $x_{n_m} (n_1 < n_2 < \cdots)$ of the sequence x_n such that

$$\rho(x, x_{n_m}) < \frac{1}{m}.$$

But then the subsequence x_{n_m} clearly converges to x. Conversely, if x_n has a subsequence converging to x, then x is obviously a limit point of x_n, by the very definition of a limit point. ∎

This leads at once to the following alternative definition of a limit point: *A point x is said to be a limit point of the sequence x_n if x_n has a subsequence converging to x.*

3.43. The second definition of a limit point rests entirely on the concept of a convergent sequence. Since a homeomorphism between two metric spaces M and M' preserves convergent sequences, we conclude that *if $x \in M$ is a limit point of a sequence $x_n \in M$ and if M' is homeomorphic to M, then the point $x' \in M'$ corresponding to the point $x \in M$ (under the homeomorphism) is a limit point of the sequence $x_n' \in M'$ corresponding to the sequence x_n.* In particular, if two distinct but homeomorphic metrics ρ and ρ' are defined on the same set M and if x is a limit point of a sequence x_n with respect to the metric ρ, then x is also a limit point of x_n with respect to the metric ρ'.

3.44. a. A point x is a metric space M (equipped with a metric ρ) is said to be a *limit point* of a given subset $A \subset M$ if every neighborhood $U_x(\varepsilon) = \{y \in M: \rho(x, y) < \varepsilon\}$ of x contains a point $y \in A$ distinct from x itself. The definition of the limit point of a subset differs somewhat from the definition of a limit point of a sequence. This is explained by the fact that the concept of a sequence of points in M does not reduce to that of a subset of M, since points can "repeat themselves" in a sequence but not in a subset (see Sec. 2.81). For example, the points 0 and 1 are limit points of the sequence $0, 1, 0, 1, \ldots$ but not of the set consisting of the two points 0 and 1.

Nevertheless, the results of Secs. 3.42 and 3.43 pertaining to limit points of a sequence, carry over to the case of limit points of a subset. Thus if x is a limit point of a set $A \subset M$, we can select a sequence of distinct points of A converging to x (the method of proof is the same as that of Theorem 3.42), while conversely, if we can select a sequence of distinct points of A converging to a point $x \in M$, then x is a limit point of A. It follows, just as in Sec. 3.43, that limit points of a set A are preserved under homeomorphisms of M.

b. THEOREM. *Let A be a set on the real line which is bounded from above, and suppose $\xi = \sup A$ does not belong to A. Then ξ is a limit point of A.*

Proof. By the very definition of the least upper bound, there is a point $x_n \in A$ such that

$$\xi - \frac{1}{n} < x_n \leqslant \xi$$

for every $n = 1, 2, \ldots$, where $x_n \neq \xi$ since $\xi \notin A$. But then every neighborhood of ξ contains a point of A distinct from ξ itself. ∎

The theorem remains true if A is bounded from below and $\xi = \inf A$ instead.

3.45. It is desirable to have general principles allowing us to determine the existence of limit points of a wide class of sets or sequences. A useful result of this type is the following

THEOREM (**Bolzano-Weierstrass principle**). *Every infinite set of points in a closed interval $[a,b] \subset R$ has at least one limit point.*

Proof. Suppose an infinite set of points E lies in some interval $[\alpha, \beta]$. Then at least one of the two halves

$$\left[\alpha, \frac{\alpha + \beta}{2} \right], \qquad \left[\frac{\alpha + \beta}{2}, \beta \right]$$

of $[\alpha, \beta]$ contains an infinite subset of E. Starting from the given interval $\Delta_1 = [a,b]$ and using this obvious fact, we construct a sequence of nested closed intervals $\Delta_1 \supset \Delta_2 \supset \cdots$, where each interval is half of the preceding interval and contains an infinite subset of the given set A. It follows from Theorem 1.82 that the intersection of all the intervals $\Delta_1, \Delta_2, \ldots$ consists of a single point x_0. To see that x_0 is a limit point of A, let $U = U_{x_0}(\varepsilon) = \{x \in R : |x - x_0| < \delta\}$ be any neighborhood of x_0, and let n be such that the length of the interval Δ_n is less than δ. Then Δ_n contains x_0 and is entirely contained in U, so that U, like Δ_n, contains infinitely many points of A, i.e., x_0 is a limit point of A. ∎

3.46. There is an analogous result, valid for sequences:

THEOREM (**Bolzano-Weierstrass principle for sequences**). *Every sequence of points in a closed interval $[a,b] \subset R$ has at least one limit point.*

Proof. Repeat the proof of Theorem 3.45, taking account of the fact that points of a sequence can repeat themselves. ∎

3.47. There are sets on the real line R without limit points, e.g., the set $\{1,2,...,n,...\}$. However, every infinite set on the extended real line $\bar{R}$, metrized in accordance with Theorem 3.35d, has a limit point. This follows from the fact that $\bar{R}$ is isometric to the interval $[-1,1]$ with the ordinary metric (see Sec. 3.35f), and the property of being a limit point is preserved under an isometry (or, for that matter, under a homeomorphism, as shown in Sec. 3.43).

3.48. Next we prove a sufficient condition for the *absence* of limit points:

THEOREM. *Suppose the distance between any two points x_m and x_n of a sequence x_1, x_2,... is bounded from below by a positive constant, so that*

$$\rho(x_m, x_n) \geqslant C \qquad (m,n = 1,2,...). \tag{3}$$

Then the sequence $x_1, x_2,...$ has no limit points.

Proof. Suppose, to the contrary, that x is a limit point of the sequence. Then the sequence certainly contains points x_m and x_n ($m \neq n$) such that

$$\rho(x, x_m) < C/2, \qquad \rho(x, x_n) < C/2.$$

But this is impossible, since then

$$\rho(x_m, x_n) \leqslant \rho(x_m, x) + \rho(x, x_n) < C,$$

contrary to (3). The proof now follows by contradiction. ∎

3.5. Closed Sets

3.51. A set F in a metric space M is said to be *closed* (in M) if it contains all its limit points. Thus the interval $a \leqslant x \leqslant b$ on the real line is closed, but not the interval $a \leqslant x < b$ which does not contain its limit point b.

THEOREM. *The closed ball*

$$V = \{x \in M : \rho(x, x_1) \leqslant r\}$$

is a closed set.†

Proof. Let x_0 be any point not in V, so that

$$\rho(x_0, x_1) = r_1 > r. \tag{1}$$

Then there are no points of V in the ball of radius $\frac{1}{2}(r_1 - r)$ centered at x_0.

† This, of course, explains why V is called a *closed* ball in the first place.

In fact, if there were such a point, say z, then

$$\rho(x_0,x_1) \leqslant \rho(x_0,z) + \rho(z,x_1) \leqslant \tfrac{1}{2}(r_1 - r) + r = \tfrac{1}{2}(r_1 + r) < r_1,$$

contrary to (1). Hence x_0 cannot be a limit point of the set V, i.e., V contains all its limit points. ∎

3.52. There is an intimate connection between closed sets and open sets in a metric space, as shown by the following

THEOREM. *Given a metric space M, the complement G of a closed set $F \subset M$ is open, while the complement F of an open set $G \subset M$ is closed.*

Proof. Let G be the complement of a closed set F, and let x_0 be any point in G. To prove that G is open, we must show that G contains some open ball centered at x_0. If there were no such ball, then every ball centered at x_0 would contain points of F. But then x_0 would be a limit point of F, and hence x_0 would have to belong to F, since F is closed. This contradicts $x_0 \in G$, thereby proving that G is open.

Next let F be the complement of an open set G, and let x_0 be any point of G. Then G contains x_0 together with some ball centered at x_0, so that x_0 cannot be a limit point of F. Hence any limit point of F can only be a point of F itself, i.e., F is closed. ∎

3.53. a. THEOREM. *Every closed set F on the real line is obtained by deleting a finite or countable collection of nonintersecting intervals from the line.*

Proof. An immediate consequence of Theorems 3.23 and 3.52. ∎

The deleted intervals, namely the component intervals of the open set G complementary to F, are said to be *adjacent* to F.

b. THEOREM. *Let F be a closed set on the real line which is bounded from above, and let $\xi = \sup F$. Then ξ belongs to F.*

Proof. If $\xi \notin F$, then, by Theorem 3.44b, ξ would be a limit point of F not belonging to F, which is impossible, since F is closed. ∎

The theorem remains true if F is bounded from below, and $\xi = \inf F$ instead.

3.54. THEOREM. *The union of any **finite** collection of closed sets and the intersection of any collection of closed sets are themselves closed sets.*

Proof. Given closed sets F_v (where the parameter v ranges over a finite "index set"), let G_v be the complementary open sets, as in Theorem 3.52. Then

$$\mathbf{C} \sum_v F_v = \prod_v \mathbf{C} F_v = \prod_v G_v,$$

by formula (2), p. 28. But $\prod_\nu G_\nu$ is open, by Theorem 3.22, and hence $\sum_\nu F_\nu$ is closed, by Theorem 3.52.

Next let ν range over an arbitrary (not necessarily finite) index set. Then

$$\mathbf{C}\prod_\nu F_\nu = \sum_\nu \mathbf{C}F_\nu = \sum_\nu G_\nu,$$

by formula (2'), p. 29. But $\sum_\nu G_\nu$ is open, again by Theorem 3.22, and hence $\prod_\nu F_\nu$ is closed, again by Theorem 3.52. ∎

3.6. Dense Sets and Closures

3.61. Definition. A set A in a metric space M is said to be (*everywhere*) *dense relative to* another set $B \subset M$ if every point of B is either a point of A or a limit point of A. In other words, we say that A is dense relative to B if every ball centered at a point of B contains a point of A. If A is dense relative to B and if, in addition, A is a subset of B, we say that A is *dense in* B. For example, the set of all rational points is dense in the real line $-\infty < x < \infty$, while the set of all points $(r_1, \ldots, r_n)$ with rational coordinates is dense in n-dimensional Euclidean space.

3.62. The property of being dense is "transitive" in the following sense:

THEOREM. *If a set A is dense relative to a set B, and if B is in turn dense relative to a set C, then A is dense relative to C.*

Proof. Given any $\varepsilon > 0$ and any $z \in C$, we can find first a point $y \in B$ such that $\rho(y, z) < \varepsilon/2$ and then a point $x \in A$ such that $\rho(x, y) < \varepsilon/2$. Hence, given any $\varepsilon > 0$ and any $z \in C$, we can find a point $x \in A$ such that

$$\rho(x, z) \leqslant \rho(x, y) + \rho(y, z) < \varepsilon.$$

But then A is dense relative to C. ∎

3.63. Definition. Given a set A in a metric space M, by the *closure* of A, denoted by $\bar{A}$, we mean the set consisting of all points of A together with all limit points of A. Obviously $\bar{A} \supset A$ in general, where $\bar{A} = A$ if and only if every limit point of A is a point of A, i.e., if and only if A is closed. If $\bar{A} \subset A$, then, since $\bar{A} \supset A$ always holds, we have $\bar{A} = A$, so that A is closed.

3.64. THEOREM. *The closure $\bar{A}$ of any set A is a closed set.*

Proof. First we note that every set is dense in its own closure, by the very definition of closure. Thus A is dense in $\bar{A}$, and $\bar{A}$ is dense in $\bar{\bar{A}}$. It follows from Theorem 3.62 that A is dense in $\bar{\bar{A}}$, i.e., every point of the set $\bar{\bar{A}}$ is either a point of A or a limit point of A. But then $\bar{\bar{A}} \subset \bar{A}$, so that $\bar{A}$ is closed. ∎

3.65. THEOREM. *The closure $\bar{A}$ of a bounded set A is itself a bounded set, and moreover* diam $\bar{A}=$ diam A.

Proof. Since $A \subset \bar{A}$, we have

$$\text{diam } A \leqslant \text{diam } \bar{A} \tag{1}$$

(recall Sec. 3.12a). Given any two points $\bar{x}, \bar{y} \in \bar{A}$ and any $\varepsilon > 0$, choose two points $x, y \in A$ such that

$$\rho(x,\bar{x}) < \frac{\varepsilon}{2}, \qquad \rho(y,\bar{y}) < \frac{\varepsilon}{2}.$$

Then

$$\rho(\bar{x},\bar{y}) \leqslant \rho(\bar{x},x) + \rho(x,y) + \rho(y,\bar{y}) < \frac{\varepsilon}{2} + \text{diam } A + \frac{\varepsilon}{2},$$

which implies

$$\text{diam } \bar{A} = \sup_{\bar{x},\bar{y} \in \bar{A}} \rho(\bar{x},\bar{y}) \leqslant \varepsilon + \text{diam } A,$$

and hence

$$\text{diam } \bar{A} \leqslant \text{diam } A, \tag{2}$$

since $\varepsilon > 0$ is arbitrary (in particular, $\bar{A}$ is bounded). Comparing (1) and (2), we find that diam $\bar{A}=$ diam A. ∎

3.66. THEOREM. *Given a closed set F and an open set $G \supset F$, there exists another open set $H \supset F$ such that*

$$F \subset H \subset \bar{H} \subset G.$$

Proof. Let $d(x)$ be the distance from the point $x \in G$ to the complement of G, i.e., the quantity

$$d(x) = \inf_{y \in M - G} \rho(x,y),$$

where M is the underlying metric space. Then $d(x)$ is positive for every $x \in G$, since G contains an entire neighborhood of x. With each $x \in F$ we now associate the open ball of radius $\frac{1}{2}d(x)$ centered at x. Let H be the union of all these balls. Then H is open, by Theorem 3.22, and obviously $H \supset F$. To prove that $\bar{H} \subset G$, suppose the contrary holds. Then there exists a point $y \in \bar{H} \cap (M - G)$. Thus we can find a sequence $z_n \in H$ such that $z_n \to y$ and a sequence $x_n \in F$ such that

$$\rho(x_n, z_n) \leqslant \frac{1}{2}d(x_n).$$

It follows that

$$d(x_n) \leqslant \rho(x_n, y) \leqslant \rho(x_n, z_n) + \rho(z_n, y) \leqslant \frac{1}{2} d(x_n) + \rho(z_n, y),$$

and hence

$$d(x_n) \leqslant 2\rho(z_n, y), \qquad \rho(x_n, y) \leqslant 2\rho(z_n, y),$$

so that

$$y = \lim_{n \to \infty} x_n \in F \subset G.$$

This contradicts the assumption $y \in M - G$, thereby proving that $\overline{H} \subset G$. ∎

3.67. In particular, if F is bounded (Sec. 3.12a), we can estimate the function $d(x)$ for $x \in F$ as

$$d(x) = \inf_{y \in M - G} \rho(x, y) \leqslant \inf_{y \in M - G} [\rho(x, x_0) + \rho(x_0, y)] \leqslant \operatorname{diam} F + d(x_0), \qquad (3)$$

where x_0 is a fixed point in F. Hence the set H constructed in Theorem 3.66 is also bounded, since the definition of H together with (3) implies

$$\operatorname{diam} H \leqslant \operatorname{diam} F + 2 \sup_{x \in F} d(x) \leqslant 3 \operatorname{diam} F + 2d(x_0).$$

3.7. Complete Metric Spaces

3.71. a. Definition. A sequence of points $x_1, x_2, \ldots$ in a metric space M is called a *fundamental sequence* (or a *Cauchy sequence*) if, given any $\varepsilon > 0$, there exists an integer $N > 0$ such that $\rho(x_m, x_n) < \varepsilon$ for all $m, n > N$.

b. THEOREM. *Every convergent sequence is fundamental.*

Proof. Let x_n be a convergent sequence, with limit x. Then, given any $\varepsilon > 0$, there is an N such that

$$\rho(x, x_n) < \varepsilon/2$$

for all $n > N$. But then

$$\rho(x_m, x_n) \leqslant \rho(x_m, x) + \rho(x, x_n) < \varepsilon$$

for all $m, n > N$, i.e., x_n is a fundamental sequence. ∎

c. THEOREM. *Every fundamental sequence is bounded.*

Proof. Given a fundamental sequence x_n, choose N such that $\rho(x_m, x_n) < 1$ for all $m, n > N$. Then

$$\rho(x_m, x_{N+1}) \leqslant \begin{cases} \max\{\rho(x_1, x_{N+1}), \ldots, \rho(x_N, x_{N+1})\} & \text{if } m \leqslant N, \\ 1 & \text{if } m > N, \end{cases}$$

so that the distance from every point $x_1, x_2, \ldots$ to the point x_{N+1} is bounded by a fixed constant. ∎

d. Definition. A metric space M is said to be *complete* if every fundamental sequence in M converges to an element of M. Otherwise, M is said to be *incomplete*.

3.72. a. THEOREM. *The real line R, equipped with the usual metric $\rho(x,y) = |x-y|$, is a complete metric space.*

Proof. Let x_n be a fundamental sequence in R. Then x_n is bounded, by Theorem 3.71c. Let

$$a_m = \inf_{n \geqslant m} x_n, \qquad b_m = \sup_{n \geqslant m} x_n \qquad (m = 1, 2, \ldots).$$

Then, obviously, $a_m \leqslant a_{m+1}$, $b_m \geqslant b_{m+1}$, so that $[a_m, b_m] \supset [a_{m+1}, b_{m+1}]$. Hence, by Theorem 1.81, there is a point $p \in R$ contained in all the intervals $[a_m, b_m]$ $(m = 1, 2, \ldots)$. We now show that

$$p = \lim_{n \to \infty} x_n, \tag{1}$$

which incidentally proves that the intersection of all the $[a_m, b_m]$ consists of the single point p (cf. Theorem 3.33a). Given any $\varepsilon > 0$, choose N such that $|x_m - x_n| < \varepsilon$ for all $m, n > N$. Then hold $m = N+1$ fixed, and let $n = N+1$, $N+2, \ldots$ Since all the x_n with $n \geqslant N+1$ are no further than ε from x_{N+1}, the same is true of the numbers a_{N+1} and b_{N+1}. Moreover, p belongs to the interval $[a_{N+1}, b_{N+1}]$, and hence

$$|p - x_n| \leqslant b_{N+1} - a_{N+1} = (b_{N+1} - x_{N+1}) + (x_{N+1} - a_{N+1}) \leqslant 2\varepsilon$$

for all $n > N$. Since $\varepsilon > 0$ is arbitrary, this implies (1). ∎

b. THEOREM. *A numerical sequence x_n is convergent if and only if it is fundamental, i.e., if and only if the following condition, called the **Cauchy convergence criterion**, is satisfied: Given any $\varepsilon > 0$, there exists an integer $N > 0$ such that $|x_m - x_n| < \varepsilon$ for all $m, n > N$.*

Proof. An immediate consequence of Theorems 3.71b and 3.72a. ∎

c. The interval $(0,1)$ is a metric space when equipped with the usual metric of the real line. The sequence

$$1, \frac{1}{2}, \ldots, \frac{1}{n}, \ldots$$

is fundamental in this space, but has no limit in the space. Therefore $(0,1)$ *is not a complete space*. In particular, this shows that the property of being a complete space is critically dependent on the choice of a metric, and may not be preserved on going over to a homeomorphic metric. In fact, the real line equipped with the usual metric

$$\rho(x,y) = |x - y|$$

is complete, as shown in Theorem 3.72a, while the real line R equipped with the homeomorphic metric

$$r(x,y) = |f(x) - f(y)|,$$

as in Theorem 3.35d, is isometric to the interval $(-1,1)$ equipped with the usual metric (see Sec. 3.35f), and hence fails to be complete,† as just shown.

d. THEOREM. *The m-dimensional real space R_m of all points $x = (\xi_1, \ldots, \xi_m)$, $y = (\eta_1, \ldots, \eta_m), \ldots$, equipped with the usual metric*

$$\rho(x,y) = \sqrt{\sum_{k=1}^{m} (\xi_k - \eta_k)^2},$$

is complete.

Proof. Let

$$x^{(n)} = (\xi_1^{(n)}, \ldots, \xi_m^{(n)}) \qquad (n = 1, 2, \ldots)$$

be a fundamental sequence of vectors in R_m. Given any $\varepsilon > 0$, let N be such that

$$\rho(x_p, x_q) = \sqrt{\sum_{k=1}^{m} (\xi_k^{(p)} - \xi_k^{(q)})^2} < \varepsilon$$

for all $p, q > N$. Then, by Theorem 3.14b,

$$|\xi_k^{(p)} - \xi_k^{(q)}| \leqslant \sqrt{\sum_{k=1}^{m} (\xi_k^{(p)} - \xi_k^{(q)})^2} < \varepsilon \qquad (k = 1, \ldots, m)$$

for all $p, q > N$. Thus each numerical sequence $\xi_k^{(p)}$ $(k = 1, \ldots, m)$ is fundamental on the real line. It follows from Theorem 3.72b that each of the limits

$$\xi_k = \lim_{p \to \infty} \xi_k^{(p)} \qquad (k = 1, \ldots, m)$$

exists.

Now consider the vector

---

† The property of being complete (or not) is obviously preserved under isometry.

$$x = (\xi_1, \ldots, \xi_m) \in R_m. \tag{2}$$

By Theorem 3.14b again, we have

$$\rho(x, x_p) = \sqrt{\sum_{k=1}^{m} (\xi_k - \xi_k^{(p)})^2} \leqslant \sqrt{m} \max_{1 \leqslant k \leqslant m} |\xi_k - \xi_k^{(p)}|. \tag{3}$$

Given any $\varepsilon > 0$, choose N such that the m inequalities

$$|\xi_k - \xi_k^{(p)}| < \frac{\varepsilon}{\sqrt{m}} \qquad (k = 1, \ldots, m)$$

hold simultaneously for all $p > N$. Then (3) implies $\rho(x, x_p) < \varepsilon$ for all $p > N$. Hence the sequence of vectors $x^{(n)}$ converges to the vector (2), and the space R_m is complete. ∎

3.73. a. THEOREM. *Let M be a complete metric space contained in another metric space P (with the same metric). Then M is closed in P.*

Proof. Let $y \in P$ be a limit point of the set M, and let x_n be a sequence of points in M converging to y (see Sec. 3.44a). Since the sequence x_n is fundamental, by Theorem 3.71b, and since the space M is complete, it follows that x_n converges to a limit

$$z = \lim_{n \to \infty} x_n$$

in M. But then $y = z \in M$, by the uniqueness of the limit (Theorem 3.33a). ∎

b. THEOREM. *Let F be a closed subset of a complete metric space M. Then F is complete, regarded as a metric space itself (with the metric "borrowed" from M).*

Proof. Every fundamental sequence $y_n \in F$ converges to a limit $y \in M$, since M is complete. But y belongs to F, since F is closed. ∎

In particular, every closed interval $[a,b]$ on the real line is a complete metric space, being a closed subset (Sec. 3.51) of a complete metric space (Theorem 3.72a).

3.74. For the real line we have the principle of nested intervals (Theorem 1.81), which asserts that a system of nested closed intervals always has a nonempty intersection. We now consider various analogues of this principle, valid for any complete metric space.

a. A set Q of nonempty subsets of a set M with the property that given any two subsets $A, B \in Q$, either $A \subset B$ or $B \subset A$, is called a *system of nested subsets*.

LEMMA. *Let Q be any system of nested subsets of a complete metric space M, and suppose Q contains subsets of arbitrarily small diameter (Sec. 3.12a). Then there is* *a unique point $p \in M$ such that every neighborhood*

$$U_\varepsilon(p) = \{x \in M : \rho(x,p) < \varepsilon\} \tag{4}$$

of p contains some set $A \in Q$.

Proof. By hypothesis, given any $m = 1,2,\ldots$, there is a subset $A_m \in Q$ such that

diam $A_m < \dfrac{1}{m}$.

Let x_m be any point of A_m. If $n > m$, then

$$\rho(x_m, x_n) < \frac{1}{m},$$

since either $A_m \subset A_n$ or $A_n \subset A_m$. Thus the sequence x_n is fundamental, with limit

$$p = \lim_{n \to \infty} x_n.$$

The point $p \in M$ satisfies the condition of the lemma. In fact, given any $\varepsilon > 0$, we need only choose n such that both inequalities

$$\rho(p, x_n) < \frac{\varepsilon}{2}, \qquad \text{diam } A_n < \frac{\varepsilon}{2}$$

hold. Then

$$\rho(p,x) \leqslant \rho(p,x_n) + \rho(x_n, x) < \frac{\varepsilon}{2} + \frac{\varepsilon}{2} = \varepsilon$$

for every $x \in A_n$, so that $A_n \subset U_\varepsilon(p)$, as required.

To prove the uniqueness of p, suppose $q \neq p$ is another point satisfying the condition of the lemma, and let $\rho(p,q) = 2\varepsilon > 0$. Then the neighborhoods $U_\varepsilon(p)$ and $U_\varepsilon(q)$ do not intersect, and moreover there are subsets $A, B \in Q$ such that $A \subset U_\varepsilon(p)$, $B \subset U_\varepsilon(q)$. But then it is impossible for one of the subsets A and B to contain the other. This contradiction shows that $q = p$. ∎

b. THEOREM. *Let Q be any system of nested* **closed** *subsets of a complete metric space M, and suppose Q contains subsets of arbitrarily small diameter. Then there is a unique point $p \in M$ such that every neighborhood* (4) *of p contains some set $A \in Q$. Moreover, p belongs to every set in Q.*

Proof. The first assertion follows at once from the lemma. Suppose p does not belong to some set $B \in Q$. Then, since B is closed, there is an $\varepsilon > 0$ such that the neighborhood $U_\varepsilon(p)$ does not intersect B. By the first assertion, there is a set $A \in Q$ entirely contained in $U_\varepsilon(p)$. But then A cannot intersect B. This contradicts the fact that either $A \subset B$ or $B \subset A$, thereby proving the second assertion. ∎

c. As a special case of Theorem 3.74b, we have the following **principle of nested balls:** *Let*

$$V_n = \{ y \in M : \rho(x,x_n) \leqslant \varepsilon_n \} \qquad (n = 1,2,\dots)$$

be a sequence of nested closed balls in a complete metric space M such that $\varepsilon_n \to 0$ as $n \to \infty$. Then the intersection of all the balls V_n consists of a single point x_0.

d. Remark. A sequence of nested closed intervals on the real line always has a nonempty intersection, whether or not the lengths of the intervals approaches zero (see Theorem 1.81). However, in a metric space (even in a *complete* metric space) there can exist sequences of nested closed balls with an empty intersection. For example, consider the space consisting of a countable sequence of points $x_1, x_2, \dots$ equipped with the metric

$$\rho(x_n, x_{n+p}) = 1 + \frac{1}{n} \qquad (n, p = 1,2,\dots),$$

where $\rho(x_n, x_n) = 0$ by definition. This space satisfies all the axioms of a metric space. Moreover, the space is complete, since it has no nonconvergent fundamental sequences (in fact, there are no fundamental sequences at all consisting of distinct points). The closed ball V_n of radius $1 + (1/n)$ centered at x_n contains the points $x_n, x_{n+1}, \dots$ and no other points, and hence

$$V_1 \supset V_2 \supset \cdots \supset V_n \supset \cdots.$$

Nevertheless, the intersection

$$\bigcap_{n=1}^{\infty} V_n$$

is empty!

3.75. a. THEOREM (**Baire**). *Suppose a complete metric space M is the union of a countable number of closed subsets $F_1, F_2, \dots \subset M$. Then at least one subset F_n contains a closed ball in M.*

Proof. Suppose to the contrary that none of the sets $F_1, F_2, \dots$ contains a closed ball, and let x_1 be a point not belonging to F_1. Since F_1 is closed, there is a closed ball

$$V_{\varepsilon_1} \{ x \in M : \rho(x,x_1) \leqslant \varepsilon_1 \}$$

which does not intersect F_1. The ball $V_{\varepsilon_1/2}(x_1)$ contains a point x_2 not belonging to F_2 (why?). Moreover, there is a closed ball $V_{\varepsilon_2}(x_2)$ which does not intersect F_2, where it can be assumed that

$$V_{\varepsilon_1}(x_1) \supset V_{\varepsilon_2}(x_1), \qquad \varepsilon_2 < \varepsilon_1/2.$$

Continuing this construction indefinitely, we get a sequence of nested closed balls

$$V_{\varepsilon_1}(x_1) \supset V_{\varepsilon_2}(x_2) \supset \cdots$$

such that $V_{\varepsilon_n}(x_n)$ does not intersect the set F_n, and moreover $\varepsilon_n \to 0$ as $n \to \infty$.†
It follows from the principle of closed balls (Sec. 3.74c) that the intersection of all the balls $V_1, V_2, \ldots$ consists of a single point x_0 which does not belong to any of the sets $F_1, F_2, \ldots$. This contradicts the condition

$$x_0 \in M = \bigcup_{n=1}^{\infty} F_n,$$

thereby proving the theorem. ∎

b. Example. The set Z of all irrational points of the interval $M = [a,b]$ cannot be represented as a countable union of closed subsets of M. In fact, if we had

$$Z = \bigcup_{n=1}^{\infty} F_n,$$

where $F_1, F_2, \ldots \subset M$ are closed sets, then the whole interval M, which is a complete space (Sec. 3.73b), could be represented as a countable union of closed subsets of M (namely, the countable collection of sets $F_1, F_2, \ldots$, together with the countable collection of all one-element sets containing a single rational point each). But this would contradict Baire's theorem, since none of these subsets can contain a closed interval.

c. In Theorem 2.41 we used the principle of nested intervals to prove that the set of all points in the unit interval $[0,1]$ is uncountable. We now prove a related result valid for a large class of complete metric spaces. First we introduce the following definition: A point x_0 of a metric space M is said to be *isolated* if there is some ball $\{x \in M : \rho(x,x_0) < \delta\}$ which contains no points of M other than the point x_0 itself. For example, let M be a set of points on the real line equipped with the usual metric. Then $x_0 \in M$ is an isolated point if and only if there is an open interval centered at x_0 containing no points of M other than x_0 itself.

---

† Note that
$$\varepsilon_n < \frac{1}{2}\varepsilon_{n-1} < \frac{1}{4}\varepsilon_{n-2} < \cdots < \frac{1}{2^{n-1}}\varepsilon_1.$$

LEMMA. *Let M be a complete metric space consisting of only countably many points. Then M contains an isolated point.*

Proof. Every one-element subset of M is closed (why?). Applying Baire's theorem, we see that some subset $\{x_0\} \subset M$ contains a closed ball $V_\varepsilon(x_0)$. But this is possible only if x_0 is an isolated point. ∎

d. THEOREM. *Every complete metric space without isolated points is uncountable.*

Proof. An immediate consequence of the above lemma. ∎

This theorem ceases to be true if we drop the condition that M have no isolated points. Consider, for example, any countable closed set on the line (e.g., any convergent sequence of points together with its limit), regarded as a metric space in its own right.

3.8. Completion of a Metric Space

3.81. The concept of a complete metric space plays a key role in the theorems of Secs. 3.73–3.75, and will continue to figure prominently in our subsequent considerations. As we now show, every incomplete metric space can be "embedded" in a certain complete space.

THEOREM (**Hausdorff**). *Let M be a metric space, in general incomplete. Then there exists a complete metric space M, called the **completion** of M, with the following properties:*
(1) M is isometric to a subset $M_1 \subset \overline{M}$;
(2) M_1 is dense in $\overline{M}$.
Moreover, every pair of spaces $\overline{M}$ and $\overline{\overline{M}}$ satisfying Properties 1 and 2 are isometric.

Proof. The proof will be given in several steps (Secs. 3.82–3.87).

3.82. *Step 1.* Two fundamental sequences x_n and y_n in the space M are said to be *cofinal* (with each other) if

$$\lim_{n \to \infty} \rho(x_n, y_n) = 0,$$

where ρ is the metric of M. For example, any two sequences in M converging to the same limit are cofinal, while two sequences converging to different limits are noncofinal. Two fundamental sequences which are cofinal with a third sequence are clearly cofinal with each other. Hence the set of all fundamental sequences consisting of elements of M can be partitioned into classes such that all sequences in the same class are cofinal, while any sequence not in a given class is noncofinal with every sequence in the class.

We now use these classes, denoted by $X, Y, \ldots$, to construct a new metric space $\overline{M}$, defining the distance between two classes X and Y by the formula

$$\rho(X,Y) = \lim_{n \to \infty} \rho(x_n, y_n), \tag{1}$$

where x_n is any fundamental sequence from the class X and y_n is any fundamental sequence from the class Y. First of all, we must verify that the limit (1) exists and is independent of the choice of the sequences x_n and y_n from the classes X and Y. It follows from the quadrilateral inequality (1), p. 53 that

$$|\rho(x_n, y_n) - \rho(x_{n+p}, y_{n+p})| \leqslant \rho(x_n, x_{n+p}) + \rho(y_n, y_{n+p}),$$

and hence the numbers $\rho(x_n, y_n)$ satisfy the Cauchy convergence criterion, i.e., are themselves a fundamental sequence on the real line R. Thus the limit (1) indeed exists, by Theorem 3.72b. Moreover, if x_n' and y_n' are other fundamental sequences from the classes X and Y, respectively, then, by the quadrilateral inequality again,

$$|\rho(x_n, y_n) - \rho(x_n', y_n')| \leqslant \rho(x_n, x_n') + \rho(y_n, y_n') \to 0$$

as $n \to \infty$, so that the sequence $\rho(x_n', y_n')$ has the same limit as the sequence $\rho(x_n, y_n)$. Thus the definition (1) of the distance between two given classes is indeed independent of the particular choice of fundamental sequences from the classes.

3.83. *Step 2.* Next we verify that the quantity (1) satisfies Axioms a–c of Sec. 3.11, thereby confirming that (1) is actually a metric in $\overline{M}$. First we note that $\rho(X,X) = 0$, as follows at once by setting $x_n = y_n$ in (1). Moreover, suppose $\rho(X,Y) = 0$. This means that

$$\lim_{n \to \infty} \rho(x_n, y_n) = 0$$

for any fundamental sequence x_n in X and any fundamental sequence y_n in Y. But then x_n and y_n are cofinal sequences, and hence the classes X and Y must coincide. It follows that $\rho(X,Y) = 0$ implies $X = Y$, or equivalently that $\rho(X,Y) > 0$ if $X \neq Y$. Together with $\rho(X,X) = 0$, this establishes Axiom a. The validity of Axiom b is an immediate consequence of the symmetry of distance in M, since $\rho(y_n, x_n) = \rho(x_n, y_n)$ obviously implies

$$\rho(Y,X) = \lim_{n \to \infty} \rho(y_n, x_n) = \lim_{n \to \infty} \rho(x_n, y_n) = \rho(X,Y).$$

As for Axiom c, let x_n, y_n, and z_n be fundamental sequences from the classes X, Y, and Z, respectively. Then, by the triangle inequality in M,

$$\rho(x_n, z_n) \leqslant \rho(x_n, y_n) + \rho(y_n, z_n). \tag{2}$$

Using Theorems 3.32e to take the limit of (2) as $n \to \infty$, we get the triangle inequality

$$\rho(X,Z) \leqslant \rho(X,Y) + \rho(Y,Z)$$

in $\overline{M}$, thereby verifying Axiom c.

3.84. *Step 3.* We now show that $\overline{M}$ contains a subset M_1 isometric to the space M. Suppose that with each element $x \in M$ we associate the class $X \in \overline{M}$ containing the sequence $x,x,\ldots,x,\ldots$, i.e., the class of all sequences converging to x. Then the set M_1 of all such classes X is a subset of $\overline{M}$ isometric to the original space M. In fact, if X and Y are the classes in M_1 corresponding to the elements x and y in M, we have

$$\rho(X,Y) = \lim_{n \to \infty} \rho(x,y) = \rho(x,y),$$

thereby verifying Property 1.

3.85. *Step 4.* To verify Property 2, i.e., that M_1 is dense in $\overline{M}$, we argue as follows. Given any class $X \in \overline{M}$, let $x_1,x_2,\ldots,x_n,\ldots$ be any fundamental sequence in X. Consider the sequence of classes $X_1,X_2,\ldots,X_n,\ldots$, where X_n corresponds to the sequence $x_n,x_n,\ldots,x_n,\ldots$, i.e., corresponds to the element x_n under the mapping of Sec. 3.84. Given any $\varepsilon > 0$, let N be such that $\rho(x_m,x_n) < \varepsilon$ for all $m,n > N$. Then

$$\rho(X,X_n) = \lim_{m \to \infty} \rho(x_m,x_n) \leqslant \varepsilon$$

for all $n > N$. But this means that the class X is the limit of the sequence of classes $X_1,X_2,\ldots,X_n,\ldots$ Since every X_n belongs to M_1, it follows that M_1 is dense in $\overline{M}$.

3.86. *Step 5.* Next we show that the space $\overline{M}$ is complete. Let $X_1,X_2,\ldots,$ $X_n,\ldots$ be a fundamental sequence of elements of $\overline{M}$. Since M_1 is dense in $\overline{M}$, for every class X_n we can find a class $Y_n \in M_1$ such that

$$\rho(X_n,Y_n) < \frac{1}{n}.$$

Let y_n be the element of M corresponding to the class Y_n under the mapping of Sec. 3.84, i.e., the common limit of all the sequences in Y_n. Then the sequence $y_1,y_2,\ldots,y_n,\ldots$ is fundamental in the space M, since

$$\rho(y_m,y_n) = \rho(Y_m,Y_n) \leqslant \rho(Y_m,X_m) + \rho(X_m,X_n) + \rho(X_n,Y_n)$$
$$\leqslant \rho(X_m,X_n) + \frac{1}{m} + \frac{1}{n} \to 0$$

as $m,n \to \infty$. The fundamental sequence $y_1, y_2, \ldots, y_n, \ldots$ determines a class $Y \in \overline{M}$. But this class is just the limit in $\overline{M}$ of the sequence $X_1, X_2, \ldots, X_n, \ldots$ In fact, given any $\varepsilon > 0$, we have

$$\rho(Y, X_n) \leqslant \rho(Y, Y_n) + \rho(Y_n, X_n) = \lim_{m \to \infty} \rho(y_m, y_n) + \frac{1}{n} < \varepsilon$$

for all sufficiently large $n > N$. It follows that every fundamental sequence $X_1, X_2, \ldots, X_n, \ldots$ of elements of $\overline{M}$ has a limit in $\overline{M}$, i.e., that $\overline{M}$ is complete.

3.87. *Final Step.* To complete the proof, we show that any metric space $\widetilde{M}$ with the Properties 1 and 2 is isometric to the space $\overline{M}$. Let M_1 and M_2 be the subsets of the spaces $\overline{M}$ and $\widetilde{M}$ isometric to the space M and hence isometric to each other. Given any element $X \in \overline{M}$, let $X_n \in M_1$ be a sequence converging to X. The corresponding sequence $Y_n \in M_2$ is certainly fundamental. In fact, because of the isometry between M_1 and M_2, the distance between any two elements of the sequence Y_n is the same as the distance between the elements of the sequence X_n with the same indices. Since the space $\widetilde{M}$ is complete, it contains the element

$$Y = \lim_{n \to \infty} Y_n.$$

We now associate this element $Y \in \widetilde{M}$ with the original element $X \in \overline{M}$. This uniquely determines Y, since cofinal sequences in M_1 correspond to cofinal sequences in M_2, and replacing X_n by a cofinal sequence leads to replacing Y_n by a cofinal sequence. The indicated correspondence is clearly one-to-one and exhausts all the elements of $\overline{M}$ and $\widetilde{M}$. It only remains to show that $\overline{M}$ and $\widetilde{M}$ are isometric. Let X, X' be elements of $\overline{M}$, and let Y, Y' be the corresponding elements of $\widetilde{M}$. Suppose that

$$X = \lim_{n \to \infty} X_n, \quad X' = \lim_{n \to \infty} X_n' \qquad (X_n, X_n' \in M_1),$$

and let Y_n, Y_n' be the elements of M_2 corresponding to the elements X_n, X_n'. Then

$$\rho(X_n, X_n') = \rho(Y_n, Y_n'),$$

and hence, by Theorem 3.32b,

$$\rho(Y, Y') = \lim_{n \to \infty} \rho(Y_n, Y_n') = \lim_{n \to \infty} \rho(X_n, X_n') = \rho(X, X'). \quad \blacksquare$$

3.88. Let M be a metric space which is a subset of a complete metric space M^*. Then $\overline{M}$, the closure of M (relative to M^*), can be chosen as the completion of M. In fact, $\overline{M}$ is complete, being a closed subset of a complete space (see Theorem 3.73b), and moreover M is obviously a dense subset of

$\overline{M}$. Thus $\overline{M}$ satisfies Properties 1 and 2 of Theorem 3.81, and hence can serve as the completion of M.

3.9. Compactness

3.91. a. Definition. A metric space M is said to be *compact* if every sequence of points in M has a limit point in M. A compact metric space M is often called a *compactum*. A metric space M is said to be *locally compact* if every point of M has a neighborhood whose closure is compact.

b. THEOREM. *A metric space M is compact if and only if every infinite subset $E \subset M$ has a limit point in M.*

Proof. Suppose M is compact, and let E be any infinite subset of M. Then E contains a sequence x_n of distinct points. This sequence has a limit point in M, which is clearly also a limit point of the set E.

Conversely, suppose every infinite subset $E \subset M$ has a limit point, and let x_n be any sequence of (not necessarily distinct) points of M. If the sequence contains only finitely many distinct points of M, at least one of these points must "repeat itself infinitely often," and this point is then a limit point of the sequence. On the other hand, if the sequence contains infinitely many distinct points of M, then the infinite set consisting of these points has a limit point, and this limit point is also a limit point of the sequence x_n. In either case, M is compact. ∎

This leads at once to the following alternative definition of compactness: *A metric space M is said to be compact if every infinite subset $E \subset M$ has a limit point in M.*

c. Examples. Any closed interval $[a,b]$ on the real line is a compactum, by the Bolzano-Weierstrass principle (Theorem 3.45). The whole real line R is not a compact space, since, for example, the sequence $1,2,\dots,n,\dots$ has no limit points. However, R is a locally compact space, since every point $x \in R$ has a neighborhood whose closure is compact, namely any closed interval centered at x. The extended real line $\overline{R}$ with the metric $r(x,y)$ of Theorem 3.35d is compact. The set of all rational points in the interval $[a,b]$ is neither compact nor locally compact.

d. As noted in Sec. 3.43, the property of being a limit point of a given sequence does not change if we equip the given metric space with a new metric homeomorphic to the original metric. More concisely, *the property of being a limit point is invariant under transformation to a new metric homeomorphic to the original metric.* Since the definitions of Sec. 3.91a involve only the notion of a limit point, we conclude that *the property of being compact or locally compact is invariant under transformation to a new metric homeomorphic to the original metric.* Thus the real line R is locally compact when equipped with either the ordinary metric $\rho(x,y) = |x-y|$ or the metric $r(x,y)$ of Theorem 3.35d.

3.92. a. THEOREM. *Every compact metric space M is complete.*

Proof. Given any fundamental sequence x_n of points in M, let $x \in M$ be the limit point of x_n guaranteed by the compactness of M. Then, as we now show, x_n converges to x, thereby proving the completeness of M. In fact, given any $\varepsilon > 0$, we first choose N such that $\rho(x_m,x_n) < \varepsilon/2$ for all $m,n > N$, and afterwards choose $p > N$ such that $\rho(x_p,x) < \varepsilon/2$. Then

$$\rho(x_n,x) \leqslant \rho(x_n,x_p) + \rho(x_p,x) < \varepsilon$$

for all $n > N$, i.e., $x_n \to x$ as $n \to \infty$. ∎

b. THEOREM. *Every compact subset M of a metric space P is closed in P.*

Proof. An immediate consequence of Theorems 3.92a and 3.73a. ∎

c. THEOREM. *Given a compact subset M of a metric space P and an open set G such that $M \subset G \subset P$, let the (open) set M_δ be the union of all open balls of radius $\delta > 0$ centered at points of M. Then $M_\delta \subset G$ for a suitable value of δ.*

Proof. Suppose to the contrary that given any $n = 1,2,\ldots$, there are two points $x_n \in M$ and $y_n \in P - G$ such that $\rho(x_n,y_n) < 1/n$. The sequence x_n lies in the compactum M, and hence has a limit point $x \in M$. Moreover, by Theorem 3.42, the sequence x_n has a subsequence x_{n_m} converging to x, i.e., such that $x_{n_m} \to x$ as $m \to \infty$. Since

$$\rho(y_{n_m},x) \leqslant \rho(y_{n_m},x_{n_m}) + \rho(x_{n_m},x) \leqslant \frac{1}{n_m} + \rho(x_{n_m},x),$$

we also have $y_{n_m} \to x$ as $m \to \infty$. But the set $P - G$ is closed. Hence $x \in P - G$, which is incompatible with $x \in M$. The proof now follows by contradiction. ∎

3.93. a. We now introduce a somewhat larger class of spaces, including compact spaces as a subclass. A metric space M is said to be *precompact* if every sequence of points in M contains a fundamental subsequence. If M is complete, this fundamental subsequence converges to a point in M, so that *a complete precompact space is necessarily compact.* Conversely, every compact space is complete (Theorem 3.92a) and obviously precompact. The open interval (a,b) on the real line is a simple example of a space which is precompact but not compact.

b. THEOREM. *Every precompact space M is bounded.*

Proof. It suffices to show that an unbounded space M is necessarily nonprecompact. If M is unbounded, then, by Sec. 3.12c, given any $C > 0$ and any point $a \in M$, we can find a point $x \in M$ such that $\rho(a,x) > C$. Noting this, we fix a point $x_1 \in M$ and then inductively construct a sequence of points $x_2, x_3, \ldots$ in M such that

$$\rho(x_1, x_2) > 1,$$

$$\rho(x_n, x_{n+1}) \geqslant \sum_{k=1}^{n-1} \rho(x_k, x_{k+1}) + 1 \qquad (n = 2, 3, \ldots).$$

It then follows from Theorem 3.11a that

$$\begin{aligned}
\rho(x_m, x_n) &\geqslant \rho(x_{n-1}, x_n) - \rho(x_m, x_{n-1}) \\
&\geqslant \rho(x_{n-1}, x_n) - [\rho(x_m, x_{m+1}) + \cdots + \rho(x_{n-2}, x_{n-1})] \\
&\geqslant \rho(x_{n-1}, x_n) - \sum_{k=1}^{n-2} \rho(x_k, x_{k+1}) > 1
\end{aligned}$$

for all $n > m$. Hence the sequence x_n cannot contain a fundamental subsequence, and the space M is nonprecompact. ∎

c. To test a metric space M for precompactness, it is sometimes convenient to embed M isometrically in a larger metric space P. We then call a set $B \subset P$ an *ε-net* for the set $M \subset P$ if the distance from every point $x \in M$ to some point $y \in B$ (in general, depending on x) does not exceed ε. In this case, the union of all closed balls of radius ε centered at the points of B contains the whole set M.

More generally, suppose the union of all the sets E_α indexed by a parameter α contains a set M. Then the sets E_α are said to *cover* M or to form a *covering* of M. Thus a set B is said to be an *ε-net* for a set M if the set of all closed balls of radius ε centered at points of B covers the set M.†

THEOREM (**Hausdorff's criterion**). *A subset M of a metric space P is precompact (in the metric ρ of P) if and only if, given any $\varepsilon > 0$, P contains a* **finite** *ε-net for M.*

Proof. Suppose M is precompact. Then, given any $\varepsilon > 0$, we construct an ε-net for M as follows. Choose any point $x_1 \in M$. If every point $x \in M$ is such that $\rho(x_1, x) \leqslant \varepsilon$, the point x_1 is itself an ε-net for M, and the construction is finished. However, if there are points of M whose distance from x_1 exceeds ε, we choose one of these points as x_2. If now every point $x \in M$ is such that either $\rho(x_1, x) \leqslant \varepsilon$ or $\rho(x_2, x) \leqslant \varepsilon$, the points x_1 and x_2 form a finite ε-net for M, and the construction is finished. Otherwise, we continue the construction,

† "Suppose a lamp illuminating a ball of radius ε is placed at every point of a set B which is an ε-net for a set M. Then the whole set M will be illuminated." (L. A. Liusternik)

noting that the distance from each new point x_n to each of the preceding points $x_1, x_2, \ldots, x_{n-1}$ exceeds ε. Hence if the construction fails to terminate after a finite number of steps, we would get a sequence of distinct points $x_1, x_2, \ldots, x_n, \ldots$ in M which certainly contains no fundamental subsequence, thereby contradicting the precompactness of M. It follows that the construction must terminate after a finite number of steps, resulting in a finite ε-net for the set M.

Conversely, suppose that given any $\varepsilon > 0$, P contains a finite ε-net for M, and let A be any infinite subset of M (in particular, a sequence containing infinitely many distinct points of M). Then we can select a fundamental sequence from A as follows: Let any point $x_0 \in A$ be the first point. Then, choosing $\varepsilon = 1$ in the finite ε-net condition, we cover A with a finite number of closed balls of radius 1. One of these balls, say V_1, must contain an infinite subset $A_1 \subset A$. Let x_1 be any point in A_1 distinct from x_0. Choosing $\varepsilon = \frac{1}{2}$, we then cover A_1 in turn with a finite number of closed balls of radius $\frac{1}{2}$. One of them, say V_2, must contain an infinite subset $A_2 \subset A_1$. Let x_2 be any point of A_2 distinct from x_0 and x_1. Continuing this process indefinitely, we get a sequence of infinite subsets

$$A \supset A_1 \supset A_2 \supset \cdots \supset A_n \supset \cdots$$

(where each A_n is contained in a closed ball of radius $1/n$) and a sequence of distinct points $x_0, x_1, x_2, \ldots, x_n, \ldots$ with $x_n \in A_n$. This sequence is fundamental. In fact, if $m < n$, then

$$V_m \supset A_m \supset A_n$$

and hence

$$\rho(x_m, x_n) \leqslant 2/m,$$

where the right-hand side approaches zero as $m \to \infty$. It follows that M is precompact. ∎

3.94. THEOREM. *Every bounded subset M of the n-dimensional Euclidean space R_n is precompact.*

Proof. Being bounded, M is contained in some closed ball $V \subset R_n$ (Sec. 3.14c), and every such ball contains only finitely many points of the form $k/2^m$ where k and $m > 0$ are integers. But, given any $\varepsilon > 0$, the set of all such points is obviously an ε-net for M provided m is sufficiently large. ∎

3.95. THEOREM. *A subset M of a metric space P is precompact if, given any $\varepsilon > 0$, there is a (possibly infinite) precompact set $B_\varepsilon \subset P$ which is an ε-net for M.*

Proof. Let Z be a finite $(\varepsilon/2)$-net for the set $B_{\varepsilon/2}$ (Z exists, since $B_{\varepsilon/2}$ is precompact). Then Z is a finite ε-net for the set M. In fact, given any point $x \in M$, there is a point $y \in B_{\varepsilon/2}$ such that $\rho(x,y) \leqslant \varepsilon/2$ and a point $z \in Z$ such that $\rho(y,z) \leqslant \varepsilon/2$, which together imply

$$\rho(x,z) \leqslant \rho(x,y) + \rho(y,z) \leqslant \varepsilon.$$

Thus, given any $\varepsilon > 0$, M has a finite ε-net and hence is precompact. ∎

3.96. a. THEOREM. *The completion $\overline{M}$ of any precompact metric space M is a compactum.*

Proof. Given any $\varepsilon > 0$, the set M is an ε-net for $\overline{M}$, being dense in $\overline{M}$. But M is precompact, and hence by Theorem 3.95, so is $\overline{M}$. Moreover, being complete, $\overline{M}$ is compact as well (Sec. 3.93a). ∎

b. THEOREM. *The closure $\overline{M}$ of any precompact subset M of a complete metric space P is compact.*

Proof. An immediate consequence of Theorem 3.96a and of the fact that the closure of M in P can be chosen as the completion of M (see Sec. 3.88). ∎

c. THEOREM. *A precompact subset M of a complete metric space P is a compactum if and only if M is closed in P.*

Proof. If M is closed in P, then M is itself a complete metric space, by Theorem 3.73b. But then, being precompact, M is compact as well (Sec. 3.93a). The converse assertion follows from Theorem 3.92b. ∎

d. THEOREM. *A subset M of a complete metric space P is a compactum if and only if M is closed in P and, given any $\varepsilon > 0$, P contains a finite ε-net for M.*

Proof. An immediate consequence of Theorem 3.93a (Hausdorff's criterion) and Theorem 3.96c. ∎

e. THEOREM. *A subset M of the space R_n is compact if and only if M is closed and bounded in R_n.*

Proof. If M is compact, then M is closed, by Theorem 3.92b, and bounded, by Theorem 3.93b. Conversely, if M is bounded in R_n, then M is precompact, by Theorem 3.94, and hence, by the completeness of R_n (Theorem 3.72d), the fact that the precompact set M is closed implies that M is compact (Theorem 3.96c). ∎

f. THEOREM. *Every compact set M on the real line is bounded and contains its greatest lower and least upper bounds.*

Proof. The set M is closed and bounded, by the preceding theorem. Now use Theorem 3.53b (and the subsequent remark). ∎

3.97. THEOREM (**Finite covering theorem**). *Suppose a compact subset K of a metric space P is covered by a family $\mathscr{B} = \{B_\alpha\}$ of open subsets of P. Then K can also be covered by some finite subfamily $B_1, \ldots, B_m$ of subsets of $\mathscr{B}$.*

Proof. Suppose to the contrary that no finite subfamily of $\mathscr{B}$ covers K. Since K is compact, given any $\varepsilon > 0$, there is a finite number of closed balls $V_1, \ldots, V_{m_1}$ of radius ε covering K (by Theorem 3.93c). If each ball V_j ($j = 1, \ldots, m_1$) could be covered by a finite subfamily of $\mathscr{B}$, then all these subfamilies taken together would give a new finite subfamily of $\mathscr{B}$ covering K itself, contrary to hypothesis. Therefore at least one of the balls V_j cannot be covered by a finite subfamily of $\mathscr{B}$. Choosing $\varepsilon = 1/n$, we see that for each $n = 1, 2, \ldots$ there is a closed ball $V_{1/n}(x_n)$ of radius $1/n$ centered at some point $x_n \in K$ which cannot be covered by a finite subfamily of $\mathscr{B}$. Let x be a limit point of the sequence x_n (since K is compact, x exists). Then there is a set $B_\alpha \in \mathscr{B}$ containing both x and some neighborhood of x, i.e., some open ball $U_\rho(x)$ of radius ρ centered at x (here we use the fact that B_α is open). But then $U_\rho(x)$ contains balls $V_{1/n}(x_n)$ with arbitrarily large values of n, so that each of these balls is covered by the single set B_α, contrary to construction. This contradiction shows that some finite subfamily of $\mathscr{B}$ must cover K. ∎

3.98. In Sec. 3.74 we showed how the principle of nested intervals (Theorem 1.81) can be generalized from the real line to an arbitrary complete metric space, provided we change nested closed intervals to nested closed subsets of arbitrarily small diameter. It will be recalled that the intersection of all the nested closed subsets may well be empty if the proviso "of arbitrarily small diameter" is dropped (see Sec. 3.74d). The following theorem shows that this cannot happen if the subsets are compact:

THEOREM. *Let Q be a system of nested (nonempty) compact subsets of a metric space M. Then the intersection of all the sets of Q is nonempty.*

Proof. Suppose the theorem is false, and choose any $K_0 \in Q$. Then for each $x \in K_0$ there is a set $K_x \in Q$ such that $x \notin K_x$. Since K_x is closed (by Theorem 3.92b), there is a neighborhood U_x of x which does not intersect K_x. The set of all such $U_x (x \in K_0)$ covers K. Hence, by Theorem 3.97, K can be covered by a finite number of these neighborhoods, say $U_1, \ldots, U_n$. Let $K_1, \ldots, K_n \in Q$ be sets which do not intersect $U_1, \ldots, U_n$, respectively. Then the intersection $K_1 \cdots K_n$ has no points in common with any of the neighborhoods $U_1, \ldots, U_n$, and hence does not intersect the original set K_0. It follows that the intersection $K_0 K_1 \cdots K_n$ is empty. On the other hand, since Q is a system of nested sets, every finite intersection of sets of Q is again a set of Q, and in particular cannot be nonempty. This contradiction shows that the theorem is true. ∎

Problems

1. Let A' denote the set of all limit points of a given subset A of a metric space M, and let

$$A^{(n)} = (A^{(n-1)})' \qquad (n = 1,2,\ldots)$$

$(A^{(0)} = A,\ A^{(1)} = A')$. Given any n, construct a set A on the real line such that $A^{(n)}$ is nonempty while $A^{(n+1)}$ is empty.

2. Given any $A \subset M$, prove that the set A' is closed.

3. Given any set A on the real line such that $A^{(n)}$ is countable for some n, prove that A itself is countable.

4. A point x on the real line is called a *condensation point* of an uncountable set A if every neighborhood of x contains uncountably many points of A. Prove that every uncountable set A has condensation points. More exactly, prove that "almost all" points of an uncountable set A (i.e., all points of A with the possible exception of at most countably many points) are condensation points.

5. Suppose a set A on the real line is covered by an arbitrary family $\mathscr{B}$ of open intervals. Prove that A can also be covered by a subfamily of $\mathscr{B}$ containing no more than countably many intervals.

6. Given any subset A of a metric space M, the quantity

$$\rho(x,A) = \inf_{y \in A} \rho(x,y)$$

is called the *distance from the point x to the set A*. Prove that the relations $\rho(x,A) = 0$ and $x \in A$ are equivalent if A is closed, but not if A fails to be closed.

7. Given any subset A of a metric space M, prove that the set

$$\{x \in M : \rho(x,A) < \varepsilon\}$$

is open, while the set

$$\{x \in M : \rho(x,A) \leqslant \varepsilon\}$$

is closed.

8. Given two nonintersecting closed subsets F_1 and F_2 of a metric space M, construct nonintersecting open sets G_1 and G_2 such that $G_1 \supset F_1$, $G_2 \supset F_2$.

9. Prove that the set of all closed subsets $A,B,\ldots$ of a bounded metric space M is itself a metric space when equipped with the metric

$$\rho(A,B) = \sup_{\substack{x \in A \\ y \in B}} \{\rho(x,B), \rho(y,A)\}.$$

Prove that this space is complete if the original space M is complete, and compact if M is compact.

10. Given a metric space M consisting of n points, prove that if $n < 4$ there exists a metric space M' isometric to M and contained in the Euclidean space R_{n-1}. Prove that this assertion is in general false if $n \geqslant 4$.

11. Let $\bar{R}_n$ denote the Euclidean space R_n together with an extra point ∞ (the "point at infinity"). Show that a metric r can be introduced in $\bar{R}_n$ such that

(a) r is homeomorphic to the usual metric ρ of Sec. 3.14a on R_n;

(b) Every sequence $x_m \in R_n$ which is unbounded in the usual metric has ∞ as a limit point.

12. Solve the analogue of Problem 11 for the case where R_n is replaced by an arbitrary unbounded metric space M.

13. Suppose that in Problem 11 we use lines going through the center of the sphere S_n instead of the lines considered in the hint. What choice of "elements at infinity" now guarantees the existence of a limit point (in the new metric) for every sequence of points $x_m \in R_n$?

4 Limits

4.1. Basic Concepts

4.11. Given any set E, a system S of nonempty subsets $A, B, \ldots$ of E is called a *direction* (on E) if

(1) Either $A \subset B$ or $B \subset A$ for every pair of sets $A, B \in S$;

(2) The intersection of all the sets $A \subset S$ is empty.

4.12. Let $f(x)$ be a function defined on a set E, taking values in a metric space M equipped with a distance ρ (Sec. 3.11). Then we say that $f(x)$ *approaches the limit p in the direction S* if, given any $\varepsilon > 0$, there exists a set $A \in S$ such that

$$\rho(p, f(x)) < \varepsilon$$

for all $x \in A$.† This fact is expressed by writing

$$p = \lim_{S} f(x) \tag{1}$$

(more concisely, $f(x) \underset{S}{\rightarrow} p$ or just $f(x) \rightarrow p$). To say that $f(x)$ *has a limit in the direction S* means that there is some point $p \in M$ such that $f(x) \underset{S}{\rightarrow} p$.

We now give examples illustrating these definitions (Secs. 4.13–4.16).

4.13. Let E be the set of all positive integers $1, 2, \ldots$, and let S be the system of all subsets $A_n \subset E$ of the form

$$A_n = \{n, n+1, n+2, \ldots\} \qquad (n = 1, 2, \ldots).$$

Then obviously either $A_m \subset A_n$ or $A_n \subset A_m$ for every pair of sets $A_m, A_n \in S$, while the intersection of all the sets A_n $(n = 1, 2, \ldots)$ is empty. Therefore S is a direction, which we denote by $n \rightarrow \infty$. Here a function $y = f(x)$ defined on E is just a sequence y_n of points in the metric space M. Thus, according to the definition of Sec. 4.12, we say that y_n approaches the limit p in the direction S, i.e., as $n \rightarrow \infty$,‡ if, given any $\varepsilon > 0$, there exists an integer $N > 0$ such that

$$\rho(p, y_n) < \varepsilon$$

for all $n > N$. This definition clearly agrees with the one already given in Sec. 3.31. In the present case, (1) takes the form

$$p = \lim_{n \rightarrow \infty} y_n.$$

4.14. a. Let $E = R_a^+$ be the real half-line $\{x : x \geqslant a\}$, and let S be the system of

---

† The uniqueness of the limit p is proved in Theorem 4.22.

‡ For brevity, we say "as $n \rightarrow \infty$" instead of "in the direction $n \rightarrow \infty$," and similarly elsewhere.

all subsets $A_\xi \subset R_a^+$ of the form

$$A_\xi = \{x : x \geqslant \xi\} \qquad (\xi \geqslant a).$$

Then S is obviously a direction, which we denote by $x \to +\infty$ (or simply $x \to \infty$). Applying the general definition of Sec. 4.12 to this case, we say that a function $f(x)$ approaches the limit p in the direction S, i.e., as $x \to +\infty$, if, given any $\varepsilon > 0$, there exists a number ξ such that

$$\rho(p, f(x)) < \varepsilon$$

for all $x \geqslant \xi$. Here the appropriate version of (1) is just

$$p = \lim_{x \to \infty} f(x).$$

b. If $M = R$ is the real line, $f(x)$ becomes a numerical function and we get the following definition: A numerical function $f(x)$ is said to approach the limit p as $x \to \infty$ if, given any $\varepsilon > 0$, there exists a number ξ such that

$$|p - f(x)| < \varepsilon$$

for all $x \geqslant \xi$.

c. Now let $E = R_a^-$ be the real half-line $\{x : x \leqslant a\}$, and let S be the system of all subsets $B_\xi \subset R_a^-$ of the form

$$B_\xi = \{x : x \leqslant \xi\} \qquad (\xi \leqslant a).$$

Then S is obviously a direction, which this time we denote by $x \to -\infty$. The general definition of Sec. 4.12 now takes the following form: A function $f(x)$ is said to approach the limit p as $x \to -\infty$ if, given any $\varepsilon > 0$, there exists a number ξ such that

$$\rho(p, f(x)) < \varepsilon$$

(or $|p - f(x)| < \varepsilon$ in the case of a numerical function) for all $x \leqslant \xi$. Formula (1) then becomes

$$p = \lim_{x \to -\infty} f(x).$$

d. We can also introduce a direction $x \to \pm \infty$, or equivalently $|x| \to \infty$, corresponding to the system of all subsets of the real line of the form $\{x : |x| \geqslant \xi\}$. This time formula (1) takes the form

$$p = \lim_{|x| \to \infty} f(x).$$

4.15. a. Next let E itself be a metric space, equipped with a distance ρ_0. Suppose a is a nonisolated point of E (cf. Sec. 3.75c), i.e., suppose every neighborhood

$$U_\delta(a) = \{x \in E: \rho_0(x,a) < \delta\}$$

contains points of E other than a itself. Then the system of *deleted neighborhoods*

$$U'_\delta(a) = \{x \in E: 0 < \rho_0(x,a) < \delta\} \qquad (\delta > 0),$$

each obtained by deleting the center a from an ordinary neighborhood $U_\delta(a)$, defines a direction which we denote by $x \to a$. Here the fact that a is nonisolated guarantees that every $U'_\delta(a)$ is nonempty, while the fact that $a \notin U'_\delta(a)$ guarantees that the intersection of all the $U'_\delta(a) (\delta > 0)$ is empty.†
The definition of Sec. 4.12 now reads as follows: A function $f(x)$ is said to approach the limit p as $x \to a$ if, given any $\varepsilon > 0$, there exists a number $\delta > 0$ such that

$$\rho(p, f(x)) < \varepsilon$$

for all $x \in U'_\delta(a)$, i.e., for all x satisfying the condition $0 < \rho_0(x,a) < \delta$. Correspondingly, formula (1) becomes

$$p = \lim_{x \to a} f(x). \tag{2}$$

Note that the value of the function $f(x)$ at the point $x = a$ plays no role in this definition. In fact, $f(x)$ may not even be *defined* at the point $x = a$.

b. In particular, if $E = R$, $M = R$, each deleted neighborhood $U'_\delta(a)$ reduces to the set

$$(a - \delta, a) \bigcup (a, a + \delta),$$

obtained by deleting the center a from the open interval $(a - \delta, a + \delta)$, and the definition of Sec. 4.12 then reads: A numerical function $f(x)$ is said to approach the limit p as $x \to a$ if, given any $\varepsilon > 0$, there exists a number $\delta > 0$ such that

$$|p - f(x)| < \varepsilon$$

for all $0 < |x - a| < \delta$. In this case, we continue to write formula (2).

c. The examples of Secs. 4.14a–4.14c are actually special cases of the definition of Sec. 4.15a.‡ To see this, we equip the real line R with the metric of the space $\bar{R}$, as in Theorem 3.35d. Then, in keeping with Sec. 3.35e, the

---

† Obviously, one of every pair of sets $U'_{\delta_1}(a)$, $U'_{\delta_2}(a)$ must contain the other ($U'_{\delta_1}(a) \supset U'_{\delta_2}(a)$ if $\delta_1 > \delta_2$).
‡ The same is true of the example of Sec. 4.13 (give the details).

directions $x \to -\infty$ and $x \to +\infty$ as defined in Secs. 4.14a and 4.14c are equivalent to the directions $x \to -\infty$ and $x \to +\infty$ as defined in Sec. 4.15a with $-\infty$ and $+\infty$ regarded as *points* of the space $\overline{R}$.

4.16. Partial limits. Given a set E equipped with a direction S, suppose we fix a subset $G \subset E$ and then consider the system of sets GA, where A is any set of the system S. Suppose every set GA is nonempty. Then, since the intersection of all the sets GA is empty (just like the intersection of all the sets $A \in S$), the system of sets GA defines a new direction, which we denote by GS. The limit in the direction GS might be called a "partial limit," as opposed to the "full limit" in the original direction S.

a. Let $f(x)$ be a function defined on E, taking values in a metric space M. If $\lim_S f(x)$ exists and equals p, then obviously $\lim_{GS} f(x)$ also exists and equals p. On the other hand, if $\lim_{GS} f(x)$ exists, then $\lim_S f(x)$ may or may not exist. The following theorem gives a criterion for the equivalence of full and partial limits:

THEOREM. *If G contains some set $B \in S$, then the existence of $\lim_{GS} f(x)$ implies that of $\lim_S f(x)$ and the two limits are equal. However, if G contains no set $B \in S$ and if the space M contains at least two distinct points p and $q \neq p$, then there exists a function $f(x)$ such that $\lim_{GS} f(x) = p$ while $\lim_S f(x)$ fails to exist.*

Proof. Suppose first that

$$G \supset B, \qquad B \in S, \qquad \lim_{GS} f(x) = p.$$

Then, given any $\varepsilon > 0$, we can find a set $GA \in GS$ such that

$$\rho(p, f(x)) < \varepsilon \tag{3}$$

for all $x \in GA$ (ρ is the metric of M). The set GA contains the set BA, which in turn equals either B or A and belongs to the direction S. Therefore (3) holds for all $x \in BA$. It follows that $\lim_S f(x)$ exists and equals p.

Now suppose G contains no set $B \in S$, and let H be the complement of G (relative to the whole set E). Introduce the function

$$f(x) = \begin{cases} p \text{ if } x \in G, \\ q \text{ if } x \in H, \end{cases}$$

and let $\varepsilon = \frac{1}{2}\rho(p,q)$. If $\lim_S f(x) = t$ existed, then we would have $\rho(t, f(x)) < \varepsilon$ for all x in some set $A \in S$. But both sets GA and HA are nonempty by hypothesis, and hence, choosing first $x \in GA$ and then $x \in HA$, we would get both $\rho(p,t) < \varepsilon$ and $\rho(q,t) < \varepsilon$, which together imply

$$\rho(p,q) \leqslant \rho(p,t) + \rho(t,q) < 2\varepsilon,$$

contrary to the definition of ε. It follows by contradiction that $\lim_S f(x)$ fails to exist. ∎

b. Suppose $\lim_{GS} f(x)$ exists. If the function $f(x)$ is defined only on the set G, then $\lim_S f(x)$ is meaningless as it stands. In the case where G contains some set $B \in S$, we let $f_E(x)$ denote any extension of the function $f(x)$ from G to E,† and then set

$$\lim_S f(x) = \lim_S f_E(x) \tag{4}$$

by definition. Theorem 4.16a then shows that (4) makes sense and does not depend on how $f(x)$ is extended from G to E.

In particular, the limit of a sequence $y_1, y_2, \ldots, y_n, \ldots$ (Sec. 4.13) makes sense not only when y_n is defined for all $n = 1, 2, \ldots$, but also when y_n is defined only for all n greater than some positive integer n_0. In the latter case, we can assign $y_1, \ldots, y_{n_0}$ arbitrary values without changing the value of

$$\lim_{n \to \infty} y_n.$$

Similarly, the definition of

$$\lim_{x \to \infty} f(x)$$

(Sec. 4.14a) depends only on the values of $f(x)$ for x greater than some number x_0, and is independent of the values of $f(x)$ for $x \leqslant x_0$. In just the same way, the definition of

$$\lim_{x \to a} f(x)$$

(Sec. 4.15a) involves only the values of $f(x)$ "near the point $x = a$," i.e., in some deleted neighborhood $U'_{\delta_0}(a)$, no matter how small.

c. Returning to the case where the set G intersects every $A \in S$ but contains no set $A \in S$, let H be the complement of G (relative to E), as before. Consider the system HS of sets HA, where $A \in S$. Since no HA is empty, the system HA is itself a direction, and we can talk about the existence or nonexistence of $\lim_{HS} f(x)$. If $\lim_S f(x)$ exists and equals p, then $\lim_{GS} f(x)$ and $\lim_{HS} f(x)$ both exist and equal p. However, the example given in the second part of the proof of Theorem 4.16a shows that the existence of both $\lim_{GS} f(x)$ and $\lim_{HS} f(x)$ does not imply that of $\lim_S f(x)$.

† Given a function $f(x)$ defined on a set G and a function $\varphi(x)$ defined on a larger set $E \supset G$, suppose $f(x) = \varphi(x)$ for all $x \in G$. Then $\varphi(x)$ is called an *extension* of $f(x)$, from G to E, while $f(x)$ is called the *restriction* of $\varphi(x)$, from E to G.

THEOREM. *If* $\lim_{GS} f(x)$ *and* $\lim_{HS} f(x)$ *both exist and are equal, to* p *say, then* $\lim_S f(x)$ *exists and equals* p.

Proof. If

$$p = \lim_{GS} f(x) = \lim_{HS} f(x),$$

then, given any $\varepsilon > 0$, there are sets A and B in S such that

$$\rho(p, f(x)) < \varepsilon \tag{5}$$

for all $x \in GA$, $x \in HB$. One of the two sets A and B contains the other, say $B \supset A$. Then (5) certainly holds for all $x \in GA$, $x \in HA$, and hence for all $x \in A = GA + HA$. Since ε is arbitrary, it follows that

$$p = \lim_S f(x). \quad \blacksquare$$

4.17. Behavior of limits under one-to-one mappings

a. Suppose the set E is mapped in a one-to-one fashion onto a set F, with the element $x \in E$ going into the element $y = \omega(x) \in F$. Given a direction S on E, consisting of subsets $A \subset E$, let T be the "image" of S under the mapping ω, i.e., let T be the system of all subsets $f(A) = \{y \in F : y = \omega(x), x \in A\}$ of F obtained as A "varies over" the system S. Then T is a direction on F, since the defining properties of a direction (the "nestedness" and "empty intersection" features) are obviously preserved under one-to-one mappings. Finally, let $f(x)$ be a function defined on E taking values in a metric space M equipped with a distance ρ, and let $g(y)$ be the function defined on F by the formula

$$g(y) = g(\omega(x)) = f(x).$$

THEOREM. *The function* $g(y)$ *has a limit in the direction* T *if and only if the function* $f(x)$ *has a limit in the direction* S. *If both limits exist, then*

$$\lim_S f(x) = \lim_T g(y).$$

Proof. Suppose

$$\lim_T g(y) = p \tag{6}$$

exists. Then, given any $\varepsilon > 0$, there is a set $B \in T$ such that $\rho(p, g(y)) < \varepsilon$ for all $y \in B$. But then

$$\rho(p, f(x)) = \rho(p, g(\omega(x)) < \varepsilon$$

for all x in the corresponding set $A \in S$, which implies

$$\lim_{S} f(x) = p. \tag{7}$$

Conversely, (7) implies (6) by the symmetry of the construction. ∎

b. It follows from Sec. 4.16b that the theorem remains valid if the one-to-one mapping is defined on some subset $A \in S$ rather than on the whole set E.

We now give two examples illustrating these considerations.

c. Let E be the half-line $\{x : x \geqslant a\}$, and let $y = -x$ map E onto the half-line $F = \{y : y \leqslant -a\}$. Choose the direction $x \to \infty$ on F. Then the corresponding direction on F is obviously $y \to -\infty$. It follows from Theorem 4.17a that the limits

$$\lim_{x \to \infty} f(x), \qquad \lim_{y \to -\infty} f(-y)$$

either both exist or both fail to exist, and that

$$\lim_{x \to \infty} f(x) = \lim_{y \to -\infty} f(-y) \tag{8}$$

if they exist.

d. Let E be the set $0 < |x - x_0| \leqslant 1$, and let F be the set $|y| \geqslant 1$. Then the formula

$$y = \frac{1}{x - x_0}$$

establishes a one-to-one correspondence between E and F. The direction $x \to x_0$ on E corresponds to the direction $|y| \to \infty$ on F. It follows from Theorem 4.17a that the limits

$$\lim_{|y| \to \infty} f(y), \qquad \lim_{x \to x_0} f\left(\frac{1}{x - x_0}\right)$$

either both exist or both fail to exist, and that

$$\lim_{|y| \to \infty} f(y) = \lim_{x \to x_0} f\left(\frac{1}{x - x_0}\right)$$

if they exist.

4.18. The definition of a limit given in Sec. 4.12 clearly depends on the metric ρ of the space M, a fact which can be indicated by writing

$$f(x) \underset{\rho}{\to} p.$$

However, *homeomorphic* metrics (Sec. 3.34c) lead to the same limits, as shown by the following

THEOREM. *Given two homeomorphic metrics ρ and r defined on the space M, $f(x)\underset{\rho}{\to}p$ if and only if $f(x)\underset{r}{\to}p$.*

Proof. Suppose $f(x)\underset{\rho}{\to}p$. Then to prove that $f(x)\underset{r}{\to}p$ as well, we must show that, given any $\varepsilon>0$, there is a set A in the underlying direction S such that

$$r(p,f(x))<\varepsilon \tag{9}$$

for all $x\in A$. But, according to Theorem 3.34c, given any $\varepsilon>0$ there is a $\delta>0$ such that $\rho(p,y)<\delta$ implies $r(p,y)<\varepsilon$. Having found δ, we choose $A\in S$ such that

$$\rho(p,f(x))<\delta$$

for all $x\in A$. But then (9) holds for all $x\in A$, as required. To prove that $f(x)\underset{r}{\to}p$ implies $f(x)\underset{\rho}{\to}p$, we need only reverse the roles of ρ and r in the above argument. ∎

4.19. As we now show, the criterion for the existence of the limit of a numerical sequence given in Theorem 3.72b can be carried over to the present more general context, provided we assume that the function $f(x)$ takes its values in a *complete* space.

THEOREM. *Let $f(x)$ be a function defined on a set E equipped with a direction S, taking values in a complete metric space M equipped with a distance ρ. Then $f(x)$ has a limit in the direction S if and only if the following condition, called the **Cauchy convergence criterion**, is satisfied: Given any $\varepsilon>0$, there exists a set $A\in S$ such that*

$$\rho(f(x'),f(x''))<\varepsilon$$

for all $x', x''\in A$.

Proof. Given that the Cauchy convergence criterion is satisfied, consider the system of all subsets $f(B)=\{y\in M:y=f(x),\ x\in B\}$ of M obtained as B varies over the system S. Since S is a system of nested subsets, the same is obviously true of the system of subsets $f(B)\subset M\,(B\in S)$. Moreover, according to (10), there are sets $f(B)$ of arbitrarily small diameter. Hence, by Lemma 3.74a, there is a unique point $p\in M$ such that every neighborhood

$$U_\varepsilon(p)=\{y\in M:\rho(y,p)<\varepsilon\}$$

of p contains some set $f(A)$ $(A\in S)$. But then

$$\rho(p,f(x))<\varepsilon$$

for all $x\in A$, i.e., $f(x)$ approaches p in the direction S.

Conversely, suppose $f(x)$ approaches p in the direction S. Then, given any $\varepsilon > 0$, there is a set $A \in S$ such that

$$\rho(p, f(x)) < \varepsilon/2$$

for all $x \in A$. But then

$$\rho(f(x'), f(x'')) \leqslant \rho(p, f(x')) + \rho(p, f(x'')) < \varepsilon$$

for all $x', x'' \in A$.† ∎

4.2. Some General Theorems

As before, let E be any set equipped with a direction S, and let $f(x)$ be a function defined on E, taking values in a metric space M equipped with a distance ρ.

4.21. THEOREM. *Suppose $f(x)$ is constant on E, i.e., suppose $f(x) = p$ for all $x \in E$. Then $\lim_S f(x)$ exists and equals p.*

Proof. Given any $\varepsilon > 0$, we obviously have

$$\rho(p, f(x)) = \rho(p, p) = 0 < \varepsilon$$

for all x in every set $A \in S$. ∎

4.22. THEOREM. *The limit of $f(x)$ in the direction S (if it exists) is unique.*

Proof. Suppose both $\lim_S f(x) = p$ and $\lim_S f(x) = q$ $(p, q \in M)$. Then, given any $\varepsilon > 0$, there is a set $A \in S$ such that

$$\rho(p, f(x)) < \varepsilon \tag{1}$$

for all $x \in A$ and a set $B \in S$ such that

$$\rho(q, f(x)) < \varepsilon \tag{2}$$

for all $x \in B$. Suppose $A \subset B$, say. Then both inequalities (1) and (2) hold for all $x \in A$, and hence

$$\rho(p, q) \leqslant \rho(p, f(x)) + \rho(q, f(x)) < 2\varepsilon$$

for all $x \in A$. Since $\varepsilon > 0$ is arbitrarily small, we have $\rho(p, q) = 0$ and hence $p = q$. ∎

4.23. We say that a function $f(x)$ *belongs asymptotically to a set* $G \subset M$ if there exists a set $A \in S$ such that $f(x) \in G$ for all $x \in A$.

THEOREM. *Suppose $\lim_S f(x) = p$, and let $G \subset M$ be a set containing an open ball*

---

† Note that the completeness of M plays no role in the second part of the proof.

centered at p. Then $f(x)$ belongs asymptotically to G.

Proof. Suppose G contains the ball

$$U = \{ y \in M : \rho(p, y) < \varepsilon \},$$

and let $A \in S$ be such that

$$\rho(p, f(x)) < \varepsilon$$

for all $A \in S$. Then $f(x) \in U \subset G$ for all $x \in A$. ∎

4.3. Limits of Numerical Functions

4.31. a. In Secs. 4.3–4.6 below we will construct a theory of limits of numerical functions, i.e., functions taking values on the real line (extended or not). The special nature of such functions is determined by the presence of a particular kind of metric on the real line, together with arithmetic operations and order relations. As a matter of fact, we know two metrics on the real line, the ordinary metric $\rho(x, y) = |x - y|$ defined on the set R of all finite numbers, and the metric $r(x, y)$ of Theorem 3.35d defined on the extended real line $\bar{R}$. But these two metrics are homeomorphic on R (see Theorem 3.35d), and hence, by Sec. 4.18, the existence or nonexistence of a finite limit

$$p = \lim_{S} f(x) \tag{1}$$

is independent of whether (1) is defined by using ρ or by using r.

b. Given two functions $f(x)$ and $g(x)$ defined on a set E with values in $\bar{R}$, we write $f(x) \leqslant g(x)$ if $f(x_0) \leqslant g(x_0)$ for all $x_0 \in E$. A function $f(x)$ is said to be *bounded from above* on E if there exists a finite number C such that $f(x) \leqslant C$ for all $x \in E$. A function $f(x)$ is said to be *bounded from below* on E if there exists a finite number C such that $f(x) \geqslant C$ for all $x \in E$. A function $f(x)$ is said to be *bounded (from both sides)* or *bounded in absolute value* on E if there exists a finite number C such that $|f(x)| \leqslant C$ for all $x \in E$.

At every point $x_0 \in E$ where $f(x)$ and $g(x)$ are finite, we define the *sum* $f(x) + g(x)$ as the sum of the corresponding values $f(x_0)$ and $g(x_0)$. The difference, product, and quotient of $f(x)$ and $g(x)$ are defined similarly, with the proviso that division is only possible at points x_0 where the denominator is nonvanishing.

4.32. Pursuing the considerations of Sec. 4.23, we now introduce a number of further definitions together with some new notation ($\pm \infty$ as limits, the symbols O and o).

A numerical function $f(x)$ is said to be *nonnegative* (or *positive*) *in the direction S* if there exists a set $A \in S$ such that $f(x)$ is nonnegative (or positive) for all $x \in A$. A numerical function $f(x)$ is said to be *bounded* (*bounded from above, bounded from below*) *in the direction S* if there exists a set $A \in S$ on which $f(x)$ is bounded (bounded from above, bounded from below). In the latter case, we write

$$f(x) = O(1).$$

Similarly, if $f(x)$ approaches zero in the direction S,† i.e., if

$$\lim_S f(x) = 0,$$

we write

$$f(x) = o(1).$$

Moreover, if given any $C \in R$, there exists a set $A \in S$ such that $f(x) > C$ for all $x \in A$, we write

$$\lim_S f(x) = +\infty,‡$$

while if there exists a set $A \in S$ such that $f(x) < C$ for all $x \in A$, we write

$$\lim_S f(x) = -\infty.$$

4.33. LEMMA. *The formula*

$$\lim_S f(x) = p$$

is equivalent to

$$\lim_S [f(x) - p] = 0.$$

Proof. An immediate consequence of the definition of a limit and the nature of the metric in R or $\bar{R}$. ∎

4.34. a. THEOREM. *If $\lim_S f(x) = p$, then, given any $\varepsilon > 0$, there exists a set $A \in S$ such that, for all $x \in A$,*

$$|f(x) - p| < \quad \varepsilon \quad \text{if } p \text{ is finite,}$$
$$f(x) > \quad 1/\varepsilon \text{ if } p = +\infty,$$
$$f(x) < -1/\varepsilon \text{ if } p = -\infty.$$

† So that, given any $\varepsilon > 0$, there exists a set $A \in S$ such that $|f(x)| < \varepsilon$ for all $x \in A$.
‡ Or simply $\lim_S f(x) = \infty$.

Proof. An immediate consequence of Lemma 4.33 and the nature of the metric in R or $\bar{R}$. ∎

b. THEOREM. *If*

$$\lim_S f(x) = p > 0 \quad or \quad \lim_S f(x) = +\infty,$$

then $f(x)$ is positive in the direction S, while if

$$\lim_S f(x) = p < 0 \quad or \quad \lim_S f(x) = -\infty,$$

then $f(x)$ is negative in the direction S.

Proof. The theorem follows at once from Theorem 4.34a if $p = \pm\infty$. If $p(\neq 0)$ is finite, we first find an $\varepsilon > 0$ such that the interval $(p-\varepsilon, p+\varepsilon)$ does not contain the number 0 and hence consists entirely of numbers of the same sign as p itself. We then use Theorem 4.34a to find a set $A \in S$ on which

$$|f(x) - p| < \varepsilon. \quad ∎$$

c. The following proposition is a kind of converse of Theorem 4.34b:

THEOREM *If $f(x)$ is nonnegative in the direction S and if $\lim_S f(x) = p$, then $p \geqslant 0$.*

Proof. If we had $p < 0$, then by Theorem 4.34b, there would exist a set $A \in S$ on which $f(x) < 0$. On the other hand, since $f(x) \geqslant 0$ in the direction S, there exists a set $B \in S$ on which $f(x) \geqslant 0$. But this is impossible, since AB is nonempty. Hence $p \geqslant 0$, by contradiction. ∎

The theorem obviously remains true if we replace "nonnegative" by "nonpositive" and $p \geqslant 0$ by $p \leqslant 0$.

d. It should be emphasized that the strict inequality > 0 (or < 0, with equality excluded) is preserved in arguing from the behavior of the limit, as in Theorem 4.34b, while only the weaker inequality $\geqslant 0$ (or $\leqslant 0$, with equality permitted) is preserved in inferring the behavior of the limit, as in Theorem 4.34c. If it is known only that $p \geqslant 0$, we can draw no conclusions at all about the behavior of $f(x)$ on the sets $A \in S$, and in fact $f(x)$ can take both positive and negative values on these sets. Moreover, even if $f(x)$ is positive in the direction S, we can only infer that $p \geqslant 0$ and not that $p > 0$.

e. We say that $f(x) \leqslant g(x)$ *in the direction S* if the difference $g(x) - f(x)$ is nonnegative in the direction S. Similarly, we say that $f(x) < g(x)$ *in the direction S* if $g(x) - f(x)$ is positive in the direction S.

f. THEOREM. *If $f(x) \leqslant g(x)$ in the direction S and if*

$$\lim_S f(x) = p, \qquad \lim_S g(x) = q,$$

then $p \leqslant q$.

Proof. An immediate consequence of Theorem 4.34c. ∎

g. THEOREM. *If*

$$\lim_S f(x) = p, \qquad \lim_S g(x) = q$$

and if $p < q$, then $f(x) < g(x)$ in the direction S.

Proof. An immediate consequence of Theorem 4.34b. ∎

h. THEOREM *If $f(x) \leqslant g(x) \leqslant h(x)$ in the direction S and if*

$$\lim_S f(x) = \lim_S h(x) = p,$$

then

$$\lim_S g(x) = p.$$

Proof. We need only consider the case $p = 0$, otherwise changing $f(x)$ and $g(x)$ to $f(x) - p$ and $g(x) - p$ (cf. Lemma 4.33). Given any $\varepsilon > 0$, there exists a set $A \in S$ on which $|f(x)| < \varepsilon$, $|h(x)| < \varepsilon$. But then $|g(x)| < \varepsilon$ on A. ∎

4.35. a. THEOREM. *Suppose $f(x)$ and $g(x)$ are bounded in the direction S. Then so are $f(x) + g(x)$ and $f(x)g(x)$.*

Proof. By hypothesis, there are sets $A, B \in S$ such that $|f(x)| \leqslant C_1$ for all $x \in A$ and $|g(x)| \leqslant C_2$ for all $x \in B$. If $A \subset B$, say, then

$$|f(x) + g(x)| \leqslant C_1 + C_2, \qquad |f(x)g(x)| \leqslant C_1 C_2$$

for all $x \in A$. ∎

b. THEOREM. *If*

$$\lim_S f(x) = \lim_S g(x) = 0,$$

then

$$\lim_S [f(x) + g(x)] = 0.$$

Proof. By hypothesis, there are sets $A, B \in S$ such that $|f(x)| < \varepsilon/2$ for all

$x \in A$ and $|g(x)| < \varepsilon/2$ for all $x \in B$. If $A \subset B$, say, then

$$|f(x) + g(x)| < \varepsilon$$

for all $x \in A$. ∎

c. THEOREM. *If $f(x)$ is bounded in the direction S and if*

$$\lim_S g(x) = 0,$$

then

$$\lim_S f(x) g(x) = 0.$$

Proof. By hypothesis, there is a set $A \in S$ and a finite number $C > 0$ such that $|f(x)| \leqslant C$ for all $x \in A$. Then, given any $\varepsilon > 0$, we can find a set $B \in S$ such that $|g(x)| < \varepsilon/C$ for all $B \in S$. It follows that

$$|f(x) g(x)| < \varepsilon$$

for all $x \in AB$. ∎

4.36. a. THEOREM. *If*

$$\lim_S f(x) = p, \qquad \lim_S g(x) = q,$$

then

$$\lim_S [f(x) + g(x)] = p + q.$$

Proof. The functions $f(x) - p$ and $g(x) - q$ both approach zero in the direction S, and hence, by Theorem 4.35b, so does the function

$$f(x) + g(x) - (p + q) = [f(x) - p] + [g(x) - q].$$

Now use Lemma 4.33. ∎

b. THEOREM. *If*

$$\lim_S f(x) = p, \qquad \lim_S g(x) = q,$$

then

$$\lim_S f(x) g(x) = pq.$$

Proof. The functions $f(g - q)$ and $q(f - p)$ both approach zero in the direction S,† by Theorem 4.35c (the boundedness of f in the direction S follows from Theorem 4.34a), and hence so does the function

$$fg - pq = f(g - q) + q(f - p). \quad ∎$$

† For brevity, we often omit arguments of functions here and below.

c. THEOREM. *If*

$$\lim_{S} f(x) = p \neq 0,$$

then

$$\frac{1}{f(x)}$$

is bounded in the direction S.

Proof. Let $A \in S$ be a set on which

$$|f(x)| \geqslant \frac{p}{2}$$

(see Theorem 4.34a). Then

$$\frac{1}{|f(x)|} \leqslant \frac{2}{p}$$

for all $x \in A$.   ∎

d. THEOREM. *If*

$$\lim_{S} f(x) = p \neq 0,$$

then

$$\lim_{S} \frac{1}{f(x)} = \frac{1}{p}.$$

Proof. Using the preceding theorem, Lemma 4.33, and Theorem 4.35c, we see that

$$\frac{1}{p} - \frac{1}{f} = \frac{f-p}{pf} = \frac{1}{pf}(f-p)$$

approaches zero in the direction S.   ∎

e. THEOREM. *If*

$$\lim_{S} f(x) = p \neq 0, \qquad \lim_{S} g(x) = q,$$

then

$$\lim_{S} \frac{g(x)}{f(x)} = \frac{q}{p}.$$

Proof. An immediate consequence of Theorems 4.36b and 4.36d.   ∎

4.37. a. THEOREM. *If*

$$\lim_{s} f(x) = 0,$$

then

$$\lim_{s} \frac{1}{|f(x)|} = \infty,$$

while if

$$\lim_{s} |f(x)| = \infty,$$

then

$$\lim_{s} \frac{1}{f(x)} = 0.$$

Proof. Note that $|f(x)| < \varepsilon$ implies

$$\frac{1}{|f(x)|} > \frac{1}{\varepsilon},$$

while $|f(x)| > C$ implies

$$\frac{1}{|f(x)|} < \frac{1}{C}. \quad \blacksquare$$

b. THEOREM. *If*

$$\lim_{s} f(x) = +\infty, \tag{2}$$

then

$$\lim_{s} [-f(x)] = -\infty.$$

Moreover, suppose (2) *holds and*

$$\lim_{s} g(x) = p \neq 0.$$

Then

$$\lim_{s} f(x)g(x) = +\infty$$

if $p > 0$, *while*

$$\lim_{s} f(x)g(x) = -\infty$$

if $p < 0$.

Proof. An immediate consequence of the definitions of Sec. 4.32. ∎

4.38. a. Given two functions $f(x)$ and $g(x)$ defined on the same set E equipped with a direction S, we write†

$$f(x) = o(g(x)) \tag{3}$$

if

$$\lim_{S} \frac{f(x)}{g(x)} = 0.$$

The functions $f(x)$ and $g(x)$ are said to be *equivalent (in the direction S)* if

$$\lim_{S} \frac{f(x)}{g(x)} = 1.$$

b. THEOREM. *If $f(x) = o(g(x))$, then‡*

$$\lim_{S} \left| \frac{g(x)}{f(x)} \right| = \infty. \tag{4}$$

Proof. An immediate consequence of Theorem 4.37a. ∎

c. THEOREM. *If $f(x)$ and $g(x)$ are equivalent and if*

$$\lim_{S} \frac{g(x)}{h(x)} = p,$$

where $h(x)$ is another function defined on E, then

$$\lim_{S} \frac{f(x)}{h(x)} = p.$$

Proof. It follows from Theorem 4.36b that

$$\lim_{S} \frac{f(x)}{h(x)} = \lim_{S} \frac{f(x)}{g(x)} \frac{g(x)}{h(x)} = \lim_{S} \frac{f(x)}{g(x)} \lim_{S} \frac{g(x)}{h(x)} = \lim_{S} \frac{g(x)}{h(x)} = p. ∎$$

Thus in calculating the limit of a quotient, we can replace the numerator (or, for that matter, the denominator) by an equivalent function.

4.39. The symbol E

a. By the symbol $E(x)$, or briefly E, we mean any function with limit 1 in a

---

† If $g(x) \equiv 1$, i.e., if $g(x)$ is identically equal to 1, then (3) reduces to $f(x) = o(1)$, as in Sec. 4.32." If $\lim_S g(x) = 0$, we describe (3) by saying that "$f(x)$ approaches zero faster than $g(x)$" (in the direction S).

‡ If $\lim_S |f(x)| = \infty$, we describe (4) by saying that "$g(x)$ approaches infinity faster than $f(x)$" (in the direction S).

given direction. In keeping with Sec. 4.23, E might be called an "asymptotic unit." The product or quotient of two asymptotic units is clearly another asymptotic unit.

b. THEOREM. *If*

$$\lim_{S} u(x) = 0,$$

then

$$[1 + u(x)]^p = 1 + pEu(x)$$

for every $p = 1, 2, \ldots$

Proof. By the binomial theorem,

$$[1 + u(x)]^p = 1 + pu(x) + \frac{p(p-1)}{1 \cdot 2} u^2(x) + \cdots + u^p(x)$$

$$= 1 + pu(x)\left[1 + \frac{p-1}{1 \cdot 2} u(x) + \cdots + \frac{1}{p} u^{p-1}(x)\right]$$

$$= 1 + pEu(x),$$

since the limit in brackets equals 1 (Theorems 4.36a and 4.36b). ∎

c. Asymptotic units are often useful in computing limits. For example, to evaluate

$$\lim_{x \to 0} \frac{(1+x)^p - (1+x+x^2)^q}{(1+2x)^r - (1-2x+x^3)^s},$$

where p, q, r, and s are positive integers, we can use the above theorem to write

$$(1+x)^p = 1 + pxE_1, \qquad x + x^2 = x(1+x) = xE_2,$$
$$(1+x+x^2)^q = (1+xE_2)^q = 1 + qxE_2E_3 = 1 + qxE_4,$$
$$(1+2x)^r = 1 + 2xrE_5, \qquad 2x - x^3 = 2x(1 - \tfrac{1}{2}x^2) = 2xE_6,$$
$$(1-2x+x^3)^s = (1 - 2xE_6)^s = 1 - 2xsE_6E_7 = 1 - 2xsE_8.$$

It follows that

$$\frac{(1+x)^p - (1+x+x^2)^q}{(1+2x)^r - (1-2x+x^3)^s} = \frac{(1+pxE_1) - (1+qxE_4)}{(1+2xrE_5) - (1-2xsE_8)} = \frac{pE_1 - qE_4}{2rE_5 + 2sE_8},$$

and hence finally

$$\lim_{x \to 0} \frac{(1+x)^p - (1+x+x^2)^q}{(1+2x)^p - (1-2x+x^3)^s} = \frac{p-q}{2(r+s)}.$$

d. Consider the rational function

$$R(x) = \frac{a_0 x^n + a_1 x^{n-1} + \cdots + a_n}{b_0 x^m + b_1 x^{m-1} + \cdots + b_m} \qquad (a_0 \neq 0,\ b_0 \neq 0).$$

Clearly

$$R(x) = \frac{a_0 x^n \left(1 + \dfrac{a_1}{a_0}\dfrac{1}{x} + \cdots + \dfrac{a_n}{a_0}\dfrac{1}{x^n}\right)}{b_0 x^m \left(1 + \dfrac{b_1}{b_0}\dfrac{1}{x} + \cdots + \dfrac{b_m}{b_0}\dfrac{1}{x^m}\right)} = \frac{a_0 x^n E_1}{b_0 x^m E_2} = \frac{a_0}{b_0} x^{n-m} E_3$$

as $|x| \to \infty$. It follows from the rules of Secs. 4.36 and 4.37 that

$$\lim_{|x| \to \infty} R(x) = \begin{cases} 0 & \text{if } n < m, \\ \dfrac{a_0}{b_0} & \text{if } n = m. \end{cases}$$

For $n > m$ the function $R(x)$ approaches an infinite limit whose sign depends on the choice of direction ($x \to +\infty$ or $x \to -\infty$), the sign of a_0/b_0, and the evenness or oddness of the number $n - m$. The reader should examine the various possibilities as an exercise.

4.4. Upper and Lower Limits

4.41. Let $f(x)$ be a numerical function defined on a set E equipped with a direction S, taking values in the extended real number system $\bar{R}$ (Sec. 1.9), and let

$$a_A = \inf \{ f(x) : x \in A \}, \qquad b_A = \sup \{ f(x) : x \in A \}$$

for every $A \in S$. Then both a_A and b_A exist in $\bar{R}$ and $a_A \leqslant b_A$. Since S is a system of nested subsets, the set of all intervals $[a_A, b_A]$ ($A \in S$) is a system of nested closed intervals. Let

$$\xi = \sup_{A \in S} a_A, \qquad \eta = \inf_{A \in S} b_A.$$

Then $\xi \leqslant \eta$ and

$$[\xi, \eta] = \prod_{A \in S} [a_A, b_A], \tag{1}$$

by the generalized principle of nested intervals (Sec. 1.94). The number $\xi \in \bar{R}$ is called the *lower limit of $f(x)$ in the direction S*, denoted by

$$\xi = \varliminf_{S} f(x),$$

while the number $\eta \in \bar{R}$ is called the *upper limit of $f(x)$ in the direction S*, denoted by

$$\eta = \overline{\lim_{S}} f(x).$$

4.42. THEOREM. *Given any $\varepsilon > 0$, there exists a set $A \in S$ such that†*

$$a_A = \inf \{ f(x) : x \in A \} \geqslant \xi - \varepsilon,$$
$$b_A = \sup \{ f(x) : x \in A \} \leqslant \eta + \varepsilon.$$

Proof. An immediate consequence of (1). ∎

4.43. THEOREM. *If $\xi = \eta$, then $f(x)$ has a limit (possibly infinite) in the direction S, equal to $\xi = \eta$.*

Proof. An immediate consequence of the preceding theorem. ∎

4.44. THEOREM. *If*

$$\lim_{S} f(x) = p, \tag{2}$$

then the interval (1) *reduces to the single point $\xi = \eta = p$.*

Proof. It follows from (2) and Theorem 4.19 that, given any $\varepsilon > 0$, there is a set $A \in S$ such that $|f(x') - f(x'')| < \varepsilon$ for all x', $x'' \in A$. Hence there is a set $A \in S$ for which $b_A - a_A$ can be made arbitrarily small. But then $\xi = \eta$ ($= p$, by Theorem 4.43), since otherwise $b_A - a_A$ could not be made smaller than $\eta - \xi$. ∎

4.45. To clarify the above concepts, we introduce another definition (cf. Sec. 3.41). A number $y \in \bar{R}$ is said to be a *limit point of $f(x)$ in the direction S* if, given any $\varepsilon > 0$ and any set $A \in S$, there exists a point $x \in A$ such that

$$|f(x) - y| < \quad \varepsilon \quad \text{if } y \text{ is finite,}$$
$$f(x) > \quad 1/\varepsilon \text{ if } y = +\infty,$$
$$f(x) < -1/\varepsilon \text{ if } y = -\infty.$$

If $\lim_{S} f(x) = p$, then p is a limit point of $f(x)$ in the direction S, but the converse is in general not true.

4.46. a. THEOREM. *Every limit point y of $f(x)$ in the direction S lies in the interval $[\xi, \eta]$.*

Proof. Clearly

$$a_A = \inf \{ f(x) : x \in A \} \leqslant y \leqslant \sup \{ f(x) : x \in A \} = b_A$$

† If $\xi \in \bar{R} - R$, we set $\xi - \varepsilon = \xi$, while if $\eta \in \bar{R} - R$, we set $\eta + \varepsilon = \eta$.

for every $A \in S$, and hence

$$y \in \prod_{A \in S} [a_A, b_A] = [\xi, \eta]. \quad \blacksquare$$

b. THEOREM. *The points ξ and η are themselves limit points of $f(x)$ in the direction S.*

Proof. If ξ is finite, then, given any $\varepsilon > 0$, there is a set $A_0 \in S$ such that

$$0 \leqslant \xi - a_{A_0} < \varepsilon/2$$

(see Theorem 4.42). Hence, for every $A \subset A_0$, we have

$$a_{A_0} \leqslant a_A \leqslant \xi,$$

which implies

$$0 \leqslant \xi - a_A < \varepsilon/2.$$

Moreover, since $a_A = \inf \{ f(x) : x \in A \}$, there exists a point $x \in A$ such that

$$0 \leqslant f(x) - a_A < \varepsilon/2$$

and hence

$$|f(x) - \xi| < \varepsilon.$$

Every set $B \supset A_0$ also contains such a point x, which can be chosen from the set A_0. Thus ξ is a limit point of $f(x)$ in the direction S. The proof for the case where $\xi \in \bar{R} - R$, as well as for the point η, is along the same lines (give the details). $\quad \blacksquare$

4.47. THEOREM. *The set of all limit points of $f(x)$ in the direction S is a nonempty subset of $\bar{R}$, with greatest lower bound $\xi = \underline{\lim}_S f(x)$ and least upper bound $\eta = \overline{\lim}_S f(x)$.*

Proof. An immediate consequence of the preceding two theorems. $\quad \blacksquare$

4.48. Note in particular that ξ, the lower limit of $f(x)$, is just the *smallest* limit point of $f(x)$, while η, the upper limit of $f(x)$, is just the *largest* limit point of $f(x)$. Note also that $f(x)$ approaches the limit p in the direction S if and only if p $(= \xi = \eta)$ is the unique limit point of $f(x)$ in the direction S.

4.5. Nondecreasing and Nonincreasing Functions

4.51. a. A numerical sequence $y_1, y_2, \ldots, y_n, \ldots$ is said to be *nondecreasing* if

$$y_1 \leqslant y_2 \leqslant \cdots \leqslant y_n \leqslant y_{n+1} \leqslant \cdots.$$

We now introduce an analogous concept for a numerical function $y = f(x)$

defined on an arbitrary set E equipped with a direction S.

b. Definition. A numerical function $f(x)$ is said to be *nondecreasing in the direction S* if $A \subset B$ (where $A, B \in S$) implies

$$\sup \{ f(x) : x \in B - A \} \leqslant \inf \{ f(x) : x \in A \}. \tag{1}$$

4.52. Examples

The following examples show that the general definition of Sec. 4.51b is consistent with the usual meaning of the term "nondecreasing."

a. If $E = \{1, 2, \dots\}$, then $y = f(x)$ is a numerical sequence $y_1 = f_1, y_2 = f_2, \dots$ As in Sec. 4.13, let S be the direction on E specified by the sets

$$A_n = \{n, n+1, n+2, \dots\} \qquad (n = 1, 2, \dots).$$

Then the sequence $f_1, f_2, \dots$ is nondecreasing in the sense of Sec. 4.51b if and only if it is nondecreasing in the usual sense (as in Sec. 4.51a). In fact, suppose $f_1, f_2, \dots$ is nondecreasing in the sense of Sec. 4.51b, and choose

$$A = A_{n+1} = \{n+1, n+2, \dots\}, \qquad B = A_n = \{n, n+1, \dots\}.$$

Then $B - A = \{n\}$, and the condition (1) implies

$$f_n \leqslant \inf \{ f_{n+1}, f_{n+2}, \dots \} \leqslant f_{n+1} \qquad (n = 1, 2, \dots),$$

so that $f_1, f_2, \dots$ is nondecreasing in the usual sense. Conversely, suppose $f_1, f_2, \dots$ is nondecreasing in the usual sense, and let

$$A = A_n = \{n, n+1, \dots\} \in S, \qquad A \subset B = \{k, k+1, \dots\} \in S \qquad (k < n).$$

Then

$$\sup \{ f(x) : x \in B - A \} = \sup \{ f_k \dots, f_{n-1} \} = f_{n-1} \leqslant f_n = \inf \{ f(x) : x \in A \},$$

so that $f_1, f_2, \dots$ is nondecreasing in the sense of Sec. 4.51b.

b. Let $E = R_a^+$ be the real half-line $\{x : x \geqslant a\}$, and let S be the system of all subsets $A_\xi \subset R_a^+$ of the form

$$A_\xi = \{x : x \geqslant \xi\} \qquad (\xi \geqslant a),$$

as in Sec. 4.14a. Then $f(x)$ is nondecreasing in the sense of Sec. 4.51b if and only if $f(x)$ is nondecreasing in the usual sense, i.e., if and only if $a \leqslant y < z$ implies $f(y) \leqslant f(z)$. In fact, suppose $f(x)$ is nondecreasing in the sense of Sec. 4.51b. Then if $A_z \subset A_y$ $(y \neq z)$, i.e., $y < z$, we have

$$f(y) \leqslant \sup \{ f(x) : y \leqslant x < z \} \leqslant \inf \{ f(x) : x \geqslant z \} \leqslant f(z),$$

so that $f(x)$ is nondecreasing in the usual sense. Conversely, let $f(x)$ be nondecreasing in the usual sense, and let $x' \in A_y - A_z$, $x'' \in A_z$. Then

$$y \leqslant x' < z, \quad x'' \geqslant z, \quad f(x') \leqslant f(x''),$$

and hence

$$\sup \{ f(x') : x' \in A_y - A_z \} \leqslant \inf \{ f(x'') : x'' \in A_z \},$$

by Theorem 1.62b, so that $f(x)$ is nondecreasing in the sense of Sec. 4.51b.

4.53. THEOREM. *If the function $f(x)$ is nondecreasing and bounded from above in the direction S, then $f(x)$ has a finite limit in the direction S. This limit is given by*

$$p = \sup \{ f(x) : x \in A \}, \tag{2}$$

where $A \in S$ is any set on which $f(x)$ is bounded from above.

Proof. Suppose $f(x)$ is bounded from above on the set $A \in S$, and let p be the quantity (2). Then, given any $\varepsilon > 0$, there is a point $x_0 \in A$ such that

$$p - \varepsilon < f(x_0) \leqslant p.$$

Since the intersection of all the sets of the system S is empty, there is a set $B \in S$ which does not contain x_0. Obviously, of the two possibilities $A \subset B$, $B \subset A$, we have $B \subset A$ in the present case. Therefore

$$p - \varepsilon < f(x_0) \leqslant \sup \{ f(x) : x \in A - B \} \leqslant \inf \{ f(x) : x \in B \},$$

since $f(x)$ is nondecreasing in the direction S. On the other hand, $B \subset A$ implies

$$\sup \{ f(x) : x \in B \} \leqslant \sup \{ f(x) : x \in A \} = p$$

(see Theorem 1.62a). It follows that

$$p - \varepsilon < f(x) \leqslant p$$

for all $x \in B$, and hence $\lim_S f(x) = p$. ∎

If $f(x)$ is a nondecreasing function which approaches the limit p in the direction S, we write $f(x) \nearrow_S p$ or simply $f(x) \nearrow p$.

4.54. Definition. A numerical function $f(x)$ is said to be *nonincreasing in the direction S* if $A \subset B$ (where $A, B \in S$) implies

$$\inf \{ f(x) : x \in B - A \} \geqslant \sup \{ f(x) : x \in A \}.$$

For nonincreasing functions we have the following analogue of Theorem 4.53:

THEOREM. *If the function $f(x)$ is nonincreasing and bounded from below in the direc-*

tion S, then f(x) has a finite limit in the direction S. This limit is given by

$$p = \inf \{ f(x) : x \in A \},$$

where $A \in S$ is any set on which $f(x)$ is bounded from below.

Proof. The function $-f(x)$ is nondecreasing and bounded from above in the direction S, and hence, by Theorem 4.53,

$$\lim_{S} [-f(x)] = \sup \{ -f(x) : x \in A \},$$

where $A \in S$ is any set on which $-f(x)$ is bounded from below. But then

$$\lim_{S} f(x) = -\sup \{ -f(x) : x \in A \} = \inf \{ f(x) : x \in A \},$$

by Theorem 1.61, where $A \in S$ is any set on which $f(x)$ is bounded from above. ∎

If $f(x)$ is a nonincreasing function which approaches the limit p in the direction S, we write $f(x) \searrow_S p$ or simply $f(x) \searrow p$.

4.55. THEOREM. *If the function $f(x)$ is nondecreasing and unbounded from above in the direction S, then*

$$\lim_{S} f(x) = +\infty. \tag{3}$$

Proof. Given any number C and any set $B \in S$, there is a point $x_0 \in B$ such that $f(x_0) > C$, by the unboundedness of $f(x)$. Holding B fixed, we now find a set $A \subset B$ ($A \in S$) which does not contain the point x_0. We then have

$$C \leqslant \sup \{ f(x) : x \in B - A \} \leqslant \inf \{ f(x) : x \in A \},$$

since $f(x)$ is nondecreasing in the direction S. But then $f(x) \geqslant C$ for all $x \in A$, which implies (3). ∎

4.56. For nonincreasing functions we have the following analogue of the above theorem:

THEOREM. *If the function $f(x)$ is nonincreasing and unbounded from below in the direction S, then*

$$\lim_{S} f(x) = -\infty.$$

Proof. Apply Theorem 4.55 to the function $-f(x)$, as in the proof of Theorem 4.54. ∎

4.6. Limits of Numerical Sequences

4.61. We now apply the general theory of Secs. 4.3–4.5 to the case of numerical sequences. It will be recalled from Sec. 3.22a that a numerical sequence $x_n \in R$ is said to converge to a limit $p \in R$ if, given any $\varepsilon > 0$, there exists an integer $N > 0$ such that $|p - x_n| < \varepsilon$ for all $n > N$, a fact expressed by writing

$$p = \lim_{n \to \infty} x_n.$$

If the numbers x_n are regarded as points of the extended real line $\bar{R}$, the definition of a finite limit remains the same, but then the possibility of infinite limits arises. A sequence x_n converging in $\bar{R}$ to the point $+\infty$ is divergent in R. As opposed to other kinds of divergent sequences, such a sequence is said to *diverge to* $+\infty$. Similarly, a sequence x_n converging in $\bar{R}$ to $-\infty$ is said to *diverge to* $-\infty$.

4.62. For numerical sequences the general *Cauchy convergence criterion* of Theorem 4.19 reduces to Theorem 3.72b, i.e., *a numerical sequence x_n is convergent if and only if, given any $\varepsilon > 0$, there exists an integer $N > 0$ such that $|x_m - x_n| < \varepsilon$ for all $m, n > N$.*

4.63. a. In keeping with Sec. 4.32, a sequence x_n is said to be *bounded* (*bounded from above, bounded from below*) as $n \to \infty$ if there exist numbers C and N such that $|x_n| \leqslant C$ $(x_n \leqslant C,\ x_n \geqslant C)$ for all $n > N$. Only a finite number of integers $1, 2, \ldots, N$ fail to satisfy the condition $n > N$, and the set $\{x_1, x_1, \ldots, x_N\}$ is obviously bounded. Hence in the definition of boundedness there is no need to invoke the number N or, for that matter, to write "as $n \to \infty$."

b. Moreover, in keeping with Secs. 4.51 and 4.54, a sequence x_n is said to be *nondecreasing* as $n \to \infty$ if there exists an N such that $x_n \leqslant x_{n+1}$ for all $n > N$. Similarly, x_n is said to be *nonincreasing* as $n \to \infty$ if there exists an N such that $x_n \geqslant x_{n+1}$ for all $n > N$.

c. As applied to sequences, Theorem 4.53 asserts that *every nondecreasing sequence which is bounded from above has a finite limit as $n \to \infty$*. According to Theorem 4.54, the same is true of every nonincreasing sequence which is bounded from below. However, a nondecreasing sequence which is unbounded from above diverges to $+\infty$, while a nonincreasing sequence which is unbounded from below diverges to $-\infty$ (see Theorems 4.55 and 4.56).

d. Example. Consider the sequence of positive numbers

$$u_n = \left(1 + \frac{1}{n}\right)^n \qquad (n = 1, 2, \ldots).$$

By the binomial theorem,

$$u_n = \left(1 + \frac{1}{n}\right)^n = 1 + 1 + \frac{1}{2!}\frac{n(n-1)}{n^2} + \frac{1}{3!}\frac{n(n-1)(n-2)}{n^3} + \cdots$$

$$= 1 + 1 + \frac{1}{2!}\left(1 - \frac{1}{n}\right) + \frac{1}{3!}\left(1 - \frac{1}{n}\right)\left(1 - \frac{2}{n}\right) + \cdots, \tag{1}$$

so that u_n is a nondecreasing sequence, since both the number of terms and the size of all the terms except the first two increase as $n \to \infty$. Moreover, replacing $1 - (k/n)$ by 1, we get

$$u_n < 1 + 1 + \frac{1}{2!} + \frac{1}{3!} + \cdots + \frac{1}{n!}$$

$$< 1 + 1 + \frac{1}{2} + \frac{1}{2^2} + \cdots + \frac{1}{2^n} = 1 + \frac{1 - (\frac{1}{2})^{n+1}}{\frac{1}{2}} < 1 + 2 = 3, \tag{2}$$

so that the sequence u_n is bounded from above. Hence, by Sec. 4.63, u_n has a finite limit as $n \to \infty$. Denoting this limit by e, we have

$$e = \lim_{n \to \infty} \left(1 + \frac{1}{n}\right)^n,$$

where $2 < e \leqslant 3$, because of (1) and (2). A more exact calculation gives $e = 2.71828\ldots$ It can be shown that the number e is irrational and in fact transcendental (*Hermite's theorem*).†

4.64. Next we specialize the considerations of Sec. 4.4 to the case of numerical sequences. A point $y \in \bar{R}$ is said to be a *limit point* of the sequence x_n (as $n \to \infty$) if, given any $\varepsilon > 0$ and any $N > 0$, there exists an $n > N$ such that

$$|x_n - y| < \quad \varepsilon \quad \text{if } y \text{ is finite,}$$
$$x_n > \quad 1/\varepsilon \text{ if } y = +\infty,$$
$$x_n < -1/\varepsilon \text{ if } y = -\infty$$

(cf. Sec. 4.45). Let

$$a_n = \inf\{x_n, x_{n+1}, \ldots\} \in \bar{R}, \qquad b_n = \sup\{x_n, x_{n+1}, \ldots\} \in \bar{R}$$

for every $n = 1, 2, \ldots$ Then the set of all intervals $[a_n, b_n]$ is a system of nested closed intervals with intersection $[\xi, \eta]$, where

$$\xi = \sup\{a_1, a_2, \ldots\}, \qquad \eta = \inf\{b_1, b_2, \ldots\}$$

(cf. Sec. 4.41). Every limit point of the sequence x_n lies in the interval $[\xi, \eta]$,

---

† See e.g., G. M. Fichtenholz, *The Definite Integral* (translated by R. A. Silverman), Gordon and Breach, N. Y. (1973), Sec. 12.

and the numbers $\xi, \eta \in \bar{R}$ are themselves limit points of x_n (cf. Sec. 4.46). The number ξ is called the *lower limit* of x_n (as $n \to \infty$), denoted by

$$\xi = \varliminf_{n \to \infty} x_n,$$

while the number η is called the *upper limit* of x_n, denoted by†

$$\eta = \varlimsup_{n \to \infty} x_n.$$

Note that ξ and η are just the *smallest* and *largest* limit points of the sequence x_n, respectively. Note also that $x_n \to p$ if and only if p $(= \xi = \eta)$ is the unique limit point of the sequence x_n.

4.65. Limit of a sequence versus limit of a function. If a function $f(x)$ defined for all $x \geq x_0$ has a limit as $x \to \infty$, equal to p say, then, by Sec. 4.16a, the sequence of numbers $y_n = f(n)$, where n takes integral values greater than x_0, approaches the same limit p (as $n \to \infty$). The converse assertion is not true: The existence of the limit as $x \to \infty$ of a function $f(x)$ defined for $x \geq x_0$ cannot be inferred from the existence of the limit of the sequence $f(n)$. For example, the function $f(x) = (x)$, where (x) is the fractional part of x (defined in Sec. 1.71) has no limit as $x \to \infty$, although the sequence $f(n) = (n) = 0$ has the limit 0. Nevertheless we have the following

THEOREM. *A function $f(x)$ defined for all $x \geq x_0$ has a limit as $x \to \infty$ if and only if every sequence $f(x_n)$ has a limit, where $x_n \geq x_0$ is any sequence diverging to ∞.*

Proof. If $f(x)$ approaches a limit p as $x \to \infty$, then by Sec. 4.16a again, p is the limit of every sequence $f(x_n)$, where $x_n \to \infty$. If $f(x)$ fails to approach a limit as $x \to \infty$, then $f(x)$ does not satisfy the Cauchy convergence criterion of Theorem 4.19. Hence for some $\varepsilon > 0$ and any $n = 1, 2, \ldots$, there are points $x_n' > n$, $x_n'' > n$ such that

$$|f(x_n') - f(x_n'')| \geq \varepsilon. \tag{3}$$

The sequence

$$x_1', x_1'', x_2', x_2'', \ldots, x_n', x_n'', \ldots$$

obviously diverges to ∞, while the sequence

$$f(x_1'), f(x_1''), f(x_2'), f(x_2''), \ldots, f(x_n'), f(x_n''), \ldots$$

of corresponding values of $f(x)$ has no limit, since, as shown by (3), it does not satisfy the Cauchy convergence criterion. ∎

† For simplicity, we often omit $n \to \infty$ in the expressions for ξ and η, as in Problem 2.

Of course, these considerations do not prevent us from using further properties of $f(x)$ to infer the existence of $\lim_{x \to \infty} f(x)$ from that of $\lim_{n \to \infty} f(n)$ in special cases.

4.7. Limits of Vector Functions

4.71. We now consider "vector functions," namely, functions taking values in the n-dimensional real space R_n (Sec. 2.61). Since R_n is a metric space, we can introduce the notion of the limit in a given direction S, where the limit has the properties indicated in Sec. 4.2. Moreover, as we shall see in a moment, a number of properties of numerical functions carry over to the case of functions taking values in R_n ($n \geqslant 2$), namely, those which involve certain arithmetic operations but make no use of order relations.

4.72.a. Addition of vector functions. Let $f(x)$ and $g(x)$ be two functions defined on a set E, taking values in R_n. Then by the *sum* $f(x) + g(x)$ we mean the function whose value at every point $x_0 \in E$ equals $f(x_0) + g(x_0)$. Clearly $f(x) + g(x)$ is itself a function on E taking values in R_n (see Sec. 2.62).

b. Multiplication of a vector function by a real function. Let $f(x)$ be a function on E taking values in R_n, while $\alpha(x)$ is a function on E taking real values. Then by the *product* $\alpha(x)f(x)$ we mean the function whose value at every point $x_0 \in E$ equals $\alpha(x_0)g(x_0)$. Clearly $\alpha(x)f(x)$ is itself a function on E taking values in R_n (see Sec. 2.63).

c. In the case $n = 2$, where the vector functions $f(x)$ and $g(x)$ can be regarded as complex-valued, the product $f(x)g(x)$ and the quotient

$$\frac{f(x)}{g(x)} \qquad (g(x) \neq 0)$$

are defined with the help of the usual rules for multiplying and dividing complex numbers (see Sec. 2.71) as the functions whose values at every point $x_0 \in E$ equal $f(x_0)g(x_0)$ and $f(x_0)/g(x_0)$, respectively.

4.73. a. THEOREM. *The formula*

$$\lim_S f(x) = p \in R_n$$

is equivalent to

$$\lim_S [f(x) - p] = 0$$

or

$$\lim_{S} |f(x) - p| = 0,$$

where $|f(x) - p|$ *is the norm of the vector* $f(x) - p$.

Proof. An immediate consequence of the definition of a limit and the nature of the metric in R_n (see Sec. 3.14a). ■

b. A function $f(x)$ with values in R_n is said to be *bounded in the direction S* if there exists a finite number C and a set $A \in S$ such that $|f(x)| \leqslant C$ for all $x \in A$ (cf. Sec. 4.32).

THEOREM. *If* $f(x)$ *and* $g(x)$ *are bounded in the direction S, then so is the sum* $f(x) + g(x)$.

Proof. A slight generalization of the proof of Theorem 4.35a. ■

c. THEOREM. *If*

$$\lim_{S} f(x) = p \in R_n, \qquad \lim_{S} g(x) = q \in R_n,$$

then

$$\lim_{S} [f(x) + g(x)] = p + q \in R_n.$$

Proof. A slight generalization of the proof of Theorem 4.36a. ■

d. THEOREM. *If*

$$\lim_{S} f(x) = p \in R_n, \qquad \lim_{S} \alpha(x) = c \in R,$$

then

$$\lim_{S} \alpha(x) f(x) = cp.$$

Proof. A slight generalization of the proof of Theorem 4.36b. ■

e. For the case $n = 2$, where the vector functions $f(x)$ and $g(x)$ can be regarded as complex-valued, so that products and quotients of functions are defined, we have the following

THEOREM. *If*

$$\lim_{S} f(x) = p \in C, \qquad \lim_{S} g(x) = q \in C,$$

then

$$\lim_{s} f(x)g(x) = pq$$

and

$$\lim_{s} \frac{f(x)}{g(x)} = \frac{p}{q} \qquad (q \neq 0).$$

Proof. A slight generalization of the proofs of Theorems 4.36b and 4.36e. ∎

f. In the field C of all complex numbers $x + iy$ we define the direction $z \to \infty$ as the system of all sets $A_r \subset C$ of the form

$$A_r = \{z \in C : |z| > r\}$$

(verify that $z \to \infty$ is a direction). Then, given any function $f(z)$ defined for $z \geqslant r_0$, we can talk about the limit

$$\lim_{z \to \infty} f(z).$$

For example, choosing

$$f(z) = \frac{1}{z} \qquad (z \neq 0),$$

we have

$$\lim_{z \to \infty} f(z) = \lim_{z \to \infty} \frac{1}{z} = 0, \tag{1}$$

since, given any $\varepsilon > 0$, the inequality

$$|f(z)| = \frac{1}{|z|} < \varepsilon$$

holds on the set

$$A_{1/\varepsilon} = \{z \in C : |z| > 1/\varepsilon\}.$$

More generally, let

$$f(z) = \frac{a_0}{z^n} + \frac{a_1}{z^{n-1}} + \cdots + a_n$$

be any polynomial in $1/z$ with complex coefficients $a_0, a_1, \ldots, a_n$. Then

$$\lim_{z \to \infty} f(z) = \lim_{z \to \infty} \left(\frac{a_0}{z^n} + \frac{a_1}{z^{n-1}} + \cdots + a_n \right) = a_n,$$

by (1) together with Theorems 4.73c and 4.73e.

4.74. THEOREM. *A function $f(x)$ with values in R_n has a limit in the direction S if and only if the following condition, called the* **Cauchy convergence criterion**, *is satisfied: Given any $\varepsilon > 0$, there exists a set $A \in S$ such that*

$$|f(x') - f(x'')| < \varepsilon$$

for all $x', x'' \in A$.

Proof. Specialize Theorem 4.19, observing that R_n is complete (Theorem 3.72d). ∎

4.75. A function $f(x)$ with values in R_n can be written in the form

$$f(x) = (f_1(x), \ldots, f_n(x)),$$

where the functions $f_1(x), \ldots, f_n(x)$ are numerical functions, being components of the vector function $f(x)$.

THEOREM. *A function $f(x)$ with values in R_n has a limit in the direction S if and only if each component function $f_1(x), \ldots, f_n(x)$ has a limit in the direction S.*

Proof. Use the fact that

$$\max_{1 \leq k \leq n} |f_k(x') - f_k(x'')| \leqslant \sqrt{\sum_{k=1}^{n} [f_k(x') - f_k(x'')]^2} \leqslant \sum_{k=1}^{n} |f_k(x') - f_k(x'')|$$

(see Theorem 3.14b), together with the Cauchy convergence criterion. ∎

Problems

1. Prove that if the sequence $x_n \in R$ is convergent, so is the sequence $|x_n|$. Is the converse true?

2. Given arbitrary real sequences a_n and b_n, prove that

$$\underline{\lim}\, a_n + \overline{\lim}\, b_n \leqslant \overline{\lim}\, (a_n + b_n) \leqslant \overline{\lim}\, a_n + \overline{\lim}\, b_n,$$
$$\underline{\lim}\, a_n + \overline{\lim}\, b_n \geqslant \underline{\lim}\, (a_n + b_n) \geqslant \underline{\lim}\, a_n + \underline{\lim}\, b_n.$$

3. Prove that if a sequence a_n converges to the limit p, then so does *any* of its rearrangements

$$a_{n_1}, a_{n_2}, \ldots, a_{n_k}, \ldots$$

Does the convergence of the sequence follow from the convergence of *one* of its rearrangements?

4. Prove that if the sequence a_n is convergent, then

$$\overline{\lim_{n \to \infty}}\, (a_n + b_n) = \lim_{n \to \infty} a_n + \overline{\lim_{n \to \infty}}\, b_n.$$

5. Prove that if

$$\varlimsup_{n\to\infty} (a_n+b_n) = \varlimsup_{n\to\infty} a_n + \varlimsup_{n\to\infty} b_n$$

for a given sequence a_n and an arbitrary sequence b_n, then a_n is convergent.

6. Given that

$$x_1 = a, \; x_2 = b, \; x_3 = \tfrac{1}{2}(a+b), \ldots, \; x_n = \tfrac{1}{2}(x_{n-1}+x_{n-2}), \ldots,$$

find $\lim_{n\to\infty} x_n$.

7. Suppose $a > 0$, $x_0 > 0$ and

$$x_{n+1} = \frac{1}{2}\left(x_n + \frac{a}{x_n}\right) \qquad (n=0,1,2,\ldots).$$

Prove that

$$\lim_{n\to\infty} x_n = \sqrt{a}.$$

8. Suppose $a > b > 0$ and

$$x_1 = a, \; y_1 = b, \ldots, \; x_{n+1} = \sqrt{x_n y_n}, \; y_{n+1} = \tfrac{1}{2}(x_n+y_n), \ldots$$

Prove that the sequences x_n and y_n have the same limit. (Gauss)

9. Given that

$$\max \{p_1,\ldots,p_m\} = p_1 \qquad (p_1 > 0,\ldots,p_m > 0),$$

prove that

$$\lim_{n\to\infty} \sqrt[n]{\sum_{k=1}^{m} p_k^n} = p_1.$$

10. A straight line $y = kx + b$ is called an *asymptote* of a curve $y = f(x)$ defined for all sufficiently large x if

$$\lim_{x\to\infty} [f(x) - (kx+b)] = 0.$$

Prove that the curve $y = f(x)$ has an asymptote as $x \to \infty$ if and only if both limits

$$k = \lim_{x\to\infty} \frac{f(x)}{x}, \quad b = \lim_{x\to\infty}\left[f(x) - x \lim_{x\to\infty} \frac{f(x)}{x}\right]$$

exist.

11. Let $f(x)$ be a function defined on a metric space M equipped with a distance ρ_0, taking values in a metric space P equipped with a distance ρ. Suppose we define the limit of $f(x)$ as $x \to a$ by using "full neighborhoods"

$U_a(\delta) = \{x \in M : \rho_0(x,a) < \delta\}$ of the point a (which include the point a itself) rather than the deleted neighborhoods $U_a'(\delta)$ of Sec. 4.15a. In other words, overlooking the fact that the sets $U_a(\delta)(\delta > 0)$ do not constitute a direction, suppose we say that $f(x) \to p$ as $x \to a$ in the new sense if, given any $\varepsilon > 0$, there exists a number $\delta > 0$ such that $\rho(p, f(x)) < \varepsilon$ for all $x \in U_\delta(a)$, i.e., for all x satisfying the inequality $\rho_0(x,a) < \delta$.

Prove that $f(x) \to p$ as $x \to a$ in this new sense if and only if $f(a) = p$ and $f(x) \to p$ as $x \to a$ in the sense of Sec. 4.15a.†

12. Let $y(x)$ be a function defined on a set X equipped with a direction S, taking values in a metric space Y, and let $z(y)$ be a function defined on Y, taking values in a metric space Z. Then the "composite function" $z(x) = z(y(x))$ is defined on the set X and takes its values in the space Z. Suppose the limits

$$p = \lim_S y(x) \in Y, \quad q = \lim_{y \to p} z(y) \in Z$$

both exist. Does $\lim_S z(x)$ necessarily exist? If so, does it equal p?

13. With the same notation as in the preceding problem, prove that

(a) If there exists a set $A \in S$ on which $y(x)$ does not take the value p, then $z(x)$ has the limit q;

(b) If there exists a set $A \in S$ on which $y(x)$ is identically equal to p, then $z(x)$ has the limit $z(p)$;

(c) If $y(x)$ takes both values equal to p and values unequal to p on every set $A \in S$, then $z(x)$ has a limit if and only if $q = z(p)$,‡ in which case

$$\lim_S z(x) = q.$$

† In particular, $f(a)$ must now be *defined*, unlike the situation in Sec. 4.15a.
‡ Thereby making p the limit of $z(x)$ in the sense of Problem 11.

5 Continuous Functions

5.1. Continuous Functions on a Metric Space

5.11. Let $f(x)$ be a function defined on a metric space M equipped with a distance ρ_0, taking values in a metric space P equipped with a distance ρ. As in Sec. 4.15a, let $x \to a$ be the direction corresponding to the deleted neighborhoods

$$U'_\delta(a) = \{x \in M: 0 < \rho_0(x,a) < \delta\} \qquad (a \in M,\ \delta > 0).$$

a. Definition. The function $f(x)$ is said to be *continuous* at $x = a$, and the point a is then called a *continuity point* of $f(x)$, if

$$\lim_{x \to a} f(x) = f(a).$$

In other words, $f(x)$ is said to be continuous at $x = a$ if, given any $\varepsilon > 0$, there exists a $\delta > 0$ such that $\rho_0(x,a) < \delta$ implies

$$\rho(f(x), f(a)) < \varepsilon. \tag{1}$$

Naturally, for a numerical function $f(x)$ the inequality (1) takes the form

$$|f(x) - f(a)| < \varepsilon.$$

Note that every isolated point $a \in M$ is automatically a continuity point of $f(x)$.

b. By an obvious modification of Theorem 4.65, a function $f(x)$ is continuous at $x = a$ if and only if $f(x_n) \to f(a) \in P$ for every sequence $x_n \to a \in M$.

c. Definition. A point $a \in M$ is called a *discontinuity point* of $f(x)$ if it is not a continuity point of $f(x)$. A function continuous at every point of a set $E \subset M$ is said to be *continuous on E*.

d. Naturally, the definition of continuity depends on the metrics of the spaces M and P. But since the definition can be formulated in terms of convergent sequences (Sec. 5.11b), the property of a function being continuous at a given point or on a given set is preserved if the metrics in M and P are replaced by homeomorphic metrics (Sec. 3.34c).

5.12. a. An obvious example of a continuous function with domain M and range P is the constant function†

$$f(x) \equiv y_0,$$

where y_0 is a fixed point of the space P (cf. Theorem 4.21).

b. As another example, consider the distance $\rho(a,x)$ from a fixed point

---

† The symbol $\equiv$ means "is identically equal to."

$a \in M$ to a variable point $x \in M$. Clearly $\rho(a,x)$ is a numerical function on the metric space M. The fact that $\rho(a,x)$ is continuous at every point $x = x_0$ of M follows from Theorem 3.32b.

5.13. Continuous numerical functions

a. THEOREM. *If the numerical functions $f(x)$ and $g(x)$ are continuous at $x = x_0$, then so is their sum $f(x) + g(x)$.*

Proof. An immediate consequence of Theorem 4.36a. ∎

b. THEOREM. *If the numerical functions $f(x)$ and $g(x)$ are continuous at $x = x_0$, then so is their product $f(x)g(x)$.*

Proof. An immediate consequence of Theorem 4.36b. ∎

c. THEOREM. *If the numerical functions $f(x)$ and $g(x)$ are continuous at $x = x_0$, then so is their quotient $f(x)/g(x)$, provided that $g(x_0) \neq 0$.*

Proof. An immediate consequence of Theorem 4.36e. ∎

d. We are now in a position to construct wide classes of continuous functions. For example, the numerical function $y = x$ defined on the real line R is obviously continuous on R. Hence it follows from the above theorems that every polynomial $a_0 x^n + a_1 x^{n-1} + \cdots + a_n$ is continuous on R while every rational function

$$\frac{a_0 x^n + a_1 x^{n-1} + \cdots + a_n}{b_0 x^m + b_1 x^{m-1} + \cdots + b_m}$$

is continuous at every point $x \in R$ where its denominator is nonvanishing.

e. Let $y = \xi_k$, where ξ_k is the kth component of the vector $x = (\xi_1, \ldots, \xi_n) \in R_n$. Then y is obviously a continuous numerical function on R_n. It follows from Theorems 5.13a–5.13c that every polynomial in the components of the vector x is continuous on R and that every rational function in the components of x is continuous at every point of R_n where its denominator is nonvanishing.

5.14. Let $f(x)$ be a continuous function on a metric space M equipped with a distance ρ_0, taking values in a metric space P equipped with a distance ρ. Moreover, let G be some subset of P, and let $H = \{x \in M : f(x) \in G\}$.

a. THEOREM. *If G is open, then H is open.*

Proof. Given $x_0 \in H$, $f(x_0) \in G$, let

$$V = \{y \in P : \rho(y, f(x_0))\} < \varepsilon$$

be an open ball centered at $f(x_0)$ and contained in G. Let $\delta > 0$ be such that $\rho_0(x,x_0) < \delta$ implies $\rho(f(x),f(x_0)) < \varepsilon$. Then the ball

$$U = \{x \in M : \rho_0(x,x_0) < \delta\}$$

is contained in H. ∎

b. THEOREM. *If G is closed, then H is closed.*

Proof. The complement of the set H relative to M is the open set $\{x \in M : f(x) \in P - G\}$. But the latter set is open, by the preceding theorem, and hence H is closed. ∎

c. THEOREM. *If $f(x)$ is a continuous numerical function and $c \in \overline{R}$ an arbitrary real number, then the sets*

$$\{x \in M : f(x) < c\}, \qquad \{x \in M : f(x) > c\}$$

are open, while the sets

$$\{x \in M : f(x) \leqslant c\}, \qquad \{x \in M : f(x) \geqslant c\}, \qquad \{x \in M : f(x) = c\}$$

are closed.

Proof. An immediate consequence of Theorems 5.14a and 5.14b. ∎

d. THEOREM. *If $f_1(x),\ldots,f_m(x)$ are continuous numerical functions and $a_1,\ldots,a_m$, $b_1,\ldots,b_m$ ($a_1 < b_1,\ldots, a_m < b_m$) are arbitrary real numbers in $\overline{R}$, then the set*

$$\{x \in M : a_1 < f_1(x) < b_1,\ldots, a_m < f_m(x) < b_m\} \tag{2}$$

is open, while the set

$$\{x \in M : a_1 \leqslant f_1(x) \leqslant b_1,\ldots, a_m \leqslant f_m(x) \leqslant b_m\} \tag{3}$$

is closed.

Proof. An immediate consequence of the theorems on intersection of open and closed sets (Theorems 3.22 and 3.54). ∎

e. Using the preceding theorem, we see that many figures of elementary geometry, described by systems of equations like those in (2) and (3) are open or closed sets. In n-dimensional space, for example, a set of the form

$$\{x \in R_n : a_1 < x_1 < b_1,\ldots, a_n < x_n < b_n\}$$

is open, while a set of the form

$$\{x \in R_n : a_1 \leqslant x_1 \leqslant b_1,\ldots, a_n \leqslant x_n \leqslant b_n\}$$

is closed. The first set is called an *open block*, while the second is called a *closed block*.

Similarly, the polygonal figure in the $x_1 x_2$-plane described by m "linear inequalities" of the form

$$a_j x_1 + b_j x_2 < c_j \qquad (j = 1, \ldots, m)$$

is an open set, called an *open m-gon*, while the figure described by the same inequalities with $<$ replaced by $\leqslant$ is a closed set, called a *closed m-gon*. In the same way, we can introduce *open* and *closed polyhedra* in n-dimensional space ($n > 2$) by using inequalities involving linear functions of the coordinates of a variable point $x \in R_n$. Closed figures of this type are compacta, provided they are bounded (see Theorem 3.96e).

5.15. Continuity of a composite function

a. Given three metric spaces M, N, and P equipped with distances ρ_M, ρ_N, and ρ_p, let $y = f(x)$ be a function defined on M and taking values in N, while $z = g(y)$ is a function defined on N and taking values in P. Then the "composite function"

$$z = h(x) \equiv g(f(x)) \tag{4}$$

is defined on M and takes its values in P.

THEOREM. *Suppose the function $y = f(x)$ is continuous at $x = x_0$, while the function $z = g(y)$ is continuous at $y = y_0 = f(x_0)$. Then the composite function (4) is continuous at $x = x_0$.*

Proof. Given any $\varepsilon > 0$, we can find a $\tau > 0$ such that $\rho_N(y, y_0) < \tau$ implies $\rho_P(g(y), g(y_0)) < \varepsilon$, since

$$\lim_{y \to y_0} g(y) = g(y_0).$$

Having found τ, we can then find a $\delta > 0$ such that $\rho_M(x, x_0) < \delta$ implies

$$\rho_N(y, y_0) \equiv \rho_N(f(x), f(x_0)) < \tau,$$

since

$$\lim_{x \to x_0} f(x) = f(x_0).$$

But then $\rho_M(x, x_0) < \delta$ implies

$$\rho_P(h(x), h(x_0)) \equiv \rho_P(g(f(x)), g(f(x_0))) = \rho_P(g(y), g(y_0)) < \varepsilon,$$

so that

$$\lim_{x \to x_0} h(x) = h(x_0). \quad \blacksquare$$

b. **THEOREM.** *Let $f(x)$ be a continuous function on a metric space M taking values in another metric space N, and let b be a fixed point of N. Then*

$$h(x) = \rho_N(f(x), b)$$

is a continuous numerical function on M.

Proof. An immediate consequence of Sec. 5.12b and the preceding theorem. ∎

In particular, if $f(x)$ is a continuous numerical function on a metric space M, then

$$|f(x)| = \rho_N(f(x), 0)$$

is also continuous on M, since in this case N is just the real line R. ∎

5.16. Continuous functions on compacta. Continuous functions on a compact metric space (Sec. 3.91a) have certain special properties which we consider here and in Sec. 5.17.

a. THEOREM. *The set of all values of a continuous function f (x) on a compact metric space M is itself compact.*

Proof. Given an arbitrary sequence $f(x_1),\ldots,f(x_n),\ldots$ of values of $f(x)$, use the compactness of M to choose a convergent subsequence $x_{n_1},\ldots,x_{n_k},\ldots$ from the sequence $x_1,\ldots,x_n,\ldots$ Suppose $x_{n_k} \to a$, say. Then $f(x_{n_k}) \to f(a)$, since $f(x)$ is continuous at $x = a$. Thus we have found a convergent subsequence of the original sequence $f(x_1),\ldots,f(x_n),\ldots$, thereby proving the compactness of the set $\{f(x): x \in M\}$. ∎

b. THEOREM (**Weierstrass**). *Every continuous numerical function $f(x)$ defined on a compact metric space M (with values in R) is bounded. Moreover, $f(x)$ achieves its greatest lower and least upper bounds on M, i.e., if*

$$\alpha = \inf_{x \in M} f(x), \qquad \beta = \sup_{x \in M} f(x),$$

then there exist points p and q in M such that $f(p) = \alpha, f(q) = \beta$.

Proof. An immediate consequence of the preceding theorem and the properties of a compact set on the real line (see Theorem 3.96f). ∎

c. Let the space M in Weierstrass' theorem be the closed interval $[a,b]$, and suppose $f(a) = f(b)$. Then at least one of the points p and q must be an *interior* point of $[a,b]$, i.e., must lie in the *open* interval (a,b). In fact, if $f(x)$ is a constant equal to $f(a)$, we can choose any point of the interval (a,b) as the point $p = q$. Suppose, on the other hand, that $f(x)$ is nonconstant, with values greater than $f(a)$, say. Then $\beta = \sup f(x) > f(a)$. By Weierstrass' theorem, there exists a point $q \in [a,b]$ such that $f(q) = \beta$. But $f(a) = f(b) < \beta$, and hence $q \in (a,b)$, as claimed. If $f(x)$ takes values less than $f(a)$, then a similar argument shows that the point p such that $f(p) = \alpha = \inf f(x)$ must belong to (a,b). ∎

5.17. Uniform continuity

a. Definition. A function $f(x)$ defined on a metric space M equipped with a distance ρ_0, taking values in a metric space P equipped with a distance ρ, is said to be *uniformly continuous* on M if, given any $\varepsilon > 0$, there exists a $\delta > 0$ such that $\rho_0(x', x'') < \delta$ implies $\rho(f(x'), f(x'')) < \varepsilon$ *for arbitrary points* x', $x'' \in M$.

Obviously, every uniformly continuous function on M is continuous at every point of M. In general, however, continuity of $f(x)$ on M does not imply uniform continuity of $f(x)$ on M. For example, the function $y = x^2$ is continuous on the half-line $0 \leqslant x < \infty$ (Sec. 5.13d). However, $y = x^2$ is not uniformly continuous on $0 \leqslant x < \infty$, since

$$x'^2 - x''^2 = (x' - x'')(x' + x''),$$

so that the condition $|x' - x''| < \delta$ certainly does not imply that $|x'^2 - x''^2|$ is bounded by any constant.

b. The situation is different in the case where M is compact, as shown by the following

THEOREM (**Heine**). *If $f(x)$ is continuous on M and if M is compact, then $f(x)$ is uniformly continuous on M.*

Proof. Suppose to the contrary that $f(x)$ is not uniformly continuous on M. Then, given any $\varepsilon > 0$ and any $\delta > 0$, there is a pair of points x', $x'' \in M$ such that $\rho_0(x', x'') < \delta$ but $\rho(f(x'), f(x'')) \geqslant \varepsilon$. Thus, given any $\varepsilon > 0$ and any $n = 1, 2, \ldots$, there is a pair of points x_n', $x_n' \in M$ such that $\rho_0(x_n', x_n'') < 1/n$ but

$$\rho(f(x_n'), f(x_n'')) \geqslant \varepsilon \tag{5}$$

(choose $\delta = 1/n$). Since M is compact, the sequence x_n' contains a convergent subsequence $x_{n_k}' \to x_0 \in M$. Moreover, since $\rho(x_{n_k}', x_{n_k}'') < 1/n_k$, we have

$$\rho(x_{n_k}'', x_0) \leqslant \rho(x_{n_k}'', x_{n_k}') + \rho(x_{n_k}', x_0) \to 0,$$

so that $x_{n_k}'' \to x_0$ as well. The function $f(x)$ is continuous at x_0, and hence, given any $\varepsilon > 0$, there is a $\delta_0 > 0$ such that $\rho_0(x, x_0) < \delta_0$ implies

$$\rho(f(x), f(x_0)) < \varepsilon/2.$$

Therefore

$$\rho(f(x_{n_k}'), f(x_0)) < \varepsilon/2, \quad \rho(f(x_{n_k}''), f(x_0)) < \varepsilon/2$$

for all sufficiently large k, since $x_{n_k}' \to x_0$, $x_{n_k}'' \to x_0$. But then

$$\rho(f(x'_{n_k}),f(x''_{n_k})) \leqslant \rho(f(x'_{n_k}),f(x_0)) + \rho(f(x''_{n_k}),f(x_0)) < \varepsilon$$

for all sufficiently large k, contrary to (5). This contradiction shows that $f(x)$ is uniformly continuous on M. ∎

c. Modulus of continuity. Again let $f(x)$ be a function with domain M and range P, where M and P are metric spaces equipped with distances ρ_0 and ρ, respectively. Consider the function

$$\omega_f(\delta) = \sup_{\rho_0(x',x'') \leq \delta} \rho(f(x'),f(x'')),$$

whose argument is the positive number δ. Clearly, $f(x)$ is uniformly continuous on M if and only if

$$\lim_{\delta \to 0} \omega_f(\delta) = 0.$$

The function $\omega_f(\delta)$ is called the *modulus of continuity* of $f(x)$ on the space M.

d. For numerical functions $y=f(x)$, i.e., for functions taking values on the real line R equipped with the usual metric $\rho(y',y'') = |y' - y''|$, the function $\omega_f(\delta)$ has the following properties:
(1) $\omega_{f+g}(\delta) \leqslant \omega_f(\delta) + \omega_g(\delta)$;
(2) $\omega_{cf}(\delta) = c\omega_f(\delta)$ for constant $c > 0$;
(3) $\omega_{fg}(\delta) \leqslant \sup_{x \in M} |f(x)| \cdot \omega_g(\delta) + \omega_f(\delta) \sup_{x \in M} |g(x)|$.

The first two properties are an immediate consequence of the definition of the modulus of continuity, while the third property follows from the inequality

$$|f(x')g(x') - f(x'')g(x'')|$$
$$= |f(x')g(x') - f(x')g(x'') + f(x')g(x'') - f(x'')g(x'')|$$
$$\leqslant |f(x')||g(x') - g(x'')| + |f(x') - f(x'')||g(x'')|$$
$$\leqslant \sup_{x' \in M} |f(x')| \cdot |g(x') - g(x'')| + |f(x') - f(x'')| \sup_{x'' \in M} |g(x'')|.$$

5.18. Continuous functions of two variables. In analysis, one often encounters continuous functions of two (or more) variables. Let M_1 and M_2 be two metric spaces, equipped with distances ρ_1 and ρ_2. Then a function $f(x_1, x_2)$ with arguments $x_1 \in M_1$, $x_2 \in M_2$, taking values in a metric space P equipped with a distance ρ,† is said to be *continuous in x_1 and x_2* (*jointly*) *at $x_1 = x_1^0, x_2 = x_2^0$* if, given any $\varepsilon > 0$, there exists $\delta > 0$ such that

$$\rho_1(x_1, x_1^0) < \delta, \quad \rho_2(x_2, x_2^0) < \delta \tag{6}$$

† In the language of Sec. 2.81, $y=f(x_1, x_2)$ is a rule associating an element $y \in P$ with every pair of points $x_1 \in M_1$, $x_2 \in M_2$.

implies $\rho(f(x_1, x_2), f(x_1^0, x_2^0)) < \varepsilon$. It is easy to see that this definition of continuity is essentially nothing new, and in fact agrees with the definition of continuity of the function $f(x_1, x_2)$ regarded as a function $f(x)$ of a single argument $x = (x_1, x_2)$, namely a variable point of the direct product $M = M_1 \times M_2$ of the metric spaces M_1 and M_2. To see this, we recall from Sec. 3.16 that a metric ρ_0 can be introduced in the space $M = M_1 \times M_2$ by setting

$$\rho_0(x, y) = \max\{\rho_1(x_1, y_1), \rho_2(x_2, y_2)\}, \tag{7}$$

where $x = (x_1, x_2), y = (y_1, y_2)$. Thus continuity of the function $f(x)$ at the point $x^0 = (x_1^0, x_2^0)$ means that given any $\varepsilon > 0$, there exists a $\delta > 0$ such that

$$\rho_0(x, x^0) = \max\{\rho_1(x_1, x_1^0), \rho_2(x_2, x_2^0)\} < \delta \tag{8}$$

implies $\rho(f(x), f(x^0)) < \varepsilon$. But the inequality (8) is equivalent to the pair of inequalities (6), and hence the two definitions of continuity are equivalent.

In Sec. 3.16 we indicated two other ways of defining a metric on the direct product $M = M_1 \times M_2$. But the alternative metrics are both homeomorphic to the metric (7), as shown in Sec. 3.34d. Hence, by Sec. 5.11d, all three metrics lead to precisely the same set of continuous functions on $M = M_1 \times M_2$ (with values in P).

5.2. Continuous Numerical Functions on the Real Line

5.21. a. In Secs. 5.2–5.7 we will consider numerical functions $f(x)$ with domain $E \subset \bar{R}$ and values in $\bar{R}$, where $\bar{R}$ is the extended real line equipped with the metric $r(x, y)$ of Theorem 3.35d. The metric $r(x, y)$ is homeomorphic to the usual metric $\rho(x, y) = |x - y|$ on the ordinary real line R. Hence, in problems involving the continuity of $f(x)$ at a point x_0, we can replace r by ρ on X if x_0 is finite and on Y if $f(x_0)$ is finite.

b. To illustrate these remarks, consider the rational function

$$f(x) = \frac{a_0 x^n + a_1 x^{n-1} + \cdots + a_n}{b_0 x^m + b_1 x^{m-1} + \cdots + b_m} \qquad (a_0 \neq 0, b_0 \neq 0),$$

defined at every point of R where the denominator is nonvanishing. Suppose we complete the definition of $f(x)$ at the points $\pm \infty$ by setting

$$f(-\infty) = \lim_{x \to -\infty} f(x), \qquad f(+\infty) = \lim_{x \to +\infty} f(x)$$

(the existence of these limits in $\bar{R}$ was proved in Sec. 4.39d). Then the resulting function, with values in $\bar{R}$, is continuous at the points $\pm \infty$ as well.

c. LEMMA. *If a numerical function $f(x)$ with values in $\overline{R}$ is continuous at a point $x = a \in \overline{R}$, then, given any $\varepsilon > 0$, there exists an open ball U_δ centered at a, i.e., a set of the form*

$\{x : |x - a| < \delta\}$ *if a is finite,*

$\{x : x > 1/\delta\}$ *if $a = +\infty$,*

$\{x : x < -1/\delta\}$ *if $a = -\infty$*

$(\delta > 0)$ *such that, for all $x \in U_\delta$,*

$|f(x) - f(a)| < \quad \varepsilon \quad$ *if $f(a)$ is finite,*

$f(x) > \quad 1/\varepsilon$ *if $f(a) = +\infty$,*

$f(x) < -1/\varepsilon$ *if $f(a) = -\infty$.*

In particular, U_δ can be chosen in such a way that $f(x)$· has the same sign as $f(a)$ at every point $x \in U_\delta$ (provided that $f(a) \neq 0$).

Proof. An immediate consequence of Theorems 4.34a and 4.34b, the definition of the metric in $\overline{R}$ (in particular, see Sec. 3.35e), and the definition of continuity. ∎

5.22. THEOREM (**Bolzano**). *Suppose a numerical function $f(x)$ is continuous on a closed interval $[a, b]$ and takes values with opposite signs at the end points of $[a, b]$. Then there exists a point $c \in (a, b)$†at which $f(c) = 0$.*

Proof. Let $f(a) < 0$, $f(b) > 0$, say. Then, by the above lemma, $f(x) < 0$ holds for all x sufficiently near a, while $f(x) > 0$ holds for all x sufficiently near b. Hence the point

$$c = \sup \{x \in [a, b] : f(x) < 0\}$$

is distinct from both a and b. By the definition of the least upper bound, $f(x') \geqslant 0$ for $x' > c$, while for every $\delta > 0$ there is an $x'' > c - \delta$, i.e., every neighborhood of c contains points x' and x'' such that $f(x') \geqslant 0$, $f(x'') < 0$. But this is impossible if $f(c) \neq 0$, by the lemma again. It follows that $f(c) = 0$.‡ ∎

5.23. THEOREM (**Intermediate value theorem**). *Suppose a numerical function $f(x)$ is continuous on a closed interval $[a, b]$ and takes distinct values $A = f(a)$, $B = f(b)$ at the end points of $[a, b]$. Then, given any number C between A and B, there exists a point $c \in (a, b)$ at which $f(c) = C$.*

Proof. The function $f(x) - C$ satisfies the conditions of Bolzano's theorem,

---

† That is, an *interior* point of $[a, b]$.
‡ To treat the case $f(a) > 0$, $f(b) < 0$, either consider the function $-f(x)$ or else let $c = \sup\{x \in [a, b] : f(x) > 0\}$.

and hence vanishes at some point $c \in (a, b)$.   ∎

5.24. One-sided continuity. We now define two further directions on the set R of all finite real numbers, besides the direction $x \to a$ of Sec. 4.15b, consisting of all deleted neighborhoods

$$U'_\delta = \{x: 0 < |x - a| < \delta\} \qquad (\delta > 0).$$

The first of these directions, denoted by $x \nearrow a$, consists of all intervals of the form

$$U^l_\delta = \{x: 0 < a - x < \delta\} \qquad (\delta > 0)$$

(l for "left"), while the second, denoted by $x \searrow a$, consists of all intervals of the form

$$U^r_\delta = \{x: 0 < x - a < \delta\} \qquad (\delta > 0)$$

(r for "right").† If a function $f(x)$ has a limit p in the direction $x \nearrow a$, we write

$$p = \lim_{x \nearrow a} f(x) = f(a - 0),$$

while if $f(x)$ has a limit p in the direction $x \searrow a$, we write

$$p = \lim_{x \searrow a} f(x) = f(a + 0).$$

Note that these limits are meaningful even if $f(x)$ fails to be defined at the point $x = a$ itself (cf. Sec. 4.15a).

Now let S be the direction $x \to a$, and let

$$G = \{x: x < a\}, \qquad H = \{x: x > a\}.$$

Then, in the notation of Sec. 4.16, the direction $x \nearrow a$ is just GS, while the direction $x \searrow a$ is just HS. Thus if $\lim_{x \to a} f(x)$ exists and equals p, then $\lim_{x \nearrow a} f(x)$ and $\lim_{x \searrow a} f(x)$ both exist and equal p, as in Sec. 4.16a. However, the existence of the "one-sided limits" $\lim_{x \nearrow a} f(x)$ and $\lim_{x \searrow a} f(x)$ does not imply the existence of $\lim_{x \to a} f(x)$ *unless the two limits are equal*, to p say, in which case $\lim_{x \to a} f(x)$ exists and equals p, by Theorem 4.16c.

Suppose now that $f(x)$ is defined at the point $x = a$. Then, according to Sec. 5.11a, we say that $f(x)$ is continuous at $x = a$ if

$$\lim_{x \to a} f(x) = f(a).$$

---

† $U_\delta{}^l$ and $U_\delta{}^r$ can be regarded as "one-sided neighborhoods" of the point a.

In the same way, we say that $f(x)$ is *continuous from the left* at $x = a$ if

$$\lim_{x \nearrow a} f(x) = f(a)$$

and *continuous from the right* at $x = a$ if

$$\lim_{x \searrow a} f(x) = f(a).$$

Suppose $f(x)$ is continuous both from the left and from the right at $x = a$. Then clearly $f(x)$ is continuous at $x = a$.

5.3. Monotonic Functions

5.31. Definition. Let $f(x)$ be a numerical function defined on a set $E \subset \bar{R}$, taking values in $\bar{R}$. Then $f(x)$ is said to be *increasing* on E if $x < y$ $(x, y \in E)$ implies $f(x) < f(y)$, *nondecreasing* on E if $x < y$ $(x, y \in E)$ implies $f(x) \leqslant f(y)$, *decreasing* on E if $x < y$ $(x, y \in E)$ implies $f(x) > f(y)$, and *nonincreasing* on E if $x < y$ $(x, y \in E)$ implies $f(x) \geqslant f(y)$. In any of these four cases, $f(x)$ is said to be *monotonic* on E.

Obviously, if an increasing (or decreasing) function takes a value C, then it can take this value at only one point of E. On the other hand, a nonincreasing or nondecreasing function can take the same value at several points of E.

5.32. Let $f(x)$ be nondecreasing on a closed interval $E = [a,b]$, and let $x_0 \in E$ be a point distinct from a. Then $f(x)$ is nondecreasing in the direction $x \nearrow x_0$, in the sense of Sec. 4.51.† Since $f(x)$ is bounded from above, by $f(x_0)$ say, it follows from Theorem 4.53 that the limit

$$f(x_0 - 0) = \lim_{x \nearrow x_0} f(x)$$

exists and equals

$$f(x_0 - 0) = \sup \{f(x) : x < x_0, x \in E\}.$$

Similarly, if $x_0 \in E$ is a point distinct from b, it follows from Theorem 4.54 that the limit

$$f(x_0 + 0) = \lim_{x \searrow x_0} f(x)$$

exists and equals

† More exactly, $f(x)$ is nondecreasing in the direction $E\{x \nearrow x_0\}$, where $E\{x \nearrow x_0\}$ denotes the intersection of the set E and the sets of the direction $x \nearrow x_0$. However, since all sufficiently small intervals of the direction $x \nearrow x_0$ are contained in E, we can replace $E\{x \nearrow x_0\}$ by $x \nearrow x_0$ (cf. Sec. 4.16b).

$f(x_0+0) = \inf \{f(x) : x > x_0, x \in E\}.$

At the points a and b, we set

$f(a-0) = f(a), \quad f(b+0) = f(b),$

by definition.

5.33. Again let $f(x)$ be nondecreasing on a closed interval $E = [a,b]$. Since $f(x') \leqslant f(x_0) \leqslant f(x'')$ if $x' \leqslant x_0 \leqslant x''$, it follows from Theorem 1.62b that

$f(x_0-0) \leqslant f(x_0) \leqslant f(x_0+0).$

By Sec. 5.24, the function $f(x)$ is continuous from the left at $x = x_0$ if $f(x_0-0) = f(x_0)$ and continuous from the right at $x = x_0$ if $f(x_0+0) = f(x_0)$. The condition $f(x_0-0) = f(x_0+0)$ (which implies $f(x_0-0) = f(x_0) = f(x_0+0)$) is necessary and sufficient for the nondecreasing function $f(x)$ to be continuous at $x = x_0$.

If $f(x_0-0) < f(x_0+0)$, then x_0 is certainly a discontinuity point of the function $f(x)$. The values of $f(x)$ cannot exceed $f(x_0-0)$ to the left of x_0 and cannot be less than $f(x_0+0)$ to the right of x_0, while the value of $f(x_0)$ must lie in the interval

$I = [f(x_0-0), f(x_0+0)].$

It follows that only one number in the interval I can be a value of $f(x)$.

5.34. The above observation leads to the following sufficient condition for continuity of a nondecreasing function:

THEOREM. *If $f(x)$ is nondecreasing on the interval $[a,b]$ and takes every value in the interval $[f(a),f(b)]$, then $f(x)$ is continuous on $[a,b]$.*

Proof. Suppose $f(x_0-0) < f(x_0+0)$ at some point $x_0 \in [a,b]$. Then $f(x)$ cannot take every value in $[f(a),f(b)]$. Hence $f(x_0-0) = f(x_0+0)$ for every $x_0 \in [a,b]$, i.e., $f(x)$ is continuous on $[a,b]$. ∎

The condition that $f(x)$ take every value in the interval $[f(a),f(b)]$ is also necessary, by the intermediate value theorem (Theorem 5.23).

As an exercise, the reader should state and prove the analogues of the results of Secs. 5.32–5.34 for *nondecreasing* functions.

5.35. The inverse function

a. Definition. Let $f(x)$ be a function with domain E and range $F = \{y : y = f(x), x \in E\}$. Then we say that $f(x)$ *has an inverse* (on E) if there exists a (single-valued) function $\varphi(y)$ with domain F such that $\varphi(f(x)) = x$ for all $x \in E$. Clearly, $f(x)$ has an inverse on E if and only if $f(x)$ is a one-to-one function on E. The function $\varphi(y)$, called the *inverse (function)* of $f(x)$, is then unique and satisfies the condition $f(\varphi(y)) = y$ for all $y \in F$.

b. THEOREM. *Let the numerical function $f(x)$ be continuous and increasing on a closed interval $[a,b] \subset \bar{R}$, and let*

$$A = f(a), \quad B = f(b) \qquad (A, B \in \bar{R}).$$

Then $f(x)$ has an inverse $\varphi(y)$ which is continuous and increasing on the closed interval $[A,B]$.

Proof. Let $\varphi(a) = A$, $\varphi(B) = b$ by definition. By the intermediate value theorem (Sec. 5.23), given any C such that $A < C < B$, there is a point $c \in (a,b)$ such that $f(c) = C$. Moreover c is unique (see Sec. 5.31). Let $\varphi(C) = c$ for all such C. The function $\varphi(y)$ is now defined on the whole interval $[f(a), f(b)]$, and is obviously the inverse of $f(x)$ since $\varphi(f(x)) \equiv x$ by construction. Moreover, $y_1 < y_2$ implies $x_1 < x_2$, i.e., $\varphi(y_1) < \varphi(y_2)$, so that $\varphi(y)$ is increasing. The continuity of $\varphi(y)$ follows from Theorem 5.34, since $\varphi(y)$ takes all values in the interval $[a,b]$. ∎

c. For the case of an open interval, this theorem takes the following form:

THEOREM. *Let the numerical function $f(x)$ be continuous and increasing on an open interval (a,b), and let*

$$A = \lim_{x \searrow a} f(x), \quad B = \lim_{x \nearrow b} f(x) \qquad (A, B \in \bar{R}).$$

Then $f(x)$ has an inverse $\varphi(y)$ which is continuous and increasing on the open interval (A,B).

Proof. Extend the definition of $f(x)$ to the closed interval $[a,b]$ by setting $f(a) = A$, $f(b) = B$. Then $f(x)$ is continuous and increasing on $[a,b]$. It follows from the preceding theorem that $f(x)$ has an inverse $\varphi(y)$ which is continuous and increasing on $[A,B]$. To complete the proof, we need only restrict $f(x)$ to the open interval, thereby restricting $\varphi(y)$ to (A,B). ∎

As an exercise, the reader should state and prove the analogues of Theorems 5.35b and 5.35c for decreasing functions.

d. The graph of an inverse function. Let $y = f(x)$ be a one-to-one function on a closed interval $[a, b]$. Then, as shown in Figure 2, the function $\varphi(y)$ has the graph obtained from the graph of $f(x)$ by reflecting the latter in the line $y = x$ (the bisector of the first quadrant). In fact, the reflection simply carries the point $(x, y = f(x))$ into the point $(y, x = \varphi(y))$.

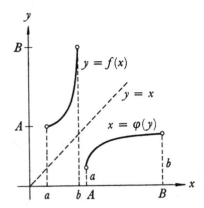

Figure   2

5.4. The Logarithm

5.41. THEOREM. *There exists a unique function $f(x)$ $(x>0)$ satisfying the following conditions:*
(a) $f(xy) = f(x) + f(y)$ *for all positive x and y;*
(b) $f(a) = 1$ *for a given $a>1$;*
(c) $f(x)$ *is an increasing function, i.e., $x<y$ implies $f(x)<f(y)$.*

Proof (after J. Dieudonné). First we construct $f(x)$, assuming that $f(x)$ actually exists. It follows from Condition a that $f(1)=2f(1)$, and hence $f(1)=0$. Then Conditions a and b imply

$$f(a^n) = nf(a) = n \qquad (n=1,2,\ldots).$$

By Theorem 1.72, given any $n=1,2,\ldots$, there is an integer m such that

$$a^m \leqslant x^n < a^{m+1}. \tag{1}$$

Since $f(x)$ is increasing, (1) implies

$$f(a^m) = m \leqslant f(x^n) = nf(x) < f(a^{m+1}) = m+1.$$

Therefore, given any $n=1,2,\ldots$, there is an integer m such that

$$\frac{m}{n} \leqslant f(x) < \frac{m+1}{n}.$$

In particular, choosing

$$n_1 = 1, \, n_2 = 2, \ldots, \, n_k = 2^{k-1}, \ldots,$$

we can find an integer m_k such that

$$a^{m_k} \leqslant x^{n_k} < a^{m_k+1} \qquad (k = 1, 2, \ldots), \tag{2}$$

thereby determining an interval

$$\left[\frac{m_k}{n_k}, \frac{m_k+1}{n_k} \right]. \tag{3}$$

The interval (3) contains the next interval

$$\left[\frac{m_{k+1}}{n_{k+1}}, \frac{m_{k+1}+1}{n_{k+1}} \right].$$

In fact, (2) implies

$$a^{2m_k} \leqslant x^{2n_k} = x^{n_{k+1}} < a^{2(m_k+1)},$$

and hence $m_{k+1} \geqslant 2m_k$, $m_{k+1} + 1 \leqslant 2(m_k+1)$,

$$\frac{m_k}{n_k} = \frac{2m_k}{2n_k} \leqslant \frac{m_{k+1}}{n_{k+1}} < \frac{m_{k+1}+1}{n_{k+1}} \leqslant \frac{2(m_k+1)}{2n_k} = \frac{m_k+1}{n_k}.$$

Thus the set of all intervals (3) with $k = 1, 2, \ldots$ is a system of nested closed intervals. The intersection of all the intervals of this system is nonempty, by Theorem 1.81, and in fact consists of a single point, by Theorem 1.82, since the length of the kth interval approaches 0 as $k \to \infty$. We now choose the value of $f(x)$ to be the number corresponding to this unique point. Since the point is unique, so is the resulting function $f(x)$ $(x > 0)$.

Next we show that the function $f(x)$ just constructed satisfies Conditions a–c. Given any positive numbers x and y, let m_k and p_k be integers such that

$$a^{m_k} \leqslant x^{n_k} < a^{m_k+1}, \qquad a^{p_k} \leqslant y^{n_k} < a^{p_k+1},$$

so that

$$\frac{m_k}{n_k} \leqslant f(x) \leqslant \frac{m_k+1}{n_k}, \qquad \frac{p_k}{n_k} \leqslant f(y) \leqslant \frac{p_k+1}{n_k}. \tag{4}$$

Then

$$a^{m_k+p_k} \leqslant (xy)^{n_k} < a^{m_k+p_k+2},$$

which implies

$$\frac{m_k+p_k}{n_k} \leqslant f(xy) \leqslant \frac{m_k+p_k+2}{n_k}.$$

On the other hand, it follows from (4) that

$$\frac{m_k+p_k}{n_k} \leqslant f(x)+f(y) \leqslant \frac{m_k+p_k+2}{n_k},$$

and hence

$$|f(x)+f(y)-f(xy)| \leqslant \frac{2}{n_k}.$$

Since n_k can be made arbitrarily large, this immediately implies Condition a, by Theorem 1.57. The fact that $f(a)=1$ is obvious from the construction of $f(x)$. Moreover, given any $x>1$, there is an integer n such that $x^n \geqslant a$, and hence, by the definition of $f(x)$, such that $f(x) \geqslant 1/n$. In particular, $f(x)>0$ for any $x>1$. But then $f(xy)=f(x)+f(y)>f(x)$ for every $y>1$, so that $f(x)$ is increasing. ∎

5.42. The function $f(x)$ figuring in Theorem 5.41 is called the *logarithm of x to the base a* and is denoted by $\log_a x$. In this notation, Conditions a–c of Theorem 5.41 take the form

$$\log_a (xy) = \log_a x + \log_a y, \tag{5}$$

$$\log_a a = 1, \tag{6}$$

$$\log_a x < \log_a y \text{ if } x < y. \tag{7}$$

In particular, (5) implies

$$\log_a x^n = n \log_a x \qquad (n=1,2,\ldots). \tag{8}$$

Choosing $x=1$, $n=2$ in (8), we get $\log_a 1 = 2 \log_a 1$ and hence

$$\log_a 1 = 0. \tag{9}$$

Moreover, since

$$\log_a x\frac{1}{x} = \log_a 1 = \log_a x + \log_a \frac{1}{x} = 0,$$

we have

$$\log_a \frac{1}{x} = -\log_a x.$$

It follows that

$$\log_a \frac{x}{y} = \log_a x - \log_a y. \tag{10}$$

In particular, (8) holds for any integer n (not necessarily positive).

5.43. THEOREM. *The function* $\log_a x$ $(x > 0)$ *is continuous.*

Proof. Given any $\varepsilon > 0$, choose $n > 1/\varepsilon$ and let $h = \sqrt[n]{a}$. Since $h^n = a$, we have

$$n \log_a h = \log_a h^n = \log_a a = 1,$$

and therefore

$$\log_a h = \frac{1}{n}, \quad \log_a \frac{1}{h} = -\frac{1}{n}.$$

Hence the inequality

$$-\varepsilon < -\frac{1}{n} \leqslant \log_a x \leqslant \frac{1}{n} < \varepsilon$$

holds in the interval $1/h \leqslant x \leqslant h$, thereby proving the continuity of $\log_a x$ at $x = 1$.

Next, given any $y_0 > 0$, let $y = x y_0$, where $1/h < x < h$. Then, since

$$\log_a y = \log_a x + \log_a y_0,$$

we have

$$|\log_a y - \log_a y_0| = |\log_a x| < \varepsilon.$$

Since the interval $(y_0/h, h y_0)$ contains an interval of the form $(y_0 - \delta, y_0 + \delta)$, it follows that $\log_a x$ is continuous at $x = y_0$. ∎

5.44. We now establish the relation between logarithms to different bases a and b. The function

$$f(x) = \frac{\log_a x}{\log_a b}$$

obviously satisfies Conditions a and c of Theorem 5.41, and moreover

$$f(b) = \frac{\log_a b}{\log_a b} = 1.$$

Hence, by the uniqueness of the logarithm, we have $f(x) = \log_b x$, thereby proving the formula

$$\log_a x = \log_a b \cdot \log_b x \tag{11}$$

for arbitrary $a > 1$, $b > 1$, $x > 0$.

5.45. Finally we note that

$$\lim_{x \to \infty} \log_a x = \infty, \quad \lim_{x \searrow 0} \log_a x = -\infty. \tag{12}$$

The first formula follows from Theorem 4.55 and the fact that $\log_a x$ is increasing and unbounded from above (since, for example, $\log_a a^n = n$ for arbitrary n) as $x \to \infty$. The second formula is an immediate consequence of the first, since

$$\lim_{x \searrow 0} \log_a x = \lim_{y \to \infty} \log_a \frac{1}{y} = -\lim_{y \to \infty} \log_a y = -\infty.$$

Using (12), we can complete the definition of $\log_a x$ by setting

$$\log_a 0 = -\infty, \quad \log_a \infty = \infty.$$

Then the function $\log_a x$ (with values in $\bar{R}$) is continuous on the interval $0 \leqslant x \leqslant \infty$.

5.5. The Exponential

5.51. According to Theorem 5.35c, the continuous increasing function $\log_a x$ has an inverse function, which we denote by a^x and call the *exponential* (*to the base a*).† The exponential is defined and positive for all real x (see Sec. 5.45), and clearly satisfies the identities

$$a^{\log_a x} = x, \tag{1}$$

$$\log_a a^x = x. \tag{2}$$

Comparing the formula

$$\log_a \underbrace{(a \cdots a)}_{n \text{ times}} = \underbrace{\log_a a + \cdots + \log_a a}_{n \text{ times}} = n$$

with (2), we see that for $n = 1, 2, \ldots$ the function a^x reduces to

$$a^n = \underbrace{a \cdot a \cdots a}_{n \text{ times}},$$

as previously defined in Sec. 1.46. Choosing $x = 1/n$ in formula (2), we get

$$\log_a a^{1/n} = \frac{1}{n},$$

which implies

$$n \log_a a^{1/n} = \log_a (a^{1/n})^n = 1,$$

so that

$$(a^{1/n})^n = a,$$

† The expression a^x is often called "a raised to the xth power."

i.e., $a^{1/n}$ is the nth root of a as defined in Theorem 1.63.

5.52. The exponential $y = a^x$ is increasing and continuous for all x, again by Theorem 5.35c. Moreover

$$a^{x+y} = a^x a^y, \tag{3}$$

$$(a^x)^y = a^{xy} \tag{4}$$

for arbitrary real x and y. In fact, if $x = \log_a \xi$, $y = \log_a \eta$, then

$$a^{x+y} = a^{\log_a \xi + \log_a \eta} = a^{\log_a \xi \eta} = \xi \eta = a^x a^y,$$

while

$$(a^x)^y = (a^{\log_a \xi})^y = a^{y \log_a \xi} = a^{yx} = a^{xy}.$$

Replacing x by b^x in formula (11), p. 148, we get

$$\log_a b^x = \log_a b \cdot \log_b b^x = x \log_a b. \tag{5}$$

The same formula (11) can be used to extend the definition of the logarithm and exponential to the case where the base b lies in the interval $(0,1)$. In fact, let $b = 1/a$ where $a > 1$, and then set

$$\log_b x = \frac{\log_a x}{\log_a b} = -\log_a x = -\log_{1/b} x,$$

by definition. The resulting function $\log_b x$ decreases as x increases and clearly satisfies the conditions

$$\log_b (xy) = \log_b x + \log_b y,$$
$$\log_b b = 1.$$

Moreover,

$$b^x = \frac{1}{a^x} = a^{-x},$$

and hence

$$b^{\log_b x} = a^{-\log_b x} = a^{\log_a x} = x,$$
$$\log_b b^x = -\log_a a^{-x} = x,$$

so that formulas (1) and (2), as well as their implications (3) and (4), continue to hold for $b \in (0,1)$.

5.53. By interchanging the roles of a and x in the exponential $y = a^x$, we get the closely related function

$$y = x^a \qquad (x > 0)$$

("x raised to an arbitrary real power a"). In terms of exponentials and logarithms, we have the formula

$$y = x^a = (b^{\log_b x})^a = b^{a \log_b x} \qquad (b > 1), \tag{6}$$

which, together with Theorem 5.15a, immediately implies the continuity of $y = x^a$. Replacing x by pq $(p, q > 0)$ in (6), we get

$$(b^{\log_b pq})^a = (b^{\log_b p + \log_b q})^a = b^{a \log_b p + a \log_b q}$$
$$= (b^{a \log_b p})(b^{a \log_b q}) = (b^{\log_b p})^a (b^{\log_b q})^a,$$

and hence

$$(pq)^a = p^a q^a. \tag{7}$$

5.54. THEOREM. *The functions a^x, $\log_a x$, and x^a have the following "limiting properties"*:

$$\lim_{x \to \infty} a^x = \begin{cases} \infty & \text{if } a > 1, \\ 0 & \text{if } a < 1, \end{cases} \tag{8}$$

$$\lim_{x \to -\infty} a^x = \begin{cases} 0 & \text{if } a > 1, \\ \infty & \text{if } a < 1, \end{cases} \tag{9}$$

$$\lim_{x \to \infty} \log_a x = \begin{cases} \infty & \text{if } a > 1, \\ -\infty & \text{if } a < 1, \end{cases} \tag{10}$$

$$\lim_{x \searrow 0} \log_a x = \begin{cases} -\infty & \text{if } a > 1, \\ \infty & \text{if } a < 1, \end{cases} \tag{11}$$

$$\lim_{x \to \infty} x^a = \begin{cases} \infty & \text{if } a > 0, \\ 0 & \text{if } a < 0, \end{cases} \tag{12}$$

$$\lim_{x \searrow 0} x^a = \begin{cases} 0 & \text{if } a > 0, \\ \infty & \text{if } a < 0. \end{cases} \tag{13}$$

Proof. The first formula of the pair (8) follows from the fact that a^x $(a > 1)$ is increasing and unbounded as $x \to \infty$ (the values of a^x fill the domain of definition of $\log_a x$, i.e., the whole half-line $0 < x < \infty$), and the second formula is then obtained from the first by replacing a by $1/a$. To get the formulas (9) we need only replace x by $-x$ in (8). The first of the formulas (10) was proved in Sec. 5.45, and the second is obtained from the first by using the definition of $\log_a x$ for $a < 1$ (Sec. 5.52). Replacing x by $1/x$ in (10) then gives (11). To get (12) and (13), we use (8)–(11) and the fact that $x^a = b^{a \log_b x}$. ∎

Theorem 5.54 helps us draw the graphs of the functions a^x, $\log_a x$, and x^a, as shown in Figures 3–5.

Using (8) and (9), we can complete the definition of a^x by setting

$$a^{-\infty} = 0, \quad a^{+\infty} = \infty$$

if $a > 1$. The function a^x (with values in $\bar{R}$) is then continuous on $\bar{R}$. The same is true if $a < 1$, provided we set

$$a^{-\infty} = \infty, \quad a^{+\infty} = 0.$$

5.55. LEMMA. *Let $a > 1$, $-\infty < r < \infty$. Then*

$$\lim_{n \to \infty} \frac{a^n}{n^r} = \infty. \tag{14}$$

Proof. Let b be a number in the interval $(1,a)$. Using (7), we have

$$\frac{a^{n+1}}{(n+1)^r} \frac{n^r}{a^n} = \frac{a}{\left(1 + \dfrac{1}{n}\right)^r}. \tag{15}$$

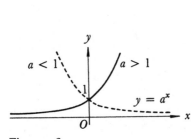

Figure 3

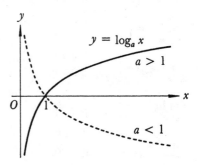

Figure 4

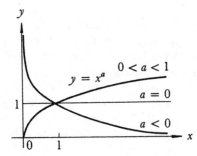

Figure 5

By the continuity of x^r at the point $x=1$, the right-hand side of (15) approaches a as $n \to \infty$, and hence exceeds b for all n starting with some integer N. Therefore

$$\frac{a^{N+1}}{(N+1)^r} \geqslant \frac{a^N}{N^r} b,$$

$$\frac{a^{N+2}}{(N+2)^r} \geqslant \frac{a^{N+1}}{(N+1)^r} b \geqslant \frac{a^N}{N^r} b^2,$$

$$\cdots$$

$$\frac{a^{N+k}}{(N+k)^r} \geqslant \frac{a^N}{N^r} b^k.$$

But

$$\lim_{k \to \infty} b^k = \infty,$$

by (8), which immediately implies (14). ∎

5.56. THEOREM. *Let $a > 1$, $p > 0$, $-\infty < r < \infty$. Then*

$$\lim_{x \to \infty} \frac{a^x}{x^r} = \infty, \tag{16}$$

$$\lim_{x \to \infty} \frac{\log_a x}{x^p} = 0, \tag{17}$$

$$\lim_{x \to 0} x^p \log_a x = 0, \tag{18}$$

$$\lim_{x \to 0} x^{x^p} = 1, \tag{19}$$

$$\lim_{x \to \infty} x^{1/x^p} = 1. \tag{20}$$

Proof. Given any $x > 0$, let n be such that $n \leqslant x < n+1$. Then

$$\frac{a^x}{x^r} \geqslant \frac{a^n}{(n+1)^r} = \frac{1}{a} \frac{a^{n+1}}{(n+1)^r},$$

which gives (16) after using the lemma. To get (17), we change x to $\log_a x$ in (16) and raise both sides to the power $p = 1/r$, assuming that $r > 0$. Replacing x by $1/x$, we get (18). Finally (19) is obtained by taking powers, while (20) is obtained from (19) by replacing x by $1/x$. ∎

According to (16) and (17), as $x \to \infty$ the exponential a^x ($a > 1$) "grows more rapidly" than any power of x while the logarithm $\log_a x$ ($a > 1$) "grows more slowly" than any positive power of x.

5.57. If a function $f(x)$ approaches a limit as $x \to \infty$, the sequence $f(n)$ $(n = 1, 2, \ldots)$ approaches the same limit as $n \to \infty$ (see Sec. 4.65). This observation leads to the following limit formulas involving sequences:

$$\lim_{n \to \infty} a^n = \begin{cases} \infty & \text{if } a > 1, \\ 0 & \text{if } a < 1, \end{cases} \tag{21}$$

$$\lim_{n \to \infty} \log_a n = \begin{cases} \infty & \text{if } a > 1, \\ -\infty & \text{if } a < 1, \end{cases} \tag{22}$$

$$\lim_{n \to \infty} n^a = \begin{cases} \infty & \text{if } a > 0, \\ 0 & \text{if } a < 0, \end{cases} \tag{23}$$

$$\lim_{n \to \infty} \frac{a^n}{n^r} = \infty \qquad (a > 1, \ -\infty < r < \infty) \tag{24}$$

$$\lim_{n \to \infty} \frac{\log_a n}{n^p} = 0 \qquad (a > 1, \ p > 0), \tag{25}$$

$$\lim_{n \to \infty} \sqrt[n]{n} = 1. \tag{26}$$

Formulas (21)–(25) follow from (8), (10), (12), (16), and (17), while (26) follows from (20) with $p = 1$.

5.58. The exponential and the number e

a. First we recall from Sec. 4.63d that

$$\lim_{n \to \infty} \left(1 + \frac{1}{n}\right)^n = e = 2.71828\ldots \tag{27}$$

THEOREM. *The function*

$$f(x) = \left(1 + \frac{1}{x}\right)^x \qquad (x \neq 0)$$

approaches e as $x \to \infty$.

Proof. This is not an immediate consequence of (27), but can be proved by using the special character of $f(x)$.† Given any $x > 1$, let $n = n(x)$ be the integer such that $n \leqslant x < n + 1$. Then $n \to \infty$ as $x \to \infty$, and

$$\left(1 + \frac{1}{x}\right)^x \leqslant \left(1 + \frac{1}{n}\right)^{n+1} = \left(1 + \frac{1}{n}\right)^n \left(1 + \frac{1}{n}\right),$$

$$\left(1 + \frac{1}{x}\right)^x \geqslant \left(1 + \frac{1}{n+1}\right)^n = \left(1 + \frac{1}{n+1}\right)^{n+1} \frac{1}{1 + \dfrac{1}{n+1}} \ .$$

† Cf. Sec. 4.65, especially the concluding remark.

The right-hand sides of both inequalities approach the same limit e as $x \to \infty$. Hence, given any $\varepsilon > 0$, there is an integer N such that both right-hand sides differ from e by less than ε for all $n > N$. But then $f(x)$ differs from e by less than 2ε for all $x \geqslant N+1$, i.e.,

$$\lim_{x \to \infty} \left(1 + \frac{1}{x}\right)^x = e. \quad \blacksquare \tag{28}$$

b. Besides (28), we also have

$$\lim_{x \to -\infty} \left(1 + \frac{1}{x}\right)^x = e. \tag{29}$$

In fact,

$$\left(1 - \frac{1}{y}\right)^{-y} = \left(\frac{y-1}{y}\right)^{-y} = \left(\frac{y}{y-1}\right)^y$$

$$= \left(1 + \frac{1}{y-1}\right)^{y-1}\left(1 + \frac{1}{y-1}\right) \to e \tag{30}$$

as $y \to \infty$, by (28). Replacing y by $-x$, we get (29).† In particular, (30) implies

$$\lim_{n \to \infty} \left(1 - \frac{1}{n}\right)^n = \frac{1}{e},$$

a result which can easily be deduced directly from (27).

c. THEOREM.

$$\lim_{z \to 0} \frac{\log_a(1+z)}{z} = \log_a e. \tag{31}$$

Proof. Noting that

$$\frac{\log_a(1+z)}{z} = \log_a(1+z)^{1/z}$$

and replacing z by $1/x$, we get

$$\lim_{z \to 0} \frac{\log_a(1+z)}{z} = \lim_{x \to \pm\infty} \log_a\left(1 + \frac{1}{x}\right)^x = \log_a e,$$

where in the last step we use the continuity of the logarithm, together with (28) and (29). $\quad \blacksquare$

Formula (31) becomes particularly simple if we choose the number e as the base a. Logarithms to the base e are called *natural logarithms* and are

† Cf. formula (8), p. 105.

denoted by

$\log_e x = \ln x.$

For natural logarithms, (31) reduces to

$$\lim_{z \to 0} \frac{\ln (1+z)}{z} = 1. \tag{32}$$

In terms of the asymptotic unit E (see Sec. 4.39), we can write (32) in the form

$$\ln (1+z) = zE \qquad (E \to 1 \text{ as } z \to 0). \tag{33}$$

d. THEOREM.

$$\lim_{h \to 0} \frac{a^h - 1}{h} = \frac{1}{\log_a e}. \tag{34}$$

Proof. Let

$$a^h - 1 = \frac{1}{x}.$$

Then

$$a^h = 1 + \frac{1}{x}, \quad h = \log_a \left(1 + \frac{1}{x} \right)$$

and

$$\frac{a^h - 1}{h} = \frac{\dfrac{1}{x}}{\log_a \left(1 + \dfrac{1}{x} \right)} = \frac{1}{\log_a \left(1 + \dfrac{1}{x} \right)^x}.$$

Now use the continuity of the logarithm, together with (28) and (29). ∎

Choosing $a = e$ in (34), we get

$$\lim_{h \to 0} \frac{e^h - 1}{h} = 1, \tag{35}$$

or equivalently

$$e^h = 1 + Eh \qquad (E \to 1 \text{ as } h \to 0) \tag{36}$$

in terms of the asymptotic unit E.

e. Although the number e does not appear explicitly in the statement of the next theorem, it would be hard to prove the theorem without making tacit use of limits involving e.

THEOREM.

$$\lim_{z \to 0} \frac{(1+z)^p - 1}{z} = p \qquad (-\infty < p < \infty). \tag{37}$$

Proof. Using (33) and (36), we get in turn

$$\ln (1+z) = zE_1, \quad 1+z = e^{zE_1},$$
$$(1+z)^p = e^{pzE_1} = 1 + pzE_1E_2 = 1 + pzE_1, \tag{38}$$

which is equivalent to (37). ∎

It is clear from (38) that Theorem 4.39, previously proved for $p = 1, 2, \ldots$, is valid for arbitrary real p.

f. Choosing $p = -1$ in (38), we get

$$\frac{1}{1+z} = 1 - zE. \tag{39}$$

More generally, we have the useful formula

$$\frac{1 + \alpha z E_1}{1 + \beta z E_2} = 1 + (\alpha - \beta) zE \qquad (\alpha \neq \beta), \tag{40}$$

where $E_1, E_2, E \to 1$ as $z \to 0$. In fact, by (39),

$$\frac{1 + \alpha z E_1}{1 + \beta z E_2} = (1 + \alpha z E_1)(1 - \beta z E_2 E_3)$$
$$= 1 + \alpha z E_1 - \beta z E_2 E_3 - \alpha \beta z^2 E_1 E_2 E_3$$
$$= 1 + z(\alpha E_1 - \beta E_2 E_3 - \alpha \beta z E_1 E_2 E_3).$$

But this implies (40), since the expression in parentheses on the right obviously approaches $\alpha - \beta$ as $z \to 0$.

5.6. Trigonometric Functions

5.61. In elementary trigonometry the functions $\sin x$ and $\cos x$ are defined geometrically, and it is then shown that they satisfy the formulas

$$\sin^2 x + \cos^2 x = 1, \tag{1}$$

$$\sin (x+y) = \sin x \cos y + \cos x \sin y, \tag{2}$$

$$\cos (x+y) = \cos x \cos y - \sin x \sin y, \tag{3}$$

$$0 < \sin x < x < \frac{\sin x}{\cos x}, \tag{4}$$

where (4) holds for all sufficiently small positive x, say for $0 < x < \varepsilon_0$. This approach will not be used here, since we wish to introduce $\sin x$ and $\cos x$ purely analytically, starting from the properties of the real numbers, without recourse to extraneous geometrical considerations. Instead we *define* $\sin x$ and $\cos x$ as the functions satisfying (1)–(4) for all real x. Thus, strictly speaking, the considerations that follow are provisional, and should read typically "If there exist functions $\sin x$ and $\cos x$ satisfying (1)–(4), then..." However, this circumlocution is hardly necessary, since the existence and uniqueness of such functions $\sin x$ and $\cos x$ will in fact be proved at a later stage of the game (in Secs. 8.54 and 8.66–8.68).

5.62. Setting $x = y = 0$ in (1)–(3), we get

$$\sin^2 0 + \cos^2 0 = 1, \tag{5}$$

$$\sin 0 = 2 \sin 0 \cos 0, \tag{6}$$

$$\cos 0 = \cos^2 0 - \sin^2 0. \tag{7}$$

It follows from (6) that either $\sin 0 = 0$ or $\cos 0 = \frac{1}{2}$. But (5) and (7) imply

$$1 + \cos 0 = 2 \cos^2 0, \tag{8}$$

from which it is apparent that $\cos 0 = \frac{1}{2}$ is impossible. Therefore $\sin 0 = 0$, and (5) then implies $\cos 0 = \pm 1$. Formula (8) shows that of these two values, only $\cos 0 = 1$ is possible. Thus, finally,

$$\sin 0 = 0, \quad \cos 0 = 1. \tag{9}$$

5.63. Next we replace y by $-x$ in (2) and (3), afterwards using (9). This gives

$$0 = \sin x \cos (-x) + \cos x \sin (-x),$$
$$1 = \cos x \cos (-x) - \sin x \sin (-x).$$

Solving this linear system for $\sin (-x)$ and $\cos (-x)$, we get

$$\sin (-x) = -\sin x, \quad \cos (-x) = \cos x. \tag{10}$$

Suppose now that from (2) we subtract the same equation with y replaced by $-y$, and then use (10). The result is

$$\sin (x+y) - \sin (x-y) = 2 \cos x \sin y.$$

Doing the same with (3), we get

$\cos (x+y) - \cos (x-y) = -2 \sin x \sin y.$

Replacing $x+y$ by α and $x-y$ by β, so that

$$x = \frac{\alpha+\beta}{2}, \quad y = \frac{\alpha-\beta}{2},$$

we can write the last two formulas as

$$\sin \alpha - \sin \beta = 2 \cos \frac{\alpha+\beta}{2} \sin \frac{\alpha-\beta}{2}, \tag{11}$$

$$\cos \alpha - \cos \beta = -2 \sin \frac{\alpha+\beta}{2} \sin \frac{\alpha-\beta}{2}. \tag{12}$$

5.64. a. THEOREM. *The functions* $\sin x$ *and* $\cos x$ *are continuous for all* x.

Proof. First we note that (1) implies

$$|\sin x| \leqslant 1, \quad |\cos x| \leqslant 1 \tag{13}$$

for all x. If $x > y$ and if $x-y$ is sufficiently small, then, by (4), (11), and (13),

$$|\sin x - \sin y| = 2 \sin \frac{x-y}{2} \left| \cos \frac{x+y}{2} \right| \leqslant 2 \sin \frac{x-y}{2} \leqslant 2 \frac{x-y}{2} = x-y,$$

which proves the continuity of $\sin x$ (why?). Similarly, by (4), (12), and (13),

$$|\cos x - \cos y| = 2 \left| \sin \frac{x+y}{2} \right| \leqslant 2 \sin \frac{x-y}{2} \leqslant x-y,$$

which proves the continuity of $\cos x$. ∎

b. THEOREM.

$$\lim_{x \to 0} \frac{\sin x}{x} = 1. \tag{14}$$

Proof. By (4),

$$\frac{1}{\cos x} < \frac{\sin x}{x} < 1$$

if $0 < x < \varepsilon_0$, where the left and right-hand sides both approach 1 as $x \searrow 0$. Hence, by Theorem 4.34h,

$$\lim_{x \searrow 0} \frac{\sin x}{x} = 1.$$

But, because of (10),

$$\lim_{x \nearrow 0} \frac{\sin x}{x} = \lim_{(-x) \searrow 0} \frac{\sin (-x)}{-x} = 1.$$

Combining the last two formulas, we get (14). ∎

5.65. a. Next we show that the function $y = \cos x$ vanishes at certain points $x > 0$. Let

$$\beta = \inf_{x > 0} \cos x,$$

and suppose $\beta > 0$. Then, by (11),

$$\sin (m+1)x - \sin mx \geq 2\beta \sin \frac{x}{2} > 0$$

for sufficiently small x and all $m = 1, 2, \ldots$, which implies that $\sin mx$ increases without limit as m increases, contrary to (13). This contradiction shows that $\beta = 0$. There must now be a point x_0 at which $\cos x_0 = 0$. In fact, if $\cos x > 0$ for all $x \geq 0$, then (11) implies that $\sin x$ is increasing and hence, by (1), that $\cos x$ is decreasing. Since $\beta = 0$, it follows that there is a point z_0 such that $x > z_0$ implies

$$\cos x < \frac{5}{13}$$

and hence

$$\sin x = \sqrt{1 - \cos^2 x} > \frac{12}{13}.$$

But then, setting $y = x > z_0$ in (2), we get

$$\frac{12}{13} < \sin 2x = 2 \sin x \cos x < 2 \cdot 1 \cdot \frac{5}{13} = \frac{10}{13},$$

which is impossible. Therefore we cannot have $\cos x > 0$ for all $x \geq 0$, i.e., $\cos x$ vanishes at some point x_0.

b. Now let

$$\frac{\pi}{2} = \inf \{x > 0 : \cos x = 0\}. \tag{15}$$

The number $\pi/180$ is called a *degree*, so that $\pi/2 = 90$ degrees (written $90°$). It follows from (15) and the continuity of $\cos x$ that

$$\cos \frac{\pi}{2} = 0, \quad \sin \frac{\pi}{2} = 1. \tag{16}$$

Moreover, $\sin x$ is increasing on the interval $[0, \pi/2]$, by (11), and hence $\cos x$ is decreasing on $[0, \pi/2]$. Using (2), (3), and (16), we get the formulas

$$\sin\left(x + \frac{\pi}{2}\right) = \sin x \cos\frac{\pi}{2} + \cos x \sin\frac{\pi}{2} = \cos x, \tag{17}$$

$$\cos\left(x + \frac{\pi}{2}\right) = \cos x \cos\frac{\pi}{2} - \sin x \sin\frac{\pi}{2} = -\sin x, \tag{18}$$

which in turn imply

$$\sin(x + \pi) = \cos\left(x + \frac{\pi}{2}\right) = -\sin x, \tag{19}$$

$$\cos(x + \pi) = -\sin\left(x + \frac{\pi}{2}\right) = -\cos x. \tag{20}$$

c. Using (19) and (20), we see at once that

$$\sin(x + 2\pi) = -\sin(x + \pi) = \sin x, \tag{21}$$

$$\cos(x + 2\pi) = -\cos(x + \pi) = \cos x. \tag{22}$$

A function $f(x)$ such that

$$f(x + T) = f(x)$$

for all x and some T is said to be *periodic*, with *period* T. Thus (21) and (22) show that the functions $\sin x$ and $\cos x$ are periodic, with period 2π. The behavior of the signs of these functions in an interval of length 2π (one period) can easily be deduced from (17)–(20), and is shown in the following table:

	$0 < x < \dfrac{\pi}{2}$	$\dfrac{\pi}{2} < x < \pi$	$\pi < x < \dfrac{3\pi}{2}$	$\dfrac{3\pi}{2} < x < 2\pi$
$\sin x$	+	+	−	−
$\cos x$	+	−	−	+

Note that given any x such that $\sin x \neq 0$, $\cos x \neq 0$, the signs of $\sin x$ and $\cos x$ uniquely determine which of the four subintervals (of length $\pi/2$) contains x. The graphs of $\sin x$ and $\cos x$ are shown in Figure 6.† It turns out that $\pi = 3.14159\ldots$ (cf. Chapter 9, Problem 15).

---

† For a justification of the "concavity" exhibited by the graphs of $\sin x$ and $\cos x$, see Example 7.53c.

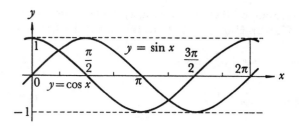

Figure   6

5.66. LEMMA. *If x and u are such that*

$$\sin (x+u) = \sin u,$$
$$\cos (x+u) = \cos u,$$ (23)

then $x = 2k\pi$, *where k is an integer.*

Proof. Solving the system

$$\sin (x+u) = \sin x \cos u + \cos x \sin u = \sin u,$$
$$\cos (x+u) = \cos x \cos u - \sin x \sin u = \cos u$$ (23')

for $\sin x$ and $\cos x$, we get

$$\sin x = 0, \quad \cos x = 1.$$ (24)

By Sec. 5.65, the only solution of (24) in the interval $0 \leqslant x < 2\pi$ is given by $x = 0$. Hence, since $\sin x$ and $\cos x$ are periodic with period 2π, the only solutions of (24) on the whole real line are given by

$$x = 2k\pi \qquad (k = 0, \pm 1, \pm 2, \ldots). \quad \blacksquare$$

5.67. As already noted, the function $\sin x$ is increasing on the interval $[0, \pi/2]$. Hence, because of the identity

$$\sin (-x) = -\sin x,$$

$\sin x$ is also increasing on the larger interval $[-\pi/2, \pi/2]$, taking values from -1 to 1. It follows from Theorem 5.35b that $\sin x$ has an inverse on the interval $[-1,1]$. This inverse function, which we denote by arc $\sin x$, is continuous and increasing on $[-1,1]$, taking values from $-\pi/2$ to $\pi/2$.

By the same token, the function

$$\cos x = \sin \left(x + \frac{\pi}{2} \right)$$ (25)

is increasing on the interval $[-\pi, 0]$, taking values from -1 to 1, and hence has an inverse on the interval $[-1,1]$, again by Theorem 5.35b. This inverse function, which we denote by arc cos x, is continuous and increasing on $[-1,1]$, taking values from $-\pi$ to 0. Letting u denote both sides of (25), we have

$$x = \text{arc cos } u, \quad x + \frac{\pi}{2} = \text{arc sin } u,$$

which together imply the formula

$$\text{arc cos } u = \text{arc sin } u - \frac{\pi}{2}, \tag{26}$$

valid for all $u \in [-1,1]$.

Alternatively, we could have defined another function arc cos u by using the "increasing branch" of the function cos x on the interval $[0,\pi]$. This time the inverse function is continuous and decreasing on $[-1,1]$, taking values from π to 0. Instead of formula (25) relating the functions cos x and sin x on the intervals $[-\pi,0]$ and $[-\pi/2,\pi/2]$, respectively, we now have

$$\cos x = \sin\left(\frac{\pi}{2} - x\right). \tag{25'}$$

Correspondingly, this formula leads to

$$\text{arc cos } u = \frac{\pi}{2} - \text{arc sin } u, \tag{26'}$$

instead of (26). To avoid possible confusion, we can designate the first of these inverse functions of cos x by arc $\cos_i u$ (i for "increasing") and the second by arc $\cos_d u$ (d for "decreasing"). The functions arc sin u, arc $\cos_i u$, and arc $\cos_d u$ have the graphs shown in Figure 7. There are, of course, other functions on $[-1,1]$ which might equally well be regarded as inverses of sin x and cos x, but we confine ourselves to the three functions just described.

5.68. Finally, let

$$\tan x = \frac{\sin x}{\cos x}.$$

The function tan x is defined everywhere except at the points where cos $x = 0$, namely, the points

$$x = \frac{\pi}{2} + k\pi \qquad (k = 0, \pm 1, \pm 2, \ldots).$$

Clearly,

$$\lim_{x \nearrow \pi/2} \tan x = +\infty, \qquad \lim_{x \searrow \pi/2} \tan x = -\infty.$$

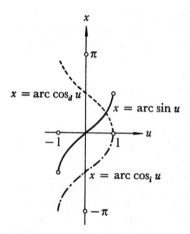

Figure   7

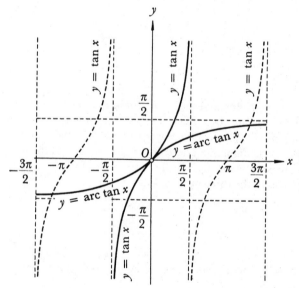

Figure   8

Moreover, (19) and (20) show that the function tan x is periodic, with period π. It follows from the behavior of sin x and cos x on the interval $(-\pi/2,\pi/2)$ that tan x is continuous and increasing on $(-\pi/2,\pi/2)$, taking all real values. By Theorem 5.35c, the function tan x has an inverse on the whole real line $(-\infty,\infty)$. This inverse function, denoted by arc tan x, is continuous and increasing on $(-\infty,\infty)$, taking values from $-\pi/2$ to $\pi/2$. The graphs of the functions tan x and arc tan x are shown in Figure 8.

The functions sin x, cos x, and tan x are called *trigonometric functions*. As an exercise, the reader should study the other trigonometric functions

$$\csc x = \frac{1}{\sin x}, \quad \sec x = \frac{1}{\cos x}, \quad \cot x = \frac{1}{\tan x}$$

and their inverses.

5.7. Applications of Trigonometric Functions

5.71. Polar coordinates in the plane. Let (x,y) be any point in the real plane R_2 such that $x^2+y^2 \neq 0$. Then obviously

$$\frac{x^2}{x^2+y^2} + \frac{y^2}{x^2+y^2} = 1.$$

Suppose we define the number θ in the interval $0 \leqslant \theta < 2\pi$ by the conditions

$$\cos\theta = \frac{x}{\sqrt{x^2+y^2}}, \quad \sin\theta = \frac{y}{\sqrt{x^2+y^2}}, \tag{1}$$

where the square roots are positive, as always. The number θ exists and is unique. For $x>0, y>0$ this follows from the continuity and monotonicity of cos θ and sin θ on the interval $(0,\pi/2)$, together with the intermediate value theorem (Theorem 5.23) and formula (1), p. 157. In the other cases, it follows from formulas (17)–(20), p. 161 and the sign rules of Sec. 5.65c. The number θ is called the *polar angle* of the point (x,y). Note that

$$\theta = \begin{cases} 0 & \text{if } x>0, y=0, \\ \pi/2 = 90° & \text{if } x=0, y>0, \\ \pi = 180° & \text{if } x<0, y=0, \\ 3\pi/2 = 270° & \text{if } x=0, y<0. \end{cases}$$

Making use of the periodicity of the functions sin θ and cos θ, we might also regard θ not only as the unique solution of the system (1) in the interval $[0,2\pi)$, but also as any real number differing from this solution by an integral multiple of 2π.

The number

$$r = \sqrt{x^2 + y^2} \tag{2}$$

is called the *radius vector*† of the point (x, y). In terms of r and θ, we can write the formulas (1) as

$$x = r \cos \theta, \quad y = r \sin \theta. \tag{3}$$

The numbers r and θ are called the *polar coordinates* of the point (x, y). They are defined everywhere except at the point $x = 0, y = 0$, where $r = 0$ but θ becomes meaningless. The curve $r = \text{const}$ is just the circle

$$x^2 + y^2 = r^2,$$

while the curve $\theta = \text{const}$ is the ray

$$\frac{y}{x} = \tan \theta.$$

5.72. Trigonometric form of the complex numbers. Let $z = x + iy \neq 0$ be a complex number, as in Sec. 2.7. Then, using (3), we can represent z in the *trigonometric form*

$$z = r(\cos \theta + i \sin \theta), \tag{4}$$

where θ and r are determined by (1) and (2). In this context, r is called the *absolute value* or *modulus* of z, written $r = |z|$, while θ is called the *argument*‡ of z, written $\theta = \arg z$.

Suppose we use formulas (2) and (3), p. 157 to form the product $z_1 z_2$ of two complex numbers

$$z_1 = r_1(\cos \theta_1 + i \sin \theta_1), \quad z_2 = r_2(\cos \theta_2 + i \sin \theta_2),$$

written in trigonometric form. The result is just

$$\begin{aligned} z_1 z_2 &= [(\cos \theta_1 \cos \theta_2 - \sin \theta_1 \sin \theta_2) + i(\sin \theta_1 \cos \theta_2 + \cos \theta_1 \sin \theta_2)] \\ &= r_1 r_2 [\cos (\theta_1 + \theta_2) + i \sin (\theta_1 + \theta_2)]. \end{aligned} \tag{5}$$

Thus *multiplication of two complex numbers leads to multiplication of their absolute values and addition of their arguments.*

5.73. In particular, (5) implies

$$z^m = r^m(\cos m\theta + i \sin m\theta) \tag{6}$$

for every $z = x + iy \neq 0$. Using (6), we now solve the equation

---

† A better (but nonstandard) term would be "polar distance."
‡ Not to be confused with the concept of the argument of a function (Sec. 2.81).

$$z^m = w, \tag{7}$$

where w is a fixed complex number. For $w = 0$, (7) has the obvious (unique) solution $z = 0$. Hence we assume $w \neq 0$ and write w in the form

$$w = R(\cos \omega + i \sin \omega).$$

Suppose we express z in the form (4), where r and θ are as yet undetermined. It follows from (6) that

$$r^m(\cos m\theta + i \sin m\theta) = R(\cos \omega + i \sin \omega),$$

and hence

$$r^m \cos m\theta = R \cos \omega, \quad r^m \sin m\theta = R \sin \omega. \tag{8}$$

Squaring these equations and adding them, we get the equation

$$r^{2m} = R^2,$$

whose unique positive solution is

$$r = \sqrt[m]{R}.$$

Dividing the equations (8) by $r^m = R$, we now have

$$\cos m\theta = \cos \omega, \quad \sin m\theta = \sin \omega,$$

which take the form

$$\cos(x + u) = \cos u, \quad \sin(x + u) = \sin u$$

if we set

$$u = \omega, \quad x + u = m\theta, \quad x = m\theta - \omega.$$

Hence, by Lemma 5.66,

$$m\theta - \omega = 2k\pi \quad (k = 0, \pm 1, \pm 2, \ldots),$$

i.e.,

$$\theta = \frac{\omega}{m} + \frac{2k\pi}{m} \quad (k = 0, \pm 1, \pm 2, \ldots).$$

Choosing $k = 0, 1, \ldots, m - 1$, we get m distinct solutions of equation (7):

$$z_k = \sqrt[m]{R}\left[\cos\left(\frac{\omega}{m} + \frac{2k\pi}{m}\right) + i \sin\left(\frac{\omega}{m} + \frac{2k\pi}{m}\right)\right]. \tag{9}$$

The remaining values of k lead to values of θ which differ by an integral multiple of 2π from one of the values already found, and hence give no new

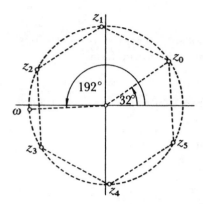

Figure   9

solutions of (7). The numbers (9), called the *complex mth roots of w* (cf. Sec. 1.63) lie at the vertices of a regular *m*-gon centered at the origin, as illustrated schematically by Figure 9 for the case $m=6$, $R=1$, $\omega=192°$.

5.74. Angle between two spatial vectors

a. Definition. Given two vectors

$$x=(x_1,\dots,x_n), \quad y=(y_1,\dots,y_n)$$

in the *n*-dimensional Euclidean space R_n, let

$$(x,y) = \sum_{k=1}^{n} x_k y_k$$

be their scalar product, as in Sec. 3.14a. Then, recalling the Schwarz inequality

$$|(x,y)| \leqslant |x||y|$$

(formula (7), p. 57), we can find a unique number ω in the interval $0 \leqslant \omega \leqslant \pi$ such that

$$\cos \omega = \frac{(x,y)}{|x||y|}, \tag{10}$$

provided of course that $x \neq 0, y \neq 0$. The number ω, denoted by $\langle x,y \rangle$, is called the *angle between the vectors x and y*.

b. Suppose $\omega=0$, so that

$$(x,y)=|x||y|. \tag{11}$$

Then the vectors x and y are linearly dependent (see Sec. 2.64). In fact, if (11) holds, the equation

$$0 = (x - \lambda y, x - \lambda y) = (x, x) - 2\lambda(x, y) + \lambda^2(y, y)$$
$$= |x|^2 - 2\lambda|x||y| + \lambda^2|y|^2 = (|x| - \lambda|y|)^2$$

has the solution

$$\lambda = \frac{|x|}{|y|},$$

and hence $x = \lambda y$ for this value of λ. Similarly, if $\omega = \pi$, i.e., if $(x, y) = -|x||y|$, the vectors x and y are again linearly dependent, this time with constant of proportionality $-\lambda$.

c. If $\omega = \pi/2$, i.e., if $(x, y) = 0$, then the vectors x and y are said to be *orthogonal*. We continue to use the condition $(x, y) = 0$ to define orthogonality even in the case where one or both of the vectors x, y equals 0. Thus *the zero vector is regarded as orthogonal to every vector*.

d. Let $x = (x_1, x_2), y = (y_1, y_2)$ be two vectors in the plane R_2, with polar coordinates r, φ and ρ, ψ, respectively, so that

$$x_1 = r \cos \varphi, \quad x_2 = r \sin \varphi, \quad y_1 = \rho \cos \psi, \quad y_2 = \rho \sin \psi.$$

Then, by definition,

$$\cos \langle x, y \rangle = \frac{(x, y)}{|x||y|} = \frac{r\rho(\cos \varphi \cos \psi + \sin \varphi \sin \psi)}{r\rho} = \cos (\varphi - \psi). \quad (12)$$

Thus the angle $\langle x, y \rangle$ between the vectors x and y is just one of the numbers

$$\varphi - \psi + 2k\pi, \quad \psi - \varphi + 2k\pi \quad (k = 0, \pm 1, \pm 2, \ldots),$$

namely the one lying in the interval $[0, \pi]$.

5.75. Rotations

a. By a *rotation through the angle* θ in the plane R_2 we mean the mapping or "transformation" of R_2 into itself which leaves the origin fixed and carries the vector $x \neq 0$ with polar coordinates r, φ into the vector x' with polar coordinates $r, \varphi + \theta$. To describe this transformation in rectangular coordinates, let $x = (x_1, x_2), x' = (x_1', x_2')$. Then

$$x_1' = r \cos (\varphi + \theta) = r (\cos \varphi \cos \theta - \sin \varphi \sin \theta) = x_1 \cos \theta - x_2 \sin \theta,$$
$$x_2' = r \sin (\varphi + \theta) = r (\sin \varphi \cos \theta + \cos \varphi \sin \theta) = x_1 \sin \theta + x_2 \cos \theta. \quad (13)$$

By its very definition, a rotation does not change lengths of vectors. Moreover, a rotation does not change angles between vectors. In fact, if φ and ψ are the polar angles of two vectors x and y, then the rotation carries x and y into vectors x' and y' with polar angles $\varphi + \theta$ and $\psi + \theta$. But the angle ω between the vectors x and y is determined by the condition

$$\cos \omega = \cos (\varphi - \psi)$$

(cf. (12)), while the angle ω' between the vectors x' and y' is determined by the condition

$$\cos \omega' = \cos [(\varphi + \theta) - (\psi + \theta)] = \cos (\varphi - \psi).$$

Hence $\omega = \omega'$, since ω and ω' both lie in the interval $[0,\pi]$.

b.† More generally, we can define a rotation in the n-dimensional Euclidean space R_n as follows. Let A be an automorphism of R_n, carrying the vector $x = (x_1,\ldots,x_n)$ into the vector $x' = A(x) = (x'_1,\ldots,x'_n)$. Then, as in Sec. 2.67, the effect of A on the components of x is described by the system of "linear relations"

$$x'_1 = a_{11}x_1 + \cdots + a_{1n}x_n,$$
$$\cdots \tag{14}$$
$$x'_n = a_{n1}x_1 + \cdots + a_{nn}x_n,$$

or more concisely,

$$x'_j = \sum_{k=1}^{n} a_{jk}x_k \qquad (j=1,\ldots,n), \tag{14'}$$

in terms of certain real numbers a_{jk} whose meaning is clear from formula (17) below. Suppose the "transformation" A does not change the lengths of vectors or the angles between them, and suppose further that the determinant of the matrix $\|a_{jk}\|$, denoted by det $\|a_{jk}\|$, equals 1.‡ Then A is called a *rotation* or (*proper*) *orthogonal transformation* in R_n.

THEOREM. *The transformation A is a rotation if and only if the coefficients a_{jk} satisfy the condition*§

---

† This section, like Sec. 2.67, presupposes a passing acquaintance with linear algebra.
‡ Note that the matrix of the transformation (13) has determinant

$$\begin{vmatrix} \cos \theta & -\sin \theta \\ \sin \theta & \cos \theta \end{vmatrix} = \cos^2 \theta + \sin^2 \theta = 1.$$

As an exercise, the reader should verify that when $n=2$ the definition of a rotation in R_n reduces to the definition of a rotation in the plane given in Sec. 5.75a.
§ The useful symbol δ_{ij} is called the *Kronecker delta*.

$$\sum_{k=1}^{n} a_{ki}a_{kj} = \delta_{ij} = \begin{cases} 1 \text{ if } i=j, \\ 0 \text{ if } i \neq j \end{cases} \tag{15}$$

(*besides* det $\|a_{jk}\| = 1$).

Proof. Let e_k be the kth vector of the basis

$$e_1 = (1,0,\ldots,0),\ldots, e_n = (0,0,\ldots,1)$$

(cf. Sec. 2.65), with 1 as its kth component and 0 for all its other components, and let $e'_k = A(e_k)$. Then, if A is a rotation, the vectors $e'_1,\ldots,e'_n$ must be orthogonal and normalized (Sec. 3.14a), like the vectors $e_1,\ldots,e_n$ themselves, i.e., we must have

$$(e'_i,e'_j) = \delta_{ij}. \tag{16}$$

But, by (14),

$$e'_k = (a_{1k},\ldots,a_{nk}) \tag{17}$$

(cf. p. 42), so that

$$(e'_i,e'_j) = \sum_{k=1}^{n} a_{ki}a_{kj},$$

which, together with (16), gives (15).

Conversely, suppose (15) holds, and let $x' = A(x), y' = A(y)$. Then, by (14),

$$\begin{aligned}
(x',y') &= \sum_{k=1}^{n} x'_k y'_k = \sum_{k=1}^{n} \left(\sum_{i=1}^{n} a_{ki}x_i\right)\left(\sum_{j=1}^{n} a_{kj}x_j\right) \\
&= \sum_{i=1}^{n}\sum_{j=1}^{n} \left(\sum_{k=1}^{n} a_{ki}a_{kj}\right)x_i y_j = \sum_{i=1}^{n}\sum_{j=1}^{n} \delta_{ij}x_i y_j \\
&= \sum_{i=1}^{n} x_i y_i = (x,y),
\end{aligned}$$

so that A preserves lengths of vectors and angles between them. Hence A is a rotation. ∎

5.76. Polar coordinates in space. Let the basis vector e_k be the same as above, and let α_k be the angle between e_k and the vector (or "point") $x = (x_1,\ldots,x_n) \in R_n$. Then, by (10),

$$\cos \alpha_k = \cos \langle x,e_k \rangle = \frac{(x,e_k)}{|x||e_k|} = \frac{x_k}{|x|}, \tag{18}$$

so that

$$x_k = |x| \cos \alpha_k \qquad (k=1,\ldots,n).$$

The "polar angles" $\alpha_k = \langle x, e_k \rangle$, together with the "radius vector" $|x|$, are called the *polar coordinates* of the vector x. Note that the angles α_k satisfy the condition

$$\sum_{k=1}^{n} \cos^2 \alpha_k = \sum_{k=1}^{n} \frac{x_k^2}{|x|^2} = 1.$$

Suppose two vectors x and y have polar angles $\alpha_1,...,\alpha_n$ and $\beta_1,...,\beta_n$, respectively. Then, clearly,

$$\cos \langle x, y \rangle = \frac{(x, y)}{|x||y|} = \cos \alpha_1 \cos \beta_1 + \cdots + \cos \alpha_n \cos \beta_n.$$

If $\omega = (\omega_1,...,\omega_n)$ is a unit vector, so that $|\omega| = 1$, then (18) gives

$$\omega_k = \cos \langle \omega, e_k \rangle \qquad (k = 1,...,n),$$

i.e., the numbers $\omega_1,...,\omega_n$ are just the cosines of the angles between ω and the corresponding basis vectors. In particular, suppose the transformation (14') is a rotation. Then the coefficients a_{jk}, being the components of the unit vectors e_k', have a simple geometrical interpretation: a_{jk} is the cosine of the angle between the basis vector e_k' and the basis vector e_j.

5.8. Continuous Vector Functions of a Vector Variable

5.81. By a *vector function of a vector variable* we mean a function $f(x)$ with domain $E \subset R_m$ and range R_n (synonymously, a *mapping of* $E \subset R_m$ *into* R_n), where R_m and R_n are the familiar Euclidean spaces (of dimensions m and n, respectively) defined in Sec. 3.14a. Since R_m and R_n are metric spaces when equipped with the usual distance,† the general definition of Sec. 5.11a takes the following form: The function $f(x)$ is said to be *continuous* at the point $x = a \in E$ if

$$\lim_{x \to a} f(x) = f(a),$$

i.e., if given any $\varepsilon > 0$, there exists a $\delta > 0$ such that $|x - a| < \delta$ implies

$$|f(x) - f(a)| < \varepsilon.$$

Here, of course, $|\cdots|$ denotes the length of the vector written between the vertical bars, calculated in R_m or R_n as the case may be.

5.82. THEOREM. *Suppose the vector function* $f(x)$ *is written in the form*

$$f(x) = (f_1(x),...,f_n(x)),$$

† Cf. formula (4), p. 56.

involving n numerical functions

$$f_j(x) = f_j(x_1,\ldots,x_m) \qquad (j=1,\ldots,n).$$

Then $f(x)$ is continuous at $x = a \in E$ if and only if each component function $f_1(x),\ldots,$ $f_n(x)$ is continuous at $x = a$.

Proof. An immediate consequence of Theorem 4.75. ∎

5.83. a. THEOREM. *If the vector functions $f(x)$ and $g(x)$ are continuous at $x = a \in E$, then so is their sum $f(x) + g(x)$.*

Proof. An immediate consequence of Theorem 4.73c. ∎

b. THEOREM. *If the vector function $f(x)$ and the numerical function $\alpha(x)$ are continuous at $x = a \in E$, then so is their product $\alpha(x)f(x)$.*

Proof. An immediate consequence of Theorem 4.73d. ∎

c. THEOREM. *If the vector function $f(x)$ is continuous at $x = a \in E$, then so is the numerical function $|f(x)|$.*

Proof. An immediate consequence of Theorem 5.15b and the subsequent remark. ∎

5.84. Now let $m = n = 2$, and consider a function $w = f(z)$ with domain $E \subset R_2$ and range R_2. In this case we can regard $x = (x,y)$ and $w = (u,v)$ as complex numbers, and $w = f(z)$ can correspondingly be called a *complex function of a complex variable*. According to Sec. 5.81, we now say that $f(z)$ is *continuous* at the point $z = a \in E$ if

$$\lim_{z \to a} f(z) = f(a),$$

i.e., if, given any $\varepsilon > 0$, there exists a $\delta > 0$ such that $|z - a| < \delta$ implies

$$|f(z) - f(a)| < \varepsilon.$$

Moreover, according to Theorem 5.82, the function $f(z) = u(z) + iv(z)$ is continuous at $z = a$ if and only if both real functions $u(z)$ and $v(z)$ are continuous at $z = a$. Invoking the results of Sec. 5.83, we see that if the complex function $f(z)$ is continuous at $z = a$, then so is its absolute value $|f(z)|$, as well as the product $\alpha(z) f(z)$, where $\alpha(z)$ is any real function continuous at $z = a$, while if two complex functions $f(z)$ and $g(z)$ are continuous at $z = a$, then so is their sum $f(z) + g(z)$.

Since products and quotients of complex functions are readily defined (see Sec. 4.72c), we have the following additional

THEOREM. *If two complex functions $f(z)$ and $g(z)$ are continuous at $z = a$, then so are* *their product $f(z)g(z)$ and quotient*

$$\frac{f(z)}{g(z)} \qquad (g(z) \neq 0).$$

Proof. An immediate consequence of Theorem 4.73e. ∎

In particular, since $f(z) = z$ is obviously continuous for all $z = x + iy$, it follows that every polynomial

$$P(z) = a_0 z^n + a_1 z^{n-1} + \cdots + a_n \qquad (a_0 \neq 0, a_1, \ldots, a_n \in C) \tag{1}$$

is continuous on the whole complex field C, while every rational function

$$R(z) = \frac{a_0 z^n + a_1 z^{n-1} + \cdots + a_n}{b_0 z^m + b_1 z^{m-1} + \cdots + b_m} \qquad (a_0 \neq 0, \ldots, a_n, b_0 \neq 0, \ldots, b_m \in C)$$

is continuous at every point $z \in C$ where its denominator is nonvanishing.

5.85. The fundamental theorem of algebra. By a *root* (or *zero*) of the polynomial (1) we mean any number z_0 (real or complex) such that $P(z_0) = 0$. There are polynomials (even polynomials with real coefficients) with no real roots at all, e.g., $P(z) = z^2 + 1$. However, as we now show, every nonconstant polynomial has at least one complex root, a result known as the *fundamental theorem of algebra*.

a. LEMMA. *Let $P(z)$ be the polynomial* (1). *Then, given any $A > 0$, there exists an $R > 0$ such that $|P(z)| \geq A$ for all $|z| \geq R$.*

Proof. Clearly

$$P(z) = a_0 z^n + a_1 z^{n-1} + \cdots + a_n = a_0 z^n \left[1 + \frac{a_1}{a_0 z} + \cdots + \frac{a_n}{a_0 z^n} \right]$$

if $z \neq 0$. But the expression in brackets approaches 1 as $z \to \infty$ (see Sec. 4.73f). Hence there is a number R_1 such that

$$\left| 1 + \frac{a_1}{a_0 z} + \cdots + \frac{a_n}{a_0 z^n} \right| \geq \frac{1}{2}$$

for all $|z| \geq R_1$. It follows that

$$|P(z)| = |a_0||z^n| \left| 1 + \frac{a_1}{a_0 z} + \cdots + \frac{a_n}{a_0 z^n} \right| \geq \frac{1}{2} |a_0||z|^n$$

for all $|z| \geq R_1$. Now let

$$R > \max \left\{ R_1, \sqrt[n]{\frac{2A}{|a_0|}} \right\}.$$

Then

$$|P(z)| \geqslant \tfrac{1}{2}|a_0||z^n| \geqslant \tfrac{1}{2}|a_0|R^n > A$$

for all $|z| \geqslant R$. ∎

b. LEMMA. *Let $P(z)$ be a polynomial (1) of degree $n \geqslant 1$, and suppose $P(z_0) \neq 0$. Then, given any $\delta_0 > 0$, there exists a point z_1 such that $|z_1 - z_0| \leqslant \delta_0$ and $|P(z_1)| < |P(z_0)|$.*

Proof. First consider the case $z_0 = 0$, $P(0) = 1$. Let a_k be the first of the coefficients $a_{n-1}, \ldots, a_0$ which is nonvanishing. Then

$$P(z) = 1 + a_k z^k + \cdots + a_0 z^n = 1 + a_k z^k [1 + H(z)],$$

where

$$H(z) = \frac{a_{k-1}}{a_k} z + \cdots + \frac{a_0}{a_k} z^{n-k}.$$

Since obviously $H(0) = 0$, it follows from the continuity of the polynomial $H(z)$ that

$$|H(z)| < \tfrac{1}{2}$$

for every point z in some disk $|z| \leqslant \delta \leqslant \delta_0$, where δ is small enough to make $\delta^k |a_k| < 1$.

Let z_1 be a solution of the equation

$$z_1^k = -\delta^k \frac{|a_k|}{a_k},$$

so that obviously $|z_1| = \delta$ (the existence of z_1 follows from Sec. 5.73). Then

$$
\begin{aligned}
|P(z_1)| &= |1 + a_k z_1^k + a_k z_1^k H(z_1)| \\
&= |1 - \delta^k |a_k| - \delta^k |a_k| H(z_1)| \leqslant 1 - \delta^k |a_k| + \delta^k |a_k||H(z_1)| \\
&\leqslant 1 - \delta^k |a_k| + \tfrac{1}{2}\delta^k |a_k| = 1 - \tfrac{1}{2}\delta^k |a_k| < 1,
\end{aligned}
$$

as required.

In the general case, where $P(z_0) \neq 0$, we use the formula

$$P(z) = \sum_{k=0}^{n} a_k z^{n-k} = \sum_{k=0}^{n} a_k [(z - z_0) + z_0]^{n-k}$$

to expand $P(z)$ in powers of $z - z_0$, obtaining

$$P(z) = \sum_{k=0}^{n} b_k (z - z_0)^{n-k}, \quad P(z_0) = b_n \neq 0.$$

Replacing $z - z_0$ by a new variable ζ, we then get

$$P(z) = b_n P_0(\zeta),$$

where

$$P_0(\zeta) = \frac{b_0}{b_n} \zeta^n + \frac{b_1}{b_n} \zeta^{n-1} + \cdots + 1.$$

Since

$$|P_0(0)| = 1, \quad |P(z)| = |b_n| |P_0(\zeta)|,$$

this case reduces to the one already considered. ∎

c. We are now in a position to prove our main result:

THEOREM (**Fundamental theorem of algebra**). *Every polynomial* (1) *of degree* $n \geqslant 1$ *has at least one complex root.*

Proof. Let

$$\alpha = \inf_{z \in C} |P(z)| \geqslant 0.$$

By the first lemma, there is an R such that

$$|P(z)| \geqslant \alpha + 1$$

for all $|z| \geqslant R$. Hence the values of $P(z)$ outside the disk $|z| \leqslant R$ play no role in the calculation of α, and we can write

$$\alpha = \inf_{|z| \leqslant R} |(P(z)|.$$

But the disk $K = \{z : |z| \leqslant R\}$ is compact (by Theorem 3.96e), unlike the whole plane C. Hence, by Weierstrass' theorem (Sec. 5.16b), the continuous function $|P(z)|$ achieves its greatest lower bound at some point $z_0 \in K$:

$$|P(z_0)| = \alpha.$$

Suppose $P(z_0) \neq 0$. The point z_0 is an interior point of K, since $|P(z)| \geqslant \alpha + 1$ on the boundary of K, i.e., for $|z| = R$. Therefore some small disk $|z - z_0| \leqslant \delta_0$ lies entirely inside K. By the second lemma, there is a point z_1 in this disk such that $|P(z_1)| < |P(z_0)|$. But then $\alpha = |P(z_0)|$ cannot be the greatest lower bound of the function $|P(z)|$ on the disk K. This contradiction shows that $P(z_0) = 0$ (and hence $\alpha = 0$). ∎

5.86. Factorization of polynomials. Given any two polynomials $P(z)$ and $Q(z) \not\equiv 0$, there are two other polynomials $S(z)$ and $R(z)$ such that

$$P(z) = Q(z)S(z) + R(z), \tag{2}$$

where either $R(z) = 0$ or the degree of $R(z)$ is less than that of $Q(z)$.[†] In this context, we call $P(z)$ the *dividend*, $Q(z)$ the *divisor*, $S(z)$ the *quotient*, and $R(z)$ the *remainder*. The sum of the degrees of the divisor and the quotient equals the degree of the dividend. Here z is simply an "unknown," not as yet assigned any numerical value, and the various polynomials are formal expressions involving z and certain numerical coefficients (subject to the usual rules for addition and multiplication). Substituting some complex number for z changes the identity (2) into an equality between two numbers.

Suppose, in particular, that $Q(z) = z - z_1$. Then $R(z)$ must be of degree zero, i.e., $R(z)$ must reduce to a constant. If z_1 is a root of the polynomial $P(z)$, this constant equals zero, since

$$R(z) \equiv R(z_1) = P(z_1) - S(z_1)(z_1 - z_1) = 0.$$

The representation (2) then takes the form

$$P(z) = S(z)(z - z_1),$$

where $S(z)$ is a polynomial of degree $n - 1$. Applying the same process in turn to the polynomial $S(z)$, we get

$$S(z) = S_1(z) = S_2(z)(z - z_2),$$

where z_2 is a root of the polynomial $S_1(z)$ and hence of the original polynomial $P(z)$, while $S_2(z)$ is a polynomial of degree $n - 2$. Repeating this process, we can eventually lower the degree of the successive quotients to zero, thereby obtaining the factorization

$$P(z) = S_0(z - z_1) \cdots (z - z_n), \tag{3}$$

where S_0 is a constant and the numbers $z_1, \ldots, z_n$ are the roots of the polynomial $P(z)$. These roots may not be distinct. Hence, combining expressions involving the same root, we can write the factorization (3) in the final form

$$P(z) = S_0(z - z_1)^{r_1} \cdots (z - z_k)^{r_k}, \tag{4}$$

where the numbers $z_1, \ldots, z_k$ are now distinct. The exponents $r_1, \ldots, r_k$ are called the *multiplicities* (or *orders*) of the corresponding roots. Comparing the two sides of (4), we see that S_0 is just the coefficient of z^n in the polynomial $P(z)$.

† See e.g., G. Birkhoff and S. MacLane, *A Survey of Modern Algebra* (third edition), The Macmillan Co., N. Y. (1965), p. 65. The polynomials $S(z)$ and $R(z)$ can be found by formal "long division" of $P(z)$ by $Q(z)$.

5.87. Partial fraction expansions of rational functions. Given any two polynomials $P_1(z)$ and $P_2(z)$, we can construct their *greatest common divisor* $D(z)$, namely, the polynomial of highest degree which divides both $P_1(z)$ and $P_2(z)$. It can be shown that $D(z)$ has the representation

$$D(z) = P_1(z)S_1(z) + P_2(z)S_2(z),$$

where $S_1(z)$ and $S_2(z)$ are themselves polynomials.† Obviously, every root of $D(z)$ is a root of both $P_1(z)$ and $P_2(z)$. Therefore, if $P_1(z)$ and $P_2(z)$ have no common roots, their greatest common divisor is just a constant, which can be taken to equal 1.

Now consider the rational function

$$\frac{Q(z)}{P_1(z)P_2(z)},$$

where $P_1(z)$ and $P_2(z)$ have no common roots and $Q(z)$ is another polynomial. Then, as just noted, there are polynomials $S_1(z)$ and $S_2(z)$ such that

$$P_1(z)S_1(z) + P_2(z)S_2(z) \equiv 1,$$

and hence

$$Q(z) = Q(z)P_1(z)S_1(z) + Q(z)P_2(z)S_2(z). \tag{5}$$

Dividing both sides of (5) by $P_1(z)P_2(z)$, we get

$$\frac{Q(z)}{P_1(z)P_2(z)} = \frac{Q(z)S_1(z)}{P_1(z)} + \frac{Q(z)S_2(z)}{P_2(z)}.$$

Clearly, this process can be continued if either $P_1(z)$ or $P_2(z)$ is itself a product of polynomials with no common roots. For example, suppose the polynomial $P(z)$ has the factorization (4). Then the rational function $Q(z)/P(z)$ can be written as a sum

$$\frac{Q(z)}{P(z)} = \frac{Q(z)}{S_0(z-z_1)^{r_1}\cdots(z-z_k)^{r_k}} = \sum_{j=1}^{k} \frac{Q_j(z)}{(z-z_j)^{r_j}}. \tag{6}$$

Expanding each numerator $Q_j(z)$ in powers of $z - z_j$, we can reduce (6) further to the form

$$\frac{Q(z)}{P(z)} = T(z) + \sum_{j=1}^{k} \left\{ \sum_{m=1}^{r_j} \frac{A_{jm}}{(z-z_j)^m} \right\}, \tag{7}$$

where $T(z)$ is a polynomial and the A_{jm} are numbers. Formula (7) is called the *partial fraction expansion* of the rational function $Q(z)/P(z)$. The expansion (7) is unique to within order of terms, a fact which justifies the use of the

† Birkhoff and MacLane, *Survey of Modern Algebra*, p. 71.

"method of undetermined coefficients" to construct (7).†

5.88. Polynomials with real coefficients. *Let*

$$P(z) = a_0 z^n + a_1 z^{n-1} + \cdots + a_n$$

be a polynomial with real coefficients, and suppose $\bar{z}_0$ is the complex conjugate of the number z_0. Then, by Sec. 2.73,

$$\overline{P(z_0)} = \overline{a_0 z^n + a_1 z^{n-1} + \cdots + a_n} = a_0 \bar{z}^n + a_1 \bar{z}^{n-1} + \cdots + \bar{a}_n = P(\bar{z}_0). \tag{8}$$

In particular, let z_0 be a root of $P(z)$, so that $P(z_0) = 0$. Then (8) shows that $P(\bar{z}_0) = 0$, i.e., $\bar{z}_0$ is also a root of $P(z)$.

Now let

$$z_0 = x_0 + iy_0, \qquad \bar{z}_0 = x_0 - iy_0$$

be a pair of complex conjugate roots of $P(z)$. Then

$$(z - z_0)(z - \bar{z}_0) = (x - x_0 - iy_0)(z - x_0 + iy) = (z - x_0)^2 + y_0^2,$$

and dividing $P(z)$ by the quadratic polynomial $(z - x_0)^2 + y_0^2$ with real coefficients, we get

$$P(z) = [(z - x_0)^2 + y_0^2] P_1(z),$$

where $P_1(z)$ is again a polynomial with real coefficients. Thus, combining all pairs of complex conjugate roots in the factorization (4), we get the following representation of $P(z)$ as a product of factors with real coefficients:

$$P(z) = S_0 [(z - x_1)^2 + y_1^2]^{r_1} \cdots [(z - x_p)^2 + y_p^2]^{r_p} (z - x_{p+1})^{r_{p+1}} \cdots (z - x_q)^{r_q}.$$

Here $x_1 \pm iy_1, \ldots, x_p \pm iy_p$ are the nonreal roots and $x_{p+1}, \ldots, x_q$ the real roots of $P(z)$, with corresponding multiplicities $r_1, \ldots, r_p, r_{p+1}, \ldots, r_q$. Note that the complex conjugate roots $x_j + iy_j$ and $x_j - iy_j$ have the same multiplicity r_j.

Similarly, suppose both polynomials $Q(z)$ and $P(z)$ have real coefficients. Then we can write the partial fraction expansion (7) in the form

$$\frac{Q(z)}{P(z)} = T(z) + \sum_{j=1}^{p} \left\{ \sum_{m=1}^{r_j} \frac{A_{jm} + B_{jm}z}{[(z - x_j)^2 + y_j^2]^m} \right\} + \sum_{j=p+1}^{q} \left\{ \sum_{m=1}^{r_j} \frac{A_{jm}}{(z - x_j)^2} \right\}$$

(give the details).

5.9. Sequences of Functions

5.91. Let

$$f_1(x), f_2(x), \ldots, f_n(x), \ldots \tag{1}$$

---

† See e.g., G. M. Fichtenholz, *The Indefinite Integral* (translated by R. A. Silverman), Gordon and Breach, N. Y. (1972), Sec. 8, esp. p. 49ff.

be a sequence of functions, all defined on the same set E and taking values in a metric space M equipped with a distance ρ. Then the sequence (1) is said to be *convergent on E* if every sequence of points

$$f_1(x_0), f_2(x_0), \ldots, f_n(x_0), \ldots \qquad (x_0 \in E)$$

has a limit in M. This limit, which clearly depends on the point x_0, will be denoted by $f(x_0)$. The function $f(x)$, equal to $f(x_0)$ at every point $x_0 \in E$, is called the *limit of the sequence of functions* (1). We then say that (1) *converges* to $f(x)$, a fact expressed by writing

$$f(x) = \lim_{n \to \infty} f_n(x). \tag{2}$$

More exactly, (2) means that given any $\varepsilon > 0$ and any $x_0 \in E$, there exists an integer $N > 0$ such that

$$\rho(f(x_0), f_n(x_0)) < \varepsilon$$

for all $n > N$.

5.92. Examples

a. Given a numerical function $\varphi(x)$, the sequence of functions

$$f_n(x) = \frac{1}{n} \varphi(x)$$

converges to zero on the domain of $\varphi(x)$.

b. The sequence of functions

$$f_n(x) = x^n$$

is convergent on the interval $-1 < x \leqslant 1$. Its limit is the function

$$f(x) = \begin{cases} 0 \text{ if } -1 < x < 1, \\ 1 \text{ if } x = 1. \end{cases}$$

c. The sequence of functions

$$f_n(x) = \begin{cases} n \sin nx \text{ if } 0 \leqslant x \leqslant \pi/n, \\ 0 \qquad \text{ if } \pi/n \leqslant x \leqslant \pi \end{cases}$$

converges to zero at every point $x \in [0, \pi]$. Note however that the maximum deviation of the function $f_n(x)$ from the limit function $f(x) \equiv 0$ equals n, and hence does not approach zero (but rather becomes arbitrarily large) as $n \to \infty$.

5.93. a. Generally speaking, the limit function of a sequence of functions

$f_n(x)$ does not have the properties of the functions $f_n(x)$ themselves. Thus, in Example 5.92b, the functions $f_n(x)$ are continuous on $(-1,1]$, but the limit function is discontinuous on $(-1,1]$. To say something definite about the limit function, it is usually necessary to impose some conditions on the character of the convergence. The following definition is one of the most important conditions of this kind:

b. Definition. A sequence of functions $f_n(x)$ is said to *converge uniformly on E* to the limit function $f(x)$ if, given any $\varepsilon > 0$, there exists an integer $N > 0$ such that

$$\rho(f(x), f_n(x)) < \varepsilon \tag{3}$$

for all $n > N = N(\varepsilon)$ and all $x \in E$. The difference between this definition of uniform convergence and the definition of ordinary convergence given in Sec. 5.91 consists in the fact that here *one and the same number N works for all $x \in E$ simultaneously*, while in ordinary convergence the number N still depends on the choice of the point x. Hence, instead of requiring that (3) hold for all $n > N$ and all $x \in E$, we could just as well have required that

$$\sup_{x \in E} \rho(f(x), f_n(x)) < \varepsilon \tag{4}$$

hold for all $n > N = N(\varepsilon)$. If the metric space M is the m-dimensional Euclidean space R_m, so that the functions $f_n(x)$ are vector-valued, the inequalities (3) and (4) become

$$|f(x) - f_n(x)| < \varepsilon \tag{3'}$$

and

$$\sup_{x \in E} |f(x) - f_n(x)| < \varepsilon, \tag{4'}$$

where $|\cdots|$ denotes the length of the vector (function) written between the vertical bars. Note that the convergence is uniform in Example 5.92a, but not in Examples 5.92b and 5.92c.

The utility of the notion of uniform convergence will be apparent in the theorems that follow:

5.94. THEOREM. *If the sequence $f_n(x)$ converges uniformly on E and if every function $f_n(x)$ is bounded on E, i. e., if*

$$\rho(a, f_n(x)) \leqslant C_n \quad (n = 1, 2, \ldots)$$

*for some fixed point $a \in M$ and all $x \in E$, then the limit function $f(x)$ is also **bounded** on E.*

Proof. Given any $\varepsilon > 0$, say $\varepsilon = 1$, choose N such that

$$\rho(f_n(x), f(x)) < \varepsilon = 1$$

for all $n > N$ and all $x \in E$. Then

$$\rho(a, f(x)) \leqslant \rho(a, f_{N+1}(x)) + \rho(f_{N+1}(x), f(x)) \leqslant C_{N+1} + 1,$$

so that $f(x)$ is bounded on E. ∎

5.95. a. THEOREM. *If the sequence $f_n(x)$ converges uniformly on E, where E is a metric space with distance ρ_0, and if every function $f_n(x)$ is continuous at a point $x = a \in E$, then the limit function $f(x)$ is also continuous at $x = a$.*

Proof. Given any $\varepsilon > 0$, choose N such that

$$\rho(f_n(x), f(x)) < \varepsilon/3 \tag{5}$$

for all $n > N$ and all $x \in E$. Since the function $f_{N+1}(x)$ is continuous at $x = a$, there exists a $\delta > 0$ such that

$$\rho(f_{N+1}(x), f_{N+1}(x_0)) < \varepsilon/3 \tag{6}$$

for all x in the ball $U = \{x : \rho_0(x,a) < \delta\}$. Writing the inequality (5) for the points a and $x \in U$ with $n = N+1$, we get

$$\rho(f_{N+1}(a), f(a)) < \varepsilon/3, \tag{7}$$

$$\rho(f_{N+1}(x), f(x)) < \varepsilon/3. \tag{8}$$

It follows from (6)–(8) that

$$\rho(f(x), f(a)) \leqslant \rho(f(x), f_{N+1}(x)) + \rho(f_{N+1}(x), f_{N+1}(a)) \\ + \rho(f_{N+1}(a), f(a)) < \varepsilon$$

for all $x \in U$, i.e., $f(x)$ is continuous at $x = a$. ∎

b. COROLLARY. *The limit of a uniformly convergent sequence of continuous functions on a metric space E is itself a continuous function on E.*

Proof. An immediate consequence of the preceding theorem and the fact that continuity on E means continuity at every point of E. ∎

5.96. Finally we prove a test for uniform convergence similar to the *Cauchy convergence criterion* of Theorems 3.72b and 4.19:

THEOREM. *A sequence of functions $f_n(x)$ defined on a set E and taking values in a complete metric space M (with distance ρ) is uniformly convergent on E if and only if the following condition is satisfied: Given any $\varepsilon > 0$, there exists an integer $N > 0$ such that*

$$\sup_{x \in E} \rho(f_m(x), f_n(x)) < \varepsilon \tag{9}$$

for all $m,n > N$.

Proof. Suppose $f_n(x)$ converges uniformly on E to $f(x)$. Then, given any $\varepsilon > 0$, there is an N such that

$$\rho(f(x), f_n(x)) < \varepsilon/2$$

for all $n > N$ and all $x \in E$. But then

$$\rho(f_m(x), f_n(x)) \leqslant \rho(f_m(x), f(x)) + \rho(f(x), f_n(x)) < \varepsilon$$

for all $m,n > N$ and all $x \in E$, i.e., (9) holds for all $m,n > N$.

Conversely, suppose (9) holds for all $m,n > N$. Then $f_n(x_0)$ is a fundamental sequence for every fixed $x_0 \in E$, and hence, by the completeness of M, $f_n(x_0)$ converges to a limit $f(x_0)$ in R. Let $f(x)$ be the function equal to $f(x_0)$ for all $x_0 \in E$, i.e., let $f(x)$ be the limit function of the sequence $f_n(x)$. Then, taking the limit of (9) as $m \to \infty$, and using the continuity of the distance ρ (Sec. 5.12b), we get

$$\sup_{x \in E} \rho(f(x), f_n(x)) \leqslant \varepsilon$$

for all $n > N$, so that $f_n(x)$ converges uniformly on E to $f(x)$. ∎

Problems

1. Given a numerical function $f(x)$ defined in a neighborhood of a point x_0, suppose that for every $\delta > 0$ there exists an $\varepsilon > 0$ such that $|x - x_0| < \delta$ implies $|f(x) - f(x_0)| < \varepsilon$. Is $f(x)$ continuous at $x = x_0$?

2. With the same $f(x)$ as in the preceding problem, suppose that for every $\varepsilon > 0$ there exists $\delta > 0$ such that $|f(x) - f(x_0)| < \varepsilon$ implies $|x - x_0| < \delta$. Is $f(x)$ continuous at $x = x_0$?

3. Prove the continuity of the function $y = |x|$.

4. Investigate the continuity of the functions $y = [x]$ and $y = (x)$, where $[x]$ is the integral part and (x) the fractional part of x, as in Sec. 1.71.

5. Let

$$f(x) = \begin{cases} 1/n & \text{if } x = m/n \text{ is a fraction in lowest terms,} \\ 0 & \text{if } x \text{ is irrational.} \end{cases}$$

Prove that $f(x)$ is continuous at irrational points and discontinuous at rational points.

6. Given two continuous numerical functions $f(x)$ and $g(x)$, prove that the functions

$$\max\{f(x), g(x)\}, \quad \min\{f(x), g(x)\}$$

are also continuous.

7. Let $f(x)$ be a continuous numerical function on $[a,b]$, and let $x_1, \ldots, x_n$ be arbitrary points in $[a,b]$. Prove that

$$f(x_0) = \frac{1}{n}[f(x_1) + \cdots + f(x_n)]$$

for some point $x_0 \in [a,b]$.

8. Prove that if a bounded monotonic function $f(x)$ is continuous on a finite or infinite interval (a,b), then $f(x)$ is uniformly continuous on (a,b).

9. Let $f(x)$ be a monotonic function on $(-\infty, \infty)$ satisfying the functional equation

$$f(x+y) = f(x) + f(y).$$

Prove that $f(x)$ is of the form

$$f(x) = ax.$$

10. Find the parts of the extended real line on which the sequence of functions

$$f_n(x) = \frac{1}{1+x^n}$$

is uniformly convergent.

11. Prove that formula (4), p. 158 is not a consequence of formulas (1)–(3), while formula (1) is not a consequence of formulas (2)–(4).

12. Let $f(x)$ be a function defined on $(-\infty, \infty)$ such that given any two points x_1, x_2 with $x_1 < x_2$ and any c between $f(x_1)$ and $f(x_2)$, there is a point $x_c \in (x_1, x_2)$ at which $f(x_c) = c$. Is $f(x)$ continuous?

13. Verify the following table:

x	$\sin x$	$\cos x$	$\tan x$
$\pi/6$	$1/2$	$\sqrt{3}/2$	$\sqrt{3}/3$
$\pi/4$	$\sqrt{2}/2$	$\sqrt{2}/2$	1
$\pi/3$	$\sqrt{3}/2$	$1/2$	$\sqrt{3}$

14. Let $f(x)$ be a numerical function defined on a closed interval $[a,b]$.

Prove that there are at most countably many points $c \in [a,b]$ at which $\lim_{x \to c} f(x)$ exists and is different from $f(c)$.

15. Let $t = 0.t_1 t_2 \ldots t_n \ldots$ be the decimal representation (Sec. 1.77) of the real number t ($0 \leqslant t \leqslant 1$). Let $n_1, n_2, \ldots, n_k, \ldots$ be an increasing subsequence of the sequence of natural numbers $1, 2, \ldots, n, \ldots$ Investigate the continuity of the function

$$x(t) = 0.t_{n_1} t_{n_2} \ldots t_{n_k} \ldots$$

16. Let Γ be the circle of unit radius centered at the origin of the plane R_2, and let M be the set of all points of Γ with polar angles $1, 2, \ldots, n, \ldots$ Prove that M is dense in Γ.

17. Suppose $f(x)$ is continuous on a set P which is dense in a compact metric space M. Prove that $f(x)$ has a continuous extension onto M (i.e., that there exists a function $\varphi(x)$ which is continuous on M and coincides with $f(x)$ on P) if and only if $f(x)$ is uniformly continuous on P.

18. Suppose $f(x)$ is nondecreasing on the interval (a,b). Prove that $f(x)$ can have no more than countably many discontinuity points.

19. Suppose we drop the requirement that $\det \|a_{jk}\| = 1$ in the definition of a rotation in the space R_n (see Sec. 5.75b). Prove that the resulting transformation A is either a rotation or a rotation combined with a reflection, more exactly the result of consecutive application of a rotation and the reflection $x_1' = x_1, \ldots, x_{n-1}' = x_{n-1}, x_n' = -x_n'$.

6 Series

6.1. Numerical Series

6.11. a. Basic definitions. Given a sequence of real numbers

$$a_1, a_2, \ldots, a_n, \ldots, \tag{1}$$

by the *partial sums* of (1) we mean the quantities

$$s_1 = a_1,$$
$$s_2 = a_1 + a_2,$$
$$\ldots$$
$$s_n = a_1 + a_2 + \cdots + a_n,$$
$$\ldots$$

Suppose the sequence of partial sums

$$s_1, s_2, \ldots, s_n, \ldots \tag{2}$$

converges to a finite limit s. Then the expression

$$a_1 + a_2 + \cdots + a_n + \cdots, \tag{3}$$

called a *(numerical) series*,† is said to be *convergent* (or to *converge*), with *sum*

$$s = \lim_{n \to \infty} s_n.$$

If, however, the sequence (2) diverges, then the series (3) is said to be *divergent* (or to *diverge*), with no sum at all. The numbers (1) are called the *terms* of the series (3), and every finite sum

$$a_{m+1} + \cdots + a_n$$

is called a *section* of the series. The series (3) is called *positive* if the numbers (1) are all positive, and similarly with the word "positive" replaced by "negative," "nonpositive," or "nonnegative." The series (3) is also written more concisely as

$$\sum_{n=1}^{\infty} a_n. \tag{3'}$$

b. Example. Consider the *geometric series*

$$1 + x + x^2 + \cdots + x^{n-1} + \cdots, \tag{4}$$

where x is a fixed number. The partial sums of this series are just

---

† The terms "series" and "infinite series" are synonymous, but we will consistently omit the word "infinite."

$$s_n = \sum_{k=0}^{n-1} x^k = \frac{1-x^n}{1-x},$$

by a familiar formula of high school mathematics. If $|x| < 1$, then $x^n \to 0$ as $n \to \infty$, so that

$$s = \lim_{n \to \infty} s_n = \frac{1}{1-x}. \tag{5}$$

Thus, if $|x| < 1$, the series (4) is convergent with sum (5). If $x = 1$, then obviously

$$s_n = 1 + \cdots + 1 = n,$$

so that (4) diverges. For $x = -1$, we have

$$s_1 = 1, \; s_2 = 0, \; s_3 = 1, \; s_4 = 0,\ldots,$$

so that the sequence $s_1, s_2, \ldots$, although bounded, has no limit, and the series (4) is again divergent. Finally, if $|x| > 1$, the quantity $|s_n|$ increases without limit as $n \to \infty$, so that (4) diverges once again.

Summarizing these results, we see that *the geometric series (4) converges to the sum (5) if $|x| < 1$ and diverges if $|x| \geqslant 1$.*

c. Applying the Cauchy convergence criterion (Theorem 3.72b) to the sequence of partial sums (2), and bearing in mind that

$$s_n - s_m = a_{m+1} + \cdots + a_n$$

if $n > m$, we get the following **Cauchy convergence criterion for series:** *The series (3) converges if and only if, given any $\varepsilon > 0$, there exists an integer $N > 0$ such that*

$$|a_{m+1} + \cdots + a_n| < \varepsilon$$

for all $n > m \geqslant N$.

d. In particular, if the series (3) converges, then, given any $\varepsilon > 0$, there exist an N such that

$$|a_n| < \varepsilon$$

for all $n > N$, i.e.,

$$\lim_{n \to \infty} a_n = 0.$$

Thus the sequence of terms (1) of a convergent series (3) converges to zero.

Therefore $a_n \to 0$ is a *necessary condition* for convergence of (3). However, there are also divergent series such that $a_n \to 0$ (see Sec. 6.15b), and hence $a_n \to 0$ is not a sufficient condition for convergence of a series.

e. Another necessary condition for convergence of a series follows from Theorem 3.33b, which asserts that every convergent sequence is *bounded*. Therefore *the sequence of partial sums* (2) *of a convergent series* (3) *must be bounded*. This necessary condition for convergence of a series also fails to be sufficient. For example, the partial sums of the series $1 - 1 + 1 - 1 + \cdots$ are bounded, but the series is obviously not convergent.

6.12. In Secs. 6.13–6.17 we will study nonnegative series, i.e., series all of whose terms are nonnegative. The partial sums $s_1, s_2, \ldots$ of such a series form a nondecreasing sequence. It follows from Sec. 4.63c that *if the partial sums* $s_1, s_2, \ldots$ *of a nonnegative series are bounded* (as $n \to \infty$), *then the series converges.*

6.13. a. THEOREM (**Comparison test**). *Let*

$$a_1 + a_2 + \cdots + a_n + \cdots$$

be a convergent nonnegative series, and let

$$b_1 + b_2 + \cdots + b_n + \cdots \tag{6}$$

be any nonnegative series such that

$$b_n \leqslant ca_n$$

for some constant c and all n exceeding some integer N. Then (6) *is also convergent.*

Proof. If

$$s = a_1 + a_2 + \cdots, \qquad \sigma_n = b_1 + \cdots + b_n,$$

then

$$\sigma_n = b_1 + \cdots + b_n \leqslant b_1 + \cdots + b_N + c(a_{N+1} + \cdots + a_n)$$
$$\leqslant b_1 + \cdots + b_N + cs$$

for all $n > N$. Now apply Sec. 6.12. ∎

b. Example. If the terms of a nonnegative series (6) satisfy the condition

$$b_n \leqslant c\theta^n \qquad (c > 0, \ 0 \leqslant \theta < 1, \ n > N),$$

then (6) converges. To see this, set $a_n = \theta^n$ in the comparison test and then use Example 6.11b.

6.14. The following two tests for convergence are consequences of the comparison test:

a. THEOREM (**D'Alembert's test**).† *The nonnegative series*

$$a_1 + a_2 + \cdots + a_n + \cdots \tag{7}$$

converges if

$$\varlimsup_{n \to \infty} \frac{a_{n+1}}{a_n} < 1 \tag{8}$$

and diverges if

$$\varliminf_{n \to \infty} \frac{a_{n+1}}{a_n} > 1. \tag{9}$$

Proof. If (8) holds, then

$$\frac{a_{n+1}}{a_n} \leqslant \theta$$

for some $\theta < 1$ and all n starting from some integer N. Therefore

$$a_{N+1} \leqslant \theta a_N, \; a_{N+2} \leqslant \theta a_{N+1} \leqslant \theta^2 a_N, \ldots, a_{N+k} \leqslant \theta^k a_N, \ldots,$$

and the convergence of (7) follows from Example 6.13b. On the other hand, if (9) holds, then

$$\frac{a_{n+1}}{a_n} > 1$$

for all n starting from some integer N. Therefore

$$a_{N+1} > a_N, \; a_{N+2} > a_{N+1}, \ldots,$$

so that the terms of (7) do not approach zero. But then (7) diverges, by Sec. 6.11d. ∎

b. Example. The series

$$1 + \frac{1 \cdot 2}{1 \cdot 3} + \frac{1 \cdot 2 \cdot 3}{1 \cdot 3 \cdot 5} + \frac{1 \cdot 2 \cdot 3 \cdot 4}{1 \cdot 3 \cdot 5 \cdot 7} + \cdots$$

converges, since

$$\frac{a_{n+1}}{a_n} = \frac{n+1}{2n+1} \to \frac{1}{2}.$$

c. THEOREM (**Cauchy's test**).‡ *The nonnegative series* (7) *converges if*

$$\varlimsup_{n \to \infty} \sqrt[n]{a_n} < 1 \tag{10}$$

---

† Also known as the *ratio test*.
‡ Also known as the *root test*.

and diverges if

$$\varlimsup_{n \to \infty} \sqrt[n]{a_n} > 1.\tag{11}$$

Proof. If (10) holds, then

$$\sqrt[n]{a_n} < \theta$$

for some $\theta < 1$ and all n starting from some integer N, so that

$$a_n < \theta^n \quad (n \geqslant N),$$

and the convergence of (7) follows from Example 6.13b. On the other hand, if (11) holds, then

$$\sqrt[n]{a_n} > 1$$

for all n starting from some integer N, i.e.,

$$a_n > 1 \quad (n \geqslant N),$$

so that the terms of (7) do not approach zero. But then (7) diverges, by Sec. 6.11d. ∎

d. Example. The series

$$\frac{2}{1} + \left(\frac{3}{3}\right)^2 + \left(\frac{4}{5}\right)^3 + \left(\frac{5}{7}\right)^4 + \cdots$$

converges, since

$$\sqrt[n]{a_n} = \frac{n+1}{2n-1} \to \frac{1}{2}.$$

6.15. Next we prove another useful convergence test due to Cauchy:

THEOREM. *If $a_1 \geqslant a_2 \geqslant \cdots \geqslant a_n \geqslant \cdots \geqslant 0$, then the series*

$$\sum_{n=1}^{\infty} a_n \tag{12}$$

converges if and only if the series

$$\sum_{k=1}^{\infty} 2^k a_{2^k} \tag{13}$$

converges.†

Proof. Consider the partial sum $a_1 + \cdots + a_n$, and choose k such that $2^k > n$.

† Equivalently, either both series (12) and (13) converge or both diverge.

Then

$$a_1 + \cdots + a_n \leqslant a_1 + \cdots + a_{2^k - 1}$$
$$= a_1 + (a_2 + a_3) + \cdots + (a_{2^{k-1}} + \cdots + a_{2^k - 1})$$
$$\leqslant a_1 + 2a_2 + \cdots + 2^{k-1} a_{2^{k-1}} .$$

If the series (13) converges, then the expression on the right cannot exceed the sum of (13). But then the partial sums $a_1 + \cdots + a_n$ are bounded, so that (12) converges.

Conversely, choosing k such that $2^k < n$, we have

$$a_1 + \cdots + a_n \geqslant a_1 + \cdots + a_{2^k}$$
$$= a_1 + a_2 + (a_3 + a_4) + \cdots + (a_{2^{k-1}+1} + \cdots + a_{2^k})$$
$$\geqslant \tfrac{1}{2} a_1 + a_2 + 2a_4 + \cdots + 2^{k-1} a_{2^k}$$
$$= \tfrac{1}{2}(a_1 + 2a_2 + 4a_4 + \cdots + 2^k a_{2^k}).$$

Hence if (13) diverges, so does (12). ∎

Examples

a. *The series*

$$\sum_{n=1}^{\infty} \frac{1}{n^p} \tag{14}$$

converges if $p > 1$ and diverges if $0 \leqslant p \leqslant 1$. In fact, the terms of the series are nonincreasing if $p \geqslant 0$, so that Theorem 6.15 is applicable. Here the appropriate series (13) is

$$\sum_{k=1}^{\infty} \frac{2^k}{2^{kp}} = \sum_{k=1}^{\infty} 2^{k(1-p)} = \sum_{k=1}^{\infty} \alpha^k,$$

where $\alpha = 2^{1-p}$, and the result follows from Example 6.13b together with the fact that $2^{1-p} < 1$ if $p > 1$ while $2^{1-p} \geqslant 1$ if $p \leqslant 1$. Note that

$$\frac{1}{(n+1)^p} : \frac{1}{n^p} = \frac{1}{\left(\dfrac{n+1}{n}\right)^p} \to 1$$

(as $n \to \infty$), so that D'Alembert's test is inapplicable in this case.

b. In particular, *the harmonic series*

$$\sum_{n=1}^{\infty} \frac{1}{n}$$

diverges. The designation "harmonic" stems from the fact that every term of the series starting from the second is the *harmonic mean*† of the two neighboring terms.

c. *The series*

$$\sum_{n=2}^{\infty} \frac{1}{n(\log_a n)^p} \qquad (a>1) \tag{15}$$

converges if $p>1$ *and diverges if* $0 \leqslant p \leqslant 1$. In fact, since the logarithm is increasing, the terms of (15) are nonincreasing, so that Theorem 6.15 is applicable. The appropriate series (13) is now

$$\sum_{k=1}^{\infty} \frac{2^k}{2^k(\log_a 2^k)^p} = \sum_{k=1}^{\infty} \frac{1}{(k \log_a 2)^p} = \frac{1}{(\log_a 2)^p} \sum_{k=1}^{\infty} \frac{1}{k^p},$$

and the convergence behavior of (15) follows at once from that of (14).

6.16. Using the series (14) and (15) in the comparison test (Theorem 6.13a), we can now establish the convergence or divergence of many kinds of series for which D'Alembert's test and Cauchy's test fail to work. We begin by showing that the comparison test continues to work if we compare *ratios* of terms rather than the terms themselves.

LEMMA. *Given two nonnegative series*

$$\sum_{n=1}^{\infty} u_n \tag{16}$$

and

$$\sum_{n=1}^{\infty} v_n, \tag{17}$$

suppose the inequality

$$\frac{u_{n+1}}{u_n} \leqslant \frac{v_{n+1}}{v_n} \tag{18}$$

holds for all sufficiently large n. Then if (17) *converges, so does* (16).

Proof. Suppose (18) holds for $n = N, N+1, \ldots$ Then, multiplying the left- and right-hand sides of (18) written for $n = N, N+1, \ldots, N+p$, we get

$$\frac{u_{N+p}}{u_N} \leqslant \frac{v_{N+p}}{v_N},$$

† A number a is said to be the *harmonic mean* of two numbers b and c if
$\frac{1}{a} = \frac{1}{2}\left(\frac{1}{b} + \frac{1}{c}\right)$.

or

$$u_{N+p} \leqslant \frac{u_N}{v_N} v_{N+p}. \tag{19}$$

Since (19) holds for all $p = 1, 2, \ldots$, the convergence of (16) follows from that of (17), by the ordinary form of the comparison test (Theorem 6.13a). ∎

6.17. a. THEOREM (**Raabe's test**). *Let* (16) *be a nonnegative series such that*

$$\frac{u_{n+1}}{u_n} = 1 - \frac{\delta E}{n},$$

where $E \to 1$ *as* $n \to \infty$ *(cf. Sec. 4.39). Then* (16) *converges if* $\delta > 1$ *and diverges if* $\delta < 1$.

Proof. Choosing

$$v_n = \frac{1}{n^p},$$

we have

$$\frac{v_{n+1}}{v_n} = \frac{n^p}{(n+1)^p} = \frac{1}{\left(1 + \frac{1}{n}\right)^p} = \frac{1}{1 + \frac{pE}{n}} = 1 - \frac{pE}{n},$$

by Secs. 5.59e and 5.59f. If $\delta > 1$, let $p \in (1, \delta)$. Then

$$\frac{u_{n+1}}{u_n} = 1 - \frac{\delta E}{n} < 1 - \frac{pE}{n} = \frac{v_{n+1}}{v_n}.$$

By Example 6.15a, the series (17) converges since $p > 1$, and hence so does (16), by the lemma. On the other hand, if $\delta < 1$, let $p \in (\delta, 1)$. Then

$$\frac{v_{n+1}}{v_n} = 1 - \frac{pE}{n} < 1 - \frac{\delta E}{n} = \frac{u_{n+1}}{u_n},$$

and if (16) converged, then so would (17), again by the lemma. But (17) diverges since $p < 1$, and hence so does (16). ∎

b. Example. Consider the series

$$\sum_{n=1}^{\infty} \frac{\alpha(\alpha+1)\cdots(\alpha+n-1)}{\beta(\beta+1)\cdots(\beta+n-1)} \qquad (\alpha, \beta \neq 0, -1, -2, \ldots). \tag{20}$$

In this case,

$$\frac{u_{n+1}}{u_n} = \frac{\alpha+n}{\beta+n} = \frac{1 + (\alpha/n)}{1 + (\beta/n)} = 1 - \frac{\beta-\alpha}{n} E,$$

and hence, by Raabe's test, (20) converges if $\beta - \alpha > 1$ and diverges if $\beta - \alpha < 1$.

6.2. Absolute and Conditional Convergence

6.21. We now turn to more general series, with terms that are not necessarily all of the same sign. Thus let

$$a_1 + a_2 + \cdots + a_n + \cdots \tag{1}$$

be a series whose terms are arbitrary real numbers, and consider the related nonnegative series

$$|a_1| + |a_2| + \cdots + |a_n| + \cdots. \tag{2}$$

THEOREM. *If the series* (2) *converges, then so does the series* (1).

Proof. Suppose (2) converges. Then, given any $\varepsilon > 0$, there is an N such that

$$|a_{m+1}| + \cdots + |a_n| < \varepsilon$$

for all $n > m \geqslant N$, by the Cauchy convergence criterion (Sec. 6.11c) applied to (2). But then

$$|a_{m+1} + \cdots + a_n| \leqslant |a_{m+1}| + \cdots + |a_n| < \varepsilon$$

for all $n > m \geqslant N$, and hence (1) converges, by the Cauchy convergence criterion applied to (1). ∎

6.22. Definition. A series (1) is said to be *absolutely convergent* if the series (2) converges. It may turn out that the series (1) converges while the series (2) diverges (an example is given in Sec. 6.24). We then say that the series (1) is *conditionally convergent*.

6.23. THEOREM (**Leibniz's test**). *If*

$$a_1 \geqslant a_2 \geqslant \cdots \geqslant a_n \geqslant \cdots$$

and $a_n \to 0$, *then the "alternating series"*

$$a_1 - a_2 + a_3 - a_4 + \cdots$$

converges.

Proof. Clearly,

$$s_{2n+1} = s_{2n-1} - a_{2n} + a_{2n+1} \leqslant s_{2n-1},$$
$$s_{2n+2} = s_{2n} + a_{2n+1} - a_{2n-2} \geqslant s_{2n},$$

so that the sequence $s_2, s_4, \ldots$ is nondecreasing, while the sequence $s_1, s_3, \ldots$ is nonincreasing. Moreover,

$$s_2 \leqslant s_4 \leqslant \cdots \leqslant s_{2n} \leqslant s_{2n} + a_{2n+1} = s_{2n+1} \leqslant \cdots \leqslant s_{2k+1}$$

for any k and $n \geqslant k$, so that the sequence $s_2, s_4, \ldots$ is bounded from above by any number s_{2k+1}. Therefore

$$\xi = \lim_{n \to \infty} s_{2n} \leqslant s_{2k+1}$$

for any k. But then $\lim_{n \to \infty} s_{2n}$ is a lower bound for all the s_{2k+1}. Therefore $\lim_{n \to \infty} s_{2n+1}$ exists and

$$\xi = \lim_{n \to \infty} s_{2n} \leqslant \lim_{n \to \infty} s_{2n+1} = \eta.$$

Finally

$$0 \leqslant \eta - \xi \leqslant s_{2n+1} - s_{2n} = a_{2n+1},$$

and hence, since $a_{2n+1} \to 0$,

$$\xi = \eta = \lim_{n \to \infty} s_n$$

(cf. Sec. 4.64).   ∎

6.24. Examples

a. It follows from Leibniz's test that the series

$$1 - \frac{1}{2^\alpha} + \frac{1}{3^\alpha} - \frac{1}{4^\alpha} + \cdots$$

converges for $\alpha > 0$. For $\alpha > 1$ the series is absolutely convergent (by Example 6.15a). For $0 < \alpha \leqslant 1$ it is only conditionally convergent, since the corresponding series of absolute values

$$1 + \frac{1}{2^\alpha} + \frac{1}{3^\alpha} + \frac{1}{4^\alpha} + \cdots$$

diverges (again by Example 6.15a).

b. In particular, the alternating series

$$1 - \frac{1}{2} + \frac{1}{3} - \frac{1}{4} + \cdots$$

converges. The sum of this series turns out to be ln 2 (see Example 9.104a).

c. Consider the series

$$\sum_{n=1}^{\infty} (-1)^{n-1} u_n,$$

where

$$u_n = \frac{\alpha(\alpha+1)\cdots(\alpha+n-1)}{\beta(\beta+1)\cdots(\beta+n-1)} \qquad (\alpha,\beta \neq 0, -1, -2,\ldots).$$

This series is absolutely convergent for $\beta > \alpha + 1$ (by Example 6.17b). As we now show, the series is (conditionally) convergent for $\beta > \alpha$ as well. In fact, since

$$\frac{u_{n+1}}{u_n} = 1 - \frac{\beta-\alpha}{n} E$$

(as already noted), $\beta > \alpha$ implies $u_{n+1} < u_n$, at least for sufficiently large n. Leibniz's test is now applicable, provided that $u_n \to 0$. But, by formula (33), p. 156,

$$\ln \frac{u_{n+1}}{u_n} = \ln \left(1 - \frac{\beta-\alpha}{n} E\right) = -\frac{\beta-\alpha}{n} E' \qquad (E' \to 1 \text{ as } n \to \infty),$$

so that

$$\ln \frac{u_{n+1}}{u_1} = \ln \frac{u_2}{u_1} + \cdots + \ln \frac{u_{n+1}}{u_n} = -(\beta-\alpha) \sum_{k=1}^{n} \frac{E'}{k}.$$

Since the harmonic series diverges (Example 6.15b), it follows that $\ln u_{n+1} \to -\infty$ and hence $u_{n+1} \to 0$, as required.

6.3. Operations on Series

6.31. By the *sum* of two numerical series

$$a_1 + a_2 + \cdots + a_n + \cdots, \tag{1}$$

$$b_1 + b_2 + \cdots + b_n + \cdots, \tag{2}$$

we mean the series

$$(a_1 + b_1) + (a_2 + b_2) + \cdots + (a_n + b_n) + \cdots. \tag{3}$$

Similarly, by the *product of the series* (1) *with a number* α, we mean the series

$$\alpha a_1 + \alpha a_2 + \cdots + \alpha a_n + \cdots. \tag{4}$$

These formal definitions, which say nothing about the convergence of the series in question, are the only natural ones, as shown by the following

THEOREM. *If the series* (1) *and* (2) *are convergent, with sums A and B, then the series* (3) *and* (4) *are also convergent, with sums $A + B$ and αA.*

Proof. Let

$$A_n = \sum_{k=1}^{n} a_k, \qquad B_n = \sum_{k=1}^{n} b_k.$$

Then the series (3) and (4) have partial sums

$$\sum_{k=1}^{n} (a_k + b_k) = A_n + B_n, \qquad \sum_{k=1}^{n} \alpha a_k = \alpha A_n.$$

But $A_n \to A$, $B_n \to B$, by hypothesis, and hence $A_n + B_n \to A + B$, $\alpha A_n \to \alpha A$. ∎

6.32. By the *product of two series* (1) *and* (2) we mean the series

$$a_1 b_1 + (a_1 b_2 + a_2 b_1) + (a_1 b_3 + a_2 b_2 + a_3 b_1) + \cdots, \tag{5}$$

where again nothing is said about convergence. The suitability of the definition (5) is shown by the following

THEOREM. *If the series* (1) *and* (2) *are convergent, with sums A and B, and if at least one of the series* (1) *and* (2) *is absolutely convergent, then the series* (5) *is convergent, with sum AB.*

Proof. Let

$$A_n = \sum_{k=1}^{n} a_k, \qquad B_n = \sum_{k=1}^{n} b_k,$$

$$c_n = \sum_{k=1}^{n} a_k b_{n-k}, \qquad C_n = \sum_{k=1}^{n} c_k.$$

Suppose the series (2) is absolutely convergent, so that

$$D = \sum_{k=1}^{\infty} |b_k| < \infty.$$

The numbers A_n are bounded, since $A_1, A_2, \ldots$ is a convergent sequence. Let

$$M = \sup_{n \geqslant 1} |A_n|.$$

The numbers

$$|A_n - A_m| \leqslant |A_n| + |A_m| \qquad (m, n = 1, 2, \cdots)$$

are bounded by $2M$. Given any $\varepsilon > 0$, let N be such that

$$\left| \sum_{k=m}^{n} a_k \right| < \frac{\varepsilon}{2D}, \qquad \sum_{k=N+1}^{\infty} |b_k| < \frac{\varepsilon}{4M}$$

for all $n > m \geqslant N$. Then

$$
\begin{aligned}
|A_n B_n - C_n| &= |(a_2 + \cdots + a_n)b_n + (a_3 + \cdots + a_n)b_{n-1} + \cdots + a_n b_2| \\
&\leqslant |a_2 + \cdots + a_n||b_n| + |a_3 + \cdots + a_n||b_{n-1}| + \cdots + |a_n||b_2| \\
&< 2M(|b_{N+1}| + \cdots + |b_n|) + \frac{\varepsilon}{2D}(|b_2| + \cdots + |b_N|) < \frac{\varepsilon}{2} + \frac{\varepsilon}{2} = \varepsilon
\end{aligned}
$$

for all $n > N$. It follows that

$$
\lim_{n \to \infty} C_n = \lim_{n \to \infty} A_n B_n = AB. \quad \blacksquare
$$

The theorem is no longer true if neither series (1) or (2) is absolutely convergent (see Problem 5).

6.33. We now consider series obtained by "grouping together" terms of a given series.

THEOREM. *Given a series*

$$
a_1 + a_2 + \cdots + a_n + \cdots, \tag{6}
$$

let $m_1, m_2, \ldots$ be any increasing sequence of positive integers ($m_1 < m_2 < \cdots$), and let

$$
\begin{aligned}
\alpha_1 &= a_1 + \cdots + a_{m_1}, \\
\alpha_2 &= a_{m_1+1} + \cdots + a_{m_2}, \\
&\cdots \\
\alpha_n &= a_{m_{n-1}+1} + \cdots + a_{m_n}, \\
&\cdots
\end{aligned}
$$

Suppose the series (6) is convergent, with sum A. Then the series

$$
\alpha_1 + \alpha_2 + \cdots + \alpha_n + \cdots \tag{6'}
$$

is also convergent, with the same sum A.

Proof. We need only note that

$$
\alpha_1 + \alpha_2 + \cdots + \alpha_n = a_1 + \cdots + a_{m_n},
$$

where the right-hand side approaches A as $n \to \infty$. $\quad \blacksquare$

6.34. In general, convergence of the "grouped series" (6') does not imply convergence of the original series (6). For example, the series

$$
1 - 1 + 1 - 1 + \cdots
$$

is obviously divergent, but the grouped series

$$
(1-1) + (1-1) + \cdots
$$

is convergent (with sum 0). However, under certain conditions, convergence of (6') does imply that of (6), as shown by the next two theorems:

a. THEOREM. *If the series* (6) *is nonnegative and if the series* (6') *is convergent, then the series* (6) *is also convergent.*

Proof. In this case,

$$a_1 + \cdots + a_n \leqslant a_1 + \cdots + a_n + \cdots + a_{m_n}$$
$$= \alpha_1 + \cdots + \alpha_n \leqslant \sum_{k=1}^{\infty} \alpha_k,$$

so that the partial sums of (6) are bounded. Now use the result of Sec. 6.12. ∎

b. THEOREM. *If $a_n \to 0$ and if the number of elements in each group is bounded, i.e., if $m_n - m_{n-1} \leqslant M$ for some constant $M > 0$ and all n, then convergence of the series* (6') *implies that of the series* (6).

Proof. Given any n, let m_k and m_{k+1} be such that

$$m_k < n \leqslant m_{k+1},$$

and let

$$s_n = a_1 + \cdots + a_n, \qquad \sigma_k = \alpha_1 + \cdots + \alpha_k.$$

Then clearly

$$|s_n - \sigma_k| = |a_{m_k+1} + \cdots + a_n| \leqslant M \max_{j > m_k} |a_j|, \tag{7}$$

where the right-hand side of (7) approaches zero as $k \to \infty$, i.e., the sequences of partial sums σ_k and s_n have the same limit. Therefore if (6') converges, so does (6). ∎

6.35. Next we consider series obtained by rearranging the terms of a given series.

THEOREM. *Given a series*

$$a_1 + a_2 + \cdots + a_n + \cdots, \tag{8}$$

let $m_1, m_2, \ldots$ be any rearrangement of the sequence of natural numbers $1, 2, \ldots$, and let

$$b_n = a_{m_n} \qquad (n = 1, 2, \ldots).$$

Suppose the series (8) *is nonnegative and convergent, with sum A. Then the series*

$$b_1 + b_2 + \cdots + b_n + \cdots \tag{8'}$$

is also convergent, with the same sum A.

Proof. For every partial sum

$$B_n = b_1 + \cdots + b_n \tag{9}$$

of the series (8′), there is a partial sum

$$A_{m(n)} = a_1 + \cdots + a_{m(n)} \tag{10}$$

of the series (8) which contains all the terms of (9). In turn, we can find a partial sum

$$B_{N(n)} = b_1 + \cdots + b_n + \cdots + b_{N(n)}$$

of the series (8′) which contains all the terms of (10). Since $a_n \to 0$ for all n, we have

$$B_n \leqslant A_{m(n)} \leqslant B_{N(n)}. \tag{11}$$

The first inequality in (11) implies

$$B_n \leqslant A_{m(n)} \leqslant A,$$

where A is the sum of the convergent series (8). It follows that the series (8′) is also convergent, with sum B, say. Obviously $m(n)$ and $N(n)$ are no smaller than n, and hence become arbitrarily large as $n \to \infty$. Taking the limit as $n \to \infty$ in (11), we get

$$B \leqslant A \leqslant B,$$

and hence $B = A$. ∎

6.36. THEOREM (**Dirichlet**). *If the series* (8) *is absolutely convergent, with sum* A, *then the series* (8′) *is also absolutely convergent, with the same sum* A.

Proof. The absolute convergence of (8′) follows from the preceding theorem. Given any $\varepsilon > 0$, let N be such that

$$|A_N - A| < \varepsilon/2$$

and

$$|a_{m+1} + \cdots + a_n| < \varepsilon/2$$

for all $n > m \geqslant N$. Moreover, let p_0 be such that every partial sum B_p of the series (8′) contains the first N terms of the series (8) if $p > p_0$. Then, if $p > p_0$, the difference $B_p - A_N$ contains only terms of (8) with indices greater than N. It follows that

$$|B_p - A_N| < \varepsilon/2$$

and hence

$$|B_p - A| \leqslant |B_p - A_N| + |A_N - A| < \varepsilon$$

for all $p > p_0$, which implies $B_p \to A$. ∎

6.37. Things are entirely different in the case of a conditionally convergent series, as shown by our next

THEOREM (**Riemann**). *Given a conditionally convergent series* (8), *let* α *and* β *($\alpha \leqslant \beta$) be any two numbers (in the extended real number system). Then there exists a rearrangement* (8′) *of the series* (8), *with partial sums* B_n, *such that*

$$\varliminf_{n \to \infty} B_n = \alpha, \qquad \varlimsup_{n \to \infty} B_n = \beta. \tag{12}$$

Proof. The proof will be given in several steps.

Step 1. Since the series (8) converges, its terms approach zero as $n \to \infty$. Hence $|a_n|$ can exceed any given $\varepsilon > 0$ for only a finite number of indices n, so that any set of terms of (8) has a term of largest absolute value. Separating positive and negative terms in (8) and choosing first the largest term (in absolute value), then the next largest, and so on, we form two series, the series

$$p_1 + p_2 + \cdots + p_n + \cdots \tag{13}$$

made up of all positive terms of (8), arranged in decreasing order, so that $p_1 \geqslant p_2 \geqslant \cdots$, and the series

$$q_1 + q_2 + \cdots + q_n + \cdots \tag{14}$$

made up of the absolute values of all the negative terms of (8), also arranged in decreasing order, so that $q_1 \geqslant q_2 \geqslant \cdots$.

Step 2. Both series (13) and (14) diverge. In fact, if (13) and (14) both converged to sums P and Q, respectively, then the original series (8) would be absolutely convergent, contrary to hypothesis, since no partial sum $|a_1| + \cdots + |a_n|$ could exceed $P + Q$. On the other hand, suppose one of the series (13) and (14), say (14), converged, while the other, say (13), diverged. Then, since the partial sums A_n of the series (8) include more and more terms of the series (13) and (14) as n increases, we could choose n such that the sum of the terms of (13) make an arbitrarily large contribution to A_n, while the sum of the terms of (14) make only a bounded contribution. But then $A_n \to \infty$, contrary to the assumption that (8) is (conditionally) convergent. A similar argument (with a sign change) shows that it is impossible for the series (13) to converge while the series (14) diverges. Thus both series (13) and (14) diverge, as asserted.

Step 3. We now use the terms of the series (13) and (14) to construct a new series, according to the following rule (in the case where α and β are finite): The partial sums P_n of the series (13) become arbitrarily large as $n \to \infty$, and hence we can find an n_1 such that

$$p_1 + \cdots + p_{n_1-1} \leqslant \beta < p_1 + \cdots + p_{n_1}$$

(setting $n_1 = 1$ if $p_1 > \beta$). Next we find an m_1 such that

$$p_1 + \cdots + p_{n_1} - q_1 - \cdots - q_{m_1} < \alpha \leqslant p_1 + \cdots + p_{n_1} - q_1 - \cdots - q_{m_1-1},$$

then an $n_2 > n_1$ such that

$$p_1 + \cdots + p_{n_1} - q_1 - \cdots - q_{m_1} + p_{n_1+1} + \cdots + p_{n_2-1}$$
$$\leqslant \beta < p_1 + \cdots + p_{n_1} - q_1 - \cdots - q_{m_1-1} + p_{n_1+1} + \cdots + p_{n_2},$$

then an integer $m_2 > m_1$ such that

$$p_1 + \cdots + p_{n_1} - q_1 - \cdots - q_{m_1} + p_{n_1+1} + \cdots + p_{n_2} - q_{m_1+1} - \cdots - q_{m_2}$$
$$< \alpha \leqslant p_1 + \cdots + p_{n_1} - q_1 - \cdots - q_{m_1-1} + p_{n_1+1} + \cdots + p_{n_2}$$
$$- q_{m_1+1} - \cdots - q_{m_2-1},$$

and so on indefinitely. This gives a series (8′) which is some rearrangement of the original series (8).

Step 4. The points α and β are limit points of the sequence B_n of partial sums of the series (8′) just constructed. In fact, the partial sums of (8′) with indices

$$n_1, n_2 + m_1 - 1, \ldots, n_k + m_{k-1} - 1, \ldots$$

differ from the number β by no more than $p_{n_1}, p_{n_2}, \ldots, p_{n_k}, \ldots$, respectively, where these quantities approach zero, while the partial sums of (8′) with indices

$$n_1 + m_1, n_2 + m_2, \ldots, n_k + m_k, \ldots$$

differ from α by no more than $q_{m_1}, q_{m_2}, \ldots, q_{m_k}, \ldots$, respectively, where these quantities also approach zero. Moreover, *there are no limit points of the sequence B_n outside the interval $[\alpha, \beta]$.* In fact, if $\gamma > \beta$ (say), let $\gamma - \beta = h > 0$. Then there is an index n_k such that $p_{n_k} < h/2$, so that all partial sums B_n with indices greater than $n_k + m_{k-1} - 1$ do not exceed

$$\beta + p_{n_k} < \beta + \frac{h}{2},$$

and hence do not get closer than $h/2$ to the point γ.

Final step. In the case of finite α and β, the theorem follows at once from the two italicized assertions in Step 4. In the case where one or both of the numbers α, β is infinite, only a slight modification of the proof is required (left as an exercise for the reader). ∎

COROLLARY. *Given a conditionally convergent series, let C be any number in the extended real number system. Then there exists a rearrangement* $(8')$ *of the series* (8) *with C as its sum.*†

Proof. Merely choose $\alpha = \beta = C$ in Riemann's theorem. ∎

6.4. Series of Vectors

6.41. The definitions of Sec. 6.11a for numerical series have natural analogues for the case of series whose terms are vectors of the space R_m. Thus, given a sequence of vectors

$$a_1, a_2, \ldots, a_n, \ldots, \tag{1}$$

all belonging to the space R_m, by the *partial sums* of (1) we mean the vectors

$$s_1 = a_1 \in R_m,$$
$$s_2 = a_1 + a_2 \in R_m,$$
$$\ldots$$
$$s_n = a_1 + a_2 + \cdots + a_n \in R_m.$$

$$\ldots$$

Suppose the sequence of partial sums

$$s_1, s_2, \ldots, s_n, \ldots \tag{2}$$

converges to a vector $s \in R_m$. Then the expression

$$a_1 + a_2 + \cdots + a_n + \cdots \tag{3}$$

is said to be *convergent* (or to *converge*), with *sum*

$$s = \lim_{n \to \infty} s_n.$$

If, however, the sequence of vectors (2) diverges, then the series (3) is said to be *divergent* (or to *diverge*).

† Here we allow our series to have the sum $-\infty$ or $+\infty$. Such series are, of course, divergent series of a special kind (cf. Sec. 4.61).

6.42. THEOREM. *Let*

$$a_1 = (a_{11}, a_{12}, \ldots, a_{1m}),$$
$$a_2 = (a_{21}, a_{22}, \ldots, a_{2m}),$$
...
$$a_n = (a_{n1}, a_{n2}, \ldots, a_{nm}),$$
...

Then the series (3) *is convergent with sum* $s = (s_1, s_2, \ldots, s_m)$ *if and only if*

$$a_{11} + a_{21} + \cdots + a_{n1} + \cdots = s_1,$$
$$a_{12} + a_{22} + \cdots + a_{n2} + \cdots = s_2,$$
...
$$a_{1m} + a_{2m} + \cdots + a_{nm} + \cdots = s_m,$$
...

i.e., if and only if each "component series" of (3) *is convergent, with the corresponding component of s as its sum.*

Proof. An immediate consequence of Theorem 3.32f. ∎

6.43. THEOREM (**Cauchy convergence criterion for vector series**). *The series* (3) *converges if and only if, given any* $\varepsilon > 0$, *there exists an integer* $N > 0$ *such that*†

$$|a_{m+1} + \cdots + a_n| < \varepsilon$$

for all $n > m \geqslant N$.

Proof. Use Theorem 4.74, taking account of Sec. 6.11c. ∎

In particular, if the series (3) converges, then, just as in Sec. 6.11c,

$$\lim_{n \to \infty} a_n = 0.$$

6.44. THEOREM. *If the numerical series*

$$|a_1| + |a_2| + \cdots + |a_n| + \cdots \tag{4}$$

converges, then so does the vector series

$$a_1 + a_2 + \cdots + a_n + \cdots.$$

Proof. Identical with that of Theorem 6.21. ∎

A vector series (3) is said to be *absolutely convergent* if the numerical series (4) converges. If the series (4) diverges while the series (3) converges, we say that (3) is *conditionally convergent*.

† Here, as usual, $|\cdots|$ denotes the length of the vector written between the vertical bars.

6.45. Theorem 6.31 on "term-by-term" addition of series and multiplication of a series by a number remains valid for vector series. However, Theorem 6.32 on multiplication of two series is in general meaningless, since multiplication of vectors is undefined (however, see Sec. 6.46). Theorems 6.33 and 6.34b on "grouping together" of terms of series and Theorem 6.36 on rearrangement of series continue to hold in the case of vector series. The appropriate generalization of Riemann's theorem (Theorem 6.37) requires a terminology all its own, as developed in Problems 17–22.

6.46. Series with complex terms, being vector series in R_2, are included in the present scheme. However, since multiplication of complex numbers is a meaningful operation (Sec. 2.71), we now have the appropriate generalization of Theorem 6.32, besides the results of Sec. 6.45:

THEOREM. *Given two series*

$$a_1 + a_2 + \cdots + a_n + \cdots, \tag{5}$$
$$b_1 + b_2 + \cdots + b_n + \cdots \tag{6}$$

with complex terms, suppose (5) and (6) are convergent, with sums A and B, while at least one of the series (5) and (6) is absolutely convergent. Then the series

$$a_1 b_1 + (a_1 b_2 + a_2 b_1) + (a_1 b_3 + a_2 b_2 + a_3 b_1) + \cdots$$

is convergent, with sum AB.

Proof. Identical with that of Theorem 6.32. ∎

6.47. To study nonabsolutely convergent vector series, we will use the *Abel-Dirichlet test* (Theorem 6.47c). This test is based on the following special transformation of finite vector sums:

a. THEOREM (**Abel's transformation**). *Given n numbers $\alpha_1, \ldots, \alpha_n$ and n vectors $b_1, \ldots, b_n$, let*

$$\alpha'_1 = \alpha_2 - \alpha_1, \ \alpha'_2 = \alpha_3 - \alpha_2, \ldots, \ \alpha'_{n-1} = \alpha_n - \alpha_{n-1},$$
$$B_1 = b_1, \ B_2 = b_1 + b_2, \ldots, \ B_n = b_1 + b_2 + \cdots + b_n.$$

Then

$$\sum_{k=1}^{n} \alpha_k b_k = \alpha_n B_n - \sum_{k=1}^{n-1} \alpha'_k B_k \qquad (n = 2, 3, \ldots). \tag{7}$$

Proof. The proof is based on induction. Formula (7) is obvious for $n = 1$, since it then reduces to

$$\alpha_1 b_1 + \alpha_2 b_2 = \alpha_2 (b_1 + b_2) - (\alpha_2 - \alpha_1) b_1.$$

Suppose (7) holds for an integer n. Then

$$\sum_{k=1}^{n+1} \alpha_k b_k = \sum_{k=1}^{n} \alpha_k b_k + \alpha_{n+1} b_{n+1} = \alpha_n B_n - \sum_{k=1}^{n-1} \alpha_k' B_k + \alpha_{n+1} b_{n+1},$$

and to show that (7) holds for $n+1$, we must prove that

$$\alpha_n B_n - \sum_{k=1}^{n-1} \alpha_k' B_k + \alpha_{n+1} b_{n+1} = \alpha_{n+1} B_{n+1} - \sum_{k=1}^{n-1} \alpha_k' B_k - \alpha_n' B_n, \tag{8}$$

i.e., that

$$\alpha_n B_n + \alpha_{n+1} b_{n+1} = \alpha_{n+1} B_{n+1} - \alpha_n' B_n = \alpha_{n+1} B_{n+1} - (\alpha_{n+1} - \alpha_n) B_n.$$

Dropping the term $\alpha_n B_n$ from both sides and then dividing through by α_{n+1}, we get the formula

$$b_{n+1} = B_{n+1} - B_n, \tag{9}$$

which is obviously true. To complete the proof, we need only reverse the steps leading from (8) to (9). ∎

b. COROLLARY. *If* $\alpha_1 \geqslant \alpha_2 \geqslant \cdots \geqslant \alpha_n \geqslant 0$ *and*

$$|B_k| \leqslant C \qquad (k = 1, \ldots, n),$$

then

$$\left| \sum_{k=1}^{n} \alpha_k b_k \right| \leqslant 2C\alpha_1. \tag{10}$$

Proof. Here

$$|\alpha_n B_n| \leqslant C\alpha_m \leqslant C\alpha_1, \qquad \alpha_k' \leqslant 0,$$

and

$$\left| \sum_{k=1}^{n-1} \alpha_k' B_k \right| = \left| \sum_{k=1}^{n-1} (-\alpha_k') B_k \right| \leqslant C \sum_{k=1}^{n-1} (-\alpha_k')$$
$$= C[(\alpha_1 - \alpha_2) + \cdots + (\alpha_{n-1} - \alpha_n)] = C(\alpha_1 - \alpha_n) \leqslant C\alpha_1,$$

so that (7) implies (10). ∎

c. THEOREM (**Abel-Dirichlet test**). *Given a numerical sequence* α_n *and a vector sequence* $b_n \in R_m$, *suppose* $\alpha_n \searrow 0$ *while the sequence* $B_n = b_1 + \cdots + b_n$ *is bounded, i.e.,*

$$|B_n| \leqslant C \qquad (n = 1, 2, \ldots).$$

Then the vector series

$$\sum_{n=1}^{\infty} \alpha_n b_n \tag{11}$$

is convergent.

Proof. Applying the estimate (10) to the "section"

$$\sum_{n=p}^{q} \alpha_n b_n$$

of the series (11), we get

$$\left| \sum_{n=p}^{q} \alpha_n b_n \right| \leqslant 2C_1 \alpha_p, \tag{12}$$

where

$$C_1 = \sup_{p \leq r \leq q} |b_p + \cdots + b_r| = \sup_{p \leq r \leq q} |(b_1 + \cdots + b_r) - (b_1 + \cdots + b_{p-1})|$$
$$= \sup_{p \leq r \leq q} |B_r - B_{p-1}| \leqslant 2C.$$

But the right-hand side of (12) approaches 0 as $p \to \infty$ for any $q > p$. The convergence of (11) now follows from the Cauchy convergence criterion (Theorem 6.43). ∎

d. Leibniz's test (Theorem 6.23) is a special case of the Abel-Dirichlet test, obtained by setting $R_m = R_1 = R$ $(m = 1)$ and

$$B_1 = 1, \ B_2 = 0, \ B_3 = 1, \ B_4 = 0, \ldots$$

But the Abel-Dirichlet test has a wider range of application than Leibniz's test, even for $m = 1$. For example, consider the real series

$$\sum_{n=0}^{\infty} \alpha_n \sin n\theta, \tag{13}$$

$$\sum_{n=0}^{\infty} \alpha_n \cos n\theta \tag{14}$$

and the complex series

$$\sum_{n=0}^{\infty} \alpha_n (\cos n\theta + i \sin n\theta), \tag{15}$$

where $\alpha_n \searrow 0$. To find the values of θ for which these series converge, we argue as follows. Let

$$z = \cos \theta + i \sin \theta.$$

Then $|z| = 1$ and

$$z^n = \cos n\theta + i \sin n\theta$$

(Sec. 5.73). Moreover

$$\sum_{k=0}^{n} (\cos k\theta + i \sin k\theta) = \sum_{k=0}^{n} z^k = \frac{1 - z^{n+1}}{1 - z}$$

(cf. Example 6.11b), and hence

$$\left| \sum_{k=0}^{n} (\cos k\theta + i \sin k\theta) \right| \leqslant \frac{2}{|1 - z|} = \frac{2}{|1 - \cos \theta - i \sin \theta|}$$

$$= \frac{2}{\sqrt{(1 - \cos \theta)^2 + \sin^2 \theta}} = \sqrt{\frac{2}{1 - \cos \theta}}. \qquad (16)$$

Thus the sequence

$$\sum_{k=0}^{n} (\cos k\theta + i \sin k\theta) \qquad (n = 0, 1, 2, \ldots)$$

is bounded (in absolute value), provided that $\theta \neq 0, \pm 2\pi, \pm 4\pi, \ldots$. Applying the Abel-Dirichlet test, we see that *the series* (15) *converges if* $\alpha_n \searrow 0$ *and* $\theta \neq 0$, $\pm 2\pi, \pm 4\pi, \ldots$ *The series* (13) *and* (14) *converge under the same conditions*, since they are just the real and imaginary parts of the series (15). Note that *the series* (13) *also converges if* $\theta = 0, \pm 2\pi, \pm 4\pi, \ldots$, since then its terms all vanish.

e. The estimate (16) implies two analogous estimates "in the real domain":

$$\left| \sum_{k=0}^{n} \cos k\theta \right| \leqslant \frac{2}{1 - \cos \theta},$$

$$\left| \sum_{k=0}^{n} \sin k\theta \right| \leqslant \frac{2}{1 - \cos \theta}.$$

These estimates often figure in problems involving the Abel-Dirichlet test.

f. For the special case where

$$\alpha_n = r^n \qquad (r < 1),$$

it is easy to write explicit expressions for the sums of the series (13)–(15). In fact, summing the appropriate geometric series, we get

$$\sum_{n=0}^{\infty} r^n e^{in\theta} = \frac{1}{1 - re^{i\theta}}.$$

Taking real and imaginary parts of the right-hand side, we get

$$\frac{1}{1-re^{i\theta}} = \frac{1-re^{-i\theta}}{(1-re^{i\theta})(1-re^{-i\theta})} = \frac{1-r\cos\theta + ir\sin\theta}{1-2r\cos\theta + r^2}.$$

It follows that

$$\sum_{n=0}^{\infty} r^n \cos n\theta = \operatorname{Re} \sum_{n=0}^{\infty} r^n e^{in\theta} = \frac{1-r\cos\theta}{1-2r\cos\theta + r^2},$$

$$\sum_{n=0}^{\infty} r^n \sin n\theta = \operatorname{Im} \sum_{n=0}^{\infty} r^n e^{in\theta} = \frac{r\sin\theta}{1-2r\cos\theta + r^2}.$$

6.48. Two-sided series

a. Let

$$\dots, a_{-n},\dots, a_{-1}, a_0, a_1,\dots, a_n,\dots$$

be a "two-sidedly infinite" sequence of vectors in the space R_m. Then by the sum of the "two-sided series"

$$\sum_{k=-\infty}^{\infty} a_k \tag{17}$$

we mean the limit (provided it exists) of the "two-sided partial sums"

$$\sum_{k=-\nu}^{n} a_k \qquad (\nu = 1,2,\dots; n = 0,1,2,\dots),$$

where ν and n approach infinity *independently*. More exactly, the series (17) is said to be *convergent* (or to *converge*), with *sum s*, if, given any $\varepsilon > 0$, there exists an integer $N > 0$ such that

$$\left| s - \sum_{k=-\nu}^{n} a_k \right| < \varepsilon \tag{18}$$

for all $\nu, n > N$.

b. The following theorem reduces the problem of the convergence of a two-sided series to that of the convergence of two one-sided series:

THEOREM. *The two-sided series* (17) *converges if and only if both "one-sided series"*

$$\sum_{k=0}^{\infty} a_k, \qquad \sum_{k=1}^{\infty} a_{-k} \tag{19}$$

converge. Moreover, if the series (19) *have the sums A and B, respectively, then the series* (17) *has the sum $A + B$.*

Proof. Suppose the series (19) converge, with sums A and B. Then, given any $\varepsilon > 0$, there is an N such that

$$\left| A - \sum_{k=0}^{n} a_k \right| < \frac{\varepsilon}{2}, \qquad \left| B - \sum_{k=1}^{v} a_{-k} \right| < \frac{\varepsilon}{2}$$

for all $v,n > N$. But then

$$\left| (A+B) - \sum_{k=-v}^{n} a_k \right| = \left| \left(A - \sum_{k=0}^{n} a_k \right) + \left(B - \sum_{k=1}^{v} a_{-k} \right) \right| < \varepsilon,$$

and hence (17) converges, with sum $A+B$.

Conversely, if the series (17) converges, then, by (18), there is an N such that

$$\left| \sum_{k=p+1}^{q} a_k \right| = \left| \left(s - \sum_{k=-v}^{p} a_k \right) - \left(s - \sum_{k=-v}^{q} a_k \right) \right| < 2\varepsilon$$

for all $q > p \geqslant N$ (v sufficiently large). Therefore the first of the series (19) converges, by the Cauchy convergence criterion (Theorem 6.43). A similar argument shows that the second of the series (19) also converges. ∎

c. Example. The two-sided series

$$\sum_{k=-\infty}^{\infty} c_k e^{ikt}$$

converges for all real t if

$$\sum_{k=-\infty}^{\infty} |c_k| < \infty.$$

The series converges everywhere except possibly at the points $t = 0, \pm 2\pi, \pm 4\pi, \ldots$ if $c_n \searrow 0, c_{-n} \searrow 0$ as $n \to \infty$ (cf. Sec. 6.47d).

6.49. Symmetric summation of two-sided series

a. The two-sided vector series (17) is said to be *symmetrically summable* if its "symmetric partial sums"

$$s_n = \sum_{k=-n}^{n} a_k \qquad (n = 0,1,2,\ldots) \tag{20}$$

approach a limit s as $n \to \infty$. The limit s (if it exists) is called the *symmetric sum* of the series (17).

If the series (17) converges in the sense of Sec. 6.48a, then it is obviously symmetrically summable, and its symmetric sum coincides with its ordinary sum. But not every symmetrically summable series converges in the sense of Sec. 6.48a. For example, the two-sided series with terms

$$a_k = \begin{cases} 1 \text{ if } k > 0, \\ 0 \text{ if } k = 0, \\ -1 \text{ if } k < 0, \end{cases}$$

is symmetrically summable, with symmetric sum 0, but does not converge in the sense of Sec. 6.48a.

b. Symmetric summation of two-sided series also reduces to ordinary "one-sided convergence":

THEOREM. *The two-sided series* (17) *is symmetrically summable if and only if the one-sided series*

$$a_0 + \sum_{k=1}^{\infty} (a_k + a_{-k}) \tag{21}$$

converges. Moreover, if (21) *converges, then its sum is the symmetric sum of* (17).

Proof. An immediate consequence of the obvious equality of the nth partial sum of the series (21) and the nth symmetric partial sum (20) of the series (17). ∎

c. Symmetric summation is often used to study trigonometric series of the form

$$\sum_{k=-\infty}^{\infty} c_k e^{ikt}.$$

For example, the series

$$\cdots - \frac{e^{-int}}{n} - \cdots - \frac{e^{-2it}}{2} - \frac{e^{-it}}{1} + \frac{e^{it}}{1} + \frac{e^{2it}}{2} + \cdots + \frac{e^{nit}}{n} + \cdots \tag{22}$$

converges for all $t \neq 0, \pm 2\pi, \pm 4\pi, \ldots$ (cf. Example 6.48c), but not for $t = 0,$ $\pm 2\pi, \pm 4\pi, \ldots$ because of Theorem 6.48b, since then the first of the series (19), say, reduces to the divergent harmonic series. However, the series (22) is symmetrically summable for *all* real t, since the series

$$\sum_{k=1}^{\infty} \frac{e^{ikt} - e^{-ikt}}{k} = 2i \sum_{k=1}^{\infty} \frac{\sin kt}{k}$$

converges everywhere (see Sec. 6.47d).

6.5. Series of Functions

6.51. Series of functions have already been encountered in Secs. 6.47 and 6.49. We now discuss such series officially.

Let

$$a_1(x) + a_2(x) + \cdots + a_n(x) + \cdots \tag{1}$$

be a series of functions, all defined on some set E and taking values in the space R_m. In keeping with Sec. 5.91, we say that the series (1) is *convergent on E* if the sequence of partial sums

$$\begin{aligned}
s_1(x) &= a_1(x), \\
s_2(x) &= a_1(x) + a_2(x), \\
&\cdots \\
s_n(x) &= a_1(x) + a_2(x) + \cdots + a_n(x), \\
&\cdots
\end{aligned} \tag{2}$$

of (1) is convergent on E. The limit

$$s(x) = \lim_{n \to \infty} s_n(x)$$

is then called the *sum* of the series (1). By the same token, we say that the series (1) is *uniformly convergent on E* if the sequence of partial sums (2) is uniformly convergent on E (see Sec. 5.93).

6.52. For series of functions we have the following version of the **Cauchy convergence criterion:**

THEOREM. *The series* (1) *is uniformly convergent on E if and only if, given any $\varepsilon > 0$, there exists an integer $N > 0$ such that*

$$\sup_{x \in E} |a_{m+1}(x) + \cdots + a_n(x)| < \varepsilon$$

for all $n > m \geqslant N$.

Proof. An immediate consequence of Theorem 5.96, with M taken to be the space R_m with the usual norm $|\cdots|$. ∎

6.53. The next theorem gives a simple sufficient condition for uniform convergence of the series (1):

THEOREM (**Weierstrass' test**). *If*

$$\sup_{x \in E} |a_n(x)| = \alpha_n$$

and if the numerical series

$$\sum_{n=1}^{\infty} \alpha_n$$

converges, then the series (1) *is uniformly convergent on E.*

Proof. An immediate consequence of the preceding theorem, together with the estimate

$$\sup_{x \in E} |a_{m+1}(x) + \cdots + a_n(x)| \leqslant \alpha_{m+1} + \cdots + \alpha_n$$

and the Cauchy convergence criterion for a numerical series (Sec. 6.11c). ∎

6.54. a. THEOREM. *If the series* (1) *converges uniformly on E and if every term $a_n(x)$ is bounded on E, then the sum $s(x)$ of the series* (1) *is also bounded on E.*

Proof. An immediate consequence of Theorem 5.94. ∎

b. THEOREM. *If the series* (1) *converges uniformly on E, where E is a metric space, and if every term $a_n(x)$ is continuous on E, then the sum $s(x)$ of the series is also continuous on E.*

Proof. An immediate consequence of Theorem 5.95a and its corollary. ∎

6.6. Power Series

6.61. By a *power series* we mean a series of the form

$$a_0 + a_1(z - z_0) + a_2(z - z_0)^2 + \cdots + a_n(z - z_0)^n + \cdots, \tag{1}$$

where $z = x + iy, z_0 = x_0 + iy_0$ and the coefficients $a_0, a_1, \ldots, a_n, \ldots$ are complex numbers.† By the *region of convergence* of the power series (1) is meant the set of all values of z for which (1) converges.

6.62. The following theorem plays a key role in determining the region of convergence of a given power series:

THEOREM (**Cauchy-Hadamard**). *Let*

$$\frac{1}{\rho} = \overline{\lim_{n \to \infty}} \sqrt[n]{|a_n|},$$

where the upper limit is taken in the extended real number system. Then the series (1) *converges (in fact absolutely) for all z such that $|z - z_0| < \rho$ and diverges for all z such that $|z - z_0| > \rho$.‡*

---

† One can also consider real power series of the form
$$a_0 + a_1(x - x_0) + a_2(x - x_0)^2 + \cdots + a_n(x - x_0)^n + \cdots$$
$(x, x_0, a_0, a_1, \ldots, a_n, \ldots$ real), but the complex case is of particular interest.
‡ We set $\rho = 0$ if $\overline{\lim} \sqrt[n]{|a_n|} = \infty$ and $\rho = \infty$ if $\overline{\lim} \sqrt[n]{|a_n|} = 0$. In the latter case, the theorem asserts that (1) converges for all complex z.

Proof. Suppose ρ is finite and nonzero, and let

$$|z - z_0| < \rho = \frac{1}{\overline{\lim} \sqrt[n]{|a_n|}}.$$

Then

$$|z - z_0| \overline{\lim} \sqrt[n]{|a_n|} = \overline{\lim} \sqrt[n]{|a_n||z - z_0|^n} < 1,$$

so that the series (1) is absolutely convergent by Cauchy's test (Theorem 6.14c) and Theorem 6.44. On the other hand, if

$$|z - z_0| > \rho = \frac{1}{\overline{\lim} \sqrt[n]{|a_n|}},$$

then

$$|z - z_0| \overline{\lim} \sqrt[n]{|a_n|} = \overline{\lim} \sqrt[n]{|a_n||z - z_0|^n} > 1,$$

so that the series (1) diverges since its terms cannot approach zero as $n \to \infty$ (cf. Sec. 6.43). The proof for the cases $\rho = 0$ and $\rho = \infty$ is left as an exercise for the reader. ∎

6.63. Theorem 6.62 shows that the region of convergence G of the power series (1) is a disk of radius ρ. Hence ρ is called the *radius of convergence* of the series (1). More exactly, G contains every point inside the "circle of convergence" $\Gamma = \{z : |z - z_0| = \rho\}$ and no points outside Γ. The theorem says nothing about the convergence of the series (1) on the boundary of the disk, i.e., on Γ itself, and in fact various possibilities can occur, requiring a special investigation in each case.

6.64. Examples

a. The series

$$\sum_{n=1}^{\infty} \frac{z^n}{n^n}$$

converges for all z, since

$$\frac{1}{\rho} = \lim_{n \to \infty} \frac{1}{\sqrt[n]{n^n}} = \lim_{n \to \infty} \frac{1}{n} = 0.$$

b. The series

$$\sum_{n=1}^{\infty} n^n z^n$$

converges only for $z = 0$, since

$$\frac{1}{\rho} = \lim_{n\to\infty} \sqrt[n]{n^n} = \lim_{n\to\infty} n = \infty.$$

c. The series

$$\sum_{n=1}^{\infty} n^\alpha z^n,$$

where α is a fixed real number, has radius of convergence 1, since, by formula (26), p. 154,

$$\frac{1}{\rho} = \lim_{n\to\infty} \frac{1}{\sqrt[n]{n^\alpha}} = \lim_{n\to\infty} \left(\frac{1}{n^{1/n}}\right)^\alpha = 1.$$

However, the series does not converge at any point of the circle $|z| = 1$ if $\alpha \geqslant 0$, since its terms then fail to converge to zero. If $\alpha < -1$, the series is absolutely convergent at every point of the circle $|z| = 1$, by Example 6.15a, while if $-1 \leqslant \alpha < 0$, the series diverges for $z = 1$ but converges for all other $z \neq 1, |z| = 1$, by Sec. 6.47d.

6.65. Suppose the power series (1) has radius of convergence ρ. Then, in general, the convergence on the disk $|z - z_0| < \rho$ fails to be uniform. For example, if the geometric series

$$1 + z + z^2 + \cdots$$

converged uniformly on the disk $|z| < 1$, then its sum would be bounded on $|z| < 1$, by Theorem 6.54a. But the sum of the geometric series is the function

$$\frac{1}{1-z}$$

(cf. Example 6.11b), which is obviously unbounded on the disk $|z| < 1$.

Nevertheless we have the following

a. THEOREM. *The series* (1) *is uniformly convergent on every disk* $|z - z_0| \leqslant \rho_1 < \rho$.

Proof. For $|z - z_0| \leqslant \rho_1$ we have the estimate

$$|a_n(z - z_0)^n| \leqslant |a_n|\rho_1^n.$$

But the series with terms $|a_n|\rho_1^n$ converges, by the Cauchy-Hadamard theorem (choose $z = z_0 + \rho_1$). Therefore the series (1) is uniformly convergent on the disk $|z - z_0| \leqslant \rho_1$, by Weierstrass' test (Theorem 6.53). ∎

b. COROLLARY. *The sum* $s(z)$ *of the series* (1) *is bounded and continuous on every disk* $|z - z_0| \leqslant \rho_1 < \rho$.

Proof. An immediate consequence of the preceding theorem and the theorems of Sec. 6.54. ∎

It follows that $s(z)$ is continuous on the whole disk $|z - z_0| < \rho$, since every point of this disk lies in some smaller disk $|z - z_0| \leqslant \rho_1$. However, $s(z)$ need not be bounded on the whole disk $|z - z_0| < \rho$, as we have just seen.

6.66. As we know, a power series with radius of convergence ρ may or may not converge at points on the boundary of its region of convergence, i.e., at points of the circle $|z - z_0| = \rho$. However, *if the series converges at a boundary point z_1, then it converges uniformly on the whole segment going from the center of the circle to the point z_1.* To see this, we need only consider the case $z_0 = 0$, $z_1 = t_1 > 0$ (why?):

THEOREM. *If the power series*

$$\sum_{n=0}^{\infty} a_n t^n$$

converges at the point $t_1 > 0$, then it converges uniformly on the whole interval $0 \leqslant t \leqslant t_1$.

Proof. The theorem will be proved if, given any $\varepsilon > 0$, we succeed in finding an N such that

$$\left| \sum_{n=p+1}^{q} a_n t^n \right| < \varepsilon$$

for all $q > p \geqslant N$ and all $0 \leqslant t \leqslant t_1$. Applying Abel's transformation (Theorem 6.47a), with

$$\alpha_k = \left(\frac{t}{t_1} \right)^k, \quad b_k = a_k t_1^k \qquad (k = p+1, \ldots, q),$$

where for the time being p and q $(p < q)$ are arbitrary, we get

$$\sum_{n=p+1}^{q} a_n t^n = \sum_{n=p+1}^{q} a_n t_1^n$$
$$= \left(\frac{t}{t_1} \right)^q \sum_{n=p+1}^{q} a_n t_1^n - \sum_{n=p+1}^{q-1} \left\{ \left[\left(\frac{t}{t_1} \right)^{n+1} - \left(\frac{t}{t_1} \right)^n \right] \sum_{k=p+1}^{n} a_k t_1^k \right\}.$$

Let N be such that

$$\left| \sum_{k=p+1}^{n} a_k t_1^k \right| < \frac{\varepsilon}{2}$$

for all $n > p \geqslant N$. Then

$$\left| \sum_{n=p+1}^{q} a_n t^n \right| < \frac{\varepsilon}{2} \left(\frac{t}{t_1} \right)^q + \frac{\varepsilon}{2} \max_{0 \leqslant t \leqslant t_1} \sum_{n=p+1}^{q-1} \left| \left(\frac{t}{t_1} \right)^{n+1} - \left(\frac{t}{t_1} \right)^n \right|$$

$$\leqslant \frac{\varepsilon}{2} + \frac{\varepsilon}{2} \max_{0 \leqslant t \leqslant t_1} \left[\left(\frac{t}{t_1} \right)^{p+1} - \left(\frac{t}{t_1} \right)^{q-1} \right] \leqslant \frac{\varepsilon}{2} + \frac{\varepsilon}{2} = \varepsilon$$

for all $q > p \geqslant N$. ∎

6.67. Examples

a. The series

$$1 + \frac{z}{1^\alpha} + \frac{z^2}{2^\alpha} + \cdots + \frac{z^n}{n^\alpha} + \cdots \qquad (0 < \alpha \leqslant 1)$$

converges for all $z \neq 1, |z| = 1$, by Sec. 6.47d. By Theorem 6.66, the series converges uniformly on every radius of the disk $|z| \leqslant 1$ going to a point z_1 such that $|z_1| = 1, z_1 \neq 1$. However, it must not be thought that the series converges uniformly on the set of points belonging to all these radii!

b. The series

$$\sum_{n=1}^{\infty} \frac{\alpha(\alpha+1)\cdots(\alpha+n-1)}{\beta(\beta+1)\cdots(\beta+n-1)} z^{n-1} \qquad (\alpha, \beta \neq 0, -1, -2, \ldots)$$

converges for $z = 1$ if $\beta > \alpha + 1$, by Example 6.17b. Hence, if $\beta > \alpha + 1$, the series converges for $|z| < 1$ and converges uniformly on the interval $0 \leqslant x \leqslant 1$. If $\beta > \alpha$, the series converges for $z = -1$, by Example 6.24c, and hence converges uniformly on the interval $-1 \leqslant x \leqslant 0$.

6.68. Theorem 6.66 has the following important implication:

THEOREM (**Abel**). *Let $f(x)$ be a function continuous on the closed interval $0 \leqslant x \leqslant x_1$ such that*

$$f(x) = \sum_{n=0}^{\infty} a_n x^n$$

on the half-open interval $0 \leqslant x < x_1$. Suppose further that the series

$$\sum_{n=0}^{\infty} a_n x_1^n \tag{2}$$

converges. Then

$$f(x_1) = \sum_{n=0}^{\infty} a_n x_1^n.$$

Proof. Let

$$\sum_{n=0}^{\infty} a_n x^n = s(x)$$

for all $x \in [0,x_1]$. Since (2) converges, the series (3) is uniformly convergent on $[0,x_1]$, by Theorem 6.66. Hence $s(x)$ is continuous on $[0,x_1]$, by Theorem 6.54b. But $f(x)$ is also continuous on $[0,x_1]$, by hypothesis, and moreover $f(x) = s(x)$ for $0 \leqslant x < x_1$. It follows that

$$f(x_1) = \lim_{x \to x_1} f(x) = \lim_{x \to x_1} s(x) = s(x_1) = \sum_{n=0}^{\infty} a_n x^n. \quad \blacksquare$$

Problems

1. Prove that the series

$$\sum_{n=1}^{\infty} a_n$$

diverges if

$$a_n > 0, \qquad \frac{a_{n+1}}{a_n} = 1 - \frac{1}{n} + \frac{\delta_n}{n^2}, \qquad |\delta_n| \leqslant C \qquad (n = 1,2,\ldots).$$

(Gauss)

2. Prove that the *hypergeometric series*

$$F(\alpha,\beta,\gamma,x) = 1 + \sum_{n=1}^{\infty} \frac{\alpha(\alpha+1)\cdots(\alpha+n-1)\beta(\beta+1)\cdots(\beta+n-1)}{n!\,\gamma(\gamma+1)\cdots(\gamma+n-1)} x^n$$

$(\alpha,\beta,\gamma \neq 0, -1, -2,\ldots)$ is absolutely convergent if $|x| < 1$ and divergent if $|x| > 1$. Prove that for $x = 1$ the series is (absolutely) convergent if $\gamma > \alpha + \beta$ and divergent if $\gamma \leqslant \alpha + \beta$, while for $x = -1$ the series is absolutely convergent if $\gamma > \alpha + \beta$, conditionally convergent if $\alpha + \beta - 1 < \gamma \leqslant \alpha + \beta$, and divergent if $\gamma \leqslant \alpha + \beta - 1$.

3. Prove that if $a_n \geqslant a_{n+1} > 0$ $(n = 1,2,\ldots)$ and if the series

$$\sum_{n=1}^{\infty} a_n$$

is convergent, then a_n approaches zero faster than $1/n$, i.e.,

$$\lim_{n \to \infty} n a_n = 0.$$

Comment. The function $1/n$ cannot be replaced by any function $\varphi(n)$ approaching zero faster than $1/n$. (A. S. Nemirovski)

4. Suppose the series

$$\sum_{n=1}^{\infty} a_n \qquad (a_n > 0)$$

converges. Prove that there exists a sequence b_n satisfying the conditions

$$b_1 \leqslant b_2 \leqslant \cdots \leqslant b_n \leqslant \cdots,$$
$$\lim_{n \to \infty} b_n = \infty$$

such that the series

$$\sum_{n=1}^{\infty} a_n b_n$$

also converges.

5. Prove that multiplying the convergent series

$$1 - \frac{1}{\sqrt{2}} + \frac{1}{\sqrt{3}} - \frac{1}{\sqrt{4}} + \cdots$$

by itself gives a divergent series.

6. Prove that for the series $a_1 - a_2 + a_3 - a_4 + \cdots$ figuring in Leibniz's test (Theorem 6.23), the difference between the sum of the entire series and the sum of the first n terms is less than the $(n+1)$st term in absolute value if $a_1 > a_2 > \cdots a_n > \cdots$.

7. Prove that the radius of convergence of a power series

$$\sum_{n=0}^{\infty} a_n (z - z_0)^n$$

for which

$$\lim_{n \to \infty} \frac{a_{n+1}}{a_n} = r$$

equals $1/r$.

8. Prove that if $a_n \geqslant 0 \ (n=1,2,\dots)$ and if

$$f(t) = \sum_{n=0}^{\infty} a_n t^n \leqslant C \qquad (0 \leqslant t < 1),$$

then the function $f(t)$ has a limit as $t \to 1$, equal to

$$\sum_{n=0}^{\infty} a_n.$$

9. Given numbers $a_{nk} \geqslant 0$ and b_n such that

$$a_{nk} \leqslant b_n \qquad (n,k=1,2,\ldots),$$

$$\sum_{n=1}^{\infty} b_n < \infty,$$

$$\lim_{k\to\infty} a_{nk} = a_n \qquad (n=1,2,\ldots),$$

let

$$\sum_{n=1}^{\infty} a_{nk} = s_k \qquad (k=1,2,\ldots),$$

$$\sum_{k=1}^{\infty} a_k = s.$$

Prove that

$$s = \lim_{k\to\infty} s_k.$$

10 (*Infinite products*). Given a sequence of complex numbers $z_1, z_2, \ldots, z_n, \ldots$, we say that the "infinite product"

$$\prod_{k=1}^{\infty} z_k = z_1 z_2 \cdots z_k \cdots \tag{1}$$

converges if the sequence of "partial products"

$$P_n = \prod_{k=1}^{n} z_k = z_1 z_2 \cdots z_n \qquad (n=1,2,\ldots)$$

converges to a finite limit $P \neq 0$. The number P is then called the *value* of the product (1).

Prove that if (1) converges, then

$$\lim_{n\to\infty} z_n = 1.$$

11 (*Continuation*). Given that $x_1 > 0, x_2 > 0, \ldots, x_n > 0, \ldots$, prove that the infinite product

$$\prod_{k=1}^{\infty} x_k$$

converges if and only if the series

$$\sum_{k=1}^{\infty} \log_b x_k$$

converges (for an arbitrary base $b > 1$).

12 (*Continuation*). Let $x_k = 1 + \omega_k$, where all the ω_k ($k=1,2,\ldots$) have the same sign. Prove that the infinite product

$$\prod_{k=1}^{\infty} (1+\omega_k)$$

converges if and only if the series

$$\sum_{k=1}^{\infty} \omega_k$$

converges.

Comment. The stipulation that all the ω_k have the same sign cannot be dropped (see Chapter 8, Problem 15).

13. Prove that

$$(1+x)(1+x^2)(1+x^4)(1+x^8)\cdots(1+x^{2^{n-1}})\cdots = \frac{1}{1-x}$$

if $|x|<1$.

14. Prove that

$$\prod_{n=1}^{\infty} \frac{1}{1-(1/p_n^x)} = \sum_{n=1}^{\infty} \frac{1}{n^x},$$

where $x>1$ and $p_1,p_2,\ldots,p_n,\ldots$ is the sequence of all prime numbers greater than 1 arranged in increasing order. (Euler)

15. Prove that the series

$$\sum_{n=1}^{\infty} \frac{1}{p_n} \qquad (p_n \text{ prime})$$

diverges.

16. Let $q_1,q_2,\ldots,q_n,\ldots$ be the sequence of all natural numbers whose decimal representations (Sec. 1.77) contain no nines at all. Does the series

$$\sum_{n=1}^{\infty} \frac{1}{q_n}$$

converge?

17. Let

$$u_1 + u_2 + \cdots + u_n + \cdots \qquad (u_n \in R_m) \tag{2}$$

be a series of vectors in the Euclidean space R_m. A vector $p \in R_m$ is called a *vector of absolute convergence* of the series (2) if the numerical series

$$(p,u_1) + (p,u_2) + \cdots + (p,u_n) + \cdots$$

where (p,u_n) denotes the scalar product of the vectors p and u_n, is absolutely convergent. Prove that the series (2) is absolutely convergent if and only if every vector $p \in R_m$ is a vector of absolute convergence of (2).

18 (*Continuation*). A vector $q \in R_m$ is called a *vector of absolute divergence* of the series (2) if the terms of (2) lying in every "solid angle" containing q form an absolutely divergent series. Prove that the series (2) is absolutely divergent if and only if it has at least one vector of absolute divergence.

19 (*Continuation*). Let $j_1, \ldots, j_n, \ldots$ be an increasing sequence of natural numbers, and let $k_1, \ldots, k_n, \ldots$ be the remaining natural numbers arranged in increasing order ($j_1 < \cdots < j_n < \cdots$, $k_1 < \cdots < k_n < \cdots$). Then the two series

$$u_{j_1} + \cdots + u_{j_n} + \cdots$$

and

$$u_{k_1} + \cdots + u_{k_n} + \cdots$$

made up of terms of the series (2), are called *complementary parts* of (2). Prove that if one of the complementary parts of a convergent series (2) converges, then so does the other, and any rearrangement of the terms of (2) which leaves all the j_n in order and all the k_n in order does not change the sum of (2).

20 (*Continuation*). Prove that if the series (2) is conditionally convergent and if q is a vector of absolute divergence of (2), then there is a part $u_{j_1} + \cdots + u_{j_n} + \cdots$ of the series (2) such that the components of the terms $u_{j_1}, \ldots, u_{j_n}, \ldots$ along q form an absolutely divergent series, while the components of $u_{j_1}, \ldots, u_{j_n}, \ldots$ orthogonal to q form an absolutely convergent series.†

21 (*Continuation*). Prove that if every nonzero vector $e \in R_m$ is a vector of absolute divergence of a conditionally convergent series (2), then every vector $f \in R_m$ is the sum of a series obtained by suitably rearranging (2).

22 (*Continuation, Steinitz's theorem*). Given any conditionally convergent series (2), prove that the set of sums of all possible rearrangements of (2) is a subspace of R_m, orthogonal to the subspace A of all vectors of absolute convergence of (2) and "shifted by" the (uniquely determined) sum of the projections of the terms of (2) onto A.‡

---

† Given two vectors p, $q \in R_m$, the vector

$$p_{\parallel} = \frac{(p, q)}{|q|^2} q$$

is called the *component of p along q* (or the *projection of p onto q*), while the vector $p_{\perp} = p - p_{\parallel}$ is called the *component of p orthogonal to q*. Note that $\langle p_{\parallel}, q \rangle = 0$, $(p_{\perp}, q) = 0$.

‡ A subset $S \subset R_m$ is called a (*linear*) *subspace* of R_m if $x + y \in S$, $\alpha x \in S$ for all x, $y \in S$ and all real α. Two subspaces S, $T \subset R_m$ are said to be *orthogonal* if $(x, y) = 0$ for all $x \in S$, $y \in T$. Given any vector $p \in R_m$ and any subspace $S \in R_m$, p has a unique representation of the form $p = p_S + p_{\perp}$, where $p_S \in S$ and $p_{\perp}$ is orthogonal to S in the sense that $(p_{\perp}, x) = 0$ for all $x \in S$ (prove this). We then call p_S the *projection of p onto S*.

7 The Derivative

7.1. Definitions and Examples

7.11. Given any (finite) real function defined on an interval (a,b) and any point x_0 in (a,b), consider the "difference quotient"

$$\frac{f(x_0 + h) - f(x_0)}{h}, \tag{1}$$

where the point $x_0 + h$ also lies in (a,b). Suppose (1) approaches a (finite) limit as $h \to 0$. Then the function $f(x)$ is said to be *differentiable* at $x = x_0$, and we write

$$\lim_{h \to 0} \frac{f(x_0 + h) - f(x_0)}{h} = f'(x_0) = [f(x)]'_{x = x_0}. \tag{2}$$

The number $f'(x_0)$ is called the *derivative* of the function $f(x)$ at the point $x = x_0$, and the operation leading from $f(x)$ to $f'(x_0)$ is called *differentiation*.

Geometrically, the quantity (1) is just the slope of the chord intersecting the graph of the function $y = f(x)$ in the points with abscissas x_0 and $x_0 + h$ (see Figure 10). The expression (2) is the slope of a certain straight line going through the point (x_0, y_0). This line (shown in the figure) is called the *tangent* to the graph of the function $y = f(x)$ at the point with abscissa x_0.

7.12. THEOREM. *If the function $f(x)$ has a derivative at $x = x_0$, then $f(x)$ is continuous at $x = x_0$.*

Proof. As $h \to 0$, the quotient (1) is bounded by some constant C. Hence

$$|f(x_0 + h) - f(x_0)| \leqslant C|h|$$

for all sufficiently small $|h|$, which implies the continuity of $f(x)$ at $x = x_0$. ∎

7.13. Basic rules for calculating derivatives

THEOREM. *Suppose the functions $f(x)$ and $g(x)$ are differentiable at the point $x = x_0$, and let k be an arbitrary real number. Then the functions $f(x) + g(x)$, $kf(x)$, $f(x)g(x)$, and $f(x)/g(x)$ are also differentiable at $x = x_0$ (provided $g(x_0) \neq 0$ in the last case),*

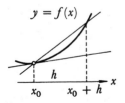

Figure 10

with derivatives

$$[kf(x)]'_{x=x_0} = kf'(x_0), \tag{3}$$

$$[f(x) + g(x)]'_{x=x_0} = f'(x_0) + g'(x_0), \tag{4}$$

$$[f(x)g(x)]'_{x=x_0} = f'(x_0)g(x_0) + f(x_0)g'(x_0), \tag{5}$$

$$\left[\frac{f(x)}{g(x)}\right]'_{x=x_0} = \frac{f'(x_0)g(x_0) - f(x_0)g'(x_0)}{g^2(x_0)}. \tag{6}$$

Proof. To prove (3) and (4), pass to the limit as $h \to 0$ in the identities

$$\frac{kf(x_0 + h) - kf(x_0)}{h} = k\frac{f(x_0 + h) - f(x_0)}{h},$$

$$\frac{[f(x_0 + h) + g(x_0 + h)] - [f(x_0) + g(x_0)]}{h} = \frac{f(x_0 + h) - f(x_0)}{h}$$
$$+ \frac{g(x_0 + h) - g(x_0)}{h},$$

using the familiar rules for calculating limits of sums and products (Theorems 4.36a and 4.36b). To prove (5), pass to the limit as $h \to 0$ in the identity

$$\frac{f(x_0 + h)g(x_0 + h) - f(x_0)g(x_0)}{h} = \frac{f(x_0 + h)g(x_0 + h) - f(x_0)g(x_0 + h)}{h}$$
$$+ \frac{f(x_0)g(x_0 + h) - f(x_0)g(x_0)}{h}$$
$$= g(x_0 + h)\frac{f(x_0 + h) - f(x_0)}{h}$$
$$+ f(x_0)\frac{g(x_0 + h) - g(x_0)}{h},$$

using the same rules and Theorem 7.12. Once having proved (5), we need only prove (6) for the case $f(x) \equiv 1$, passing to the limit as $h \to 0$ in the identity

$$\frac{1}{g(x_0 + h)} - \frac{1}{g(x_0)} = -\frac{g(x_0 + h) - g(x_0)}{g(x_0)g(x_0 + h)},$$

with the help of Theorem 7.12 and the rule for calculating the limit of a reciprocal (Theorem 4.36d). ∎

7.14. Examples

a. The constant function $f(x) \equiv c$ is obviously differentiable everywhere, with derivative 0.

b. The function $f(x) \equiv x$ is differentiable everywhere, with derivative 1.

c. Given any polynomial

$$P(x) = a_0 x^n + a_1 x^{n-1} + \cdots + a_n$$

and any rational function

$$R(x) = \frac{P(x)}{Q(x)} = \frac{a_0 x^n + a_1 x^{n-1} + \cdots + a_n}{b_0 x^m + b_1 x^{m-1} + \cdots + b_m},$$

it follows from Theorem 7.13 that $P(x)$ is differentiable at every point $x = x_0$, while $R(x)$ is differentiable at every point $x = x_0$ such that $Q(x_0) \neq 0$. The derivatives of $P(x)$ and $R(x)$ are easily calculated by using (3)–(6), together with the formula

$$(x^n)' = nx^{n-1}. \tag{7}$$

To prove (7), we use induction (see Sec. 1.43), noting that (7) obviously holds for $n = 1$, while if (7) holds for an integer n, then, by (5),

$$(x^{n+1})' = (x^n \cdot x)' = (x^n)'x + x^n(x') = nx^{n-1} \cdot x + x^n = (n+1)x^n.$$

d. Similarly, it follows from (5) and mathematical induction that the derivative of a product of n functions is given by

$$[f_1(x)f_2(x)\cdots f_n(x)]' = f_1'(x)f_2(x)\cdots f_n(x)$$
$$+ f_1(x)f_2'(x)\cdots f_n(x) + \cdots + f_1(x)f_2(x)\cdots f_n'(x). \tag{8}$$

e. Consider the determinant

$$W(x) = \begin{vmatrix} u_{11}(x) & u_{12}(x) & \cdots & u_{1n}(x) \\ u_{21}(x) & u_{22}(x) & \cdots & u_{2n}(x) \\ \cdot & \cdot & \cdots & \cdot \\ u_{n1}(x) & u_{n2}(x) & \cdots & u_{nn}(x) \end{vmatrix},$$

made up of n^2 differentiable functions $u_{jk}(x)$. By definition, $W(x)$ is the sum of $n!$ terms with appropriate signs, each term a product of n factors, one from each row and each column of the determinant.† Using the rule (8) to differentiate each term and collecting first the terms differentiated with respect to the factors in the first column, then the terms differentiated with respect to the factors in the second column, and so on, we get the formula

† G. E. Shilov, *Linear Algebra*, Sec. 1.31.

$$W'(x) = \begin{vmatrix} u'_{11}(x) & u_{12}(x) & \cdots & u_{1n}(x) \\ u'_{21}(x) & u_{22}(x) & \cdots & u_{2n}(x) \\ \cdot & \cdot & \cdots & \cdot \\ u'_{n1}(x) & u_{n2}(x) & \cdots & u_{nn}(x) \end{vmatrix}$$

$$+ \begin{vmatrix} u_{11}(x) & u'_{12}(x) & \cdots & u_{1n}(x) \\ u_{21}(x) & u'_{22}(x) & \cdots & u_{2n}(x) \\ \cdot & \cdot & \cdots & \cdot \\ u_{n1}(x) & u'_{n2}(x) & \cdots & u_{nn}(x) \end{vmatrix}$$

$$+ \cdots + \begin{vmatrix} u_{11}(x) & u_{12}(x) & \cdots & u'_{1n}(x) \\ u_{21}(x) & u_{22}(x) & \cdots & u'_{2n}(x) \\ \cdot & \cdot & \cdots & \cdot \\ u_{n1}(x) & u_{n2}(x) & \cdots & u'_{nn}(x) \end{vmatrix}.$$

7.15. a. THEOREM (**Differentiation of a composite function**). *Suppose* $y = f(x)$ *is differentiable at* $x = x_0$, *while* $z = g(y)$ *is defined on an interval containing the point* $y_0 = f(x_0)$ *and is differentiable at* $y = y_0$. *Then the composite function*

$$z = h(x) \equiv g(f(x))$$

is differentiable at $x = x_0$, *with derivative*

$$h'(x_0) = g'(y_0)f'(x_0). \tag{9}$$

Proof. By the definition of the derivative,

$$y - y_0 = f(x) - f(x_0) = [f'(x_0) + \varepsilon(x)](x - x_0),$$
$$g(y) - g(y_0) = [g'(y_0) + \delta(y)](y - y_0),$$

where $\varepsilon(x) \to 0$ as $x \to x_0$, $\delta(y) \to 0$ as $y \to y_0$, and hence

$$\begin{aligned} h(x) - h(x_0) &= g(f(x)) - g(f(x_0)) = g(y) - g(y_0) \\ &= [g'(y_0) + \delta(y)](y - y_0) \\ &= [g'(y_0) + \delta(y)][f'(x_0) + \varepsilon(x)](x - x_0). \end{aligned}$$

As $x \to x_0$, $\varepsilon(x) \to 0$ and moreover $y \to y_0$, by the continuity of $y = f(x)$ at the point $x = x_0$ (Theorem 7.12), so that $\delta(y) \to 0$. To get (9), we now take the limit as $x \to x_0$ in the identity

$$\frac{h(x) - h(x_0)}{x - x_0} = [g'(y_0) + \delta(y)][f'(x_0) + \varepsilon(x)]. \quad \blacksquare$$

b. Example. To find the derivative of the function

$$z = (1 + x^2)^{99},$$

we need only write z as a composite function

$$z = y^{99}, \qquad y = 1 + x^2.$$

We then have

$$z' = 99y^{98}, \qquad y' = 2x,$$

and hence

$$z' = 99y^{98} \cdot 2x = 198x(1 + x^2)^{98},$$

by the preceding theorem. It would be the height of folly to expand $(1 + x^2)^{99}$ by the binomial theorem and then use Theorem 7.13 directly.

7.16. THEOREM (**Differentiation of an inverse function**). Let $y = f(x)$ be continuous and increasing on an interval (a,b), with inverse $x = \varphi(y)$, and suppose $y = f(x)$ is differentiable at $x = x_0 \in (a,b)$, with derivative $f'(x_0) \neq 0$. Then the function $x = \varphi(y)$ is differentiable at $y = y_0 = f(x_0)$, with derivative

$$\varphi'(y_0) = \frac{1}{f'(x_0)}.$$

Proof. Since

$$\varphi(y) = x, \quad \varphi(y_0) = x_0, \quad y = f(x), \quad y_0 = f(x_0),$$

where y lies in some neighborhood of y_0, we have

$$\frac{\varphi(y) - \varphi(y_0)}{y - y_0} = \frac{x - x_0}{y - y_0} = \frac{1}{\dfrac{f(x) - f(x_0)}{x - x_0}}. \tag{10}$$

But $\varphi(y)$ is continuous at the point $y = y_0$ (see Theorem 5.35b), and hence $y \to y_0$ implies $x \to x_0$. Therefore, taking the limit of (10) as $y \to y_0$, we get

$$\varphi'(y_0) = \lim_{y \to y_0} \frac{\varphi(y) - \varphi(y_0)}{y - y_0} = \lim_{x \to x_0} \frac{1}{\dfrac{y - y_0}{x - x_0}} = \frac{1}{\displaystyle\lim_{x \to x_0} \frac{f(x) - f(x_0)}{x - x_0}} = \frac{1}{f'(x_0)}. \quad \blacksquare$$

7.17. Differentiation of the logarithm and related functions

a. Let $y = \log_a x$. Then, for $x_0 > 0$,

$$\frac{\log_a (x_0 + h) - \log_a (x_0)}{h} = \log_a \left(\frac{x_0 + h}{x_0} \right)^{1/h} = \log_a \left(1 + \frac{h}{x_0} \right)^{1/h}$$

$$= \frac{1}{h} \log_a \left(1 + \frac{h}{x_0} \right) = \frac{1}{x_0} \left[\frac{\log_a [1 + (h/x_0)]}{h/x_0} \right],$$

where the expression in brackets approaches $\log_a e$ as $h \to 0$ (see Theorem 5.58c). It follows that

$$(\log_a x)'_{x=x_0} = \frac{\log_a e}{x_0}.$$

This formula takes a particularly simple form if $a = e$, in which case we are dealing with natural logarithms (Sec. 5.58c):

$$(\ln x)' = \frac{1}{x}. \tag{11}$$

b. To differentiate the function $y = \ln(-x)$, defined for $x < 0$, we use Theorem 7.15a, obtaining

$$[\ln(-x)]' = \frac{1}{-x}(-x)' = -\frac{1}{x}(-1) = \frac{1}{x}. \tag{11'}$$

Formulas (11) and (11') can be combined into the single formula

$$(\ln |x|)' = \frac{1}{x}, \tag{12}$$

valid for all $x \neq 0$.

c. The exponential $x = a^y$ is the inverse of the logarithm $y = \log_a x$, and hence, by Theorem 7.16,

$$(a^y)'_{y=y_0} = \frac{1}{\log_a' x_0} = \frac{x_0}{\log_a e} = \frac{a^{y_0}}{\log_a e} = a^{y_0} \log_e a = a^{y_0} \ln a.$$

Replacing y_0 by x ($-\infty < x < \infty$), we get the formula

$$(a^x)' = a^x \ln a,$$

which becomes particularly simple for $a = e$:

$$(e^x)' = e^x. \tag{13}$$

d. The function

$$y = x^\alpha, \tag{14}$$

where α is an arbitrary real number, is defined for all $x > 0$ (Sec. 5.53). Writing (14) in the form

$$y = e^{\alpha \ln x}$$

and using Theorem 7.15a, we get

$$y' = e^{\alpha \ln x} \frac{\alpha}{x} = \alpha x^{\alpha - 1}, \tag{15}$$

with the help of (11) and (13).† For $n = 1, 2, \ldots$ this agrees with formula (7), as is to be expected.

7.18. Differentiation of trigonometric functions and their inverses

a. For the function $y = \sin x$, we have

$$\sin (x+h) - \sin x = 2 \sin \frac{h}{2} \cos\left(x + \frac{h}{2}\right)$$

(Sec. 5.63), and hence

$$\frac{\sin (x+h) - \sin x}{h} = \frac{\sin \frac{h}{2}}{\frac{h}{2}} \cos\left(x + \frac{h}{2}\right).$$

But

$$\lim_{x \to 0} \frac{\sin x}{x} = 1$$

(Theorem 5.64b), and hence

$$(\sin x)' = \lim_{h \to 0} \frac{\sin \frac{h}{2}}{\frac{h}{2}} \cos\left(x + \frac{h}{2}\right) = \lim_{h \to 0} \cos\left(x + \frac{h}{2}\right),$$

which implies

$$(\sin x)' = \cos x, \tag{16}$$

by the continuity of the function $\cos x$ (see Theorem 5.64a).

b. It follows from (16) and the formulas

$$\sin\left(\frac{\pi}{2} + x\right) = \cos x, \qquad \cos\left(\frac{\pi}{2} + x\right) = -\sin x$$

† As an exercise, the reader should calculate the derivative of the more complicated function
$$y = [f(x)]^{g(x)} = e^{g(x) \ln f(x)}.$$

(Sec. 5.65b) that

$$(\cos x)' = \left[\sin \left(\frac{\pi}{2} + x \right) \right]' = \cos \left(\frac{\pi}{2} + x \right),$$

and hence

$$(\cos x)' = -\sin x. \tag{17}$$

c. Formulas (6), (16), and (17) in turn imply

$$(\tan x)' = \left(\frac{\sin x}{\cos x} \right)' = \frac{\cos^2 x + \sin^2 x}{\cos^2 x} = \frac{1}{\cos^2 x},$$

or simply

$$(\tan x)' = \sec^2 x, \tag{18}$$

provided of course that cos x does not vanish, i.e., that

$$x \neq \frac{2k+1}{2}\pi \qquad (k=0,\pm 1,\pm 2,...).$$

d. If

$$x = \text{arc sin } u, \qquad u = \sin x \qquad (-\pi/2 < x < \pi/2, \ -1 < u < 1),$$

then, by Theorem 7.16,†

$$(\text{arc sin } u)' = \frac{1}{(\sin x)'} = \frac{1}{\cos x} = \frac{1}{\sqrt{1 - \sin^2 x}},$$

i.e.,

$$(\text{arc sin } u)' = \frac{1}{\sqrt{1 - u^2}}. \tag{19}$$

e. Similarly, if

$$x = \text{arc cos } u, \qquad u = \cos x \qquad (-\pi < x < 0, \ -1 < u < 1),$$

then

$$(\text{arc cos } u)' = \frac{1}{(\cos x)'} = -\frac{1}{\sin x} = -\frac{1}{-\sqrt{1 - \cos^2 x}},$$

i.e.,

$$(\text{arc cos } u)' = \frac{1}{\sqrt{1 - u^2}}. \tag{20}$$

† As always, the radical denotes the *positive* square root.

We can also get (20) by differentiating the formula

$$\text{arc} \cos u = \text{arc} \sin u - \frac{\pi}{2} \qquad (21)$$

(Sec. 5.67) with the help of (19).

More exactly, formula (20) pertains to the "increasing branch" of arc cos u. For the "decreasing branch" we have

$$\text{arc} \cos u = \frac{\pi}{2} - \text{arc} \sin u \qquad (21')$$

instead of (21) and

$$(\text{arc} \cos u)' = -\frac{1}{\sqrt{1-u^2}} \qquad (20')$$

instead of (20).†

f. Finally, if

$$x = \text{arc} \tan u, \qquad u = \tan x \qquad (-\pi/2 < x < \pi/2, \ -\infty < u < \infty),$$

then

$$(\text{arc} \tan u)' = \frac{1}{(\tan x)'} = \cos^2 x = \frac{1}{1+\tan^2 x},$$

i.e.,

$$(\text{arc} \tan u)' = \frac{1}{1+u^2}. \qquad (22)$$

As an exercise, the reader should find the derivatives of the remaining trigonometric functions (cot x, sec x, csc x) and their inverses.

7.19. One-sided derivatives

a. By the *left-hand derivative* of a function $f(x)$, defined on an interval $a < x \leqslant x_0$, we mean the quantity

$$f'_l(x_0) = \lim_{h \nearrow 0} \frac{f(x_0 + h) - f(x_0)}{h}$$

(l for "left"), provided it exists. Instead of $f'_l(x_0)$ we sometimes write $f'(x_0 - 0)$.

b. By the *right-hand derivative* of a function $f(x)$, defined on an interval

† See Sec. 5.67, esp. Figure 7.

$x_0 \leqslant x < b$, we mean the quantity

$$f'_r(x_0) = \lim_{h \searrow 0} \frac{f(x_0 + h) - f(x_0)}{h}$$

(r for "right"), provided it exists. Instead of $f'_r(x_0)$ we sometimes write $f'(x_0 + 0)$.

c. If a function $f(x)$ is defined on an interval (a,b) containing the point x_0, we can inquire as to the existence of the "one-sided derivatives" $f'_l(x_0)$ and $f'_r(x_0)$. Obviously, if the ordinary derivative $f'(x_0)$ exists, then both $f'_l(x_0)$ and $f'_r(x_0)$ exist and satisfy the equality

$$f'_l(x_0) = f'_r(x_0) = f'(x_0).$$

But it may well happen that $f'_l(x_0)$ and $f'_r(x_0)$ both exist, without $f'(x_0)$ existing (see Figure 11). On the other hand, if $f'_l(x_0)$ and $f'_r(x_0)$ both exist *and are equal*, then $f'(x_0)$ exists and equals the common value of $f'_l(x_0)$ and $f'_r(x_0)$. This follows at once from the corresponding theorem on "partial limits" (Theorem 4.16c).

d. The ray defined by the equation

$$y = f(x_0) + f'_l(x_0)(x - x_0) \qquad (x \leqslant x_0) \tag{23}$$

is called the *left-hand tangent* to the curve $y = f(x)$ at the point $x = x_0$. Similarly, the ray defined by the equation

$$y = f(x_0) + f'_r(x_0)(x - x_0) \qquad (x \geqslant x_0) \tag{23'}$$

is called the *right-hand tangent*† to the curve $y = f(x)$ at the point $x = x_0$.

e. Theorem 7.16 is easily generalized to the case of one-sided derivatives:

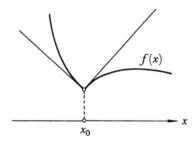

Figure   11

† The term "semitangent" can be used instead of "tangent" to emphasize that (23) and (23′) are rays (half-lines) rather than "full" straight lines.

THEOREM. *Let $y = f(x)$ be continuous and increasing on an interval (a,b), with inverse $x = \varphi(y)$, and suppose $y = f(x)$ has a left-hand derivative $f'(x_0 - 0) \neq 0$ at $x = x_0 \in (a,b)$. Then the function $x = \varphi(y)$ has a left-hand derivative*

$$\varphi'(y_0 - 0) = \frac{1}{f'(x_0 - 0)}$$

at $y = y_0 = f(x_0)$.

Proof. The same as that of Theorem 7.16, except for the restriction that $x \leqslant x_0, y \leqslant y_0$. ∎

The theorem obviously remains true if we change "left-hand" to "right-hand" and -0 to $+0$.

7.2. Properties of Differentiable Functions

7.21. We now turn to a study of general properties of differentiable functions.

THEOREM. *Given a function $y = f(x)$, consider all linear functions with the same value as $f(x)$ at the point $x = x_0$, i.e., all functions of the form*

$$y_A = A(x - x_0) + f(x_0) = Ah + f(x_0) \qquad (h = x - x_0). \tag{1}$$

Then $y = f(x)$ is differentiable at $x = x_0$, with derivative $f'(x_0)$, if and only if there exists a linear function (1) such that

$$y - y_A = \varepsilon(h)h \qquad (h = x - x_0), \tag{2}$$

where $\varepsilon(h) \to 0$ as $h \to 0$,† in which case $A = f'(x_0)$.

Proof. If (2) holds, then

$$\frac{y - y_A}{h} = \frac{f(x) - Ah - f(x_0)}{h} = \frac{f(x_0 + h) - f(x_0)}{h} - A = \varepsilon(h) \to 0$$

as $h \to 0$, and hence

$$f'(x_0) = \lim_{h \to 0} \frac{f(x_0 + h) - f(x_0)}{h}$$

exists and equals A. Conversely, if $f(x)$ is differentiable at x_0, with deriva-

---

† That is, if and only if there exists a linear function (1) whose deviation from the function $y = f(x)$ is "of a higher order of smallness than h." Note that two linear functions y_A and y_B differ from one another by a quantity "of the same order of smallness as h," in fact by the quantity $(A - B)h$ proportional to h.

tive $f'(x_0)$, then

$$\frac{f(x_0+h)-f(x_0)}{h}-f'(x_0)=\varepsilon(h)\to 0$$

as $h\to 0$, and hence

$$f(x_0+h)-[f'(x_0)h+f(x_0)]=\varepsilon(h)h. \tag{3}$$

Noting that $f(x_0+h)=f(x)=y$ and choosing

$$y_A=Ah+f(x_0)=f'(x_0)h+f(x_0),$$

we get (2). ∎

7.22. By definition (Sec. 7.11), the straight line in the xy-plane with equation

$$y=f'(x_0)(x-x_0)+f(x_0) \tag{4}$$

is the tangent to the curve $y=f(x)$ at $x=x_0$ (more exactly, at the point with abscissa x_0). According to the above theorem, given any $\varepsilon>0$, there is a $\delta>0$ such that $|h|<\delta$ implies $|\varepsilon(h)|<\varepsilon$, where $\varepsilon(h)h$ is the deviation of the linear function (4) from the curve $y=f(x)$. Thus

$$-\varepsilon h<f(x)-f'(x_0)h-f(x_0)<\varepsilon h$$

if $|h|<\delta$, i.e.,

$$[f'(x_0)h-\varepsilon]h+f(x_0)<f(x)<[f'(x_0)h+\varepsilon]h+f(x_0) \tag{5}$$

if $|h|<\delta$. This has the following geometric interpretation: If $f(x)$ is differentiable at $x=x_0$, then near $x=x_0$ the curve $y=f(x)$ lies between two straight lines making arbitrarily small angles with the tangent to the curve at $x=x_0$ (see Figure 12). It follows at once that *if $f'(x_0)<0$, there is a $\delta>0$ such that*

$$f(x_0-h)>f(x_0)>f(x_0+h) \tag{6}$$

whenever $0<h<\delta$, while if $f'(x_0)>0$, there is a $\delta>0$ such that

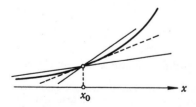

Figure   12

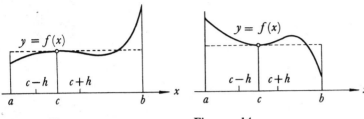

Figure 13   **Figure 14**

$$f(x_0 - h) < f(x_0) < f(x_0 + h) \tag{7}$$

whenever $0 < h < \delta$.

7.23. A function $f(x)$ is said to have a *local maximum* at a point $c \in (a,b)$ if there exists an $h > 0$ such that $f(x) \leqslant f(c)$ for all $x \in (c-h, c+h)$. Similarly, a function $f(x)$ is said to have a *local minimum* at a point $c \in (a,b)$ if there exists an $h > 0$ such that $f(x) \geqslant f(c)$ for all $x \in (c-h, c+h)$. If $f(x)$ has a local maximum or a local minimum at a point c, we say that $f(x)$ has a *local extremum* at c.

Figure 13 illustrates the geometric meaning of a local maximum, while Figure 14 illustrates that of a local minimum.

THEOREM. *If $f(x)$ is differentiable at a point $x = x_0$ where it has a local extremum, then $f'(x_0) = 0$.*

Proof. If $f'(x_0) \neq 0$, then $f(x)$ cannot have a local extremum at x_0, because of the inequalities (6) and (7). ∎

Thus to find the points at which an everywhere differentiable function $f(x)$ has local extrema, we must analyze the equation

$$f'(x) = 0, \tag{8}$$

since the points in question must be among the solutions of (8). Note, however, that $f'(x)$ may well vanish at a point c where there is no local extremum, as for example in the case of the function $f(x) = x^3$ at the point $c = 0$. A more detailed analysis of extrema will be given in Secs. 7.5 and 8.4.

7.3. The Differential

7.31. It follows from formula (3) of the preceding page that the *increment*

$$\Delta y = f(x_0 + h) - f(x_0)$$

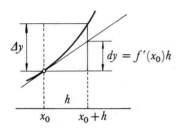

$$\Delta y \qquad\qquad dy = f'(x_0)h$$

$$h$$

$$x_0 \qquad x_0 + h$$

Figure   15

of the function $y = f(x)$, when the independent variable x changes from x_0 to $x_0 + h$, is the sum of two parts, a part $f'(x_0)h$ which is linear in the increment h of the independent variable and a part $\varepsilon(h)h$ which is "of a higher order of smallness than h." Geometrically, the first part is the increment of $f(x)$ as measured along the tangent, while the second part is the difference between the "true value" $f(x_0 + h)$ and the value as measured along the tangent (see Figure 15). The term $f'(x_0)h$ is called the *principal linear part* of the increment Δy. Thus the existence of the derivative implies the possibility of separating out a principal linear part from the increment of the function, and conversely (this is the content of Theorem 7.21).

7.32. Suppose $f(x)$ is differentiable at $x = x_0$, and let the quantity $h = x - x_0$ be denoted by dx and the quantity

$$f'(x_0)h = f'(x_0)\, dx$$

by $dy(x_0)$, or more briefly by dy or df. Then the quantities dx and dy are called *differentials*, the first the differential of the independent variable x, the second the differential of the dependent variable y (at the point $x = x_0$). Note that the differential dx is just the increment h of the independent variable x, while the differential dy is just the principal linear part of the increment of the dependent variable y. Thus dy is a linear function of dx. Using dx and dy, we can write $f'(x)$ as a ratio of differentials:

$$f'(x) = \frac{dy}{dx} = \frac{df}{dx}.$$

7.33. The rules for calculating derivatives summarized in Theorem 7.13 lead to corresponding rules for calculating differentials. In fact, multiplying both sides of formulas (3)–(6), p. 224 by dx, we get

$$d(kf) = k\, df, \tag{1}$$

$$d(f + g) = df + dg, \tag{2}$$

$$d(fg) = f\,dg + g\,df, \tag{3}$$

$$d\left(\frac{f}{g}\right) = \frac{g\,df - f\,dg}{g^2(x_0)} \qquad (g(x_0) \neq 0). \tag{4}$$

7.34. The differential of a composite function. Multiplying both sides of formula (9), p. 226 by dx, we get

$$dz = h'(x_0)\,dx = g'(y_0)f'(x_0)\,dx = g'(y_0)\,dy.$$

But if y were the independent variable rather than a function of x, the differential of the function z would take precisely the same form

$$dz = g'(y_0)\,dy.$$

Hence *the differential of a function does not depend on whether its argument is the independent variable or a function of some other independent variable.* This "invariance property" can be used to differentiate composite functions. Thus the function $(x^2 + 1)^{99}$ has the differential

$$d(x^2 + 1)^{99} = 99(x^2 + 1)^{98}\,d(x^2 + 1) = 99(x^2 + 1)^{98}2x\,dx,$$

and hence

$$[(x^2 + 1)^{99}]' = \frac{d(x^2 + 1)^{99}}{dx} = 198x(x^2 + 1)^{98},$$

just as in Example 7.15b.

7.4. Mean Value Theorems

We now establish a class of results called "mean value theorems," relating the values of a function at the end points of a closed interval to the value of its derivative at a suitable interior point of the interval.

7.41. THEOREM (**Rolle's theorem**). *Suppose a (finite) function $f(x)$ is continuous on a (possibly infinite†) closed interval $[a,b]$ and differentiable at every point of the open interval (a,b), and suppose further that $f(a) = f(b)$. Then there exists a point $c \in (a,b)$ such that $f'(c) = 0$.*

Proof. By a simple argument involving Weierstrass' theorem (see Sec. 5.16c), there is a point $c \in (a,b)$ such that

$$f(c) = \sup_{a \leq x \leq b} f(x)$$

† In this case, we regard $[a,b]$ as an interval of the extended real number system $\bar{R}$ (see Sec. 1.9).

or

$$f(c) = \inf_{a \le x \le b} f(x).$$

Hence $f(x)$ has a local extremum at the point c. But then $f'(c) = 0$, by Theorem 7.23. ∎

7.42. THEOREM. *Let $f(x)$ have the same continuity and differentiability properties as in Rolle's theorem, and suppose $f'(x) \ne 0$ for all $x \in (a,b)$. Then $f(a) \ne f(b)$.*

Proof. An obvious consequence of Rolle's theorem. ∎

7.43. THEOREM (**Cauchy's theorem**). *Suppose two (finite) functions $f(x)$ and $g(x)$ are continuous on a (possibly infinite) closed interval $[a,b]$ and differentiable at every point of the open interval (a,b), and suppose further that $g'(x)$ does not vanish at any point of (a,b). Then there exists a point $c \in (a,b)$ such that*

$$\frac{f(b) - f(a)}{g(b) - g(a)} = \frac{f'(c)}{g'(c)}. \tag{1}$$

Proof. First we note that $g(b) \ne g(a)$, by Theorem 7.42. The function

$$\varphi(x) = f(x) - Ag(x)$$

(A any constant) has the same continuity and differentiability properties as the functions $f(x)$ and $g(x)$ themselves. Let A be such that $\varphi(a) = \varphi(b)$, as in Rolle's theorem. Then A satisfies the equation

$$f(a) - Ag(a) = f(b) - Ag(b),$$

and hence

$$A = \frac{f(b) - f(a)}{g(b) - g(a)}.$$

Applying Rolle's theorem to the function $\varphi(x)$, we find that there exists a point $c \in (a,b)$ such that

$$\varphi'(c) = f'(c) - \frac{f(b) - f(a)}{g(b) - g(a)} g'(c) = 0. \tag{2}$$

But (2) is equivalent to (1). ∎

7.44. THEOREM (**Lagrange's theorem**). *If $f(x)$ is continuous on a finite closed interval $[a,b]$ and differentiable at every interior point of $[a,b]$, then there exists a point $c \in (a,b)$ such that*

$$f'(c) = \frac{f(b) - f(a)}{b - a}. \tag{3}$$

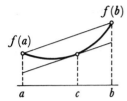

Figure   16

Proof. Choose $g(x) = x$ in Cauchy's theorem.   ∎

Another version of (3) is the "finite difference formula"

$$f(b) = f(a) + f'(c)(b - a). \tag{3'}$$

The geometric meaning of the point c, implied by (3) and shown in Figure 16, is simply that the tangent to the curve $y = f(x)$ at the point with abscissa c is parallel to the chord joining the points $(a, f(a))$ and $(b, f(b))$.

7.45. a. THEOREM. *The function $f(x)$ is decreasing on $[a,b]$ if $f'(x) < 0$ for all $x \in (a,b)$ and increasing on $[a,b]$ if $f'(x) > 0$ for all $x \in (a,b)$.*

Proof. Choosing $a = x', b = x''$ in (3'), where $[x', x''] \subset [a,b]$, we get

$$f(x'') = f(x') + (x'' - x')f'(c) < f(x')$$

if $f'(c) < 0$ and

$$f(x'') = f(x') + (x'' - x')f'(c) > f(x')$$

if $f'(c) > 0$.   ∎

b. THEOREM. *The function $f(x)$ is constant on $[a,b]$ if $f'(x)$ vanishes for all $x \in (a,b)$.*

Proof. This time

$$f(x'') = f(x') + (x'' - x')f'(c) = f(x'),$$

since $f'(c) = 0$.   ∎

c. Example. Let

$$f(x) = \sin x, \quad f'(x) = \cos x.$$

Then $\sin x$ is increasing on the intervals

$$(2k - \tfrac{1}{2})\pi < x < (2k + \tfrac{1}{2})\pi \qquad (k = 0, \pm 1, \pm 2, \ldots)$$

on which cos x is positive and decreasing on the intervals

$$(2k+\tfrac{1}{2})\pi<x<(2k+\tfrac{3}{2})\pi \qquad (k=0,\pm1,\pm2,\dots)$$

on which cos x is negative (see Sec. 5.65). Similarly, choosing

$$f(x)=\cos x, \qquad f'(x)=-\sin x,$$

we see that cos x is increasing on the intervals

$$(2k-1)\pi<x<2k\pi \qquad (k=0,\pm1,\pm2,\dots)$$

on which $-\sin x$ is positive and decreasing on the intervals

$$2k\pi<x<(2k+1)\pi \qquad (k=0,\pm1,\pm2,\dots)$$

on which $-\sin x$ is negative. These results have already been found by other means (Sec. 5.65), but the present approach is much simpler.

7.5. Concavity and Inflection Points

7.51. Given a function $f(x)$ continuous on an interval (a,b) and differentiable at a point $x_0 \in (a,b)$, let $y=T(x)$ be the tangent to the curve $y=f(x)$ at the point (with abscissa) x_0, so that

$$y=T(x)=f'(x_0)(x-x_0)+f(x_0)$$

(see Sec. 7.22). Then $f(x)$ is said to be *concave downward* at x_0 if there is a deleted neighborhood of x_0† in which $f(x)<T(x)$, i.e., in which the curve $y=f(x)$ lies below its tangent at x_0 (see Figure 17). Similarly, $f(x)$ is said to be *concave upward* at x_0 if there is a deleted neighborhood of x_0 in which

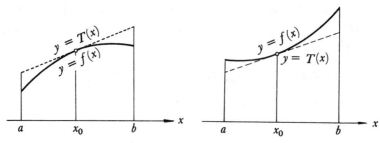

Figure   17   Figure   18

† By a *neighborhood* of a point x_0 we mean a set of the form $U_\delta=\{x:|x-x_0|<\delta\}$ and by a *deleted neighborhood* of x_0 we mean a set U_δ minus the point x_0 itself, i.e., a set of the form $U_\delta'=\{x:0<|x-x_0|<\delta\}$ (cf. Sec. 4.15a).

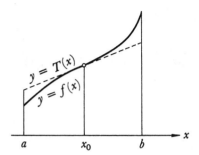

Figure 19

$f(x) > T(x)$, i.e., in which the curve $y = f(x)$ lies above its tangent at x_0 (see Figure 18). If $f(x)$ is concave downward (or upward) at every point $x \in (a,b)$, we say that $f(x)$ is concave upward (or downward) *on the interval* (a,b). Finally, a point x_0 is said to be an *inflection point* of the function $f(x)$ if the curve $y = f(x)$ lies on one side of its tangent (at x_0) if $x < x_0$ and on the other side of its tangent if $x > x_0$ (see Figure 19).

7.52. a. LEMMA. *Let C be the curve with equation*

$$y = f(x) \qquad (a < x < b),$$

and let T be the tangent to C at the point x_0. Then C lies above T to the left of x_0 $(x < x_0)$ if $f'(\xi) < f'(x_0)$ for all $\xi \in (a,x_0)$, while C lies below T to the left of x_0 if $f'(\xi) > f'(x_0)$ for all $\xi \in (a,x_0)$.

Proof. Choosing $a = x_0, b = x, c = \xi$ in the finite difference formula (3′), p. 239, we get

$$f(x) = f(x_0) + f'(\xi)(x - x_0) > f(x_0) + f'(x_0)(x - x_0) = T(x) \tag{1}$$

$(x - x_0 < 0)$ if $f'(\xi) < f'(x_0)$, while $f(x) < T(x)$ if $f'(\xi) > f'(x_0)$. ∎

b. LEMMA. *Let C and T be the same as in the preceding lemma. Then C lies above T to the right of x_0 $(x > x_0)$ if $f'(\eta) > f'(x_0)$ for all $\eta \in (x_0,b)$, while C lies below T to the right of x_0 if $f'(\eta) < f'(x_0)$ for all $\eta \in (x_0,b)$.*

Proof. Instead of (1), we now have

$$f(x) = f(x_0) + f'(\eta)(x - x_0) > f(x_0) + f'(x_0)(x - x_0) = T(x)$$

$(x - x_0 > 0)$ if $f'(\eta) > f'(x_0)$, while $f(x) < T(x)$ if $f'(\eta) < f'(x_0)$. ∎

7.53. a. THEOREM. *The function $f(x)$ is concave downward at x_0 if*

$$f'(\xi) > f'(x_0) > f'(\eta)$$

whenever $a < \xi < x_0 < \eta < b$ and concave upward at x_0 if

$$f'(\xi) < f'(x_0) < f'(\eta)$$

whenever $a < \xi < x_0 < \eta < b$. On the other hand, x_0 is an inflection point of $f(x)$ if

$$f'(\xi) > f'(x_0), \qquad f'(\eta) > f'(x_0)$$

or

$$f'(\xi) < f'(x_0), \qquad f'(\eta) < f'(x_0)$$

whenever $a < \xi < x_0 < \eta < b$.

Proof. An immediate consequence of the above two lemmas. ∎

b. THEOREM. *The function $f(x)$ is concave downward on (a,b) if $f'(x)$ is decreasing on (a,b) and concave upward on (a,b) if $f'(x)$ is increasing on (a,b).*

Proof. An immediate consequence of the preceding theorem. ∎

Further tests for concavity and inflection points, based on the use of higher derivatives of $f(x)$, will be given in Sec. 8.3.

c. Example. Let

$$f(x) = \sin x, \qquad f'(x) = \cos x.$$

Then $\sin x$ is concave upward on the intervals

$$(2k-1)\pi < x < 2k\pi \qquad (k = 0, \pm 1, \pm 2, \ldots)$$

on which $\cos x$ is increasing and concave downward on the intervals

$$2k\pi < x < (2k+1)\pi \qquad (k = 0, \pm 1, \pm 2, \ldots)$$

on which $\cos x$ is decreasing (see Example 7.45c). Similarly, choosing

$$f(x) = \cos x, \qquad f'(x) = -\sin x,$$

we see that $\cos x$ is concave downward on the intervals

$$(2k-\tfrac{1}{2})\pi < x < (2k+\tfrac{1}{2})\pi \qquad (k = 0, \pm 1, \pm 2, \ldots)$$

on which $-\sin x$ is decreasing and concave upward on the intervals

$$(2k+\tfrac{1}{2})\pi < x < (2k+\tfrac{3}{2})\pi \qquad (k = 0, \pm 1, \pm 2, \ldots)$$

on which $-\sin x$ is increasing.

7.54. THEOREM. *Suppose $f'(x_0) = 0$. Then $f(x)$ has a local maximum at x_0 if*

$$f'(\xi) > 0 > f'(\eta)$$

whenever $a < \xi < x_0 < \eta < b$ and a local minimum at x_0 if

$$f'(\xi) < 0 < f'(\eta)$$

whenever $a < \xi < x_0 < \eta < b$. On the other hand, $f(x)$ has neither a maximum nor a minimum at x_0 if

$$f'(\xi) > 0, \qquad f'(\eta) > 0$$

or

$$f'(\xi) < 0, \qquad f'(\eta) < 0$$

whenever $a < \xi < x_0 < \eta < b$.

Proof. If $f'(x_0) = 0$, then the tangent T to the curve $y = f(x)$ at the point x_0 is horizontal. Under these circumstances, if $f(x)$ is concave downward at x_0, then $f(x) < f(x_0)$ in a deleted neighborhood of x_0, so that $f(x)$ has a local maximum at x_0, while if $f(x)$ is concave upward at x_0, then $f(x) > f(x_0)$ in a deleted neighborhood of x_0, so that $f(x)$ has a local minimum at x_0. On the other hand, if x_0 is an inflection point of $f(x)$, then the curve $y = f(x)$ goes from one side of the tangent T to the other as it passes through the point x_0, so that $f(x)$ can have neither a maximum nor a minimum at x_0. The rest of the proof is now an immediate consequence of Theorem 7.53a, with $f'(x_0) = 0$. ∎

7.6. L'Hospital's Rules

The next two theorems are often useful in evaluating limits. They apply to limits of the form

$$\lim_{x \searrow a} \frac{f(x)}{g(x)},$$

where in the first case $f(x)$ and $g(x)$ both approach 0 as $x \searrow a$, leading to the "indeterminate form" $0/0$, while in the second case $f(x)$ and $g(x)$ both approach ∞ as $x \searrow a$, leading to the "indeterminate form" ∞/∞.

7.61. THEOREM (**L'Hospital's rule for 0/0**). *Suppose the (finite) functions $f(x)$ and $g(x)$ are continuous on the (possibly infinite) closed interval $[a,b]$ and differentiable on the open interval (a,b).† Suppose further that $g'(x) \neq 0$ for all $x \in (a,b)$, while $f(a) = g(a) = 0$. Then*

$$\lim_{x \searrow a} \frac{f'(x)}{g'(x)} = A \qquad (A \in \bar{R}) \tag{1}$$

† A function is said to be differentiable *on a set E* if it is differentiable at every point of E.

implies

$$\lim_{x \searrow a} \frac{f(x)}{g(x)} = A. \tag{2}$$

Proof. Suppose first that $-\infty < A < \infty$. Then, given any $\varepsilon > 0$, choose x_0 such that

$$\left| \frac{f'(x)}{g'(x)} - A \right| < \varepsilon \tag{3}$$

if $a < x < x_0$. By Cauchy's theorem (Sec. 7.43) applied to the interval $[a,b]$, there is a point $c \in (a,x)$ such that

$$\frac{f(x)}{g(x)} = \frac{f(x) - f(a)}{g(x) - g(a)} = \frac{f'(c)}{g'(c)}.$$

It follows that

$$\left| \frac{f(x)}{g(x)} - A \right| = \left| \frac{f'(c)}{g'(c)} - A \right| < \varepsilon$$

if $a < x < x_0$, which implies (2).

   In the case where A is infinite, the inequality (3) is replaced by

$$\frac{f'(x)}{g'(x)} > \frac{1}{\varepsilon}$$

or

$$\frac{f'(x)}{g'(x)} < -\frac{1}{\varepsilon},$$

depending on the sign of A, and the rest of the proof is the same. ∎

7.62. THEOREM (**L'Hospital's rule for** ∞/∞). *Suppose the functions $f(x)$ and $g(x)$ are continuous and differentiable on the (possibly infinite) open interval (a,b). Suppose further that $g'(x) \neq 0$ for all $x \in (a,b)$, while $f(a) = g(a) = \infty$. Then (1) implies (2).*

Proof. Suppose first that $-\infty < A < \infty$. As before, given any $\varepsilon > 0$, choose x_0 such that (3) holds if $a < x < x_0$. Defining a function $D(x,x_0)$ by the condition

$$\frac{f(x)}{g(x)} = \frac{f(x) - f(x_0)}{g(x) - g(x_0)} D(x,x_0),$$

we see that

$$D(x,x_0) = \frac{\dfrac{g(x)-g(x_0)}{g(x)}}{\dfrac{f(x)-f(x_0)}{f(x)}} = \frac{1-\dfrac{g(x_0)}{g(x)}}{1-\dfrac{f(x_0)}{f(x)}} \to 1 \tag{4}$$

as $x \searrow a$. By Cauchy's theorem, applied this time to the interval $[x,x_0]$, there is a point $c \in (x,x_0)$ such that

$$\frac{f(x)}{g(x)} = \frac{f'(c)}{g'(c)} D(x,x_0) = \frac{f'(c)}{g'(c)} + \frac{f'(c)}{g'(c)}[D(x,x_0)-1].$$

Hence for all x such that $|D(x,x_0)-1| < \varepsilon$ we have

$$\left|\frac{f(x)}{g(x)} - A\right| \leqslant \left|\frac{f'(c)}{g'(c)} - A\right| + \left|\frac{f'(c)}{g'(c)}\right| |D(x,x_0)-1| < \varepsilon(|A|+\varepsilon)\varepsilon \tag{5}$$

which implies (2), since ε is arbitrarily small.

In the case where A is infinite, say $A = \infty$, the inequality (3) is replaced by

$$\frac{f'(x)}{g'(x)} > \frac{1}{\varepsilon}$$

and the inequality (5) by

$$\frac{f(x)}{g(x)} \geqslant \frac{f'(c)}{g'(c)} \frac{1}{2} > \frac{1}{2\varepsilon}$$

for all x sufficiently near a, because of (4). The case $A = -\infty$ is treated similarly. ∎

There are obvious analogues of L'Hospital's rules for the case where (1) and (2) are replaced by

$$\lim_{x \nearrow b} \frac{f'(x)}{g'(x)} = A \qquad (A \in \bar{R}) \tag{1'}$$

and

$$\lim_{x \nearrow b} \frac{f(x)}{g(x)} = A. \tag{2'}$$

Problems

1. Suppose $f(x)$ is defined on $[a,b]$ and satisfies the inequality

$$|f(x_1) - f(x_2)| \leqslant C|x_1 - x_2|^{1+\alpha} \qquad (\alpha > 0)$$

for all $x_1, x_2 \in [a,b]$. Prove that $f(x)$ is a constant.

2. Let $f(x)$ be differentiable for all $x > c$, and suppose that

$$\lim_{x \to \infty} f'(x) = 0.$$

Prove that

$$\lim_{x \to \infty} [f(x+h) - f(x)] = 0$$

for any $h > 0$.

3. Suppose $f(x)$ has a continuous derivative at every point of the interval $[a,b]$. Prove that $f(x)$ is *uniformly differentiable* on $[a,b]$ in the sense that given any $\varepsilon > 0$, there exists a $\delta > 0$ such that

$$\left| f'(x_1) - \frac{f(x_2) - f(x_1)}{x_2 - x_1} \right| < \varepsilon$$

for every pair of points $x_1, x_2 \in [a,b]$ with $|x_1 - x_2| < \delta$.

4. Prove that the function

$$y = x^2 \sin \frac{1}{x}$$

has a derivative y' at every point of $[0,1]$, but y' fails to be continuous on $[0,1]$.

5 (*Darboux's theorem*). Prove that if $f(x)$ is differentiable on $[a,b]$, then $f'(x)$ takes every value between $f'(a)$ and $f'(b)$.

6. Suppose $f(x)$ is concave upward on (a,b), as defined in Sec. 7.51. Prove that

$$f(x) \leqslant \frac{f(b)(x-a) + f(a)(b-x)}{b-a}$$

for all $x \in [a,b]$, i.e., the curve $y = f(x)$ does not go above the chord joining the points of the curve with abscissas a and b.

7. Guided by the preceding problem, suppose that a function $f(x)$ defined on a closed interval $[a,b]$ has the property that

$$f(x) \leqslant \frac{f(\beta)(x-\alpha) + f(\alpha)(\beta-x)}{\beta-\alpha} \qquad (1)$$

for all x in any interval $(\alpha,\beta) \subset [a,b]$, so that the curve $y = f(x)$ does not go above the chord joining the points of the curve with abscissas α and β. Then $f(x)$ is said to be *convex* on $[a,b]$. This generalizes the notion of a function which is concave upward, since we make no assumptions about the differentiability of $f(x)$.

Prove that if $f(x)$ is convex on $[a,b]$, then

$$f\left(\sum_{j=1}^{n} \lambda_j x_j\right) \leqslant \sum_{j=1}^{n} \lambda_j f(x_j)$$

for arbitrary $x_1,\ldots,x_n$ in $[a,b]$ and arbitrary numbers $\lambda_1,\ldots,\lambda_n$ such that $0 \leqslant \lambda_j \leqslant 1$ $(j=1,\ldots,n)$ and

$$\sum_{j=1}^{n} \lambda_j = 1.$$

The notion of convexity is explored further in the next six problems.

8. Consider the function

$$p(\alpha,x) = \frac{f(x) - f(\alpha)}{x - \alpha} \qquad (a \leqslant \alpha < x < b),$$

i.e., the slope of the chord joining the points of the curve $y = f(x)$ with abscissas α and β. Prove that $f(x)$ is convex on $[a,b]$ if and only if $p(\alpha,x)$ is a nondecreasing function of x for every fixed α.

9. Prove that a function $f(x)$ convex on $[a,b]$ is continuous on (a,b) and has finite left-hand and right-hand derivatives $f_l'(x)$ and $f_r'(x)$ at every point of (a,b). Prove that $f_l'(x) \leqslant f_r'(x)$.

10. Given that $f(x)$ is convex on $[a,b]$, prove that $f_l'(x)$ and $f_r'(x)$ are nondecreasing on (a,b). Prove that

$$f_r'(\alpha) \leqslant \frac{f(\beta) - f(\alpha)}{\beta - \alpha} \leqslant f_l'(\beta)$$

whenever $a < \alpha < \beta < b$.

11. Prove that if $f(x)$ is convex on $[a,b]$, then there are no more than countably many points of $[a,b]$ at which $f(x)$ fails to have a derivative.

12. Let $f(x)$ be differentiable on $[a,b]$ and suppose that, given any two points α,β $(a \leqslant \alpha < \beta \leqslant b)$, there exists a unique point γ such that

$$f'(\gamma) = \frac{f(\beta) - f(\alpha)}{\beta - \alpha}. \tag{2}$$

Prove that either $f(x)$ or $-f(x)$ is convex on $[a,b]$.

13. Prove that if $f(x)$ is convex in a neighborhood of a point $x_0 \in (a,b)$, then its graph does not go below either the left-hand tangent or the right-hand tangent at $x = x_0$ (Sec. 7.19d), i.e.,

$$f(x) \geqslant (x - x_0) f_l'(x_0) + f(x_0) \qquad (x < x_0),$$
$$f(x) \geqslant (x - x_0) f_r'(x_0) + f(x_0) \qquad (x > x_0).$$

14. Prove that a differentiable curve $y=f(x)$ defined for all sufficiently large x has an asymptote $y=kx+b$ (see Chapter 4, Problem 10) if and only if both limits

$$k= \lim_{x \to \infty} f'(x), \qquad b= \lim_{x \to \infty} [f(x) -xf'(x)]$$

exist.

15. Given a function $f(x)$ defined on $[a,b]$, suppose $f(x)$ has a derivative $f'(x)$ at every point of (a,b), where $f'(x) \to p$ as $x \searrow a$. Prove that p is the right-hand derivative of $f(x)$ at $x=a$.

16. Suppose $f(t)$ is increasing on the interval $0 \leqslant t \leqslant b$, a with decreasing derivative $f'(t)$ on $0 < t \leqslant b$ (the derivative may become arbitrarily large as $t \searrow 0$). Prove that

$$\sum_{n=1}^{\infty} \frac{1}{n^2} f'\left(\frac{b}{n}\right) < \infty.$$

17. Following van der Waerden, let

$$\varphi_0(x) = \begin{cases} x & \text{if } 0 \leqslant x \leqslant \frac{1}{2}, \\ 1-x & \text{if } \frac{1}{2} \leqslant x \leqslant 1, \end{cases}$$

and then continue $\varphi_0(x)$ by periodicity with period 1 (see Sec. 5.65c) over the whole real line. Next let

$$\varphi_n(x) = \frac{1}{4^n} \varphi_0(4^n x),$$

so that $\varphi_n(x)$ has the period 4^{-n} and a derivative everywhere, equal to either 1 or -1, except at the points with abscissas $p/2 \cdot 4^n$ $(p=0, \pm 1, \pm 2, \ldots)$.[†] Prove that the function

$$f(x) = \sum_{n=0}^{\infty} \varphi_n(x)$$

is continuous everywhere but fails to have a derivative at every point x_0.

[†] These points correspond to "corners" of the curve $y = \varphi_n(x)$.

8 Higher Derivatives

8.1. Definitions and Examples

8.11. Let $y = f(x)$ be a function differentiable on the interval (a,b), with derivative $f'(x)$, and suppose the function $f'(x)$ is in turn differentiable on (a,b). Then the derivative of $f'(x)$ is called the *second derivative* of $f(x)$, denoted by $y'' = f''(x)$. Repeating this process, we get a sequence of functions

$$f(x), f'(x), f''(x), \ldots, f^{(n)}(x), \ldots$$

on the interval (a,b), where

$$f^{(n)}(x) = [f^{(n-1)}(x)]' \qquad (n = 1, 2, \ldots).$$

The function $f^{(n)}(x)$ is called the *nth derivative* (or the *derivative of order n*) of $f(x)$. The operation leading from $f(x)$ to $f^{(n)}(x)$ is called *n-fold differentiation* of $f(x)$. If the function $f'(x)$ exists and is continuous, we say that $f(x)$ is *smooth* (or *continuously differentiable*). The function $f(x)$ itself is regarded as its own derivative of order zero:

$$f(x) = f^{(0)}(x).$$

8.12. a. THEOREM. *If $f(x)$ and $g(x)$ have derivatives of order n at every point of (a,b), then*

$$[kf(x)]^{(n)} = kf^{(n)}(x),$$
$$[f(x) + g(x)]^{(n)} = f^{(n)}(x) + g^{(n)}(x)$$

for arbitrary real k and every $x \in (a,b)$.

Proof. Apply formulas (3) and (4), p. 224 repeatedly. ∎

b. THEOREM (**Leibniz's rule**). *If $f(x)$ and $g(x)$ have derivatives of order n at every point of (a,b), then*

$$[f(x)g(x)]^{(n)} = \sum_{k=0}^{n} C_k^n f^{(k)}(x) g^{(n-k)}(x), \tag{1}$$

in terms of the binomial coefficients†

$$C_k^n = \frac{n!}{k!(n-k)!}.$$

Proof. The proof is by induction. If (1) holds for some n, then differentiating

† By definition $0! = 1$, $n! = 1 \cdot 2 \cdots n$.

(1) again, we get

$$(fg)^{(n+1)} = \sum_{k=0}^{n} C_k^n [f^{(k)}g^{(n-k+1)} + f^{(k+1)}g^{(n-k)}]$$

$$= \sum_{k=0}^{n} C_k^n f^{(k)}g^{(n-k+1)} + \sum_{k=1}^{n+1} C_{k-1}^n f^{(k)}g^{(n-k+1)}$$

$$= fg^{(n+1)} + \sum_{k=1}^{n} (C_k^n + C_{k-1}^n)f^{(k)}g^{(n-k+1)} + f^{(n+1)}g$$

$$= \sum_{k=0}^{n+1} C_k^{n+1} f^{(k)}g^{(n+1-k)},$$

because of the easily verified formula

$$C_k^n + C_{k-1}^n = C_k^{n+1}.$$

Thus the validity of (1) for n implies its validity for $n+1$. Since (1) obviously holds for $n=1$, the induction is now complete. ∎

8.13. The following table shows the higher derivatives of various commonly encountered functions:

$f(x)$	$f'(x)$	$f''(x)$	$\cdots$	$f^{(n)}(x)$
x^α	$\alpha x^{\alpha-1}$	$\alpha(\alpha-1)x^{\alpha-2}$	$\cdots$	$\alpha(\alpha-1)\cdots(\alpha-n+1)x^{\alpha-n}$
e^{ax}	ae^{ax}	$a^2 e^{ax}$	$\cdots$	$a^n e^{ax}$
$\ln x$	$\dfrac{1}{x}$	$-\dfrac{1}{x^2}$	$\cdots$	$(-1)^{n-1}\dfrac{(n-1)!}{x^n}$
$\sin bx$	$b\cos bx$	$-b^2 \sin bx$	$\cdots$	$b^n \sin\left(bx + \dfrac{n\pi}{2}\right)$
$\cos bx$	$-b\sin bx$	$-b^2 \cos bx$	$\cdots$	$b^n \cos\left(bx + \dfrac{n\pi}{2}\right)$

Note that differentiating a polynomial has the effect of lowering its degree by 1, so that n-fold differentiation of a polynomial of degree n leads to a constant, while all derivatives of order $n+1$ and higher vanish.

8.14. Given any polynomial

$$P(x) = \sum_{k=0}^{n} a_k x^k \tag{2}$$

of degree n and any real number a, we can expand $P(x)$ in powers of $x-a$ as follows. Substitution of $x=(x-a)+a$ into (2) gives

$$P(x) = P((x-a)+a) = \sum_{k=0}^{n} a_k((x-a)+a)^k,$$

which can be written in the form

$$P(x) = b_0 + b_1(x-a) + b_2(x-a)^2 + \cdots + b_n(x-a)^n. \tag{3}$$

To find the coefficients $b_0, b_1, \ldots, b_n$, we first set $x = a$ in (3). This gives

$$P(a) = b_0.$$

We then differentiate (3) with respect to x, obtaining

$$P'(x) = b_1 + 2b_2(x-a) + 3b_3(x-a)^2 + \cdots + nb_n(x-a)^{n-1}. \tag{4}$$

Setting $x = a$ then gives

$$P'(a) = b_1.$$

Similarly, differentiating (4) and setting $x = a$, we get first

$$P''(x) = 2b_2 + 3\cdot 2b_3(x-a) + \cdots + n(n-1)b_n(x-a)^{n-2}$$

and then

$$P''(a) = 2b_2.$$

Continuing this process, we find more generally that

$$P^{(k)}(a) = k! b_k \qquad (k=0,1,\ldots,n),$$

$(0! = 1)$ or equivalently,

$$b_k = \frac{1}{k!} P^{(k)}(a). \tag{5}$$

Substituting (5) into (3), we finally get

$$P(x) = \sum_{k=0}^{n} \frac{P^{(k)}(a)}{k!}(x-a)^k. \tag{6}$$

8.2. Taylor's Formula

8.21. LEMMA. *Suppose the functions $F(x)$ and $G(x)$ are continuous on a finite interval $[a,b]$, with derivatives up to order $n+1$ inclusive on (a,b), where the functions*

$$G^{(k)}(x) \qquad (k=0,1,\ldots,n)$$

are nonvanishing on (a,b). Suppose further that the functions

$$F^{(k)}(x), \; G^{(k)}(x) \qquad (k=0,1,\ldots,n)$$

approach zero as $x \searrow a$. *Then*

$$\frac{F(b)}{G(b)} = \frac{F^{(n+1)}(c)}{G^{(n+1)}(c)} \tag{1}$$

for some point $c \in (a,b)$.

Proof. By Cauchy's theorem (Theorem 7.43), there is a point $c_1 \in (a,b)$ such that

$$\frac{F(b)-F(a)}{G(b)-G(a)} = \frac{F(b)}{G(b)} = \frac{F'(c_1)}{G'(c_1)}.$$

Applying Cauchy's theorem again to the interval (a,c_1), we find a point $c_2 \in (a,c_1)$ such that

$$\frac{F(b)}{G(b)} = \frac{F'(c_1)}{G'(c_1)} = \frac{F'(c_1)-F'(a)}{G'(c_1)-G'(a)} = \frac{F''(c_2)}{G''(c_2)}.$$

Continuing this process, we finally find at the $(n+1)$st step that

$$\frac{F(b)}{G(b)} = \frac{F^{(n)}(c_n)}{G^{(n)}(c_n)} = \frac{F^{(n)}(c_n)-F^{(n)}(a)}{G^{(n)}(c_n)-G^{(n)}(a)} = \frac{F^{(n+1)}(c)}{G^{(n+1)}(c)}$$

for some point $c \in (a,c_n) \subset (a,b)$. ∎

8.22. THEOREM. *Suppose the function* $f(x)$ *is continuous on a finite interval* $[a,b]$, *with derivatives up to order* $n+1$ *inclusive on* (a,b). *Suppose further that the functions*

$$f^{(k)}(x) \qquad (k=0,1,\dots,n)$$

approach finite limits $f^{(k)}(a)$ *as* $x \searrow a$.† *Then* **Taylor's formula**

$$f(b) = \sum_{k=0}^{n} \frac{f^{(k)}(a)}{k!}(b-a)^k + \frac{f^{(n+1)}(c)}{(n+1)!}(b-a)^{n+1} \tag{2}$$

holds for some point $c \in (a,b)$.

Proof. For $n=1$ formula (2) reduces to the finite difference formula (3′), p. 239. Moreover, $f^{(n+1)}(x) \equiv 0$ for a polynomial of degree n, and we then get formula (6) on the preceding page. More generally, let

$$F(x) = f(x) - \sum_{k=0}^{n} \frac{f^{(k)}(a)}{k!}(x-a)^k, \qquad G(x) = (x-a)^{n+1}. \tag{3}$$

Then $F(x)$ has derivatives up to order $n+1$, like $f(x)$ itself, while the function $G(x)$ has derivatives of all orders, with

$$G^{(k)}(x) > 0 \qquad (k=0,1,\dots,n)$$

† The limiting values $f^{(k)}(a)$ are actually the corresponding right-hand derivatives (see Chapter 7, Problem 15), but this fact plays no role here.

for $x > a$. Moreover,

$$\left[\sum_{k=0}^{n} \frac{f^{(k)}(a)}{k!}(x-a)^k\right]_{x=a}^{(m)} = f^{(m)}(a), \qquad [(x-a)^{n+1}]_{x=a}^{(m)} = 0$$

for $m = 0, 1, \ldots, n$, since clearly

$$[(x-a)^k]_{x=a}^{(m)} = \begin{cases} 0 & \text{if } k \neq m, \\ k! & \text{if } k = m, \end{cases}$$

and hence

$$F^{(m)}(a) = f^{(m)}(a) - f^{(m)}(a) = 0, \qquad G^{(m)}(a) = 0 \qquad (m = 0, 1, \ldots, n).$$

Thus $F(x)$ and $G(x)$ have all the properties of the functions figuring in the lemma, and furthermore

$$F^{(n+1)}(x) = f^{(n+1)}(x), \qquad G^{(n+1)}(x) = (n+1)! \qquad (4)$$

Therefore, substituting (3) and (4) into (1), we get

$$\frac{F(b)}{G(b)} = \frac{f(b) - \sum_{k=0}^{n} \frac{f^{(k)}(a)}{k!}(b-a)^k}{(b-a)^{n+1}} = \frac{f^{(n+1)}(c)}{(n+1)!},$$

which is equivalent to (2). ∎

It is tacitly assumed in Theorem 8.22 that $a < b$, but Taylor's formula remains valid in the case $b < a$ provided that $b < c < a$ (check this). Thus, regardless of the position of the point $b \neq a$, (2) holds for some point c between a and b.

8.23. Taylor's formula is often written in the form

$$f(x) = \sum_{k=0}^{n} \frac{f^{(k)}(a)}{k!}(x-a)^k + \frac{f^{(n+1)}(c)}{(n+1)!}(x-a)^{n+1} \qquad (a < c < x).$$

The first term on the right is a polynomial of degree n, called *Taylor's polynomial*, while the second term

$$R_n(x,a) = \frac{f^{(n+1)}(c)}{(n+1)!}(x-a)^{n+1} \qquad (5)$$

is called the *remainder* (in Lagrange's form). If $f^{(n+1)}(x)$ is bounded as $x \searrow a$, the remainder is "of a higher order of smallness than $(x-a)^n$," i.e.,

$$R_n(x,a) = o((x-a)^n)$$

(see Sec. 4.38a). In particular, it is easily verified that the following expansions hold for $x < 0$ or $x > 0$:

$$e^x = 1 + x + \frac{x^2}{2!} + \cdots + \frac{x^n}{n!} + o(x^n), \tag{6}$$

$$\cos x = 1 - \frac{x^2}{2!} + \frac{x^4}{4!} - \cdots + (-1)^n \frac{x^{2n}}{(2n)!} + o(x^{2n+1}), \tag{7}$$

$$\sin x = x - \frac{x^2}{3!} + \frac{x^5}{5!} - \cdots + (-1)^n \frac{x^{2n+1}}{(2n+1)!} + o(x^{2n+2}), \tag{8}$$

$$\ln (1+x) = x - \frac{x^2}{2} + \frac{x^3}{3} - \cdots + (-1)^{n-1} \frac{x^n}{n} + o(x^n), \tag{9}$$

$$(1+x)^\alpha = 1 + \alpha x + \frac{\alpha(\alpha-1)}{2!} x^2 + \cdots + \frac{\alpha(\alpha-1)\cdots(\alpha-n+1)}{n!} x^n + o(x^n). \tag{10}$$

8.24. The importance of Taylor's formula consists in the fact that it allows us to replace the function $f(x)$, which may well be very complicated, by a comparatively simple function, i.e., a polynomial, with an error (5) which in many cases can be estimated quite simply and made as small as we please. Suppose it is known that

$$\sup_{a \le x \le a+h} |f^{(n)}(x)| = M_n(h) \qquad (n = 0,1,2,\ldots).$$

Then we have the estimate

$$|R_n(x,a)| \le \frac{M_{n+1}(h)}{(n+1)!} h^{n+1}$$

for the remainder (5). Denoting the expression on the right by $\omega = \omega(n,h)$, we now pose three natural problems, each involving the determination of one of the quantities ω, n, and h in terms of the other two.

(a) *Given n and h, find ω.* In other words, find an upper bound for the error made in replacing a function $f(x)$ on the interval $(a, a+h)$ by its Taylor polynomial of degree n.

(b) *Given n and ω, find h.* In other words, find the interval $(a, a+h)$ on which the error made in replacing $f(x)$ by its Taylor polynomial of degree n is guaranteed not to exceed a given quantity ω.

(c) *Given h and ω, find n.* In other words, find the degree of the Taylor polynomial such that the error made in replacing $f(x)$ by the polynomial does not exceed a given quantity ω on a given interval $(a, a+h)$.

In concrete cases all three problems can be solved by a more or less elementary calculation (see Problem 10). The same problems can be posed for the interval $(a-h, a)$ and solved in the same way.

8.3. More on Concavity and Inflection Points

8.31. In Sec. 7.5 we used the values of the derivative $f'(x)$ in a neighborhood of the point x_0 to determine the position of the curve $y=f(x)$ relative to its tangent at x_0. We now consider another approach to the same problem, which uses the value of the second derivative at the single point $x=x_0$ rather than the whole set of values of $f'(x)$ near x_0.

THEOREM. *Suppose $f(x)$ has a first derivative $f'(x)$ in some interval (a,b) containing the point x_0 and a second derivative $f''(x_0)$ at the point x_0 itself. Then $f(x)$ is concave downward at x_0 if $f''(x_0)<0$ and concave upward at x_0 if $f''(x_0)>0$.*

Proof. If $f''(x_0)<0$, then, applying formula (6), p. 234 to the function $f'(x)$, we see that there is a $\delta>0$ such that

$$f'(x_0-h)>f'(x_0)>f'(x_0+h)$$

whenever $0<h<\delta$. But then $f(x)$ is concave downward at x_0, by Theorem 7.53a applied to the interval $(x_0-\delta,x_0+\delta)$. The case $f''(x_0)>0$ is treated similarly. ∎

8.32. The above theorem leads to new sufficient conditions for the presence of local extrema:

THEOREM. *Suppose $f(x)$ has a first derivative $f'(x)$ in some interval (a,b) containing the point x_0 and a second derivative $f''(x_0)$ at the point x_0 itself. Then $f(x)$ has a local maximum at x_0 if $f'(x_0)=0$ and $f''(x_0)<0$, while $f(x)$ has a local minimum at x_0 if $f'(x_0)=0$ and $f''(x_0)>0$.*

Proof. An immediate consequence of Theorem 8.31 and the argument given in the proof of Theorem 7.54.† ∎

8.33. What extra information about the function $f(x)$ is given by a knowledge of the values of $f''(x)$ in a neighborhood of the point x_0 as well as at the point x_0 itself? It turns out that a knowledge of $f''(x)$ allows us to determine the position of the curve $y=f(x)$ relative to the parabola

$$y=P(x)=f(x_0)+f'(x_0)(x-x_0)+\tfrac{1}{2}f''(x_0)(x-x_0)^2, \tag{1}$$

called the *osculating parabola* to the curve $y=f(x)$ at the point x_0. We begin with the analogues of Lemmas 7.52a and 7.52b.

a. LEMMA. *Let C be the curve with equation*

$$y=f(x) \qquad (a<x<b),$$

---

† Alternatively, use Theorem 7.54 and the argument given in the proof of Theorem 8.31.

and let P be the osculating parabola to C at the point x_0. Then C lies above P to the left of x_0 $(x<x_0)$ if $f''(\xi)>f''(x_0)$ for all $\xi \in (a,x_0)$, while C lies below P to the left of x_0 if $f''(\xi)<f''(x_0)$ for all $\xi \in (a,x_0)$.

Proof. By Taylor's formula with $n=1$, we have

$$f(x)=f(x_0)+f'(x_0)(x-x_0)+\tfrac{1}{2}f''(c)(x-x_0)^2$$
$$>f(x)+f'(x_0)(x-x_0)+\tfrac{1}{2}f''(x_0)(x-x_0)^2=P(x)$$

if $f''(\xi)>f''(x_0)$, while $f(x)<P(x)$ if $f''(\xi)<f''(x_0)$. ∎

b. LEMMA. *Let C and P be the same as in the preceding lemma. Then C lies above P to the right of x_0 $(x>x_0)$ if $f''(\eta)>f''(x_0)$ for all $\eta \in (x_0,b)$, while C lies below P to the right of x_0 if $f''(\eta)<f''(x_0)$ for all $\eta \in (x_0,b)$.*

Proof. The same as above, with ξ replaced by η. ∎

8.34. If $f''(x_0)=0$, the osculating parabola with equation (1) reduces to the line with equation

$$y=f(x_0)+f'(x_0)(x-x_0),$$

i.e., the tangent T to the curve C at the point x_0. This observation leads to the following analogue of Theorem 7.54:

THEOREM. *Suppose $f''(x_0)=0$. Then $f(x)$ is concave upward at x_0 if*

$$f''(\xi)>0, \qquad f''(\eta)>0$$

whenever $a<\xi<x_0<\eta<b$ and concave downward at x_0 if

$$f''(\xi)<0, \qquad f''(\eta)<0$$

whenever $a<\xi<x_0<\eta<b$. On the other hand, x_0 is an inflection point of $f(x)$ if

$$f''(\xi)>0>f''(\eta)$$

or

$$f''(\xi)<0<f''(\eta)$$

whenever $a<\xi<x_0<\eta<b$.

Proof. An immediate consequence of the above two lemmas. ∎

8.35. Treating the second derivative $f''(x)$ in the same way as the first derivative $f'(x)$ was treated earlier, we now make use of the value of the third derivative $f'''(x)$ at the single point $x=x_0$ rather than the whole set of values of $f''(x)$ near x_0. This leads to the following analogue of Theorem 8.31:

THEOREM. *Suppose* $f(x)$ *has first and second derivatives* $f'(x)$ *and* $f''(x)$ *in some interval* (a,b) *containing the point* x_0 *and a third derivative* $f'''(x)$ *at the point* x_0 *itself, and let C and P be the same as before. Then C lies above P to the left of* x_0 *and below P to the right of* x_0 *if* $f'''(x_0) < 0$, *while C lies below P to the left of* x_0 *and above P to the right of* x_0 *if* $f'''(x_0) > 0$.

Proof. If $f'''(x_0) < 0$, then, applying formula (6), p. 234 to the function $f''(x)$, we see that there is a $\delta > 0$ such that

$$f''(x_0 - h) > f''(x_0) > f''(x_0 + h)$$

whenever $0 < h < \delta$. But then, by Lemmas 8.33a and 8.33b, C lies above P to the left of x_0 and below P to the right of x_0. The case $f'''(x_0) > 0$ is treated similarly. ∎

8.36. THEOREM. *Let* $f(x)$ *be the same as in the preceding theorem. Then* x_0 *is an inflection point of* $f(x)$ *if* $f''(x_0) = 0, f'''(x_0) \neq 0$.

Proof. An immediate consequence of the preceding theorem, since P reduces to the tangent T to the curve C at the point x_0 if $f''(x_0) = 0$. ∎

If $f'''(x_0) = 0$, the behavior of $f(x)$ near x_0 is determined by the sign of the first nonvanishing derivative $f^{(n)}(x_0)$, as shown in Problem 11.

8.4. Another Version of Taylor's Formula

8.41. We now give another way of writing Taylor's formula, based on the use of the "asymptotic unit" E (see Sec. 4.39). This way of writing Taylor's formula is useful in a variety of problems involving the behavior of functions near given points. Suppose $f(x)$ has derivatives up to order $n + 1$ inclusive in a neighborhood of a point a, where $f^{(n)}(a) \neq 0$ and $f^{(n+1)}(x)$ is bounded. Then Taylor's formula (2), p. 252, can be written as

$$f(a+h) = f(a) + hf'(a) + \cdots + \frac{h^{n-1}}{(n-1)!}f^{(n-1)}(a) + \frac{h^n}{n!}f^{(n)}(a)E, \tag{1}$$

where E denotes a quantity approaching 1 as $h \to 0$. In fact, in this case, E is just

$$E = 1 + \frac{h}{n+1}\frac{f^{(n+1)}(c)}{f^{(n)}(a)}.$$

8.42. Comparing (1) with formulas (6)–(10), p. 254, and choosing the indicated values of n, we get†

$$e^x = 1 + xE \qquad\qquad (n=1), \qquad\qquad (2)$$

$$\cos x = 1 - \frac{x^2}{2}E \qquad\qquad (n=2), \qquad\qquad (3)$$

$$\sin x = xE \qquad\qquad (n=1), \qquad\qquad (4)$$

$$\ln (1+x) = xE \qquad\qquad (n=1), \qquad\qquad (5)$$

$$(1+x)^\alpha = 1 + \alpha xE \qquad\qquad (n=1). \qquad\qquad (6)$$

Choosing larger values of n, we get the following more accurate formulas:

$$e^x = 1 + x + \frac{x^2}{2}E \qquad\qquad (n=2), \qquad\qquad (2')$$

$$\cos x = 1 - \frac{x^2}{2} + \frac{x^4}{24}E \qquad\qquad (n=4), \qquad\qquad (3')$$

$$\sin x = x - \frac{x^3}{6}E \qquad\qquad (n=3), \qquad\qquad (4')$$

$$\ln (1+x) = x + \frac{x^2}{2}E \qquad\qquad (n=2), \qquad\qquad (5')$$

$$(1+x)^\alpha = 1 + \alpha x + \frac{\alpha(\alpha-1)}{2}x^2 E \qquad (n=2). \qquad\qquad (6')$$

8.43. Examples

a. Evaluate

$$\lim_{x\to 0} \frac{e^x - \cos x}{\sin x}.$$

Solution. Using (2)–(4), we find that

$$\frac{e^x - \cos x}{\sin x} = \frac{(1+xE) - (1 - \frac{1}{2}x^2 E)}{xE} = \frac{E + \frac{1}{2}xE}{E} \to 1$$

as $x\to 0$. The same result can also be obtained by using L'Hospital's rule for $0/0$ (Theorem 7.61).

b. Evaluate

$$\lim_{x\to 0} \frac{e^{x^2} - \sqrt{1-x^2+x^3}}{\ln (1+x^2)}.$$

† Formula (6) has already been found in Theorem 5.58e.

Solution. Using (2), (5), and (6), we find that

$$\frac{e^{x^2} - \sqrt{1 - x^2 + x^3}}{\ln(1 + x^2)} = \frac{1 + x^2 E - [1 - \frac{1}{2}(x^2 - x^3)E]}{x^2 E} = \frac{\frac{3}{2}E - \frac{1}{2}xE}{E} \to \frac{3}{2}$$

as $x \to 0$, a result which can again be obtained by using L'Hospital's rule for $0/0$.

c. How does the function

$$f(x) = \frac{x - \sin x}{x^3}$$

behave near the point $x = 0$?

Solution. By (4′), we have

$$f(x) = \frac{x - (x - \frac{1}{6}x^3 E)}{x^3} = \frac{1}{6}E \to \frac{1}{6}$$

as $x \to 0$. To investigate $f(x)$ further, we use the more exact expression

$$\sin x = x - \frac{x^3}{6} + \frac{x^5}{120}E,$$

implied by formula (8), p. 254. We then get

$$f(x) = \frac{1}{6} - \frac{x^2}{120}E,$$

which shows that $f(x)$ has a local maximum at $x = 0$.

8.5. Taylor Series

8.51. Let $f(x)$ be a function with derivatives of all orders at every point of an interval (a,b).† Choosing any point $x_0 \in (a,b)$, we write Taylor's formula

$$f(x) = f(x_0) + f'(x_0)(x - x_0) + \cdots + \frac{f^{(n)}(x_0)}{n!}(x - x_0)^n + R_n(x, x_0),$$

with remainder

$$R_n(x, x_0) = \frac{f^{(n+1)}(c)}{(n+1)!}(x - x_0)^{n+1} \tag{1}$$

where c lies between x_0 and x (see Sec. 8.23). Suppose (1) approaches zero

† Such a function is said to be *infinitely differentiable* on (a,b).

as $n \to \infty$, for some given value of x. Then the (infinite) series

$$f(x_0) + f'(x_0)(x-x_0) + \cdots + \frac{f^{(n)}(x_0)}{n!}(x-x_0)^n + \cdots = \sum_{n=0}^{\infty} \frac{f^{(n)}(x_0)}{n!}(x-x_0)^n$$

(2)

is convergent, with sum $f(x)$. The series (2), written for an infinitely differentiable function $f(x)$, is called the *Taylor series* of $f(x)$, regardless of whether or not the series converges and regardless of whether or not the sum of the series (if convergent) equals $f(x)$.

8.52. Clearly, the Taylor series (2) is convergent with sum $f(x)$ if and only if the remainder (1) approaches zero as $n \to \infty$, in which case $f(x)$ is the sum of its own Taylor series and we are entitled to write†

$$f(x) = \sum_{n=0}^{\infty} \frac{f^{(n)}(x_0)}{n!}(x-x_0)^n.$$

(3)

A condition guaranteeing this is given by the following

THEOREM. *If*

$$\frac{1}{n!B^n} \sup_{a < x < b} |f^{(n)}(x)| \leqslant C \qquad (n = 0,1,2,\ldots)$$

(4)

for suitable constants $B > 0$ and $C > 0$, then

$$\lim_{n \to \infty} R_n(x,x_0) = 0$$

for all $x \in (a,b)$ such that $|x - x_0| < 1/B$.

Proof. It follows from (1) and (4) that

$$|R_n(x,x_0)| \leqslant \frac{|x-x_0|^{n+1}}{(n+1)!} CB^{n+1}(n+1)! = C(B|x-x_0|)^{n+1}.$$

But the expression on the right approaches zero as $n \to \infty$ if $|x - x_0| < 1/B$. ∎

8.53. Examples

a. The inequality (4) holds true for any polynomial of degree m, since $f^{(m+1)}(x)$ and all higher derivatives vanish. Thus every polynomial has a Taylor series expansion, which reduces to the finite sum (6) already found on p. 251.

b. If $f(x) = e^x$, then $f^{(n)}(x) = e^x$, so that

$$\frac{1}{n!B^n} \sup_{|x| < a} |f^{(n)}(x)| \leqslant \frac{e^a}{n!B^n}$$

† We often call (3) the *Taylor (series) expansion* of $f(x)$ at $x = x_0$.

for all $a > 0, B > 0$. But the right-hand side approaches zero as $n \to \infty$ (with a and B fixed), and hence forms a bounded sequence.† It follows from Theorem 8.52 that the Taylor series of e^x converges to e^x on every interval $(-a, a)$ and hence for all real x:

$$e^x = 1 + x + \frac{x^2}{2!} + \cdots + \frac{x^n}{n!} + \cdots \qquad (-\infty < x < \infty). \tag{5}$$

c. If $f(x) = \sin x$, then

$$|f^{(n)}(x)| = \left| \sin\left(x + \frac{n\pi}{2} \right) \right| \leqslant 1$$

for all x (cf. Sec. 8.13), so that

$$\frac{1}{n! B^n} \sup_{|x| < a} |f^n(x)| \leqslant \frac{1}{n! B^n}$$

for all $a > 0$, $B > 0$. The right-hand side again approaches zero as $n \to \infty$ (with B fixed), and hence forms a bounded sequence. Therefore the Taylor series of $\sin x$ converges to $\sin x$ for all real x:

$$\sin x = x - \frac{x^3}{3!} + \frac{x^5}{5!} - \frac{x^7}{7!} + \cdots \qquad (-\infty < x < \infty). \tag{6}$$

d. In just the same way, we find that $\cos x$ has the Taylor expansion

$$\cos x = 1 - \frac{x^2}{2!} + \frac{x^4}{4!} - \frac{x^6}{6!} + \cdots \qquad (-\infty < x < \infty). \tag{7}$$

8.54. In our treatment of the trigonometric functions (see Sec. 5.6), we defined $\sin x$ and $\cos x$ as the functions satisfying the formulas

$$\sin^2 x + \cos^2 x = 1, \tag{8}$$

$$\sin (x + y) = \sin x \cos y + \cos x \sin y, \tag{9}$$

$$\cos (x + y) = \cos x \cos y - \sin x \sin y, \tag{10}$$

$$0 < \sin x < x < \frac{\sin x}{\cos x}, \tag{11}$$

where (11) holds for all sufficiently small positive x, say for $0 < x < \varepsilon_0$. The fact that $\sin x$ and $\cos x$ have the expansions (6) and (7) now proves the

---

† More exactly, the numerical sequence

$$\frac{e^a}{1! B}, \frac{e^a}{2! B^2}, \ldots, \frac{e^a}{n! B^n}, \ldots$$

is bounded.

uniqueness of sin x and cos x. To complete the theory, we must still prove the *existence* of functions sin x and cos x satisfying (8)–(11). This will be done in Secs. 8.66–8.68, by the simple expedient of showing that if we *define* the functions sin x and cos x by (6) and (7), then they satisfy (8)–(11).

8.55. Analytic continuation. Given that $f(x)$ is infinitely differentiable on an interval (a,b) and has a Taylor expansion

$$f(x) = \sum_{n=0}^{\infty} a_n(x-x_0)^n, \qquad a_n = \frac{f^{(n)}(x_0)}{n!} \tag{12}$$

on (a,b), suppose we consider (12) in the complex plane by replacing the real variable x by the complex variable $z = x + iy$. This gives the power series

$$\sum_{n=0}^{\infty} a_n(z-x_0)^n, \tag{13}$$

whose region of convergence is some disk G of radius ρ centered at the point x_0 (see Theorem 6.62). At the very least, ρ must be such that the whole interval (a,b) lies inside the "circle of convergence" $\Gamma = \{z : |z - x_0| = \rho\}$, since the series (13) diverges at every point outside Γ (see Sec. 6.63). By hypothesis, the sum of (13) equals $f(x)$ at every point $x \in (a,b)$. At the other points of G, the sum of (13) is some function of z which we denote by $f(z)$ and call the *analytic continuation* of $f(x)$. In particular, if (13) converges for all real x, then (13) converges for all complex z, i.e., the analytic continuation of $f(x)$ is defined in the whole complex plane.

8.6. Complex Exponentials and Trigonometric Functions

We now use analytic continuation to define the exponential and trigonometric functions "in the complex domain," i.e., for complex values of the argument.

8.61. The Taylor series

$$e^x = 1 + x + \frac{x^2}{2!} + \cdots + \frac{x^n}{n!} + \cdots \tag{1}$$

converges for all real x (Example 8.53b), and hence the series

$$e^z = 1 + z + \frac{z^2}{2!} + \cdots + \frac{z^n}{n!} + \cdots \tag{1'}$$

converges for all complex z, thereby defining the exponential function e^z on the whole complex plane (as the analytic continuation of e^x).

8.62. Similarly, the Taylor series

$$\sin x = x - \frac{x^3}{3!} + \frac{x^5}{5!} - \frac{x^7}{7!} + \cdots \tag{2}$$

and

$$\cos x = 1 - \frac{x^2}{2!} + \frac{x^4}{4!} - \frac{x^6}{6!} + \cdots \tag{3}$$

converge for all real x (Examples 8.53c and 8.53d), and hence the series

$$\sin z = z - \frac{z^3}{3!} + \frac{z^5}{5!} - \frac{z^7}{7!} + \cdots \tag{2'}$$

and

$$\cos z = 1 - \frac{z^2}{2!} + \frac{z^4}{4!} - \frac{z^6}{6!} + \cdots \tag{3'}$$

converge for all complex z, thereby defining the trigonometric functions $\sin z$ and $\cos z$ on the whole complex plane (as the analytic continuations of $\sin x$ and $\cos x$). In particular, it follows from (2') and (3') that

$$\sin(-z) = -\sin z, \tag{4}$$

$$\cos(-z) = \cos z, \tag{5}$$

i.e., $\sin z$ is *odd* and $\cos z$ is *even* for complex z (just as for real values of the argument).

8.63. The expansions (1')–(3') imply a remarkable connection between the exponential and trigonometric functions in the complex domain. In fact, replacing z by iz in (1'), we find that

$$e^{iz} = 1 + iz - \frac{z^2}{2!} - i\frac{z^3}{3!} + \frac{z^4}{4!} + i\frac{z^5}{5!} - \cdots$$

$$= \left(1 - \frac{z^2}{2!} + \frac{z^4}{4!} - \cdots\right) + i\left(z - \frac{z^3}{3!} + \frac{z^5}{5!} - \cdots\right)$$

and hence

$$e^{iz} = \cos z + i \sin z \tag{6}$$

for arbitrary complex z, a result known as *Euler's formula*. Replacing z

by $-z$, with the help of (4) and (5), we get

$$e^{-iz} = \cos z - i \sin z. \tag{7}$$

Together (6) and (7) imply the formulas

$$\cos z = \frac{e^{iz} + e^{-iz}}{2}, \tag{8}$$

$$\sin z = \frac{e^{iz} - e^{-iz}}{2i}, \tag{9}$$

which can also be obtained directly from $(1')$–$(3')$.

Letting z be a real number θ in (6), we get

$$e^{i\theta} = \cos \theta + i \sin \theta. \tag{10}$$

Thus every complex number of the form $e^{i\theta}$ (θ real) has absolute value 1. It will be recalled from Sec. 5.72 that every complex number z can be written in the *trigonometric form*

$$z = r(\cos \theta + i \sin \theta), \tag{11}$$

where z has absolute value r and argument θ. Combining (10) and (11), we can write z in the *exponential form*

$$z = re^{i\theta}. \tag{11'}$$

8.64. THEOREM. *The formula*

$$e^{z_1 + z_2} = e^{z_1} e^{z_2} \tag{12}$$

holds for arbitrary complex z_1 and z_2.

Proof. Both series

$$e^{z_1} = 1 + z_1 + \frac{z_1^2}{2!} + \cdots, \qquad e^{z_2} = 1 + z_2 + \frac{z_2^2}{2!} + \cdots$$

are absolutely convergent (why?), and hence, by the theorem on multiplication of complex series (Theorem 6.46),

$$\begin{aligned}
e^{z_1} e^{z_2} &= 1 + (z_1 + z_2) + \left(\frac{z_1^2}{2!} + z_1 z_2 + \frac{z_2^2}{2!} \right) \\
&\quad + \left(\frac{z_1^3}{3!} + \frac{z_1^2 z_2}{2!1!} + \frac{z_1 z_2^2}{1!2!} + \frac{z_2^3}{3!} \right) + \cdots \\
&= 1 + (z_1 + z_2) + \frac{(z_1 + z_2)^2}{2!} + \frac{(z_1 + z_2)^3}{3!} + \cdots = e^{z_1 + z_2}. \quad \blacksquare
\end{aligned}$$

8.65. In particular, if $z = x + iy$ (x and y real), then, by (12) and (10), we have

$$e^z = e^{x+iy} = e^x e^{iy} = e^x(\cos y + i \sin y) = e^x \cos y + ie^x \sin y, \tag{13}$$

thereby explicitly exhibiting the real and imaginary parts of e^z. Formula (13) reveals an interesting property of e^z, namely its periodicity in the complex plane. As in the real case (see Sec. 5.65c), a function $f(z)$ such that

$$f(z + T) = f(z)$$

for all complex z and some T is said to be *periodic*, with *period* T (in general, T is itself complex). According to (13),

$$e^{z+2\pi i} = e^{x+iy+2\pi i} = e^{x+i(y+2\pi)}$$
$$= e^x[\cos (y + 2\pi) + i \sin (y + 2\pi)] = e^x(\cos y + i \sin y) = e^z, \tag{14}$$

so that e^z is periodic, with the purely imaginary period $T = 2\pi i$. More generally,

$$e^{z+2k\pi i} = e^z$$

for all $k = 0, \pm 1, \pm 2, \cdots$ (why?).

On the other hand, the functions e^{iz} and e^{-iz} are periodic, with the real period $T = 2\pi$. For example, changing z to iz in (14), we get

$$e^{iz+2\pi i} = e^{i(z+2\pi)} = e^{iz},$$

and similarly for e^{-iz}. It then follows from (8) and (9) that $\sin z$ and $\cos z$ are also periodic, with period 2π.[†]

Setting $z = i\pi$ in (13), we get the interesting formula[‡]

$$e^{i\pi} = -1.$$

8.66. Next, using Theorem 8.64 to multiply formulas (6) and (7), we find that

$$e^{iz}e^{-iz} = e^0 = 1 = (\cos z + i \sin z)(\cos z - i \sin z)$$
$$= \cos^2 z + \sin^2 z, \tag{15}$$

which reduces to formula (8), p. 261 for real $z = x$. Unlike the real case, however, we cannot infer from (15) that $\sin z$ and $\cos z$ do not exceed 1 in absolute value, since $\sin z$ and $\cos z$ are now complex. In fact, setting $z = iy$

[†] The periodicity of $\sin x$ and $\cos x$ for real x (see Sec. 5.65c) has already been used in (14).
[‡] "This remarkable formula symbolizes, as it were, the unity of all mathematics, with -1 representing arithmetic, i algebra, π geometry, and e analysis." (A. N. Krylov)

(y real) in (8) and (9), we get

$$\cos iy = \frac{e^{-y} + e^y}{2},$$

$$\sin iy = \frac{e^{-y} - e^y}{2i},$$

so that $\cos iy$ and $\sin iy$ become arbitrarily large in absolute value as $y \to \pm \infty$.

8.67. Suppose we replace z by z_1 in (6) and z by z_2 in (7), where z_1 and z_2 are arbitrary complex numbers, and then use Theorem 8.64 to multiply the resulting formulas. This gives

$$\cos (z_1 + z_2) + i \sin (z_1 + z_2) = e^{i(z_1 + z_2)} = e^{iz_1} e^{iz_2}$$
$$= (\cos z_1 + i \sin z_1)(\cos z_2 + i \sin z_2)$$
$$= (\cos z_1 \cos z_2 - \sin z_1 \sin z_2) + i(\cos z_1 \sin z_2 + \sin z_1 \cos z_2),$$

and similarly,

$$\cos (z_1 + z_2) - i \sin (z_1 + z_2) = e^{-i(z_1 + z_2)} = e^{-iz_1} e^{-iz_2}$$
$$= (\cos z_1 - i \sin z_2)(\cos z_2 - i \sin z_2)$$
$$= (\cos z_1 \cos z_2 - \sin z_1 \sin z_2) - i(\cos z_1 \sin z_2 + \sin z_1 \cos z_2).$$

First subtracting and then adding these formulas, we find that

$$\sin (z_1 + z_2) = \sin z_1 \cos z_2 + \cos z_1 \sin z_2,$$
$$\cos (z_1 + z_2) = \cos z_1 \cos z_2 - \sin z_1 \sin z_2,$$

which reduce to formulas (9) and (10), p. 261 for real $z_1 = x, z_2 = y$.

8.68. In keeping with the discussion of Sec. 8.54, to complete the theory of the trigonometric functions, we need only show that the functions $\sin x$ and $\cos x$, regarded as *defined* by the series (2) and (3), satisfy the inequalities

$$0 < \sin x < x < \frac{\sin x}{\cos x} \tag{16}$$

for all sufficiently small $x > 0$. By formula (4'), p. 258,

$$\sin x = x - \frac{x^3}{6} E = x\left(1 - \frac{x^2}{6} E\right) \qquad (E \to 1 \text{ as } x \to 0). \tag{17}$$

But, for all sufficiently small $x > 0$,

$$0 < 1 - \frac{x^2}{6} E < 1$$

and hence

$0 < \sin x < x,$

because of (17). Moreover, by formulas (3) and (6), p. 258,

$$\cos x = 1 - \frac{x^2}{2}E \qquad (E \to 1 \text{ as } x \to 0),$$

$$(1+x)^{-1} = 1 - xE \qquad (E \to 1 \text{ as } x \to 0),$$

which imply

$$\frac{\sin x}{\cos x} = x\left(1 - \frac{x^2}{6}E\right)\left(1 - \frac{x^2}{2}E\right)^{-1}$$

$$= x\left(1 - \frac{x^2}{6}E\right)\left(1 + \frac{x^2}{2}E\right) = x\left(1 + \frac{x^2}{3}E\right). \tag{18}$$

This time, for all sufficiently small $x > 0$,

$$1 + \frac{x^2}{3}E > 1$$

and hence

$$\frac{\sin x}{\cos x} > x,$$

because of (18).

This completes the proof of (16), thereby accomplishing the program of Sec. 8.54.

8.7. Hyperbolic Functions

8.71. Definition. The functions defined by the series

$$\cosh z = 1 + \frac{z^2}{2!} + \frac{z^4}{4!} + \cdots, \tag{1}$$

$$\sinh z = z + \frac{z^3}{3!} + \frac{z^5}{5!} + \cdots \tag{2}$$

(which converge for all complex z), are called the *hyperbolic cosine* and the *hyperbolic sine*, respectively.

8.72. Replacing z by $-z$ in the series

$$e^z = 1 + z + \frac{z^2}{2!} + \frac{z^3}{3!} + \cdots \tag{3}$$

(Sec. 8.61), we get

$$e^{-z} = 1 - z + \frac{z^2}{2!} - \frac{z^3}{3!} + \cdots. \tag{4}$$

Comparing (1)–(4), we find that

$$e^z = \cosh z + \sinh z,$$
$$e^{-z} = \cosh z - \sinh z.$$

Solving these equations for $\cosh z$ and $\sinh z$ then gives

$$\cosh z = \frac{e^z + e^{-z}}{2}, \tag{5}$$

$$\sinh z = \frac{e^z - e^{-z}}{2}. \tag{6}$$

In particular, we have

$$\cosh x = \frac{e^x + e^{-x}}{2}, \tag{5'}$$

$$\sinh x = \frac{e^x - e^{-x}}{2} \tag{6'}$$

for real $z = x$. The functions $\cosh x$ and $\sinh x$ have the graphs shown in Figure 20.

8.73. The hyperbolic functions are intimately related to the trigonometric functions in the complex domain. In fact, replacing z by iz in (5) and (6), we get

$$\cos z = \frac{e^{iz} + e^{-iz}}{2} = \cosh iz,$$
$$i \sin z = \frac{e^{iz} - e^{-iz}}{2} = \sinh iz.$$

Then, replacing iz by z, we find that

$$\cosh z = \cos(-iz) = \cos iz,$$
$$\sinh z = i \sin(-iz) = -i \sin iz.$$

Thus, to within constant factors, the trigonometric functions are obtained from the hyperbolic functions and the hyperbolic functions from the trigonometric functions by multiplying the argument z by i. This corresponds to

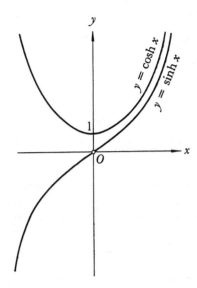

Figure 20

rotating z through a right angle in the complex plane (see Sec. 5.72).

8.74. The following formulas are all easy consequences of the results of Secs. 8.66, 8.67, and 8.73 (for arbitrary real x, y and complex z, z_1, z_2):

$$\sin (x+iy) = \sin x \cos iy + \cos x \sin iy$$
$$= \sin x \cosh y + i \cos x \sinh y, \tag{7}$$
$$\cos (x+iy) = \cos x \cos iy - \sin x \sin iy$$
$$= \cos x \cosh y - i \sin x \sinh y,$$
$$\cosh^2 z - \sinh^2 z = \cos^2 (iz) + \sin^2 (iz) = 1,$$
$$\cosh (z_1 + z_2) = \cos i(z_1 + z_2) = \cos iz_1 \cos iz_2 - \sin iz_1 \sin iz_2$$
$$= \cosh z_1 \cosh z_2 + \sinh z_1 \sinh z_2,$$
$$\sinh (z_1 + z_2) = -i \sin i(z_1 + z_2) = -i(\sin iz_1 \cos iz_2 + \cos iz_1 \sin iz_2)$$
$$= \sinh z_1 \cosh z_2 + \cosh z_1 \sinh z_2.$$

In particular, we have

$$\cosh 2z = \cosh^2 z + \sinh^2 z,$$
$$\sinh 2z = 2 \sinh z \cosh z$$

and

$$\sinh (x + iy) = \sinh x \cosh iy + \cosh x \sinh iy$$
$$= \sinh x \cos y + i \cosh x \sin y,$$
$$\cosh (x + iy) = \cosh x \cosh iy + \sinh x \sinh iy$$
$$= \cosh x \cos y + i \sinh x \sin y.$$

To differentiate $\cosh x$ and $\sinh x$ (for real x), we need only note that

$$(\cosh x)' = \tfrac{1}{2}(e^x + e^{-x})' = \tfrac{1}{2}(e^x - e^{-x}) = \sinh x,$$
$$(\sinh x)' = \tfrac{1}{2}(e^x - e^{-x})' = \tfrac{1}{2}(e^x + e^{-x}) = \cosh x.$$

8.75. Thus we see that the theory of the hyperbolic functions is in certain respects simpler than that of the trigonometric functions. The adjective "hyperbolic" is explained by the fact that the curve in the xy-plane with the parametric representation

$$x = \cosh t, \qquad y = \sinh t$$

is the hyperbola $x^2 - y^2 = 1$. The corresponding curve for the trigonometric functions has the parametric representation

$$x = \cos t, \qquad y = \sin t,$$

and is just the circle $x^2 + y^2 = 1$. Hence the trigonometric functions are sometimes called "circular" functions.

Problems

1. Prove that the function e^{-1/x^2} cannot be represented as a Taylor series in powers of x, although it has derivatives of all orders at $x = 0$.

2. Prove that the function

$$y = x^{2n} \sin \frac{1}{x}$$

has derivatives up to order n inclusive at $x = 0$, while the nth derivative $y^{(n+1)}(0)$ fails to exist.

3. Prove that if all the roots of a polynomial $P(x)$ of degree n are real, then so are all the roots of the polynomials $P'(x)$, $P''(x), \ldots, P^{(n-1)}(x)$.

4. Suppose $f(x)$ is bounded and has continuous bounded derivatives $f'(x)$ and $f''(x)$ on $(-\infty, \infty)$. Prove that

$$M_1^2 \leqslant 2 M_0 M_2,$$

where

$$M_k = \sup_{-\infty < x < \infty} |f^{(k)}(x)| \qquad (k=0,1,2).$$

5. Prove that the second derivative $f''(x)$, if it exists, can be found by taking the following limit of a quotient involving the values of $f(x)$ at three neighboring points:

$$f''(x) = \lim_{h \to 0} \frac{f(x) - 2f(x+h) + f(x+2h)}{h^2}.$$

Write an analogous formula for $f^{(n)}(x)$.

6. Prove that if $f(x)$ is convex in the sense of Chapter 2, Problem 7 and if $f(x)$ has a second derivative at $x=x_0$, then $f''(x_0) \geqslant 0$.

7. Prove that if $f''(x) > 0$ for all $x \in [a,b]$, then $f(x)$ is convex on $[a,b]$.

8. Prove that the equation $\sin z = 0$ has no roots in the complex plane other than the real roots $z=0, \pm\pi, \pm 2\pi, \ldots$

9. Prove that there is a constant $C > 0$ such that $|\sin z| \geqslant C$ on the set of all circles

$$|z| = (n+\tfrac{1}{2})\pi \qquad (n=0,1,2,\ldots).$$

10. Estimate the error ω made in replacing the function e^x on the interval $[0,1]$ by its Taylor polynomial of degree 10. On what interval $[0,h]$ does the function e^x differ from its Taylor polynomial of degree 10 by no more than 10^{-7}? For what value of n does the function e^x differ from its Taylor polynomial of degree n by no more than 10^{-7} on the interval $[0,1]$?

11. Suppose $f(x)$ has derivatives up to order n in some interval containing the point x_0 and a derivative of order $n+1$ at the point x_0 itself, where

$$f'(x_0) = \cdots = f^{(n)}(x_0) = 0, \qquad f^{(n+1)}(x_0) \neq 0.$$

Describe the position of the curve $y = f(x)$ relative to its tangent at the point x_0, assuming that
(a) $f^{(n+1)}(x_0) > 0$, n even;
(b) $f^{(n+1)}(x_0) > 0$, n odd;
(c) $f^{(n+1)}(x_0) < 0$, n even;
(d) $f^{(n+1)}(x_0) < 0$, n odd.

12. If $f(x)$ can be written in the form

$$f(a+h) = f(a) + Ah + o(h)$$

in a neighborhood of the point a, then, by Theorem 7.21, $f'(a)$ exists and equals A. Suppose $f(x)$ can be written in the form

$$f(a+h) = f(a) + f'(a)h + \tfrac{1}{2}Bh^2 + o(h^2)$$

in a neighborhood of the point a. Does $f''(a)$ necessarily exist?

13. Prove the inequality

$$e - \sum_{k=0}^{n} \frac{1}{k!} < \frac{1}{n!n}.$$

14. Use the fact that

$$n! \sum_{k=0}^{n} \frac{1}{k!}$$

is an integer to prove that e is irrational.

15. Show that the infinite product

$$\prod_{k=1}^{\infty} (1 + \omega_k) \tag{1}$$

diverges if

$$\omega_k = (-1)^k \frac{1}{\sqrt{k+1}},$$

although

$$\sum_{k=1}^{\infty} \omega_k \tag{2}$$

converges. Show that (1) converges if

$$\omega_k = e^{(-1)^k/\sqrt{k}} - 1,$$

although (2) then diverges.†

16. Prove that

$$1 - \frac{x^2}{2!} + \frac{x^4}{4!} - \cdots - \frac{x^{4n-2}}{(4n-2)!} < \cos x$$

$$< 1 - \frac{x^2}{2!} + \frac{x^4}{4!} - \cdots + \frac{x^{4n}}{(4n)!} \qquad (n = 1, 2, \ldots) \tag{3}$$

for all $x \neq 0$.

17. Deduce from (3) that $2.82 < \pi < 3.19$.

† Thus the stipulation that all the ω_k have the same sign cannot be dropped in Chapter 6, Problem 12.

18. Suppose $y = f(x)$ has derivatives up to order n inclusive at every point of an interval (a,b), and define the *differential of order n of $f(x)$* as the function†

$$d^n y = f^{(n)}(x)(dx)^n. \tag{4}$$

Prove that $d^n y = d(d^{n-1} y)$ if dx is regarded as constant. Calculate $d^n y$ for the function $y = x$. Find the second differential of a composite function $z = g(y) = g(f(x))$, and show that, unlike the first differential, the result does not agree with the expression obtained for the case where y is the independent variable, unless y is a linear function $Ax + B$, in which case

$$d^n z = g^{(n)}(y)(dy)^n.$$

Write Taylor's formula in terms of higher differentials.

† We can use (4) to interpret the common notation $f^{(n)}(x) = \dfrac{d^n y}{dx^n}$.

9 The Integral

The preceding two chapters have been devoted to one of the basic concepts of mathematical analysis, namely the *derivative*. We now turn to the equally important concept of the *integral*.

9.1. Definitions and Basic Properties

9.11. Partitions with marked points. A set of points $x_0, x_1, \ldots, x_n$ of a closed interval $[a,b]$ such that

$$a = x_0 \leqslant x_1 \leqslant \cdots \leqslant x_n = b$$

is called a *partition* of $[a,b]$ into subintervals $[x_k, x_{k+1}]$, and the points $x_0, x_1, \ldots, x_n$ themselves are called *points of subdivision* of $[a,b]$. If, in addition, a point ξ_k is chosen in each subinterval $[x_k, x_{k+1}]$, so that

$$x_k \leqslant \xi_k \leqslant x_{k+1} \qquad (k = 0, 1, \ldots, n-1),$$

our partition is said to be a *partition with marked points*. We denote such a partition by Π, or in more detail,

$$\Pi = \{a = x_0 \leqslant \xi_0 \leqslant x_1 \leqslant \xi_1 \leqslant x_2 \leqslant \cdots \leqslant x_{n-1} \leqslant \xi_{n-1} \leqslant x_n = b\}.$$

The difference $x_{k+1} - x_k$ is denoted by Δx_k, and the number

$$d(\Pi) = \max\{\Delta x_0, \Delta x_1, \ldots, \Delta x_{n-1}\}$$

is called the *parameter* of the partition Π.

9.12. Riemann sums. Let $y = f(x)$ be a (finite) function defined on $[a,b]$. Then, starting from a given partition Π with marked points, we can form the *Riemann sum*

$$S_\Pi(f) = \sum_{k=0}^{n-1} f(\xi_k) \Delta x_k. \tag{1}$$

9.13. The integral. The finite number I is called the *(Riemann) integral of the function $f(x)$ over the interval $[a,b]$* if, given any $\varepsilon > 0$, there exists a $\delta > 0$ such that

$$|I - S_\Pi(f)| < \varepsilon$$

for every partition Π with $d(\Pi) < \delta$. This definition can be examined in the light of the general definition of a limit. To this end, let E be the set of all partitions Π (with marked points) of the underlying interval $[a,b]$, and, given any $\delta > 0$, let E_δ be the subset of E consisting of all partitions of E such that $d(\Pi) < \delta$. Then, obviously, given any two sets E_{δ_1} and E_{δ_2}, either $E_{\delta_1} \subset E_{\delta_2}$ or $E_{\delta_2} \subset E_{\delta_1}$, while the intersection of all the sets E_δ is empty. Hence the sets E_δ make up a direction T on the set E (see Sec. 4.11), which can then be used to construct limits of functions defined on E. In fact, the general definition of a limit (Sec. 4.12) takes the following form when applied to the present situation: The number I is said to be the limit in the direction T of a function $S(\Pi)$ defined on the set E of all partitions Π, and we write

$$I = \lim_T S(\Pi), \tag{2}$$

if, given any $\varepsilon > 0$, there exists a set $E_\delta \in T$ such that

$$|I - S(\Pi)| < \varepsilon \tag{3}$$

for all $\Pi \in E_\delta$, or equivalently if there exists a $\delta > 0$ such that (3) holds for all Π such that $d(\Pi) < \delta$. The limit (2) will also be called *the limit of $S(\Pi)$ under arbitrary refinement of the partition.*

In particular, let $S(\Pi)$ be the function $S_\Pi(f)$ defined by (1). Then the existence of (2) is equivalent to the existence of the integral of the function $f(x)$ over $[a,b]$. The integral I is denoted by

$$\int_a^b f(x) \, dx, \tag{4}$$

an expression resembling the original Riemann sum (1). However, at the same time that the expression (1) indicates the operations involved in calculating the Riemann sum, the expression (4) is "unsplittable," at least for the time being, since there is no operation that corresponds to multiplying $f(x)$ by dx and then "integrating" the result. The suitability of this way of writing the integral, rather than just

$$\int_a^b f(x),$$

say, will be apparent later.† On the other hand, it is clear that changing the "variable of integration" has no effect on the value of the integral, so that, for example,

$$\int_a^b f(x) \, dx = \int_a^b f(\xi) \, d\xi = \int_a^b f(t) \, dt.$$

The number a is called the *lower limit of integration*, while b is called the *upper limit of integration*. An integral with limits of integration a and b is often simply called an integral "from a to b."

The integral of a given function $f(x)$ may well fail to exist. However, as will be shown in Secs. 9.14 and 9.16, continuous functions always have inte-

† See esp. Secs. 9.38 and 9.53.

grals, and so do certain simple discontinuous functions. On the other hand, "strongly discontinuous" functions like the Dirichlet function (Sec. 9.17) do not have integrals. If the integral of $f(x)$ over $[a,b]$ exists, we say that $f(x)$ is *integrable on* $[a,b]$.

9.14. We now prove that every continuous function is integrable on the interval $[a,b]$.

a. Let $\omega_f(\delta)$ denote the *modulus of continuity* of the function $f(x)$ on $[a,b]$, i.e., let

$$\omega_f(\delta) = \sup_{|x'-x''|\leq\delta} |f(x')-f(x'')|.$$

It will be recalled from Sec. 5.17c that

$$\lim_{\delta\to 0} \omega_f(\delta) = 0$$

for every continuous function on $[a,b]$.†

b. A partition

$$\Pi' = \{a=x_0' \leq \xi_0' \leq x_1' \leq \xi_1' \leq x_2' \leq \cdots \leq x_{p-1}' \leq \xi_p' \leq x_p' = b\}$$

is called *finer* than a partition

$$\Pi = \{a=x_0 \leq \xi_0 \leq x_1 \leq \xi_1 \leq x_2 \leq \cdots \leq x_{n-1} \leq \xi_n \leq x_n = b\}$$

if every point $x_0,\ldots,x_n$ is one of the points $x_0',\ldots,x_p'$ (so that the set $x_0',\ldots,x_p'$ is obtained by adding extra points of subdivision to the set $x_0,\ldots,x_n$), with no new requirements whatsoever being imposed on the marked points ξ_k, ξ_k'.

c. LEMMA. *If the partition Π is such that $d(\Pi)<\delta$, then*

$$|S_{\Pi'}(f) - S_{\Pi}(f)| \leq \omega_f(\delta)(b-a) \tag{5}$$

for every finer partition Π'.

Proof. In going from the partition Π to the partition Π', each subinterval $[x_k,x_{k+1}]$ acquires new points of subdivision and new marked points which, with a slight change of notation, can be written as follows:

$$x_k=x_{k0} \leq \xi_{k0} \leq x_{k1} \leq \xi_{k1} \leq x_{k2} \leq \cdots \leq \xi_{k,m_k-1} \leq x_{km_k}=x_{k+1}.‡$$

The part of the Riemann sum S_Π coming from $[x_k,x_{k+1}]$ is of the form

† See also Theorem 5.17b, which shows that $f(x)$ is *uniformly* continuous on $[a,b]$.
‡ Note, however, that there are actually no new points of subdivision in $[x_k,x_{k+1}]$ if $m_k=1$.

$$\sum_{j=0}^{m_k-1} f(\xi_{kj})\Delta x_{kj} \qquad (\Delta x_{kj}=x_{k,j+1}-x_{kj}).$$

The term $f(\xi_k)\Delta x_k$ in the Riemann sum S_Π can also be written as a sum

$$\sum_{j=0}^{m_k-1} f(\xi_k)\Delta x_{kj}$$

of the same form. Clearly,

$$\sum_{j=0}^{m_k-1} f(\xi_{kj})\Delta x_{kj} - \sum_{j=0}^{m_k-1} f(\xi_k)\Delta x_{kj} = \sum_{j=0}^{m_k-1} [f(\xi_{kj})-f(\xi_k)]\Delta x_{kj},$$

and hence, since $d(\Pi)<\delta$,

$$\left|\sum_{j=0}^{m_k-1} f(\xi_{kj})\Delta x_{kj} - \sum_{j=0}^{m_k-1} f(\xi_k)\Delta x_{kj}\right| \leqslant \sum_{j=0}^{m_k-1} |f(\xi_{kj})-f(\xi_k)|\Delta x_{kj} \leqslant \omega_f(\delta)\Delta x_k.$$

Summing over k, we finally get

$$\left|\sum_{k=0}^{n-1}\sum_{j=0}^{m_k-1} f(\xi_{kj})\Delta x_{kj} - \sum_{k=0}^{n-1} f(\xi_k)\Delta x_k\right| \leqslant \omega_f(\delta)(b-a). \quad \blacksquare$$

d. LEMMA. *If Π_1 and Π_2 are any two partitions such that $d(\Pi_1)<\delta$, $d(\Pi_2)<\delta$, then*

$$|S_{\Pi_1}(f)-S_{\Pi_2}(f)|\leqslant 2\omega_f(\delta)(b-a). \tag{6}$$

Proof. Let Π_3 be a new partition using all the points of subdivision of Π_1 and Π_2 as its points of subdivision, with the marked points being chosen arbitrarily. The partition Π_3 is clearly finer than both Π_1 and Π_2. Hence, by the preceding lemma,

$$|S_{\Pi_1}(f)-S_{\Pi_3}(f)|\leqslant \omega_f(\delta)(b-a),$$
$$|S_{\Pi_2}(f)-S_{\Pi_3}(f)|\leqslant \omega_f(\delta)(b-a),$$

which together imply (6). $\quad \blacksquare$

e. We are now in a position to prove the following key

THEOREM. *If $f(x)$ is continuous on $[a,b]$, then $f(x)$ is integrable on $[a,b]$.*

Proof. By Cauchy's convergence criterion for the existence of a limit in a direction (Theorem 4.19),† we need only prove that, given any $\varepsilon>0$, there exists a $\delta>0$ such that

$$|S_{\Pi_1}(f)-S_{\Pi_2}(f)|<\varepsilon$$

† Here, of course, we have in mind the direction T or $d(\Pi)\to 0$ discussed on p. 275.

whenever $d(\Pi_1) < \delta$, $d(\Pi_2) < \delta$. Since $f(x)$ is continuous (and hence uniformly continuous) on $[a,b]$, given any $\varepsilon > 0$, we can find a $\delta > 0$ such that

$$\omega_f(\delta) < \frac{\varepsilon}{2(b-a)}.$$

But then, by the preceding lemma,

$$|S_{\Pi_1}(f) - S_{\Pi_2}(f)| < 2\frac{\varepsilon}{2(b-a)}(b-a) = \varepsilon$$

for any two partitions Π_1 and Π_2 such that $d(\Pi_1) < \delta$, $d(\Pi_2) < \delta$. ∎

9.15. Before considering the integrals of a wider class of functions, we now prove a number of general properties of integrals, assuming that the integrals in question exist, without using any special properties of the functions involved.

a. THEOREM. *If the function $f(x)$ is integrable on $[a,b]$ and if k is constant, then $kf(x)$ is also integrable on $[a,b]$ and*

$$\int_a^b kf(x)\ dx = k \int_a^b f(x)\ dx, \tag{7}$$

i.e., a constant can always be brought out in front of the integral sign.†

Proof. In fact,

$$S_\Pi(kf) = \sum_{j=0}^{n-1} kf(\xi_j)\Delta x_j = k\sum_{j=0}^{n-1} f(\xi_j)\Delta x_j = kS_\Pi(f),$$

where the right-hand side clearly has a limit under arbitrary refinement of the partition Π, equal to

$$k \int_a^b f(x)\ dx.$$

But then the left-hand side has the same limit under arbitrary refinement of the partition, thereby proving (7). ∎

b. THEOREM. *If the functions $f(x)$ and $g(x)$ are integrable on $[a,b]$, then so is the function $f(x) + g(x)$ and*

$$\int_a^b [f(x) + g(x)]\ dx = \int_a^b f(x)\ dx + \int_a^b g(x)\ dx, \tag{8}$$

i.e., the integral of a sum equals the sum of the integrals of the separate terms.

† The symbol ∫ is called the *integral sign*, and the function appearing behind the integral sign (i.e., the function "being integrated") is called the *integrand*. The process leading from a function to its integral is, of course, called *integration*.

Proof. For any partition Π we have

$$S_\Pi(f+g) = \sum_{k=0}^{n-1} [f(\xi_k) + g(\xi_k)]\Delta x_k$$
$$= \sum_{k=0}^{n-1} f(\xi_k)\Delta x_k + \sum_{k=0}^{n-1} g(\xi_k)\Delta x_k = S_\Pi(f) + S_\Pi(g).$$

Taking the limit of both sides under arbitrary refinement of the partition, we deduce the integrability of $f(x) + g(x)$ and the validity of (8). ∎

c. THEOREM. *Every integrable function $f(x)$ is bounded.*

Proof. Suppose $f(x)$ is integrable but unbounded on the underlying interval $[a,b]$, and consider any partition Π of $[a,b]$. Then $f(x)$ is unbounded on at least one of the subintervals $[x_k, x_{k+1}]$, and hence, by fixing the marked points in the other subintervals and varying ξ_k in $[x_k, x_{k+1}]$, we can make $S_\Pi(f)$ arbitrarily large in absolute value. But then $S_\Pi(f)$ cannot have a finite limit under arbitrary refinement of the partition, i.e., $f(x)$ fails to be integrable. This contradiction shows that $f(x)$ must be bounded. ∎

d. THEOREM. *If $f(x)$ and $g(x)$ are integrable on $[a,b]$ and satisfy the inequality $f(x) \leqslant g(x)$, then*

$$\int_a^b f(x)\, dx \leqslant \int_a^b g(x)\, dx. \tag{9}$$

Proof. The Riemann sums $S_\Pi(f)$ and $S_\Pi(g)$ for the same partition Π with the same marked points ξ_k satisfy the inequality

$$S_\Pi(f) = \sum_{k=0}^{n-1} f(\xi_k)\Delta x_k \leqslant \sum_{k=0}^{n-1} g(\xi_k)\Delta x_k = S_\Pi(g),$$

which gives (9) after taking the limit as $d(\Pi) \to 0$. ∎

e. In particular, if $f(x)$ is integrable on $[a,b]$ and $f(x) \leqslant M$, then

$$\int_a^b f(x)\, dx \leqslant \int_a^b M\, dx = M(b-a), \tag{10}$$

while if $f(x) \geqslant m$, then

$$\int_a^b f(x)\, dx \geqslant \int_a^b m\, dx = m(b-a). \tag{11}$$

In particular, if $f(x)$ is nonnegative, we can choose $m = 0$ in (11), so that the integral of $f(x)$ is also nonnegative, while if $f(x)$ is nonpositive, we can choose $M = 0$ in (10), so that the integral of $f(x)$ is also nonpositive.

f. Let $f(x)$ be integrable on $[a,b]$ and let

$$m \leqslant f(x) \leqslant M \qquad (a \leqslant x \leqslant b).$$

Then, by (10) and (11),

$$m(b-a) \leqslant \int_a^b f(x) \, dx \leqslant M(b-a). \tag{12}$$

The quantity

$$\frac{1}{b-a} \int_a^b f(x) \, dx \tag{13}$$

is called the *(integral) mean value* of $f(x)$ on the interval $[a,b]$. It follows from (12) that

$$m \leqslant \frac{1}{b-a} \int_a^b f(x) \, dx \leqslant M. \tag{14}$$

THEOREM (**Mean value theorem for integrals**). *If $f(x)$ is continuous on $[a,b]$, then there exists a point $\xi \in [a,b]$ such that*

$$\frac{1}{b-a} \int_a^b f(x) \, dx = f(\xi).$$

Proof. Let

$$m = \inf_{a \leqslant x \leqslant b} f(x), \qquad M = \sup_{a \leqslant x \leqslant b} f(x)$$

in (14). By Weierstrass' theorem (Theorem 5.16b), there are points $p, q \in [a,b]$ such that $f(p) = m, f(q) = M$. If $p < q$, then by (14) and the intermediate value theorem (Theorem 5.23), $f(x)$ takes the value (13) at some point $\xi \in [p,q] \subset [a,b]$, and similarly if $p > q$. ∎

g. THEOREM. *If $f(x)$ and $|f(x)|$ are integrable on $[a,b]$,† then*

$$\left| \int_a^b f(x) \, dx \right| \leqslant \int_a^b |f(x)| dx. \tag{15}$$

Proof. It follows from

$$-|f(x)| \leqslant f(x) \leqslant |f(x)|$$

and Theorem 9.15d that

$$-\int_a^b |f(x)| dx \leqslant \int_a^b f(x) \, dx \leqslant \int_a^b |f(x)| dx,$$

† Actually, the integrability of $f(x)$ implies that of $|f(x)|$ (see Problem 7).

which is equivalent to (15). ∎

h. THEOREM. *If $f(x)$ is integrable on the intervals $[a,c]$ and $[c,b]$, where $a < c < b$, then $f(x)$ is integrable on the interval $[a,b]$ and*

$$\int_a^b f(x)\ dx = \int_a^c f(x)\ dx + \int_c^b f(x)\ dx. \tag{16}$$

Proof. Let

$$\Pi = \{a = x_0 \leqslant x_1 \leqslant \cdots \leqslant x_m \leqslant c \leqslant x_{m+1} \leqslant \cdots \leqslant x_n = b\}$$

be any partition of the interval $[a,b]$. The corresponding Riemann sum with marked points ξ_k $(k = 0,1,\ldots,n-1)$ can be written in the form

$$
\begin{aligned}
S_\Pi(f) &= \sum_{k=0}^{n-1} f(\xi_k)\Delta x_k = \sum_{k=0}^{m-1} f(\xi_k)\Delta x_k + f(\xi_m)\Delta x_m + \sum_{k=m+1}^{n-1} f(\xi_k)\Delta x_k \\
&= \sum_{k=0}^{m-1} f(\xi_k)\Delta x_k + f(c)(c - x_m) + f(c)(x_{m+1} - c) \\
&\qquad + \sum_{k=m+1}^{n-1} f(\xi_k)\Delta x_k + [f(\xi_m) - f(c)]\Delta x_m \\
&= S_{\Pi_1}(f) + S_{\Pi_2}(f) + [f(\xi_m) - f(c)]\Delta x_m, \tag{17}
\end{aligned}
$$

where Π_1 and Π_2 are the corresponding partitions of the intervals $[a,c]$ and $[c,b]$. Under an arbitrary refinement of Π, the left-hand side of (17) approaches

$$\int_a^b f(x)\ dx,$$

while the right-hand side approaches

$$\int_a^c f(x)\ dx + \int_c^b f(x)\ dx.$$

In fact, an arbitrary refinement of Π obviously leads to an arbitrary refinement of both Π_1 and Π_2, while at the same time causing the last term in the right-hand side of (17) to approach zero, since $f(x)$ is bounded (being integrable) and $\Delta x_m \to 0$. ∎

i. Next we prove the converse assertion:

THEOREM. *If $f(x)$ is integrable on the interval $[a,b]$, then $f(x)$ is integrable on the intervals $[a,c]$ and $[c,b]$, where $a < c < b$.*

Proof. We need only give the proof for the interval $[a,c]$, since the proof for

$[c,b]$ is completely analogous. Given two partitions

$$\Pi = \{a = x_0 \leqslant \xi_0 \leqslant x_1 \leqslant \xi_1 \leqslant \cdots \leqslant x_m = c\},$$
$$\Pi' = \{a = x_0' \leqslant \xi_0' \leqslant x_1' \leqslant \xi_1' \leqslant \cdots \leqslant x_p' = c\}$$

of the interval $[a,c]$, suppose we enlarge Π and Π' in the *same* way to form partitions

$$\overline{\Pi} = \{a = x_0 \leqslant \xi_0 \leqslant x_1 \leqslant \xi_1 \leqslant \cdots \leqslant x_m \leqslant \xi_m \leqslant x_{m+1} \leqslant \cdots \leqslant x_n = b\},$$
$$\overline{\Pi}' = \{a = x_0' \leqslant \xi_0' \leqslant x_1' \leqslant \xi_1' \leqslant \cdots \leqslant x_p' \leqslant \xi_m \leqslant x_{m+1} \leqslant \cdots \leqslant x_n = b\}.$$

Then, since $f(x)$ is integrable on $[a,b]$, given any $\varepsilon > 0$, there exists a $\delta > 0$ such that

$$|S_{\overline{\Pi}}(f) - S_{\overline{\Pi}'}(f)| < \varepsilon$$

whenever $d(\overline{\Pi}) < \delta$, $d(\overline{\Pi}') < \delta$. But obviously

$$S_{\overline{\Pi}}(f) - S_{\overline{\Pi}'}(f) = S_\Pi(f) - S_{\Pi'}(f),$$

and hence

$$|S_\Pi(f) - S_{\Pi'}(f)| < \varepsilon$$

whenever $d(\Pi) < \delta$, $d(\Pi') < \delta$. Therefore the Cauchy convergence criterion for the existence of a limit in the direction $d(\Pi) \to 0$ is satisfied for the Riemann sums of the function $f(x)$ on the interval $[a,c]$. It follows that $f(x)$ is integrable on $[a,c]$. ∎

j. THEOREM. *If $f(x)$ is integrable on the interval $[a,b]$ and if $a = c_0 < c_1 < \cdots < c_n = b$, then $f(x)$ is integrable on every interval $[c_0,c_1],\ldots,[c_{n-1},c_n]$ and*

$$\int_a^b f(x)\ dx = \int_{c_0}^{c_1} f(x)\ dx + \cdots + \int_{c_{n-1}}^{c_n} f(x)\ dx.$$

Proof. Apply the preceding two theorems repeatedly. ∎

k. Finally we prove a useful estimate for the deviation of an integral from its integral sums:

THEOREM. *If $f(x)$ is integrable on $[a,b]$ and $d(\Pi) < \delta$, then*

$$\left| \int_a^b f(x)\ dx - S_\Pi(f) \right| \leqslant \omega_f(\delta)(b-a).$$

Proof. Take the limit as $d(\Pi') \to 0$ of the inequality (5). ∎

9.16. Next we show that a number of discontinuous functions are integrable. The class of discontinuous functions to be considered here is admittedly rather small, but is quite sufficient for our subsequent purposes.

a. LEMMA. *Let $h(x)$ be a function equal to 0 for $a < x < b$ but taking arbitrary values $h(a)$ and $h(b)$. Then $h(x)$ is integrable on $[a,b]$ and*

$$\int_a^b h(x)\ dx = 0. \tag{18}$$

Proof. The Riemann sum of $h(x)$ for any partition Π reduces to

$$h(\xi_0)\Delta x_0 + h(\xi_{n-1})\Delta x_{n-1},$$

where $h(\xi_0)$ equals 0 or $h(a)$ and $h(\xi_{n-1})$ equals 0 or $h(b)$. But this expression obviously approaches 0 as $d(\Pi) \to 0$. ∎

b. LEMMA. *Given a function $f(x)$ continuous on $[a,b]$, suppose $f_1(x)$ coincides with $f(x)$ on the open interval (a,b), while taking arbitrary values $f_1(a)$ and $f_1(b)$. Then $f_1(x)$ is integrable on $[a,b]$ and*

$$\int_a^b f_1(x)\ dx = \int_a^b f(x)\ dx. \tag{19}$$

Proof. The function $h(x) = f_1(x) - f(x)$ is integrable on $[a,b]$ and satisfies the conditions of the preceding lemma. Hence $h(x)$ is integrable on $[a,b]$ and satisfies (18). But then the function $f_1(x) = f(x) + h(x)$ is also integrable on $[a,b]$ and satisfies (19), because of (18) and Theorem 9.15b. ∎

c. A function $f(x)$ is said to be *piecewise continuous* on $[a,b]$ if there exists a partition

$$a = x_0 < x_1 < x_2 < \cdots < x_n = b$$

such that $f(x)$ is continuous on every open interval $x_k < x < x_{k+1}$ and has finite limits

$$f(x_0 + 0), f(x_1 - 0), f(x_1 + 0), f(x_2 - 0), f(x_2 + 0), \ldots, f(x_n - 0)$$

(see Sec. 5.32).†

THEOREM. *If $f(x)$ is piecewise continuous on $[a,b]$, then $f(x)$ is integrable on $[a,b]$.*

Proof. Let $f_k(x)$ denote the function defined on $x_k \leqslant x \leqslant x_{k+1}$ which coincides with $f(x)$ for $x_k < x < x_{k+1}$ and takes the appropriate limiting values $f(x_k + 0)$ and $f(x_{k+1} - 0)$ at the end points of $[x_k, x_{k+1}]$. Then $f_k(x)$ is continuous on $[x_k, x_{k+1}]$ and hence integrable on $[x_k, x_{k+1}]$, by Theorem 9.14e. Moreover, by the preceding lemma, $f(x)$ is integrable on $[x_k, x_{k+1}]$ and

$$\int_{x_k}^{x_{k+1}} f(x)\ dx = \int_{x_k}^{x_{k+1}} f_k(x)\ dx.$$

† The values of $f(x)$ at the points x_k themselves can be arbitrary.

Applying Theorem 9.15h repeatedly, we see that $f(x)$ is integrable on the whole interval $[a,b]$. ∎

It follows from Theorem 9.15j that

$$\int_a^b f(x)\ dx = \sum_{k=0}^{n-1} \int_{x_k}^{x_{k+1}} f(x)\ dx = \sum_{k=0}^{n-1} h_k \int_{x_k}^{x_{k+1}} f_k(x)\ dx,$$

so that calculating the integral of a piecewise continuous function reduces to calculating integrals of continuous functions.

d. In particular, every *piecewise constant* function

$$h(x) = \begin{cases} h_0 = \text{const if } a = x_0 < x < x_1, \\ h_1 = \text{const if } x_1 < x < x_2, \\ \cdots \\ h_{n-1} = \text{const if } x_{n-1} < x < x_n = b \end{cases}$$

is integrable, with integral

$$\int_a^b h(x)\ dx = \sum_{k=0}^{n-1} \int_{x_k}^{x_{k+1}} h(x)\ dx = \sum_{k=0}^{n-1} h_k \int_{x_k}^{x_{k+1}} dx = \sum_{k=0}^{n-1} h_k(x_{k+1} - x_k).$$

Incidentally, this shows that the Riemann sum $S_\Pi(f)$ of every function $f(x)$ is the integral of a piecewise constant function $h_\Pi(x)$, namely the function taking the constant value $h_k = f(\xi_k)$ in each interval $x_k < x < x_{k+1}$.

e. THEOREM. *If $g(x)$ is a nonnegative piecewise continuous function and if*

$$\int_a^b g(x)\ dx = 0, \tag{20}$$

then $g(x)$ can be nonzero only at a finite number of points.

Proof. Suppose first that $g(x)$ is continuous, and suppose $g(x) \not\equiv 0$ on $[a,b]$. Then there exists a point $c \in [a,b]$ such that $g(c) \neq 0$, and hence, by the continuity of $g(x)$, a constant $C > 0$ and a whole interval $[\alpha,\beta]$, $a < \alpha < \beta < b$ such that $g(x) \geqslant C$ for all $x \in [\alpha,\beta]$. But then

$$\int_a^b g(x)\ dx = \int_\alpha^\beta g(x)\ dx + \int_a^\alpha g(x)\ dx + \int_\beta^b g(x)\ dx$$
$$\geqslant \int_\alpha^\beta g(x)\ dx \geqslant (\beta - \alpha)C > 0,$$

contrary to (20). It follows that $g(x) \equiv 0$.

Now suppose $g(x)$ is a piecewise continuous function, with points of discontinuity $c_0 = a < c_1 < \cdots < c_n = b$. Then, by Sec. 9.16c,

$$\int_a^b g(x)\ dx = \sum_{k=0}^{n-1} \int_{c_k}^{c_{k+1}} g(x)\ dx = 0,$$

and hence

$$\int_{c_k}^{c_{k+1}} g(x)\ dx = 0 \qquad (k = 0, 1, \ldots, n-1),$$

since each term is nonvanishing. But $g(x)$ is continuous on (c_k, c_{k+1}), and hence $g(x) = 0$ at every point of (c_k, c_{k+1}). The only points where $g(x)$ can fail to vanish are the points $c_0, c_1, \ldots, c_n$, of which there are only finitely many. ∎

9.17. We have just proved the integrability of piecewise continuous functions, a class of functions with only finitely many points of discontinuity.† We now inquire about the integrability of functions with infinitely many points of discontinuity. In this connection, consider the *Dirichlet function*

$$\chi(x) = \begin{cases} 1 \text{ if } x \text{ is rational,} \\ 0 \text{ if } x \text{ is irrational.} \end{cases}$$

Given any partition Π (of the unit interval $[0,1]$, say), we get an integral sum

$$\sum_{k=0}^{n-1} \chi(\xi_k) \Delta x_k, \tag{21}$$

equal to 1 if we choose the points ξ_k to be rational. However, if we choose the points ξ_k to be irrational (with the same partition Π), we find that the sum (21) equals 0. Hence the Riemann sums of the Dirichlet function do not approach a limit under arbitrary refinement of the partition, i.e., the Dirichlet function is *nonintegrable*.

Thus we see that, so to speak, the boundary between integrable and nonintegrable functions lies somewhere in the region of functions with infinitely many points of discontinuity. In fact, there is a theorem which asserts that a function $f(x)$ has a Riemann integral if and only if the set of all its points of discontinuity can be covered by a finite or countable set of intervals whose total length is less than δ, where $\delta > 0$ is arbitrary (see Problem 9). However, in problems involving the integration of discontinuous functions, it is best to use another kind of integral, which is more general than the Riemann integral, namely the *Lebesgue integral*.‡

---

† It can be shown that every bounded function with only finitely many points of discontinuity is integrable (see Problem 9).

‡ See, e.g., G. E. Shilov and B. L. Gurevich, *Integral, Measure and Derivative: A Unified Approach* (translated by R. A. Silverman), Dover Publications, Inc., N.Y. (1977).

9.18. Suppose $f(x)$ is integrable on the interval $a \leqslant x \leqslant b$. Then we set

$$\int_b^a f(x)\ dx = -\int_a^b f(x)\ dx, \tag{22}$$

by definition. In particular, (22) implies

$$\int_a^a f(x)\ dx = 0.$$

The formula

$$\int_\alpha^\gamma f(x)\ dx = \int_\alpha^\beta f(x)\ dx + \int_\beta^\gamma f(x)\ dx \tag{23}$$

is now valid for arbitrary $\alpha, \beta, \gamma \in [a,b]$, regardless of their relative position in the interval $[a,b]$. In fact, (23) has already been proved for the case $\alpha < \beta < \gamma$ (see Theorem 9.15h). Choosing another configuration of the points α, β, and γ, say $\gamma < \beta < \alpha$, we have

$$\int_\gamma^\alpha f(x)\ dx = \int_\gamma^\beta f(x)\ dx + \int_\beta^\alpha f(x)\ dx,$$

as already proved, and hence, by (22),

$$\int_\alpha^\gamma f(x)\ dx = -\int_\gamma^\alpha f(x)\ dx = -\int_\gamma^\beta f(x)\ dx - \int_\beta^\alpha f(x)\ dx$$
$$= \int_\alpha^\beta f(x)\ dx + \int_\beta^\gamma f(x)\ dx,$$

by (22) again. The argument is the same for any other configuration of the points α, β, and γ.

Note that (22) immediately implies

$$\frac{1}{a-b}\int_b^a f(x)\ dx = -\frac{1}{a-b}\int_a^b f(x)\ dx = \frac{1}{b-a}\int_a^b f(x)\ dx. \tag{24}$$

Hence, if $m \leqslant f(x) \leqslant M$, the estimate (14) can be written in the alternative form

$$m \leqslant \frac{1}{a-b}\int_b^a f(x)\ dx \leqslant M. \tag{25}$$

In other words, in estimating the mean value of $f(x)$, it does not matter whether $a < b$ or $b < a$.

9.2. Area and Arc Length

We now give two geometrical examples illustrating the utility of the concept of the integral in mathematical analysis. A much more detailed discussion of the applications of integration will be given in Secs. 9.6–9.9.

9.21. Area under a curve. The concept of the area of a simple geometric figure bounded by straight line segments is familiar from elementary geometry. But what is the "area" of a figure bounded by more general curves? The answer to this question is not to be found in elementary geometry.†

A natural starting point for a general theory of area is the following axiom: *The area of a figure* Φ *is a number no larger than the area* Ω *of any "elementary figure"* (by which we mean a finite plane region whose boundary consists of a finite number of simple closed polygonal lines‡) *containing* Φ *and no smaller than the area* ω *of any elementary figure contained in* Φ (see Figure 21). Since we always have $\omega \leqslant \Omega$, by the theorems of elementary geometry, it follows that

$$\sup \omega \leqslant \inf \Omega$$

(see Theorem 1.62b), where the least upper bound on the left is taken over

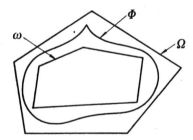

Figure   21

---

† The case of the area enclosed by a circle is no exception, since even in elementary geometry this area is calculated by nonelementary means (with the help of limits).

‡ Given a finite number of line segments $\sigma_1, \sigma_2, \ldots, \sigma_n$, suppose that

(1) The final point Q_j of σ_j coincides with the initial point P_{j+1} of σ_{j+1} ($j = 1, 2, \ldots, n$), where $\sigma_{n+1} = \sigma_1$ by definition;

(2) Two segments σ_j and σ_k have points in common (if and) only if $k = j+1$, in which case $Q_j = P_{j+1}$ is the only shared point.

Then the segments $\sigma_1, \sigma_2, \ldots, \sigma_n$ are said to form a *simple closed polygonal line*.

all elementary figures contained in Φ and the greatest lower bound on the right is taken over all elementary figures containing Φ. If sup $\omega =$ inf Ω then the only possible value of the area of Φ is the common value

$$S = \sup \omega = \inf \Omega. \tag{1}$$

Suppose we now apply these general considerations to the case of the area of the "curvilinear trapezoid" Φ bounded on the bottom by the segment $[a,b]$ of the x-axis, on top by the graph of a continuous function $y = f(x) \geqslant 0$, and on the sides by segments of the vertical lines $x = a$ and $x = b$. The area of this trapezoid is often simply called the "area under the curve $y = f(x)$" (from $x = a$ to $x = b$). Consider a partition

$$\Pi = \{a = x_0 \leqslant \xi_0 \leqslant x_1 \leqslant \xi_1 \leqslant x_2 \leqslant \cdots \leqslant x_{n-1} \leqslant \xi_{n-1} \leqslant x_n = b\},$$

where the point $\xi_k \in [x_k, x_{k+1}]$ is chosen so that

$$f(\xi_k) = M_k = \max_{x_k \leq x \leq x_{k+1}} f(x).$$

Then the corresponding Riemann sum

$$\bar{S}_\Pi(f) = \sum_{k=0}^{n-1} M_k \Delta x_k$$

can be interpreted as the area of an elementary figure containing Φ, made up of certain "circumscribed rectangles" (see Figure 22). On the other hand, if we choose ξ_k so that

$$f(\xi_k) = m_k = \min_{x_k \leq x \leq x_{k+1}} f(x),$$

then the corresponding Riemann sum

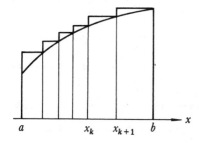

Figure   22

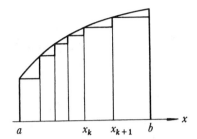

Figure   23

$$\underline{S}_\Pi(f) = \sum_{k=0}^{n-1} m_k \Delta x_k$$

can be interpreted as the area of an elementary figure contained in Φ, made up of certain "inscribed rectangles" (see Figure 23). Therefore, according to our basic axiom, every number S which can lay claim to be a value of the "area" of the figure Φ must satisfy the condition

$$\underline{S}_\Pi(f) \leqslant S \leqslant \bar{S}_\Pi(f).$$

But $f(x)$ is integrable, being continuous (see Theorem 9.14e), so that $\underline{S}_\Pi(f)$ and $\bar{S}_\Pi(f)$ approach the same limit, equal to the integral of $f(x)$ over $[a,b]$, under arbitrary refinement of the partition Π. It follows that

$$S = \int_a^b f(x)\ dx \qquad\qquad (2)$$

is the only possible value of the area of the figure Φ. The fact that S exists as defined by (1) in terms of arbitrary elementary figures (instead of just elementary figures of a special kind, made up of rectangles) can be shown with the help of the general considerations given in Appendix B (esp. Secs. B.4, B.5, and B.7).

In three-dimensional space, the concept of area goes over into that of "volume." A theory of "measure" (i.e., "generalized volume") and integration on a compact metric space is outlined in Appendix B.

9.22. Arc length. In elementary geometry, one encounters first the concept of the length of a line segment or more generally a polygonal line, and then the concept of the length of a circular arc, defined as the limit of the length of a polygonal line "inscribed" in the arc as the length of the segments making

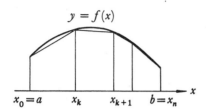

Figure   24

up the line approaches zero. Without lingering to make this latter definition more precise,† we pass at once to the problem of defining the length of an arbitrary curve L with equation

$$y=f(x) \qquad (a\leqslant x\leqslant b).$$

Consider a partition

$$\Pi = \{a=x_0 \leqslant \xi_0 \leqslant x_1 \leqslant \xi_1 \leqslant x_2 \leqslant \cdots \leqslant x_{n-1} \leqslant \xi_{n-1} \leqslant x_n = b\},$$

where the points $\xi_0, \xi_1, \ldots, \xi_{n-1}$ are as yet undetermined, and construct the polygonal line L_Π with vertices at the points (x_0, y_0), $(x_1, y_1), \ldots, (x_n, y_n)$. Then L_Π is "inscribed" in the curve L, in the sense illustrated in Figure 24.

Now let $s(L_\Pi)$ be the length of L_Π. Then clearly

$$s(L_\Pi) = \sum_{k=0}^{n-1} \sqrt{(\Delta x_k)^2 + (\Delta y_k)^2}, \tag{3}$$

where $\Delta x_k = x_{k+1} - x_k$, $\Delta y_k = y_{k+1} - y_k$. Suppose $f(x)$ is continuous and differentiable, with derivative $f'(x)$. Then, by Lagrange's theorem (Sec. 7.44), we can choose a point $\xi_k \in (x_k, x_{k+1})$ such that

$$\sqrt{(\Delta x_k)^2 + (\Delta y_k)^2} = \sqrt{1 + (\Delta y_k/\Delta x_k)^2}\,\Delta x_k = \sqrt{1 + [f'(\xi_k)]^2}\,\Delta x_k.$$

We can then write (3) in the form

$$s(L_\Pi) = \sum_{k=0}^{n-1} \sqrt{1 + [f'(\xi_k)]^2}\,\Delta x_k. \tag{4}$$

This expression is a Riemann sum of the function

$$g(x) = \sqrt{1 + [f'(x)]^2},$$

corresponding to the partition Π with marked points $\xi_0, \xi_1, \ldots, \xi_{n-1}$. If $g(x)$

† Because of the inadequacy of the high school definition of a limit, the reader should examine the definition of the length of a circular arc in the light of the more general theory of arc length given here. A reexamination of the elementary definition of the area enclosed by a circle from the standpoint of the considerations leading to formula (2) is also in order.

is integrable, then the right-hand side of (4) approaches the integral

$$\int_a^b g(x)\, dx$$

under arbitrary refinement of the partition Π. Thus, finally, we define $s(L)$, the *length* of the curve L, by the formula

$$s(L) = \int_a^b \sqrt{1 + [f'(x)]^2}\, dx. \tag{5}$$

9.23. Writing integrals like (2) and (5) would be of little value, and would in fact be hardly more than a tautology, if there were no way of calculating integrals other than that given in the direct definition of Sec. 9.13. Actually, however, there are other powerful methods for evaluating integrals (presented in Secs. 9.4 and 9.5). These methods rest on the considerations of Sec. 9.3, of great importance in their own right.

9.3. Antiderivatives and Indefinite Integrals

9.31. The integral as a function of its upper limit. Let the function $f(x)$ be integrable on the interval $[a,b]$, and consider the integral

$$F(x) = \int_a^x f(t)\, dt \qquad (a \leqslant x \leqslant b), \tag{1}$$

where we denote the variable of integration by t to avoid confusion with the upper limit of integration x. Then (1) is a new function of x, defined on the interval $[a,b]$.

a. THEOREM. *The function $F(x)$ is continuous on $[a,b]$.*

Proof. An immediate consequence of the estimate

$$|F(x') - F(x'')| = \left| \int_{x'}^{x''} f(t)\, dt \right| \leqslant (x'' - x') \sup_{a \leqslant x \leqslant b} |f(x)|$$

(see Secs. 9.15e and 9.18). ∎

b. THEOREM. *If $f(x)$ is continuous from the left (or from the right) at $x = c$, then $F(x)$ has a left-hand (or right-hand) derivative at $x = c$, equal to $f(c)$.*

Proof. Clearly

$$\frac{F(c+h) - F(c)}{h} = \frac{1}{h} \int_a^{c+h} f(t)\, dt - \int_a^c f(t)\, dt = \frac{1}{h} \int_c^{c+h} f(t)\, dt, \tag{2}$$

where the expression on the right is the mean value of $f(x)$ on the interval $[c, c+h]$.† Let

$$m_h = \inf_{x \in [c, c+h]} f(x), \qquad M_h = \sup_{x \in [c, c+h]} f(x).$$

Then

$$m_h \leqslant \frac{1}{h} \int_c^{c+h} f(t)\, dt = \frac{F(c+h) - F(c)}{h} \leqslant M_h$$

(cf. formula (14), p. 280 and formula (25), p. 286). If $f(x)$ is continuous from the left (or from the right) at $x = c$, then $m_h \to f(c)$, $M_h \to f(c)$ as $h \nearrow 0$ (or $h \searrow 0$), i.e., the limit of (2) as $h \nearrow 0$ (or $h \searrow 0$) exists and equals $f(c)$, or equivalently, $F(x)$ has a left-hand (or right-hand) derivative at $x = c$, equal to $f(c)$. ∎

COROLLARY. *If $f(x)$ is continuous (from both sides) at $x = c$, then $F(x)$ has a derivative at $x = c$, equal to $f(c)$.*

Proof. $F(x)$ has a derivative at $x = c$ if and only if $F(x)$ has equal left-hand and right-hand derivatives at $x = c$ (Sec. 7.19c). But, by the preceding theorem, these two one-sided derivatives exist and equal $f(c)$. ∎

Note that if $f(x)$ is continuous on $[a, b]$ but on no larger interval, then the derivative of $F(x)$ at the end point $x = a$ can only be a right-hand derivative, equal to $f(a)$, while the derivative of $F(x)$ at the end point $x = b$ can only be a left-hand derivative, equal to $f(b)$.

9.32. Antiderivatives. Let $f(x)$ be a piecewise continuous function on $[a, b]$. Then any continuous function $G(x)$ defined on $[a, b]$ with a derivative equal to $f(x)$ at every continuity point of $f(x)$ is said to be an *antiderivative* $f(x)$ on $[a, b]$. For example, the function x^n is an antiderivative of nx^{n-1} on every closed interval. Given any piecewise continuous function $f(x)$, it follows from the above corollary that the function $F(x)$ defined by (1) is an antiderivative of $f(x)$. If $G(x)$ is an antiderivative of $f(x)$, then obviously so is every function

$$H(x) = G(x) + C, \tag{3}$$

where C is a constant (so that, in particular, $f(x)$ has infinitely many antiderivatives). The converse is also true, as shown by the following

THEOREM. *Every antiderivative $H(x)$ of the function $f(x)$ on the interval $[a, b]$ is of the form (3), where $G(x)$ is any fixed antiderivative of $f(x)$ on $[a, b]$.*

† See Secs. 9.15 and 9.18. If $h < 0$, we set $[c, c+h] = \{x : c+h \leqslant x \leqslant c\}$, by definition.

Proof. Clearly

$$[H(x) - G(x)]' = f(x) - f(x) = 0$$

at every continuity point of $f(x)$, and hence, by Theorem 7.45b, $H(x) - G(x)$ is constant on every interval on which $f(x)$ is continuous. But $G(x)$ and $H(x)$ are continuous on the whole interval $[a,b]$, and hence so is $H(x) - G(x)$. Therefore $H(x) - G(x)$ is constant on the whole interval $[a,b]$. ∎

9.33. The following key proposition, relating the operations of differentiation and integration, is often called the "fundamental theorem of calculus":

THEOREM. *If $f(x)$ is piecewise continuous on $[a,b]$, then*

$$\int_a^b f(x)\,dx = G(b) - G(a), \tag{4}$$

where $G(x)$ is any antiderivative of $f(x)$ on $[a,b]$.

Proof. By the preceding theorem, every other antiderivative of $f(x)$, in particular the function (1), differs from $G(x)$ by a constant, so that

$$F(x) = \int_a^x f(t)\,dt = G(x) + C.$$

To determine the constant C, let $x = a$. Then obviously $F(a) = 0$, and hence $C = -G(a)$, so that

$$\int_a^x f(t)\,dt = G(x) - G(a). \tag{5}$$

To get (4), we now set $x = b$ in (5), afterwards changing the variable of integration from t to x. ∎

9.34. Indefinite versus definite integrals. The set of all antiderivatives of a function $f(x)$ is called the *indefinite integral* of $f(x)$, denoted by

$$\int f(x)\,dx$$

(without limits of integration). By contrast, the expression

$$\int_a^b f(x)\,dx$$

(defined in Sec. 9.13) is called the *definite integral* of $f(x)$. The adjectives "indefinite" and "definite" are often dropped in cases where the kind of integral under discussion is clear from the context.

9.35. Formula (4) is often written as

$$\int_a^b f(x)\ dx = G(x)\Big|_{x=a}^{x=b}$$

or simply

$$\int_a^b f(x)\ dx = G(x)\Big|_a^b,$$

where each of the expressions on the right stands for the value of the function $G(x)$ at $x=b$ minus its value at $x=a$. Thus, for example,

$$\int_0^b nx^{n-1}\ dx = x^n\Big|_0^b = b^n.$$

Geometrically, this means that the area under the curve $y=x^{n-1}$ from $x=0$ to $x=b$ equals $1/n$ times the area of the rectangle $ABCD$ shown in Figure 25 (where the points A and B have abscissas 0 and b).

9.36. Let $f(x)$ be a function continuous on $[a,b]$, and let $F(x)$ be the indefinite integral of $f(x)$, so that $F'(x)=f(x)$. Then, by the definition of the indefinite integral,

$$d\int f(x)\ dx = d[F(x)+C] = F'(x)\ dx = f(x)\ dx,$$

so that the signs d and $\int$ (in that order) cancel each other out. On the other

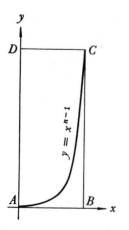

Figure 25

hand, by the definition of the differential and the indefinite integral,

$$\int dF(x) = \int F'(x)\, dx = F(x) + C, \tag{6}$$

so that the signs $\int$ and d (in that order) cancel each other out, provided we add a constant to the right-hand side.

9.37. Next we prove a number of general rules of indefinite integration, starting from the formulas

$$d(ku) = k\, du \qquad (k \text{ constant}), \tag{7}$$

$$d(u+v) = du + dv, \tag{8}$$

$$d(uv) = u\, dv + v\, du, \tag{9}$$

proved in Sec. 7.33. It follows from (6) and (7) that

$$\int k\, du = \int d(ku) = ku + C = k(u + C_1) = k \int du,$$

where C and $C_1 = C/k$ are arbitrary "constants of integration," i.e., a constant can always be brought out in front of the indefinite integral sign (cf. Theorem 9.15a). Similarly, (6) and (8) imply

$$\int (du + dv) = \int d(u+v) = u + v + C = \int du + \int dv, \tag{10}$$

where the constant of integration is absorbed into the indefinite integrals on the right, i.e., the indefinite integral of a sum equals the sum of the indefinite integrals of the separate terms (cf. Theorem 9.15b). Finally, it follows from (6), (9), and (10) that

$$\int (u\, dv + v\, du) = \int u\, dv + \int v\, du = \int d(uv) = uv + C,$$

or equivalently,

$$\int u\, dv = uv - \int v\, du, \tag{11}$$

a result known as the formula for *integration by parts* (the constant of integration has been absorbed into the second integral). This important formula allows us to determine the integral $\int u\, dv$ from a knowledge of the integral $\int v\, du$.

9.38. THEOREM. *Suppose the function $u(t)$ has a continuous derivative $u'(t)$ on $\alpha \leqslant t \leqslant \beta$ and takes values in $A \leqslant u \leqslant B$, while the function $f(u)$ is continuous on*

$A \leqslant u \leqslant B$, and let

$$F(u) = \int f(u)\ du, \qquad G(t) = \int f(u(t))u'(t)\ dt.$$

Then

$$F(u(t)) \equiv G(t) + C, \tag{12}$$

i.e., du can be replaced by u'(t) dt in the indefinite integral $\int f(u)\ du$.†

Proof. We need only prove that both sides of (12) have the same differential. But

$$dF(u) = F'(u)\ du = F'(u)u'(t)\ dt = f(u(t))u'(t)\ dt$$

(see Sec. 7.34), while

$$dG(t) = f(u(t))u'(t)\ dt. \quad \blacksquare$$

9.39. Finally we prove the converse of Theorem 9.31b. Given a function $F(x)$ continuous on $[a,b]$, suppose $F(x)$ has a derivative $F'(x)$ at all but a finite number of points of $[a,b]$, where $F'(x)$ is piecewise continuous on $[a,b]$. Then $F(x)$ is said to be *piecewise smooth* on $[a,b]$. In effect, Theorem 9.31b says that if $f(x)$ is piecewise continuous on $[a,b]$, then

$$F(x) = \int_a^x f(t)\ dt$$

is piecewise smooth on $[a,b]$. Conversely, suppose $G(x)$ is piecewise smooth on $[a,b]$, and let $f(x)$ be its piecewise continuous derivative. Then $G(x)$ is an antiderivative of $f(x)$, by Sec. 9.32, and hence

$$G(x) = \int_a^x f(t)\ dt + G(a),$$

by the same argument as in the proof of Theorem 9.33. It follows that *to within an additive constant every piecewise smooth function $G(x)$ is the integral of a piecewise continuous function* (with a variable upper limit of integration).

9.4. Technique of Indefinite Integration

9.41. Integration of polynomials

a. Applying the sign $\int$ to both sides of the formula

$$d(x^n) = nx^{n-1}dx \qquad (n = 1, 2, \ldots),$$

---

† In other words, the "invariance property" of differentials (Sec. 7.34) continues to hold for differentials appearing behind the indefinite integral sign. The importance of writing *du* in the expression $\int f(u)\ du$ is now apparent (see also Sec. 9.54).

and then changing $n-1$ to m and dividing by $n = m+1$, we get

$$\int x^m \, dx = \frac{x^{m+1}}{m+1} + C \qquad (m = 0,1,2,\ldots), \tag{1}$$

where C is an arbitrary constant.

b. Let

$$P(x) = a_0 + a_1 x + \cdots + a_n x^n$$

be any polynomial. Then, using (1) and the rules of Sec. 9.37, we find that

$$\int P(x) \, dx = a_0 x + a_1 \frac{x^2}{2} + \cdots + a_n \frac{x^{n+1}}{n+1} + C. \tag{2}$$

c. In some cases, it is more economical to use a less direct approach. For example, it would be absurd to evaluate the integral

$$\int x(x^2 + 1)^{500} \, dx$$

by first using the binomial theorem to expand the expression $(x^2 + 1)^{500}$ and then applying (2). The sensible thing to do is to use Theorem 9.38 together with the substitution $u = x^2 + 1$. This at once gives

$$\int x(x^2 + 1)^{500} \, dx = \int u^{500} \frac{1}{2} \, du = \frac{1}{2} \frac{u^{501}}{501} + C = \frac{(x^2 + 1)^{501}}{1002} + C. \tag{3}$$

d. To evaluate the integral

$$\int x^3 (x^2 + 1)^{500} \, dx,$$

we use formula (11), p. 295 to integrate by parts, setting

$$u = x^2, \qquad dv = x(x^2 + 1)^{500} \, dx.$$

Since

$$du = 2x \, dx, \qquad v = \frac{(x^2 + 1)^{501}}{1002},$$

by (3), the result is

$$\int x^3 (x^2 + 1)^{500} \, dx = uv - \int v \, du$$

$$= \frac{x^2 (x^2 + 1)^{501}}{1002} - \int \frac{(x^2 + 1)^{501} 2x \, dx}{1002}$$

$$= \frac{x^2 (x^2 + 1)^{501}}{1002} - \frac{(x^2 + 1)^{502}}{1002 \cdot 502} + C.$$

9.42. Integration of rational functions. Next we consider the problem of integrating a function of the form

$$f(x) = \frac{S(x)}{P(x)},$$

where $S(x)$ and $P(x)$ are both polynomials with real coefficients.

a. If the degree of $S(x)$ is greater than or equal to that of $P(x)$, we can carry out the division, thereby reducing $f(x)$ to the form

$$f(x) = T(x) + \frac{Q(x)}{P(x)},$$

where $T(x)$ and $Q(x)$ are new polynomials, and the degree of $Q(x)$ is now less than that of $P(x)$. Thus, having just seen how to integrate polynomials, we need only consider the problem of integrating the proper fraction

$$\frac{Q(x)}{P(x)}.$$

b. Let $x_1 \pm iy_1, \ldots, x_p \pm iy_p$ be the nonreal roots and $x_{p+1}, \ldots, x_q$ the real roots of $P(x)$, with corresponding multiplicities $r_1, \ldots, r_p$, $r_{p+1}, \ldots, r_q$.† Then the rational function $Q(x)/P(x)$ has the following partial fraction expansion (see Sec. 5.88):

$$\frac{Q(x)}{P(x)} = \frac{A_{11} + B_{11}x}{(x - x_1)^2 + y_1^2} + \cdots + \frac{A_{1r_1} + B_{1r_1}x}{[(x - x_1)^2 + y_1^2]^{r_1}} + \cdots$$

$$+ \frac{A_{p1} + B_{p1}x}{(x - x_p)^2 + y_p^2} + \cdots + \frac{A_{pr_p} + B_{pr_p}x}{[(x - x_p)^2 + y_p^2]^{r_p}}$$

$$+ \frac{A_{p+1,1}}{x - x_{p+1}} + \cdots + \frac{A_{p+1,r_{p+1}}}{(x - x_{p+1})^{r_{p+1}}} + \cdots$$

$$+ \frac{A_{q1}}{x - x_q} + \cdots + \frac{A_{qr_q}}{(x - x_q)^{r_q}}.$$

Thus our problem reduces to that of integrating the various kinds of partial fractions.

c. To evaluate the integral

$$\int \frac{dx}{(x - x_k)^n} \qquad (n > 1),$$

we integrate both sides of the formula

---

† Since $P(x)$ has real coefficients, if $z = \alpha + i\beta$ is a root of $P(x)$, then the complex conjugate $\bar{z} = \alpha - i\beta$ is also a root, with the same multiplicity (Sec. 5.88).

$$d\left(\frac{1}{(x-x_k)^{n-1}}\right) = \frac{1-n}{(x-x_k)^n}\, dx,$$

obtaining

$$\int \frac{dx}{(x-x_k)^n} = \frac{1}{(1-n)(x-x_k)^{n-1}} + C. \tag{4}$$

d. If $n=1$, we start from formula (12), p. 228, which implies

$$d\left(\ln|x-x_k|\right) = \frac{dx}{x-x_k}$$

and hence

$$\int \frac{dx}{x-x_k} = \ln|x-x_k| + C. \tag{5}$$

Note that (4) and (5) hold in any interval $[a,b]$ which does not contain the point x_k.†

e. The substitution $x-x_k = y_k u$ reduces the integral

$$\int \frac{Ax+B}{[(x-x_k)^2 + y_k^2]^n}\, dx$$

to the simpler form

$$\int \frac{Cu+D}{(u^2+1)^n}\, du.$$

If $D=0$, the integral is easily evaluated by making the further substitution $u^2+1=v$:

$$\int \frac{Cu\, du}{(u^2+1)^n} = \frac{C}{2}\int \frac{dv}{v^n} = \begin{cases} \dfrac{C}{2(1-n)v^{n-1}} + C_1 = \dfrac{C}{2(1-n)(u^2+1)^{n-1}} + C_1 & \text{if } n>1, \\[2ex] \dfrac{C}{2}\ln|v| + C_1 = \dfrac{C}{2}\ln(u^2+1) + C_1 & \text{if } n=1. \end{cases}$$

f. We must still consider the integral

$$I_n = \int \frac{du}{(u^2+1)^n}. \tag{6}$$

If $n=1$, then

$$\int \frac{du}{u^2+1} = \arctan u + C,$$

---

† If $[a,b]$ contains x_k, the integrand fails to be piecewise continuous, and the underlying theory of Secs. 9.32–9.34 is no longer applicable.

by formula (22), p. 231. Suppose $n > 1$. Then applying the formula for integration by parts to (6) with

$$v = \frac{1}{(u^2 + 1)^n}, \qquad dv = -\frac{2nu\,du}{(u^2+1)^{n+1}},$$

we get

$$\begin{aligned}
I_n &= \int \frac{du}{(u^2+1)^n} = \frac{u}{(u^2+1)^n} + 2n \int \frac{u^2\,du}{(u^2+1)^{n+1}} \\
&= \frac{u}{(u^2+1)^n} + 2n \left\{ \int \frac{(u^2+1)\,du}{(u^2+1)^{n+1}} - \int \frac{du}{(u^2+1)^{n+1}} \right\} \\
&= \frac{u}{(u^2+1)^n} + 2n(I_n - I_{n+1}),
\end{aligned}$$

and hence

$$I_{n+1} = \frac{1}{2n}\frac{u}{(u^2+1)^n} + \frac{2n-1}{2n}I_n. \tag{7}$$

This recursion formula expresses I_{n+1} in terms of I_n and a rational function. For example, setting $n = 1$ in (7), we get

$$I_2 = \int \frac{du}{(u^2+1)^2} = \frac{1}{2}\frac{u}{(u^2+1)^2} + \frac{1}{2}\arctan u + C;$$

setting $n = 2$, we find a more complicated expression for I_3, and so on.

g. Thus, finally, *the indefinite integral of every rational function can be expressed entirely in terms of rational functions, logarithms, and arc tangents.* For more on the practical technique of integrating rational functions, we refer to the literature.†

9.43. Rationalizing substitutions. Next we show how certain types of integrals can be "rationalized," i.e., reduced to integrals of rational functions, by making appropriate standard substitutions. In what follows, R denotes a rational function of one or two arguments.

a. The integral

$$\int R(e^x)\,dx$$

† See e.g., G. M. Fichtenholz, *The Indefinite Integral* (translated by R. A. Silverman), Gordon and Breach, N. Y. (1971), Chapter 2.

can be rationalized by the substitution

$$e^x = u, \qquad x = \ln u, \qquad dx = \frac{du}{u}.$$

b. The integral

$$\int R(\tan x)\, dx$$

can be rationalized by the substitution

$$\tan x = u, \qquad x = \arctan u, \qquad dx = \frac{du}{1 + u^2}. \tag{8}$$

Since the functions

$$\cos^2 x = \frac{1}{1 + \tan^2 x},$$

$$\sin^2 x = 1 - \cos^2 x,$$

$$\cos 2x = \cos^2 x - \sin^2 x,$$

$$\sin 2x = 2 \sin x \cos x = 2 \tan x \cos^2 x$$

are all rational functions of $\tan x$, the integral

$$\int R(\sin 2x,\ \cos 2x)\, dx$$

can also be rationalized by the substitution (8).

c. Unlike $\sin 2x$ and $\cos 2x$, the functions $\sin x$ and $\cos x$ are not rational functions of $\tan x$ (this can be seen from the fact that $\sin x$ and $\cos x$ fail to have the period π of the function $\tan x$). However,

$$\sin x = \frac{2 \tan (x/2)}{1 + \tan^2 (x/2)}, \qquad \cos x = \frac{1 - \tan^2 (x/2)}{1 + \tan^2 (x/2)},$$

and hence this time the integral

$$\int R(\sin x,\ \cos x)\, dx$$

is rationalized by the substitution

$$\tan \frac{x}{2} = u, \qquad x = 2 \arctan u, \qquad dx = \frac{2du}{1 + u^2}. \tag{8'}$$

9.44. Integration of irrational expressions. Consider the integral

$$\int R(x,y)\ dx,\tag{9}$$

where y is some function of x. If x and y can both be represented as rational functions

$$x = x(t),\qquad y = y(t)$$

of a new variable t, then (9) can be rationalized in an obvious way:

$$\int R(x,y)\ dx = \int R(x(t),y(t))x'(t)\ dt.$$

a. For example, suppose

$$y = \sqrt{ax+b}.$$

Then

$$ax + b = t^2,\qquad x = \frac{t^2-b}{a},\qquad y = t$$

is a rationalizing substitution, since

$$\int R(x,\sqrt{ax+b})\ dx = \int R\!\left(\frac{t^2-b}{a},t\right)\frac{2t\ dt}{a}.$$

b. The same method works for

$$y = \sqrt{\frac{\alpha x + \beta}{\gamma x + \delta}}$$

if we choose

$$\frac{\alpha x + \beta}{\gamma x + \delta} = t^2,\qquad x = \frac{\delta t^2 - \beta}{\alpha - \gamma t^2},\qquad y = t$$

(give the details).

c. However, the substitution

$$ax^2 + bx + c = t^2$$

no longer rationalizes the square root

$$y = \sqrt{ax^2 + bx + c},\tag{10}$$

since in this case x does not turn out to be a rational function of t. Here we proceed differently. If α is such that $b = -2a\alpha$, then

$$a(x-\alpha)^2 + c - a\alpha^2 = ax^2 + bx + c.$$

Now let $x - \alpha = \beta u$. Then, apart from a multiplicative constant, (10) reduces to one of the expressions

$$y_1 = \sqrt{1-u^2}, \qquad y_2 = \sqrt{1+u^2}, \qquad y_3 = \sqrt{u^2-1}$$

for a suitable choice of β.

d. In the first case ($y_1 = \sqrt{1-u^2}$), we make the substitution $u = \sin\theta$ or $u = \cos\theta$, thereby obtaining one of the integrals†

$$\int R(u, \sqrt{1-u^2})\, du = \int R(\sin\theta, \cos\theta)\cos\theta\, d\theta$$

or

$$\int R(u, \sqrt{1-u^2})\, du = -\int R(\cos\theta, \sin\theta)\sin\theta\, d\theta$$

already considered in Sec. 9.43c. For example,

$$\int \frac{du}{\sqrt{1-u^2}} = \int \frac{\cos\theta\, d\theta}{\cos\theta} = \int d\theta = \theta = \text{arc}\sin u + C_1 \tag{11}$$

if $u = \sin\theta$, or

$$\int \frac{du}{\sqrt{1-u^2}} = -\int \frac{\sin\theta\, d\theta}{-\sin\theta} = \int d\theta = \theta = \text{arc}\cos u + C_2 \tag{12}$$

if $u = \cos\theta$.‡ The difference between (11) and (12) should not bother us, since according to the general theory, the two results can differ by a constant, and this is in fact the case, as shown by the formula

$$\text{arc}\cos u = \text{arc}\sin u - \frac{\pi}{2} \tag{13}$$

(see Sec. 5.67).

† Here, of course, R is a different rational function from that appearing in (9), but one easily found from the latter.

‡ In (12) we choose the increasing branch of arc cos u (Sec. 5.67), corresponding to choosing $[-\pi, 0]$ as the domain of cos θ, so that $\sqrt{1-u^2} = -\sin\theta$ (as always, the radical denotes the positive square root). Using the decreasing branch of arc cos u corresponds to choosing $[0, \pi]$ as the domain of cos θ, so that $\sqrt{1-u^2} = \sin\theta$. We then get

$$\int \frac{du}{\sqrt{1-u^2}} = -\text{arc}\cos u + C_2$$

instead of (12), with the connection between the two results now being given by

$$\text{arc}\cos u = \frac{\pi}{2} - \text{arc}\sin u$$

(Sec. 5.67) instead of (13).

e. In the second case $(y = \sqrt{1 + u^2})$, we make the substitution

$$u = \sinh \theta, \qquad 1 + u^2 = 1 + \sinh^2 \theta = \cosh^2 \theta,$$

leading to the integral

$$\int R(u, \sqrt{1 + u^2}) \; du = \int R(\sinh \theta, \cosh \theta) \cosh \theta \; d\theta.$$

For example,

$$\int \frac{du}{\sqrt{1 + u^2}} = \int \frac{\cosh \theta \; d\theta}{\cosh \theta} = \int d\theta = \theta = \text{arc sinh } u + C,$$

where the function $\theta = \text{arc sinh } u$ is the inverse of the function $u = \sinh \theta$. We can express arc sinh u in terms of more familiar functions by solving the equation

$$u = \sinh \theta = \frac{e^\theta - e^{-\theta}}{2} = \frac{1}{2}\left(e^\theta - \frac{1}{e^\theta}\right)$$

for θ. This gives first

$$e^\theta = u + \sqrt{1 + u^2}$$

(where we drop the minus sign since $e^\theta > 0$) and then

$$\theta = \text{arc sinh } u = \ln \; (u + \sqrt{1 + u^2}).$$

f. Finally, in the third case $(y_3 = \sqrt{u^2 - 1})$, the appropriate substitution is

$$u = \cosh \theta, \qquad u^2 - 1 = \cosh^2 \theta - 1 = \sinh^2 \theta,$$

leading to the integral

$$R(u, \sqrt{u^2 - 1}) = R(\cosh \theta, \sinh \theta) \sinh \theta \; d\theta.$$

For example,

$$\int \frac{du}{\sqrt{u^2 - 1}} = \int \frac{\sinh \theta \; d\theta}{\sinh \theta} = \int d\theta = \theta = \text{arc cosh } u + C,$$

where the function $\theta = \text{arc cosh } u$ is the inverse of the function $u = \cosh \theta$. Just as before, we can express arc cosh u in terms of more familiar functions by solving the equation

$$u = \cosh \theta = \frac{e^\theta + e^{-\theta}}{2} = \frac{1}{2}\left(e^\theta + \frac{1}{e^\theta}\right)$$

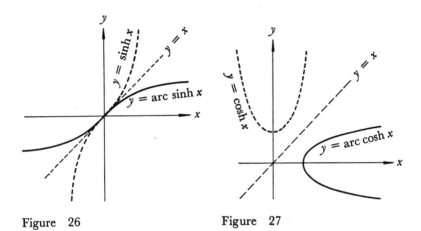

Figure 26   Figure 27

for θ, obtaining first

$$e^{\theta} = u \pm \sqrt{u^2 - 1}$$

(both signs are permissible) and then

$$\theta = \text{arc cosh } u = \ln \left(u \pm \sqrt{u^2 - 1} \right).$$

g. The graphs of the functions $y = \text{arc sinh } x$ and $y = \text{arc cosh } x$† are obtained in the usual way (Sec. 5.35d), i.e., by reflecting the graphs of the functions $y = \sinh x$ and $y = \cosh x$ in the line $y = x$ bisecting the first quadrant (see Figures 26 and 27). This construction confirms the single-valued character of arc sinh x and the double-valued character of arc cosh x (cf. Sec. 2.84).

9.45. Elliptic integrals. In the case where the polynomial $P(x)$ is of degree $n > 2$, the integral

$$\int R(x, \sqrt{P(x)})\, dx \tag{14}$$

cannot in general be expressed in terms of elementary functions.‡ The integral (14) is said to be *elliptic* if $n = 3$ or $n = 4$ and *hyperelliptic* if $n > 4$. The two

---

† For simplicity, we revert to the usual rectangular coordinates x and y.

‡ This is the *Abel-Liouville theorem*, proved, say, by N. G. Chebotarev, Uspekhi Mat. Nauk, vol. 2, no. 2 (1947), pp. 3–20. By an *elementary function* we mean any function formed from polynomials, exponentials, logarithms, trigonometric functions, and inverse trigonometric functions by a finite number of algebraic operations (addition, subtraction, multiplication, and division) and a finite number of compositions (i.e., formation of composite functions).

most important elliptic integrals are†

$$F(k,\varphi) = \int_0^{\sin\varphi} \frac{dt}{\sqrt{(1-t^2)(1-k^2t^2)}} = \int_0^\varphi \frac{d\theta}{\sqrt{1-k^2\sin^2\theta}},$$

$$E(k,\varphi) = \int_0^{\sin\varphi} \sqrt{\frac{1-k^2t^2}{1-t^2}}\, dt = \int_0^\varphi \sqrt{1-k^2\sin^2\theta}\, d\theta,$$

where $0 \leqslant k \leqslant 1$ and in each case we make the simplifying assumption $t = \sin\theta$.

For $k=0$ both integrals give the same function

$$F(0,\varphi) = \int_0^\varphi d\theta = \varphi, \qquad E(0,\varphi) = \int_0^\varphi d\theta = \varphi,$$

while for $k=1$ they can be expressed in terms of elementary functions

$$F(1,\varphi) = \int_0^{\sin\varphi} \frac{dt}{1-t^2} = \frac{1}{2}\ln\frac{1+t}{1-t}\Big|_0^{\sin\varphi} = \frac{1}{2}\ln\frac{1+\sin\varphi}{1-\sin\varphi},$$

$$E(1,\varphi) = \int_0^{\sin\varphi} dt = t\Big|_0^{\sin\varphi} = \sin\varphi.$$

There exist extensive tables of $F(k,\varphi)$ and $E(k,\varphi)$ as functions of the angle φ and another angle α, related to the "modulus" k by the formula $k = \sin\alpha$. The table below lists a few values of $F(k,\varphi)$ and $E(k,\varphi)$.

The behavior of $F(k,\varphi)$ and $E(k,\varphi)$ as functions of φ for various values of k (or α) is shown in Figure 28. Elliptic integrals come up in calculating the arc length of an ellipse (see Sec. 9.63c), which explains the adjective "elliptic." They also play a role in many other problems of analysis, and hence have been studied in great detail.

	$E(k,\varphi)$				$F(k,\varphi)$			
φ	$\alpha=90°$	$60°$	$45°$	$30°$	$\alpha=0°$	$30°$	$45°$	$60°$	$90°$
$0°$	0	0	0	0	0	0	0	0	0
$30°$	0.50	0.51	0.51	0.52	$0.52=\frac{\pi}{6}$	0.53	0.54	0.54	0.55
$45°$	0.71	0.73	0.75	0.77	$0.78=\frac{\pi}{4}$	0.80	0.82	0.85	0.88
$60°$	0.87	0.92	0.96	1.01	$1.05=\frac{\pi}{3}$	1.09	1.14	1.21	1.32
$90°$	1	1.21	1.35	1.48	$1.57=\frac{\pi}{2}$	1.69	1.85	2.16	

† $F(k,\varphi)$ are $E(k,\varphi)$ are known as *Legendre's incomplete elliptic integrals of the first and second kind*, respectively.

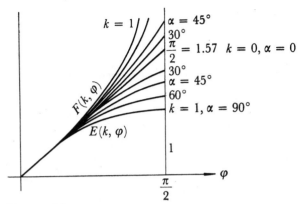

Figure 28

It turns out that the general elliptic integral

$$\int R(x,\sqrt{ax^4 + bx^3 + cx^2 + ex + f}) \, dx$$

can be expressed (to within an elementary function) in terms of the functions $F(k,\varphi)$, $E(k,\varphi)$, and the "elliptic integral of the third kind"

$$\int \frac{d\theta}{(1 + h^2 \sin^2 \theta)\sqrt{1 - k^2 \sin^2 \theta}},$$

depending on two parameters h and k.†

9.46. Other frequently encountered integrals

a. The integral

$$\int P(x)e^x \, dx,$$

where $P(x)$ is a polynomial, can be expressed as a linear combination of integrals of the form

$$I_n = \int x^n e^x \, dx.$$

To evaluate I_n, we use integration by parts (with $u = x^n$, $dv = e^x \, dx$, $du = nx^{n-1} \, dx$, $v = e^x$), obtaining the recursion formula

$$I_n = x^n e^x - n \int x^{n-1} e^x \, dx = x^n e^x - nI_{n-1}.$$

† The proof is given in Fichtenholz, *The Indefinite Integral*, Appendix.

But

$$I_0 = \int e^x \, dx = e^x + C,$$

and hence I_n can be expressed entirely in terms of elementary functions. The same is true of the integrals†

$$\int P(x) \cos x \, dx, \qquad \int P(x) \sin x \, dx, \qquad \int P(x) \ln x \, dx.$$

b. On the other hand, the *exponential integral*

$$\int \frac{e^x}{x} \, dx, \tag{15}$$

the *logarithmic integral*

$$\int \frac{du}{\ln u}$$

(obtained from (15) by making the substitution $e^x = u$), the *sine integral*

$$\int \frac{\sin x}{x} \, dx$$

and the *cosine integral*

$$\int \frac{\cos x}{x} \, dx$$

cannot be expressed in terms of elementary functions.‡

9.5. Evaluation of Definite Integrals

9.51. Integration by parts

a. Let u and v be piecewise smooth functions. Then, since uv is an antiderivative of its derivative $uv' + vu'$, it follows from Theorem 9.33 that

$$\int_a^b (uv' + vu') \, dx = \int_a^b (u \, dv + v \, du) = uv \Big|_a^b,$$

or equivalently

† Fichtenholz, *The Indefinite Integral*, Sec. 5, Problems 3 and 5.
‡ For a proof of this assertion, see G. H. Hardy, *The Integration of Functions of a Single Variable*, second edition, Cambridge University Press, London (1928).

$$\int_a^b u\,dv = uv\Big|_a^b - \int_a^b v\,du,$$

a result known as the formula for *integration by parts* (for a definite integral).

b. Example. *Evaluate the integral*

$$I_m = \int_0^{\pi/2} \sin^m x\,dx.$$

Solution. Integrating by parts, we find that

$$I_m = \int_0^{\pi/2} \sin^{m-1}x\,d(-\cos x) = -\sin^{m-1}x\,\cos x\Big|_0^{\pi/2}$$
$$+ (m-1)\int_0^{\pi/2} \sin^{m-2}x\,\cos^2 x\,dx.$$

The first term on the right vanishes and hence

$$I_m = (m-1)\int_0^{\pi/2} \sin^{m-2} x(1-\sin^2 x)\,dx$$
$$= (m-1)I_{m-2} - (m-1)I_m,$$

which implies

$$I_m = \frac{m-1}{m}I_{m-2}.$$

Thus, since

$$I_0 = \int_0^{\pi/2} dx = \frac{\pi}{2},$$

we get

$$I_{2n} = \int_0^{\pi/2} \sin^{2n} x\,dx = \frac{(2n-1)(2n-3)\cdots 3\cdot 1}{2n(2n-2)\cdots 4\cdot 2}\,\frac{\pi}{2} \qquad (1)$$

for even $m = 2n$. On the other hand, since

$$I_1 = \int_0^{\pi/2} \sin x\,dx = 1,$$

we get

$$I_{2n+1} = \int_0^{\pi/2} \sin^{2n+1} x\,dx = \frac{2n(2n-1)\cdots 4\cdot 2}{(2n+1)(2n-1)\cdots 3\cdot 1} \qquad (2)$$

for odd $m = 2n+1$.

9.52. Taylor's formula with remainder in integral form

a. Consider the integral

$$\int_a^b f(x)(b-a)^n \, dx.$$

Assuming that $f(x)$ is piecewise smooth, we integrate by parts, obtaining

$$\int_a^b f(x)(b-x)^n \, dx = -f(x)\frac{(b-x)^{n+1}}{n+1}\bigg|_a^b + \int_a^b f'(x)\frac{(b-x)^{n+1}}{n+1} \, dx$$

$$= f(a)\frac{(b-a)^{n+1}}{n+1} + \frac{1}{n+1}\int_a^b f'(x)(b-x)^{n+1} \, dx. \tag{3}$$

Suppose $f(x)$ has continuous derivatives up to order n and a piecewise continuous derivative $f^{(n+1)}(x)$ on $[a,b]$. Then, applying (3) repeatedly, we get

$$f(b) - f(a) = \int_a^b f'(x) \, dx = f'(a)(b-a) + \int_a^b f''(x)(b-x) \, dx$$

$$= f'(a)(b-a) + \frac{1}{2}f''(a)(b-a)^2 + \frac{1}{2}\int_a^b f'''(x)(b-x)^2 \, dx = \cdots$$

$$= f'(a)(b-a) + \frac{1}{2}f''(a)(b-a)^2 + \cdots + \frac{1}{n!}f^{(n)}(a)(b-a)^n$$

$$+ \frac{1}{n!}\int_a^b f^{(n+1)}(x)(b-x)^n \, dx,$$

which is equivalent to Taylor's formula

$$f(b) = f(a) + f'(a)(b-a) + \frac{1}{2}f''(a)(b-a)^2 + \cdots + \frac{1}{n!}f^{(n)}(a)(b-a)^n + R_n,$$

with the remainder

$$R_n = \frac{1}{n!}\int_a^b f^{(n+1)}(x)(b-x)^n \, dx$$

in integral form.

b. Since obviously

$$\frac{1}{n!}\min_{a \leq x \leq b} f^{(n+1)}(x) \int_a^b (b-x)^n \, dx$$

$$\leq R_n \leq \frac{1}{n!}\max_{a \leq x \leq b} f^{(n+1)}(x) \int_a^b (b-x)^n \, dx$$

and hence

$$\frac{(b-a)^{n+1}}{(n+1)!} \min_{a\leq x\leq b} f^{(n+1)}(x) \leqslant R_n \leqslant \frac{(b-a)^{n+1}}{(n+1)!} \max_{a\leq x\leq b} f^{(n+1)}(x),$$

we immediately get Lagrange's form of the remainder

$$R_n = \frac{(b-a)^{n+1}}{(n+1)!} f^{(n+1)}(c)$$

(Sec. 8.23), provided that the function $f^{(n+1)}(x)$ is continuous and hence takes every value between its minimum and maximum values at some intermediate point c (see Theorem 5.23).

c. We now use these considerations to investigate the convergence of the Taylor series (Sec. 8.51) of the function

$$f(x) = (1+x)^{\alpha}.$$

In this case, Taylor's formula (with $a=0$, $b=x$) takes the form

$$(1+x)^{\alpha} = 1 + \alpha x + \frac{\alpha(\alpha-1)}{2!}x^2 + \cdots + \frac{\alpha(\alpha-1)\cdots(\alpha-n+1)}{n!}x^n$$
$$+ \frac{\alpha(\alpha-1)\cdots(\alpha-n)}{n!}\int_0^x (1+t)^{\alpha-n-1}(x-t)^n \, dt.$$

If $0 < t < x < 1$, then obviously

$$\frac{x-t}{1+t} \leqslant \frac{x-t}{1-t} \leqslant x,$$

and hence

$$\int_0^x \frac{(x-t)^n}{(1+t)^n}(1+t)^{\alpha-1} \, dt \leqslant x^n \int_0^x (1+t)^{\alpha-1} \, dt,$$

while if $-1 < x < t < 0$, then

$$\left| \int_0^x \frac{(x-t)^n}{(1+t)^n}(1+t)^{\alpha-1} \, dt \right| \leqslant \int_{-\xi}^0 \frac{(\xi-\tau)^n}{(1-\tau)^n}(1-\tau)^{\alpha-1} \, d\tau$$
$$\leqslant \xi^n \int_{-\xi}^0 (1-\tau)^{\alpha-1} \, d\tau = |x|^n \int_0^{|x|} (1+t)^{\alpha-1} \, dt,$$

where $x = -\xi$, $t = -\tau$. Therefore, if $|x| < 1$,

$$\left| \int_0^x \frac{(x-t)^n}{(1+t)^n}(1+t)^{\alpha-1} \, dt \right| \leqslant |x|^n \int_0^{|x|} (1+t)^{\alpha-1} \, dt = C|x|^n,$$

where $C = C(x, \alpha)$. It follows that

$$|R_n| \leqslant C \frac{|\alpha(\alpha-1)\cdots(\alpha-n)|}{n!} |x|^n.$$

Denoting the quantity on the right by ω_n, we get

$$\frac{\omega_{n+1}}{\omega_n} = \frac{|\alpha-n-1|}{n+1} |x|, \qquad \lim_{n\to\infty} \frac{\omega_{n+1}}{\omega_n} = |x|,$$

so that $\omega_n \to 0$ as $n \to \infty$ if $|x| < 1$. Therefore the remainder R_n approaches 0 as $n \to \infty$ if $|x| < 1$, and hence

$$(1+x)^\alpha = 1 + \alpha x + \frac{\alpha(\alpha-1)}{2!} x^2 + \cdots + \frac{\alpha(\alpha-1)\cdots(\alpha-n+1)}{n!} x^n + \cdots \tag{4}$$

if $-1 < x < 1$. Unless α is a positive integer (in which case the right-hand side of (4) reduces to a polynomial), the series (4) diverges if $|x| > 1$, by the same argument as in the proof of D'Alembert's test (Theorem 6.14a). Hence the radius of convergence of the series (4) equals 1.

Replacing the real variable x in (4) by the complex variable $z = x + iy$, we get the analytic continuation (Sec. 8.55) of the function $(1+x)^\alpha$ into the disk $|z| < 1$:

$$(1+z)^\alpha = 1 + \alpha z + \frac{\alpha(\alpha-1)}{2!} z^2 + \cdots + \frac{\alpha(\alpha-1)\cdots(\alpha-n+1)}{n!} z^n + \cdots. \tag{5}$$

Choosing $\alpha = -1$ and changing z to $-z$, we obtain the familiar formula for the sum of a convergent geometric series

$$\frac{1}{1-z} = 1 + z + z^2 + \cdots + z^n + \cdots$$

(see Example 6.11b).

d. Writing (5) in the form

$$(1+z)^\alpha = 1 + (-\alpha)(-z) + \frac{(-\alpha)(1-\alpha)}{2!}(-z)^2 + \cdots$$
$$+ \frac{(-\alpha)(1-\alpha)\cdots(n-1-\alpha)}{n!}(-z)^n + \cdots$$

and using the fact that the series

$$\sum_{n=1}^\infty \frac{\alpha(\alpha+1)\cdots(\alpha+n-1)}{\beta(\beta+1)\cdots(\beta+n-1)} z^{n-1} \qquad (\alpha, \beta \neq 0, -1, -2, \ldots)$$

converges for $z = 1$ if $\beta > \alpha + 1$ and for $z = -1$ if $\beta > \alpha$ (see Example 6.67b), we find that (5) converges for $-z = 1$ if $1 > -\alpha + 1$ and for $-z = -1$ if

$1 > -\alpha$, i.e., for $z = -1$ if $\alpha > 0$ and for $z = 1$ if $\alpha > -1$. Then Abel's theorem (Theorem 6.68) implies the interesting formulas

$$1 - \alpha + \frac{\alpha(\alpha-1)}{2!} - \frac{\alpha(\alpha-1)(\alpha-2)}{3!} + \cdots = 0 \qquad (\alpha > 0),$$

$$1 + \alpha + \frac{\alpha(\alpha-1)}{2!} + \frac{\alpha(\alpha-1)(\alpha-2)}{3!} + \cdots = 2^\alpha \qquad (\alpha > -1).$$

For $\alpha = n = 1, 2, \ldots$, these reduce to the familiar formulas

$$C_0^n - C_1^n - C_2^n - \cdots + (-1)^n C_n^n = 0,$$
$$C_0^n + C_1^n + C_2^n + \cdots + C_n^n = 2^n$$

satisfied by the binomial coefficients

$$C_k^n = \frac{n!}{k!(n-k)!}.$$

9.53. Integration by substitution. Given an integral

$$\int_a^b f(u) \, du, \tag{6}$$

where $f(u)$ is continuous on the interval $a \leqslant u \leqslant b$, let $u = u(t)$ be a function with a continuous derivative on some interval $\alpha \leqslant t \leqslant \beta$, where $u(\alpha) = a$, $u(\beta) = b$. Then making a formal change of variables in (6), we get†

$$\int_a^b f(u) \, du = \int_\alpha^\beta f(u(t))u'(t) \, dt. \tag{7}$$

We now look for conditions guaranteeing the validity of (7). It might seem natural to assume that $u(t)$ is increasing on $[\alpha, \beta]$. However, it turns out that (7) is valid in the more general case where $u(t)$ fails to be monotonic or even takes values outside the interval $[a, b]$ on which $f(u)$ is originally defined, provided only that $f(u)$ has a continuous extension onto the (in general) larger interval‡

$$[A, B] = \{u : u = u(t), \, \alpha \leqslant t \leqslant \beta\}. \tag{8}$$

Thus we have the following

THEOREM. *Formula (7) holds (with $a = u(\alpha)$, $b = u(\beta)$) if $u(t)$ has a continuous*

---

† Here, and in the subsequent theorem, we again see the importance of writing du in the integral on the left (recall the discussion on p. 275). In fact, the formula $du = u'(t) \, dt$ for the differential of u automatically leads to the correct integral on the right (cf. Theorem 9.38).

‡ Obviously, $[A, B] \supset [a, b] = [u(\alpha), u(\beta)]$. As an exercise, verify that the set $\{u : u = u(t), \, \alpha \leqslant t \leqslant \beta\}$ is actually a closed interval, as asserted.

derivative on the interval $[\alpha,\beta]$ *and if* $f(u)$ *is continuous on the interval* (8).

Proof. Let $F(u)$ be an antiderivative of the function $f(u)$ on $[A,B]$, and let $G(t)$ be an antiderivative of the function $f(u(t))u'(t)$ on $[\alpha,\beta]$. Then, by Theorem 9.38,

$$F(u(t)) = G(t) + C,$$

and hence, by Theorem 9.33,

$$\int_a^b f(u)\ du = F(b) - F(a) = F(u(\beta)) - F(u(\alpha))$$
$$= G(\beta) - G(\alpha) = \int_\alpha^\beta f(u(t))u'(t)\ dt. \quad\blacksquare$$

It should be noted that in changing integrands in (7), we must also make a corresponding change of limits of integration. Moreover, since Theorem 9.33 is valid for the case $a > b$,† it does not matter which of the limits of integration in either side of (7) is larger.

9.54. Applications of integration by substitution

a. If $f(x)$ is defined on $[a,b]$, then the "shifted" function $f(x-h)$ is defined on the interval $[a+h, b+h]$. Making the substitution $x-h=t$, we get

$$\int_{a+h}^{b+h} f(x-h)\ dx = \int_a^b f(t)\ dt = \int_a^b f(x)\ dx.$$

Similarly, the "reflected" function $f(a+b-x)$ is defined on the original interval $[a,b]$, and the substitution $a+b-x=t$ shows that

$$\int_a^b f(a+b-x)\ dx = -\int_b^a f(t)\ dt = \int_a^b f(t)dt = \int_a^b f(x)\ dx.$$

b. THEOREM. *The integral of a periodic function $f(x)$ with period T has the same value over every interval of length T.*

Proof. Let m be the unique integer such that $a \leqslant mT < a + T$. Then

$$\int_a^{a+T} f(x)\ dx = \int_a^{mT} f(x)\ dx + \int_{mT}^{a+T} f(x)\ dx.$$

Making the substitution $x = t + (m-1)T$ in the first integral on the right and the substitution $x = t + mT$ in the second integral, and using the periodicity

† In fact, if $a > b$, then
$$\int_a^b f(x)\ dx = -\int_b^a f(x)\ dx = -[F(a) - F(b)] = F(b) - F(a)$$
(see Sec. 9.18).

to write

$$f(t+mT)\equiv f(t+(m-1)T)\equiv f(t)$$

(cf. Sec. 8.65), we get

$$\int_a^{a+T} f(x)\ dx = \int_{a-(m-1)T}^{T} f(t)\ dt + \int_0^{a-(m-1)T} f(t)\ dt$$

$$= \int_0^T f(t)\ dt = \int_0^T f(x)\ dx.\ \ \blacksquare$$

c. Example. Consider the integral

$$\int_{-\pi}^{\pi} \sin^{2\alpha} kx\ dx \qquad (\alpha>0,\ k=1,2,\ldots).$$

Using the formula

$$\sin kx = \cos\left(\frac{\pi}{2}-kx\right)$$

and the substitution

$$\frac{\pi}{2}-kx=ku,$$

we get

$$\int_{-\pi}^{\pi} \sin^{2\alpha} kx\ dx = \int_{-\pi}^{\pi} \cos^{2\alpha}\left(\frac{\pi}{2}-kx\right)dx = \int_{(\pi/2k)-\pi}^{(\pi/2k)+\pi} \cos^{2\alpha} ku\ du,$$

and hence, by the above theorem,

$$\int_{-\pi}^{\pi} \sin^{2\alpha} kx\ dx = \int_{-\pi}^{\pi} \cos^{2\alpha} kx\ dx. \tag{9}$$

For $\alpha=1$ it follows from (9) and the obvious formula

$$\int_{-\pi}^{\pi} (\sin^2 kx + \cos^2 kx)\ dx = \int_{-\pi}^{\pi} dx = 2\pi$$

that

$$\int_{-\pi}^{\pi} \sin^2 kx\ dx = \int_{-\pi}^{\pi} \cos^2 kx\ dx = \pi.$$

More generally, using formula (1), p. 309, we find that (9) equals

$$\frac{(2\alpha-1)(2\alpha-3)\cdots 3\cdot 1}{2\alpha(2\alpha-2)\cdots 4\cdot 2} 2\pi$$

if $\alpha = 1,2,\ldots$ Clearly (9) vanishes if $\alpha = \frac{1}{2}$, $\alpha = \frac{3}{2},\ldots$ (why?).

9.55. The Stieltjes integral

a. Definition. We now introduce an important generalization of the concept of a definite integral. Given two functions $f(x)$ and $g(x)$ defined on an interval $[a,b]$, let $\Pi = \{a = x_0 < x_1 < < x_n = b\}$ be a partition of $[a,b]$ with marked points $\xi_k \in [x_k, x_{k+1}]$, and let

$$d(\Pi) = \max\{\Delta x_0, \Delta x_1, \ldots, \Delta x_{n-1}\} \qquad (\Delta x_k = x_{k+1} - x_k)$$

as before. Then, introducing the *Riemann-Stieltjes sum*

$$S_\Pi(f,g) = \sum_{k=0}^{n-1} f(\xi_k)[g(x_{k+1}) - g(x_k)], \tag{10}$$

we call the finite number I the *Stieltjes integral of the function $f(x)$ with respect to the function $g(x)$ over the interval $[a,b]$* if, given any $\varepsilon > 0$, there exists a $\delta > 0$ such that

$$|I - S_\Pi(f,g)| < \varepsilon$$

for every partition Π with $d(\Pi) < \delta$. In other words, the Stieltjes integral I, denoted by

$$\int_a^b f(x)\,dg(x), \tag{11}$$

is the limit of the sum (10) under arbitrary refinement of the partition Π. Note that

$$\int_a^b dg(x) = g(b) - g(a)$$

for every function $g(x)$.

b. THEOREM. *If $f(x)$ is continuous on $[a,b]$ and if $g(x)$ has a continuous derivative on $[a,b]$, then the Stieltjes integral (11) exists and equals*

$$\int_a^b f(x)g'(x)\,dx. \tag{12}$$

Proof. By Lagrange's theorem (Sec. 7.44),

$$\Delta g(x_k) = g(x_{k+1}) - g(x_k) = g'(c_k)\Delta x_k$$

where $c_k \in (x_k, x_{k+1})$. Therefore

$$\sum_{k=0}^{n-1} f(\xi_k)\,\Delta g(x_k) = \sum_{k=0}^{n-1} f(\xi_k) g'(c_k)\Delta x_k$$

$$= \sum_{k=0}^{n-1} f(c_k) g'(c_k)\Delta x_k + \sum_{k=0}^{n-1} [f(\xi_k) - f(c_k)]g'(c_k)\Delta x_k$$

$$= \sum_{k=0}^{n-1} f(c_k) g'(c_k)\Delta x_k + \sum_{k=0}^{n-1} \varepsilon_k g'(c_k)\Delta x_k, \tag{13}$$

where, by the uniform continuity of $f(x)$ on $[a,b]$, the numbers $|\varepsilon_k|$ can all be made less than any given $\varepsilon > 0$ for a "sufficiently fine" partition Π. Thus the second term in the right-hand side of (13) satisfies the inequality

$$\left| \sum_{k=0}^{n-1} \varepsilon_k g'(c_k)\Delta x_k \right| \leqslant \varepsilon \max_{a \leq x \leq b} |g'(x)|(b-a),$$

and hence approaches zero under arbitrary refinement of Π, while the first term is an ordinary Riemann sum of the function $f(x)g'(x)$ and hence approaches the limit (12) under arbitrary refinement of Π.† $\blacksquare$

c. The following formulas involving Stieltjes integrals hold whenever the integrals on the right exist:

$$\int_a^b [\alpha_1 f_1(x) + \alpha_2 f_2(x)]\,dg(x) = \alpha_1 \int_a^b f_1(x)\,dg(x) + \alpha_2 \int_a^b f_2(x)\,dg(x),$$

$$\int_a^b f(x)\,d[\beta_1 g_1(x) + \beta_2 g_2(x)] = \beta_1 \int_a^b f(x)\,dg_1(x) + \beta_2 \int_a^b f(x)\,dg_2(x),$$

$$\int_a^b f(x)\,dg(x) = \int_a^c f(x)\,dg(x) + \int_c^b f(x)\,dg(x) \qquad (a<c<b),$$

$$\int_a^b f(x)\,dg(x) = f(x)g(x)\Big|_a^b - \int_a^b g(x)\,df(x)$$

$(\alpha_1, \alpha_2, \beta_1, \beta_2 \text{ real}).\ddagger$

9.6. More on Area

9.61. Let $f(x)$ be a continuous nonnegative function on the interval $[a,b]$. Then, according to Sec. 9.21, the area under the curve $y = f(x)$ from $x = a$ to $x = b$ (more exactly, the area of the "curvilinear trapezoid" Φ bounded on

† The existence of (12) follows from Theorem 9.14e and the assumed continuity of $f(x)$ and $g'(x)$. As an exercise, the reader should examine what happens if $f(x)$ or $g'(x)$ or both are only piecewise continuous.

‡ The details are left as an exercise. For more on the Stieltjes integral, see e.g., Shilov and Gurevich, *Integral, Measure and Derivative: A Unified Approach*, Part 2.

the bottom by the segment $[a,b]$ of the x-axis, on top by the curve $y=f(x)$, and on the sides by segments of the vertical lines $x=a$ and $x=b$) is given by the integral

$$I= \int_a^b f(x)\ dx \tag{1}$$

(see Figure 29). More generally, if Φ is a figure made up of several curvilinear trapezoids with no common interior points, then the area of Φ is just the sum of the areas of the separate trapezoids.†

Formula (1) gives the area if $f(x)\geqslant 0$. If $f(x)\leqslant 0$, as in Figure 30, then $I\leqslant 0$ and I is the "area over the curve $y=f(x)$ from $x=a$ to $x=b$"‡ *taken with the minus sign*. If $f(x)$ changes sign in the interval $[a,b]$, then I is the "area between the curve $y=f(x)$ and the x-axis," more exactly, the sum of all the areas lying under the x-axis and over the curve $y=f(x)$ (see Figure 31).

9.62. Examples

a. Let

$$f(x) = A + Bx + Cx^2 + Dx^3 \tag{2}$$

be a polynomial of degree no higher than three. Then the area between the curve $y=f(x)$ and the x-axis (see Figure 32) is given by

$$\int_a^b f(x)\ dx = \frac{1}{6}(b-a)\left[f(a) + 4f\left(\frac{a+b}{2}\right) + f(b)\right], \tag{3}$$

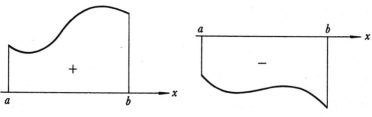

Figure   29   Figure   30

---

† See the discussion pertaining to Figure 82, p. 487, as well as the remarks on p. 486, concerning the case where Φ is a union of more general figures with area.

‡ More exactly, the area of the curvilinear trapezoid bounded on the bottom by the curve $y=f(x)$, on the top by the segment $[a,b]$ of the x-axis, and on the sides by segments of the vertical lines $x=a$ and $x=b$.

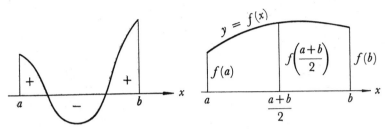

Figure 31

Figure 32

or equivalently

$$\int_{-c}^{c} f(x)\ dx = \frac{1}{3}c[f(-c) + 4f(0) + f(c)] \qquad (2c = b - a) \qquad (3')$$

after shifting the origin to the midpoint of the interval $[a,b]$. To prove $(3')$, we first note that if $(3')$ holds for $f(x)$ and $g(x)$, then it also holds for $kf(x)$ and $f(x) + g(x)$, where k is any real number. Hence to show that (2) satisfies $(3')$, we need only show that the functions

$$f_0(x) = 1, \qquad f_1(x) = x, \qquad f_2(x) = x^2, \qquad f_3(x) = x^3$$

all satisfy $(3')$. The validity of $(3')$ is obvious for $f_0(x)$ and also for $f_1(x)$ and $f_3(x)$, since both sides of $(3')$ vanish if the integrand is odd.† Moreover, $(3')$ holds for $f_2(x)$, since clearly

$$\int_{-c}^{c} x^2\ dx = \frac{1}{3}x^3 \Big|_{-c}^{c} = \frac{2}{3}c^3$$

(Sec. 9.35). It follows that $(3')$, or equivalently (3), holds for the function (2).

b. Area of a circular disk. To find the area S of the circular disk

---

† A function $f(x)$ is said to be *odd* if $f(-x) \equiv -f(x)$ and *even* if $f(-x) \equiv f(x)$. Clearly $f(0) = 0$ and

$$\int_{-c}^{c} f(x)\ dx = 0$$

if $f(x)$ is odd, while

$$\int_{-c}^{c} f(x)\ dx = 2\int_{0}^{c} f(x)\ dx$$

if $f(x)$ is even (why?).

$x^2 + y^2 \leqslant r^2$, we note that

$$S = 4 \int_0^r y\, dx = 4 \int_0^r \sqrt{r^2 - x^2}\, dx,$$

by symmetry (see Figure 33). Setting $x = r \cos \theta$, $dx = -r \sin \theta\, d\theta$, we get

$$S = -4r^2 \int_{\pi/2}^0 \sin^2 \theta\, d\theta = 4r^2 \int_0^{\pi/2} \sin^2 \theta\, d\theta = 4r^2 \int_0^{\pi/2} \frac{1 - \cos 2\theta}{2}\, d\theta$$

$$= 2r^2 \int_0^{\pi/2} d\theta - 2r^2 \int_0^{\pi/2} \cos 2\theta\, d\theta = \pi r^2 - r^2 \sin 2\theta \Big|_0^{\pi/2} = \pi r^2.$$

Thus the number $\pi/2$, which we have defined as the first positive root of the equation $\cos x = 0$ (see Sec. 5.65b) can be interpreted as half the area of a circular disk of radius 1.

c. Area of a circular sector. Let $S(\theta_0, \theta_1)$ be the area of the sector of the disk $x^2 + y^2 \leqslant r^2$ bounded by the rays $\theta = \theta_0$ and $\theta = \theta_1$, where $0 \leqslant \theta_0 < \theta_1 \leqslant \pi/2$ (see Figure 34). Then obviously

$$S(\theta_0, \theta_1) = S(0, \theta_1) - S(0, \theta_0),$$

so that we need only calculate $S(0, \theta_0)$ for arbitrary $\theta_0 \in [0, \pi/2]$. It is clear

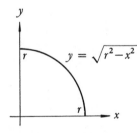

Figure   33   Figure   34

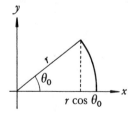

Figure   35

from Figure 35 that

$$S(0,\theta_0) = \int_0^{r\cos\theta_0} x\tan\theta_0\,dx + \int_{r\cos\theta_0}^r \sqrt{r^2-x^2}\,dx$$

$$= \frac{x^2}{2}\Big|_0^{r\cos\theta_0}\tan\theta_0 + r^2\int_0^{\theta_0}\sin^2\theta\,d\theta$$

$$= \frac{r^2\cos^2\theta_0}{2}\tan\theta_0 + r^2\int_0^{\theta_0}\frac{1-\cos 2\theta}{2}\,d\theta$$

$$= \frac{r^2}{2}\cos^2\theta_0\tan\theta_0 + \frac{r^2}{2}\theta_0 - \frac{r^2}{4}\sin 2\theta\Big|_0^{\theta_0}$$

$$= \frac{r^2}{2}\sin\theta_0\cos\theta_0 + \frac{r^2}{2}\theta_0 - \frac{r^2}{2}\sin\theta_0\cos\theta_0 = \frac{r^2}{2}\theta_0,$$

and hence

$$S(\theta_0,\theta_1) = \frac{r^2}{2}(\theta_1-\theta_0). \tag{4}$$

It is easy to see that (4) is actually valid for any circular sector of central angle $0 \leqslant \theta_1 - \theta_0 \leqslant 2\pi$.

d. Suppose we subject a curvilinear trapezoid Φ to a k-fold expansion along the y-axis, as shown in Figure 36. Then the area S_k of the resulting figure is

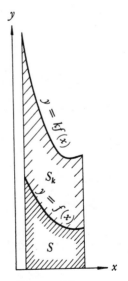

Figure   36

k times the area S of the original figure, since obviously

$$S_k = \int_a^b kf(x)\ dx = k \int_a^b f(x)\ dx = kS \tag{5}$$

(cf. Sec. B.6, p. 486). The same is true if we subject Φ to a k-fold expansion along the x-axis, as shown in Figure 37, since then

$$S_k = \int_{ka}^{kb} f\left(\frac{x}{k}\right) dx = k \int_a^b f(t)\ dt = k \int_a^b f(x)\ dx = kS \tag{5'}$$

(make the substitution $x = kt$).

e. Area enclosed by an ellipse. The ellipse

$$\frac{x^2}{a^2} + \frac{y^2}{b^2} = 1$$

is obtained from the circle $x^2 + y^2 = 1$ by an a-fold expansion along the x-axis and a b-fold expansion along the y-axis (see Figure 38). Since the area enclosed by the circle equals π, it follows from (5) and (5') that the area enclosed by the ellipse equals πab.

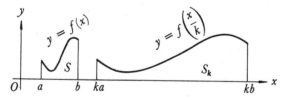

Figure   37

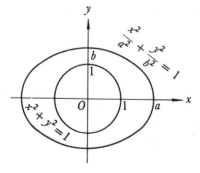

Figure   38

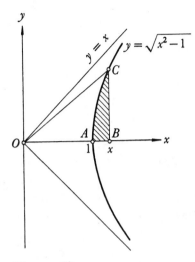

Figure 39

f. Geometric meaning of the inverse hyperbolic cosine. Consider the problem of finding the area of the sector ABC shown in Figure 39, where C lies on the right-hand branch of the hyperbola $x^2 - y^2 = 1$. This area is given by the integral

$$I = \int_1^x \sqrt{u^2 - 1}\, du,$$

which we evaluate by making the substitution $u = \cosh t$ (cf. Sec. 9.44f):

$$I = \int_1^x \sqrt{u^2 - 1}\, du = \int_0^{\text{arc cosh } x} \sinh^2 t\, dt.$$

But

$$\sinh^2 t = \frac{\cosh 2t - 1}{2},$$

since

$$\cosh 2t = \cosh^2 t + \sinh^2 t, \qquad 1 = \cosh^2 t - \sinh^2 t$$

(cf. Sec. 8.74), and hence

$$I = \int_0^{\text{arc cosh } x} \frac{\cosh 2t - 1}{2}\, dt = \left(\frac{\sinh 2t}{4} - \frac{t}{2} \right)\Bigg|_0^{\text{arc cosh } x}$$

Moreover,

$$\sinh 2t = 2\sqrt{\cosh^2 t - 1}\ \cosh t,$$

since

$$\sinh 2t = 2 \sinh t \cosh t,$$

and hence

$$I = \tfrac{1}{2}(x\sqrt{x^2-1} - \text{arc cosh } x) = \tfrac{1}{2}(x\sqrt{x^2-1} - \tfrac{1}{2}\text{ arc cosh } x).$$

The first term is the area of the triangle OBC. Therefore $\tfrac{1}{2}$ arc cosh x is the area of the sector OAC, so that

$$S = \text{arc cosh } x \tag{6}$$

is the area of the sector $OC'AC$ shown in Figure 40. Inverting (6), we get the parametric representation of the hyperbola

$$x = \cosh S, \qquad y = \sqrt{x^2 - 1} = \sinh S. \tag{7}$$

Naturally, it is the interpretation of the parameter S as the area of the sector $OC'AC$ determined by the variable point C that is at issue here, since the mere fact that (7) is a parametric representation of the hyperbola $x^2 - y^2 = 1$ is immediately obvious.

It is interesting to compare (7) with the parametric representation

$$x = \cos \theta, \qquad y = \sin \theta$$

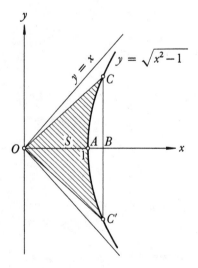

Figure   40

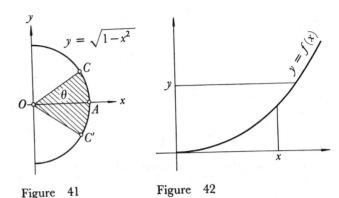

Figure 41   Figure 42

of the circle $x^2 + y^2 = 1$. Here θ is the central angle AOC shown in Figure 41, or alternatively, because of (4), the area of the sector $OC'AC$. Thus the area of a variable sector can be chosen as the basic parameter both for the circle and for the hyperbola.

g. Young's inequality. Let

$$y = f(x) \qquad (0 \leqslant x < \infty)$$

be an increasing continuous function such that $f(0) = 0, f(\infty) = \infty$, with inverse function

$$x = g(y) \qquad (0 \leqslant x < \infty),$$

which is itself increasing and continuous (see Theorem 5.35b), and let

$$F(x) = \int_0^x f(\xi) \, d\xi, \qquad G(y) = \int_0^y g(\eta) \, d\eta \qquad (x \geqslant 0, y \geqslant 0).$$

Then it is geometrically clear from Figure 42 that

$$xy \leqslant F(x) + G(y) \tag{8}$$

for all $x \geqslant 0$, $y \geqslant 0$, a result known as *Young's inequality*. Note that the inequality becomes an equality if and only if $y = f(x)$.†

Young's inequality is the starting point for a number of other important inequalities. For example, (8) implies

$$xy \leqslant \frac{x^p}{p} + \frac{y^q}{q} \qquad \left(\frac{1}{p} + \frac{1}{q} = 1 \right)$$

† As an exercise, the reader should contrive a purely analytic proof of (8).

if we choose

$$y = x^{p-1}, \qquad x = y^{1/(p-1)}$$

and

$$xy \leqslant (x+1) \ln(x+1) - x + e^y - y - 1$$

if we choose

$$y = \ln(x+1), \qquad x = e^y - 1,$$

where both inequalities hold for all $x \geqslant 0, y \geqslant 0$.

9.63. Area in polar coordinates. Given a curve with equation

$$r = f(\theta) \geqslant 0 \qquad (\alpha \leqslant \theta \leqslant \beta)$$

in polar coordinates, we now examine the problem of calculating the area S of the "curvilinear sector" OAB bounded by the two rays $\theta = \alpha$, $\theta = \beta$ and the arc of the curve between the rays (see Figure 43). Our approach is the exact analogue of the treatment of the area of a "curvilinear trapezoid" given in Sec. 9.21. Consider a partition

$$\Pi = \{\alpha = \theta_0 \leqslant \varphi_0 \leqslant \theta_1 \leqslant \varphi_1 \leqslant \theta_2 \leqslant \cdots \leqslant \theta_{n-1} \leqslant \varphi_n \leqslant \theta_n = \beta\},$$

where the point $\varphi_k \in [\theta_k, \theta_{k+1}]$ is chosen so that

$$f(\varphi_k) = M_k = \max_{\theta_k \leqslant \theta \leqslant \theta_{k+1}} f(\theta).$$

Then the corresponding Riemann sum of the function

$$g(\theta) = \frac{1}{2} f^2(\theta),$$

equal to

$$\bar{S}_\Pi(g) = \frac{1}{2} \sum_{k=0}^{n-1} M_k^2 \Delta \theta_k$$

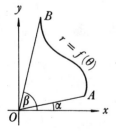

Figure   43

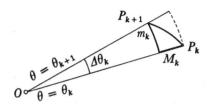

Figure 44

$(\Delta\theta_k = \theta_{k+1} - \theta_k)$, can be interpreted as the area of an "elementary figure" containing OAB, made up of n "circumscribed circular sectors" like the sector of radius M_k shown in Figure 44.† The use of $g(\theta)$ instead of the original function $f(\theta)$ is due, of course, to the fact that the area of a circular sector of radius r and central angle $\Delta\theta$ is given by $\frac{1}{2}r^2\Delta\theta$ and not by $r\Delta\theta$, as shown by formula (4). On the other hand, if we choose φ_k so that

$$f(\varphi_k) = m_k = \min_{\theta_k \leq \theta \leq \theta_{k+1}} f(\theta),$$

then the corresponding Riemann sum of $g(\theta)$, equal to

$$\underline{S}_\Pi(g) = \frac{1}{2}\sum_{k=0}^{n-1} m_k^2 \Delta\theta_k,$$

can be interpreted as the area of an "elementary figure" contained in OAB, made up of n "inscribed circular sectors," like the sector of radius m_k shown in Figure 44. Therefore any number S which can lay claim to be a value of the "area" of the figure OAB must satisfy the condition

$$\underline{S}_\Pi(g) \leqslant S \leqslant \bar{S}_\Pi(g).$$

But $g(\theta)$ is integrable, being continuous, so that $\underline{S}_\Pi(g)$ and $\bar{S}_\Pi(g)$ approach the same limit, equal to the integral of $g(\theta)$ over $[\alpha, \beta]$, under arbitrary refinement of the partition Π. It follows that

$$S = \frac{1}{2}\int_\alpha^\beta f^2(\theta)\,d\theta \tag{9}$$

is the only possible value of S. The existence of S is established in the same way as for the case of rectangular coordinates.‡

As an example of the use of formula (9), consider the curve

$$r = a\sin n\theta \geqslant 0 \qquad (n \text{ odd}), \tag{10}$$

called an *n-leaved rose* (shown in Figure 45 for the case $n = 3$). Since

† P_k is the point of the curve $r = f(\theta)$ with polar coordinates $r_k = f(\theta_k)$ and θ_k.
‡ See the discussion following formula (2), p. 289.

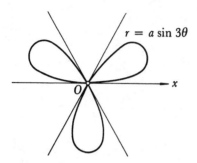

Figure   45

$$\frac{1}{2}\int_0^{\pi/n} a^2 \sin^2 n\theta \; d\theta = \frac{1}{2n}\int_0^{\pi} a^2 \sin^2 \varphi \; d\varphi$$

(make the substitution $n\theta = \varphi$), the area enclosed by one leaf of an n-leaved rose is $1/n$ times the area enclosed by the "single-leaved rose"

$$r = a \sin \theta \tag{11}$$

But (11) is simply an ordinary circle of radius $a/2$. It follows that the area enclosed by all n leaves of the curve (10) is precisely the area $\pi a^2/4$ enclosed by the circle (11).

9.7. More on Arc Length

9.71. Length of a parametric curve. It will be recalled from Sec. 9.22 that the length of the curve

$$y = f(x) \qquad (a \leqslant x \leqslant b),$$

defined as the limiting length of a polygonal line inscribed in the curve as the length of the segments making up the line approaches zero, is given by the integral

$$s = \int_a^b \sqrt{1 + [f'(x)]^2} \; dx \tag{1}$$

(provided the integral exists). In the general case, a curve L is described by parametric equations, specifying the coordinates of a variable point of L as functions of some parameter t. There are two such parametric equations

$$x = x(t), \qquad y = y(t) \qquad (a \leqslant t \leqslant b)$$

in the plane, three equations

$$x = x(t), \qquad y = y(t), \qquad z = z(t) \qquad (a \leqslant t \leqslant b)$$

in three-dimensional space, and more generally, n equations

$$x_1 = x_1(t), \qquad x_2 = x_2(t), \ldots, \qquad x_n = x_n(t) \qquad (a \leqslant t \leqslant b) \tag{2}$$

in n-dimensional space. We now derive a formula for the length of such a "parametric curve."

THEOREM. *Let L be the curve with parametric equations* (2), *where every function $x_j(t)$ is piecewise smooth on $[a,b]$.† Then the length of L, defined as the limiting length of a polygonal line "inscribed" in L, is given by the formula*

$$s = \int_a^b \sqrt{\sum_{j=1}^n [x_j'(t)]^2} \, dt. \tag{3}$$

Proof. Let

$$\Pi = \{a = t_0 \leqslant t_1 \leqslant \cdots \leqslant t_p = b\}$$

be a partition of $[a,b]$, where each parameter value t_k leads to a point P_k on the curve L, with coordinates $x_1(t_k), x_2(t_k), \ldots, x_n(t_k)$. Joining consecutive points P_k with "n-dimensional line segments," we get a polygonal line $L_\Pi = P_0 P_1 \ldots P_n$ "inscribed in L," of length

$$s(L_\Pi) = \sum_{k=0}^{p-1} \sqrt{\sum_{j=1}^n [x_j(t_{k+1}) - x_j(t_k)]^2}.$$

Since the functions $x_j(t)$ are all piecewise smooth, we have

$$\begin{aligned}
x_j(t_{k+1}) - x_j(t_k) &= \int_{t_k}^{t_{k+1}} x_j'(t) \, dt \\
&= \int_{t_k}^{t_{k+1}} x_j'(t_k) \, dt + \int_{t_k}^{t_{k+1}} [x_j'(t) - x_j'(t_k)] \, dt \\
&\leqslant x_j'(t_k) \Delta t_k + \max_{t_k \leq t \leq t_{k+1}} |x_j'(t) - x_j'(t_k)| \Delta t_k,
\end{aligned}$$

where $\Delta t_k = t_{k+1} - t_k$ and we interpret $x_j'(t_k)$ as $x_j'(t_k + 0)$ if t_k is a discontinuity point of $x_j'(t)$. Therefore

$$|x_j(t_{k+1}) - x_j(t_k) - x_j'(t_k) \Delta t_k| \leqslant \varepsilon_{jk} \Delta t_k,$$

where

$$\varepsilon_{jk} = \max_{t_k \leq t \leq t_{k+1}} |x_j'(t) - x_j'(t_k)|.$$

† As in Sec. 9.39, this means that $x_j(t)$ is continuous on $[a,b]$, with a piecewise continuous derivative $x_j'(t)$ on $[a,b]$. If the functions $x_j(t)$ are all piecewise smooth, the curve L itself is said to be *piecewise smooth*.

But then

$$\left|\sqrt{\sum_{j=1}^{n}[x_j(t_{k+1})-x_j(t_k)]^2}-\sqrt{\sum_{j=1}^{n}[x_j'(t_k)]^2\Delta t_i}\right|\leqslant\sqrt{\sum_{j=1}^{n}\varepsilon_{jk}^2}\,\Delta t_k$$

(see Sec. 3.14b), which implies

$$\left|\sum_{k=0}^{p-1}\sqrt{\sum_{j=1}^{n}[x_j(t_{k+1})-x_j(t_k)]^2}-\sum_{k=0}^{p-1}\sqrt{\sum_{j=1}^{n}[x_j'(t_k)]^2\Delta t_k}\right|$$
$$\leqslant\sum_{k=0}^{p-1}\sqrt{\sum_{j=1}^{n}\varepsilon_{jk}^2}\,\Delta t_k,$$

or equivalently,

$$\left|s(L_\Pi)-\sum_{k=0}^{p-1}\sqrt{\sum_{j=1}^{n}[x_j'(t_k)]^2}\,\Delta t_k\right|\leqslant\sum_{k=0}^{p-1}\sqrt{\sum_{j=1}^{n}\varepsilon_{jk}^2}\,\Delta t_k.$$

We can write the sum on the right in the form

$$\sum{}'\sqrt{\sum_{j=1}^{n}\varepsilon_{jk}^2}\,\Delta t_k+\sum{}''\sqrt{\sum_{j=1}^{n}\varepsilon_{jk}^2}\,\Delta t_k,$$

where the first sum consists of the terms stemming from intervals on which all the $x_j(t)$ are continuous and the second sum consists of the terms stemming from intervals containing a discontinuity point of at least one $x_j'(t)$. Since every $x_j'(t)$ is uniformly continuous on all the intervals $[t_k,t_{k+1}]$ involved in the first sum, we have

$$\varepsilon_{jk}\leqslant\omega(d(\Pi))\qquad(j=1,2,\ldots,n)$$

for all such intervals, where $\omega(u)$ approaches zero as $u\to0$ and $d(\Pi)$ has the usual meaning, i.e.,

$$d(\Pi)=\max\{\Delta t_0,\Delta t_1,\ldots,\Delta t_{n-1}\}.$$

It follows that

$$\left|s(L_\Pi)-\sum_{k=0}^{p-1}\sqrt{\sum_{j=1}^{n}[x_j'(t_k)]^2}\Delta t_k\right|\leqslant\sqrt{n}(b-a)\omega(d(\Pi))+2M\sqrt{n}\sum{}''\Delta t_k,\qquad(4)$$

where

$$M=\sup_{\substack{a\leq t\leq b\\1\leq j\leq n}}|x_j'(t)|$$

and $\sum''\,\Delta t_k$ is the sum of the lengths of all the intervals $[t_k,t_{k+1}]$ containing discontinuity points of at least one of the $x_j'(t)$. There are a fixed number of such points, and hence $\sum''\,\Delta t_k \to 0$ as $d(\Pi)\to 0$. Hence, passing to the limit as $d(\Pi)\to 0$ in (4), we find that

$$\lim_{d(\Pi)\to 0} s(L_\Pi) = \lim_{d(\Pi)\to 0} \sum_{k=0}^{p-1} \sqrt{\sum_{j=1}^{n}\,[x_j'(t_k)]^2}\,\Delta t_k.$$

But this proves (3), since the quantity on the left is just the length s of the curve L, defined as the limiting length of the inscribed polygonal line L_Π as $d(\Pi)\to 0$, while the quantity on the right, involving Riemann sums of the function

$$\sqrt{\sum_{j=1}^{n}\,[x_j'(t)]^2},$$

is just the integral

$$\int_a^b \sqrt{\sum_{j=1}^{n}\,[x_j'(t)]^2}\;dt$$

(which exists since the integrand is piecewise continuous, like the functions $x_j'(t)$ themselves). ∎

If $n=2$, formula (3) reduces to

$$s = \int_a^b \sqrt{[x'(t)]^2 + [y'(t)]^2}\;dt \qquad\qquad (3')$$

($x_1 = x$, $x_2 = y$). Choosing x as the parameter, we get formula (1).

9.72. Examples

a. It follows from (1) that the length of the catenary $y = \cosh x$ between the points with abscissas $x = 0$ and $x = b$ is just

$$s = \int_0^b \sqrt{1 + \sinh^2 x}\;dx = \int_0^b \cosh x\;dx = \sinh b$$

(see Figure 46, noting that the length of the curve AB is the same as that of the segment CD).

b. Using (3'), we find that the length of the arc of the circle

$$x = r\cos\theta, \qquad y = r\sin\theta,$$

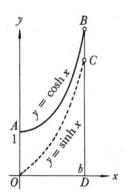

Figure   46

bounded by the rays $\theta = \theta_1$ and $\theta = \theta_2$ $(0 \leqslant \theta_1 < \theta_2 \leqslant 2\pi)$ equals

$$s = \int_{\theta_1}^{\theta_2} \sqrt{r^2 \sin^2 \theta + r^2 \cos^2 \theta} \; d\theta = r(\theta_2 - \theta_1).$$

In particular, choosing $\theta_1 = 0$, $\theta_2 = 2\pi$, we find that the circumference (length) of the whole circle is just $2\pi r$. This justifies the familiar interpretation of the number π as the ratio of the circumference of a circle to its diameter.

c. The length of the arc

$$x = a \cos t, \qquad y = b \sin t \qquad (0 \leqslant t \leqslant T \leqslant 2\pi, \; a < b)$$

of an ellipse of eccentricity

$$\varepsilon = \frac{c}{b} = \frac{\sqrt{b^2 - a^2}}{b}$$

is given by

$$\begin{aligned}
s &= \int_0^T \sqrt{a^2 \sin^2 t + b^2 \cos^2 t} \; dt = \int_0^T \sqrt{a^2 \sin^2 t + b^2(1 - \sin^2 t)} \; dt \\
&= \int_0^T \sqrt{b^2 - (b^2 - a^2) \sin^2 t} \; dt = b \int_0^T \sqrt{1 - \varepsilon^2 \sin^2 t} \; dt = bE(\varepsilon, T),
\end{aligned}$$

in terms of the elliptic integral $E(\varepsilon, T)$ introduced in Sec. 9.45.

d. The length of a "quarter cycle" of the sinusoid $y = \sin x$ is also given by an elliptic integral, since

$$s = \int_0^{\pi/2} \sqrt{1 + \cos^2 x}\, dx = \int_0^{\pi/2} \sqrt{2 - \sin^2 x}\, dx$$
$$= \sqrt{2} \int_0^{\pi/2} \sqrt{1 - \tfrac{1}{2} \sin^2 x}\, dx = \sqrt{2} E\left(\frac{1}{\sqrt{2}}, \frac{\pi}{2}\right).$$

This is hardly surprising, since the ellipse obtained by cutting an oblique cross section of a circular cylinder becomes a sinusoid when the cylinder is "unrolled" onto a plane.

e. Unlike the case of area, the length of a curve does not in general transform in a simple way under an expansion along one of the coordinate axes. In fact, even in the case of a polygonal line, each segment of the line is multiplied by its own coefficient under such an expansion, where the coefficient depends on the angle between the segment and the direction of the axis in question. However, if we carry out a k-fold expansion along all the coordinate axes *simultaneously*, thereby making a similarity transformation (see Sec. 2.67), then obviously the length of each segment of every polygonal line is multiplied by k, and hence so is the length of every piecewise smooth curve.

9.73. Arc length versus chord length. Let s_0 denote the length of the chord L_0 joining the end points of a piecewise smooth curve L, and let $L_\Pi = P_0 P_1 \ldots P_{n-1}$ be any polygonal line inscribed in L, as in the proof of Theorem 9.71. Then, by the triangle inequality (cf. Theorem 3.11a),

$$s(L_\Pi) \geqslant s_0,$$

where $s(L_\Pi)$ is the length of L_Π and s_0 is the length of L_0. Taking the limit as $d(\Pi) \to 0$, we find that

$$s \geqslant s_0,$$

or equivalently

$$\frac{s}{s_0} \geqslant 1,$$

where s is the length of the curve L.

THEOREM. *If s is the length of the curve L with parametric equations (2) and if s_0 is the length of the chord joining the end points of L, then*

$$\lim_{b \searrow a} \frac{s}{s_0} = 1,$$

provided that

$$\sum_{j=1}^{n} [x'_j(a)]^2 \neq 0. \tag{5}$$

Proof. Since b will eventually be made to approach a, there is no loss of generality in assuming that every $x'_j(t)$ is continuous on $[a,b]$.† Using (3), we have

$$\frac{s}{s_0} = \frac{\displaystyle\int_a^b \sqrt{\sum_{j=1}^{n} [x'_j(t)]^2} \, dt}{\sqrt{\displaystyle\sum_{j=1}^{n} [x_j(b) - x_j(a)]^2}}$$

$$= \frac{(b-a)\sqrt{\displaystyle\sum_{j=1}^{n} [x'_j(\xi)]^2}}{(b-a)\sqrt{\displaystyle\sum_{j=1}^{n} [x'_j(\xi_j)]^2}} = \sqrt{\frac{\displaystyle\sum_{j=1}^{n} [x'_j(\xi)]^2}{\displaystyle\sum_{j=1}^{n} [x'_j(\xi_j)]^2}} \tag{6}$$

for suitable $\xi, \xi_1, \ldots, \xi_n$ all belonging to the interval (a,b). This follows from Theorem 9.15f in the case of the numerator and Lagrange's theorem (Sec. 7.44) in the case of the denominator. But the right-hand side of (6) clearly approaches 1 as $b \searrow a$, provided that (5) holds. ∎

9.74. Arc length as parameter. Again let L be a piecewise smooth curve with parametric equations (2), and let $s(t)$ be the length of the arc of L joining the initial point of L (with parameter value a) to the variable point

$$P(t) = (x_1(t), x_2(t), \ldots, x_n(t)) \qquad (a \leqslant t \leqslant b).$$

Then

$$s(t) = \int_a^t \sqrt{\sum_{j=1}^{n} [x'_j(\tau)]^2} \, d\tau, \tag{7}$$

where $s(t)$ is itself piecewise smooth, with derivative

$$s'(t) = \sqrt{\sum_{j=1}^{n} [x'_j(t)]^2}. \tag{8}$$

Clearly (8) is nonnegative and continuous everywhere except at the discontinuity points of $x'_j(t)$. It follows that $s(t)$ is nondecreasing on $[a,b]$.

Now suppose the functions $x'_1(t), \ldots, x'_n(t)$ do not vanish simultaneously on $[a,b]$, so that $s(t)$ is increasing.‡ Then $s = s(t)$ has a continuous increasing

---

† We interpret $x_j'(a)$ as $x_j'(a+0)$ if $x_j'(t)$ is discontinuous at $t=a$.
‡ At discontinuity points we replace this condition by the requirement that the quantities $x_j'(t+0)$ do not vanish simultaneously, and similarly for the $x_j'(t-0)$.

inverse $t = t(s)$ on the interval $[0,l]$, where l is the length of L (see Theorem 5.35b). Moreover, let $[\alpha,\beta] \subset [a,b]$ be an interval on which the functions $x'_j(t)$ are all continuous. Then, by Theorem 7.16, $t = t(s)$ has a derivative on $[s(\alpha), s(\beta)]$ equal to

$$t'(s) = \frac{1}{s'(t)}.$$

Hence $t(s)$ is piecewise smooth, like $s(t)$ itself (why?). But then every function $x_j(t)$ can be expressed as a continuous function of s, with a piecewise continuous derivative. In particular, if the arc length s is chosen as the original parameter t, it follows from (8) that

$$\sum_{j=1}^{n} [x'_j(s)]^2 = 1 \qquad (0 \leqslant s \leqslant l).$$

Points at which all the functions $x'_j(t)$ vanish simultaneously are called *singular points* of the curve L. Thus we see that a parametric representation of L with the arc length as parameter is possible on parts of the curve which contain no singular points.

9.8. Area of a Surface of Revolution

By a *surface of revolution* we mean the surface Ω generated by rotating a curve L about a straight line lying in its plane. For example, rotating the curve with equation

$$y = f(x) \geqslant 0 \qquad (a \leqslant x \leqslant b) \tag{1}$$

about the x-axis generates the surface of revolution Ω shown in Figure 47. More generally, rotating the curve with parametric equations

$$x = x(t), \qquad y = y(t) \geqslant 0 \qquad (a \leqslant x \leqslant b) \tag{2}$$

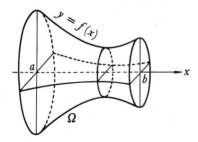

Figure 47

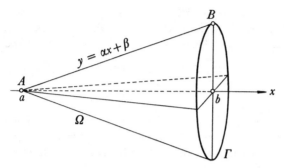

Figure   48

about the x-axis also generates a surface of revolution Ω. We now consider the problem of defining and calculating the area of Ω.

9.81. Area of a cone. We begin with the case where the curve (1) is a straight line with equation

$$y = \alpha x + \beta. \tag{3}$$

Rotating the line (3) about the x-axis then generates the surface of revolution Ω shown in Figure 48, i.e., a *conical surface* (or simply a *cone*). The line (3) is called the *generator* of the cone, the point A in which the generator intersects the x-axis is called the *vertex* of the cone, and the circle Γ obtained by rotating the point B about the x-axis is called the *directrix* of the cone.

To define the "area" of the cone Ω we argue as follows. Choose consecutive points $B_1,\ldots,B_n$ on the circle Γ, and then join each point to the next (and the point B_n to the point B_1), thereby obtaining a polygonal line $\lambda = B_1 B_2 \ldots B_n B_1$ and a set of isosceles triangles $B_1 A B_2,\ldots,B_n A B_1$ which together make up a *pyramidal surface* Φ_λ "inscribed" in P, as shown in Figure 49. The area of Φ_λ is defined in the obvious way as the sum $S(\Phi_\lambda)$ of the areas of the triangles $B_1 A B_2,\ldots,B_n A B_1$. Let the length of the segment $B_k B_{k+1}$ be $2\delta_k$ ($k=1,\ldots,n$), and let

$$d(\lambda) = \max \{\delta_1,\ldots,\delta_n\}.$$

Defining the area of the cone Ω as the limit

$$S = \lim_{d(\lambda) \to 0} S(\Phi_\lambda), \tag{4}$$

we now derive a formula for S (this incidentally proves the existence and uniqueness of S):

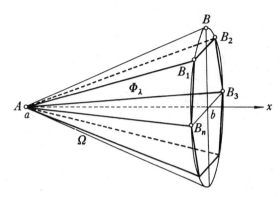

Figure   49

THEOREM. *The limit* (4) *equals* $\pi l r$, *where* l *is the "slant height" of the cone and* r *the radius of its base.*†

Proof. The altitude of the triangle $B_k A B_{k+1}$ equals

$$\sqrt{l^2 - \delta_k^2} = \sqrt{1 - \frac{\delta_k^2}{l^2}} = l\left(1 - \frac{\delta_k^2 E}{2l^2}\right) = l - \frac{\delta_k^2 E}{2l},$$

where $E \to 1$ as $\delta_k \to 0$ (see Theorem 5.59e). Hence the area of $B_k A B_{k+1}$ equals

$$\delta_k \sqrt{l^2 - \delta_k^2} = \delta_k l - \frac{\delta_k^3 E}{2l},$$

and the area of the inscribed pyramidal surface Φ_λ is just

$$S(\Phi_\lambda) = \sum_{k=1}^{n} \left(\delta_k l - \frac{\delta_k^3 E}{2l}\right) = \frac{1}{2} l s(\lambda) - \frac{1}{2l} \sum_{k=1}^{n} \delta_k^3 E, \qquad (5)$$

where $s(\lambda)$ is the length of the polygonal line $\lambda = B_1 B_2 \ldots B_n B_1$. As $d(\lambda) \to 0$, $s(\lambda)$ approaches the circumference $2\pi r$ of the circle Γ (Example 9.72b), and hence $\frac{1}{2} l s(\lambda) \to \pi l r$. Moreover, if the δ_k are all less than ε, then

$$\frac{1}{2l} \sum_{k=1}^{n} \delta_k^3 E \leqslant \frac{1}{2l} 2\varepsilon^2 s(\lambda) \qquad (6)$$

(for sufficiently small δ_k), so that (6) approaches zero as $d(\lambda) \to 0$. It follows that (5) approaches $\pi l r$ as $d(\lambda) \to 0$. ∎

9.82. Area of a conical band. Let Ω be the "conical band" shown in

† Thus l is the length of the generator AB and r the radius of the directrix Γ.

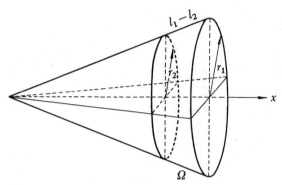

Figure   50

Figure 50, generated by rotating a segment of the line (3) about the x-axis. Then we define the area of Ω in the natural way, i.e., as the difference between the area of the "larger cone," of slant height l_1 and base radius r_1, and the area of the "smaller cone," of slant height l_2 and base radius r_2. By the above theorem,

$$S = \pi(l_1 r_1 - l_2 r_2).$$

But, by similarity,

$$\frac{l_1}{l_2} = \frac{r_1}{r_2},$$

or equivalently,

$$l_1 r_2 = l_2 r_1,$$

and hence

$$S = \pi(l_1 - l_2)(r_1 + r_2),$$

or finally

$$S = \pi l(r_1 + r_2), \tag{7}$$

where $l = l_1 - l_2$ is just the "slant height" of the conical band Ω.

9.83. Area of a general surface of revolution. Let Ω be the surface of revolution generated by rotating the curve L with parametric equations (2) about the x-axis, and let L_Π be the polygonal line inscribed in L, corresponding to the partition

$$\Pi = \{a = t_0 \leqslant t_1 \leqslant \cdots \leqslant t_n = b\},$$

with parameter

$$d(\Pi) = \max \{\Delta t_0, \Delta t_1, \ldots, \Delta t_{n-1}\} \qquad (\Delta t_k = t_{k+1} - t_k).$$

Let Ω_Π be the surface of revolution generated by rotating L_Π about the x-axis. Since Ω_Π is made up of conical bands, we define its area $S(\Omega_\Pi)$ in the obvious way as the sum of the areas of these bands. Defining the area of Ω itself as the limit

$$S = \lim_{d(\Pi) \to 0} S(\Omega_\Pi), \tag{8}$$

we now derive a formula for S (this incidentally proves the existence and uniqueness of S).

THEOREM. *The limit* (8) *equals*

$$S = 2\pi \int_a^b y(t) \sqrt{[x'(t)]^2 + [y'(t)]^2} \, dt \tag{9}$$

if $x(t)$ *and* $y(t)$ *have continuous derivatives on* $[a,b]$.

Proof. The vertices of L_Π are at the points $P_k = (x_k, y_k)$, where

$$x_k = x(t_k), \qquad y_k = y(t_k) \qquad (k = 0, 1, \ldots, n),$$

and the length of the segment $P_k P_{k+1}$ is just

$$\Delta l_k = \sqrt{(\Delta x_k)^2 + (\Delta y_k)^2}.$$

Rotating $P_k P_{k+1}$ about the x-axis generates a conical band of area

$$\pi(y_k + y_{k+1})\Delta l_k,$$

by (7), and hence the area of Ω_Π is given by

$$S_\Pi = S(\Omega_\Pi) = \pi \sum_{k=0}^{n-1} (y_k + y_{k+1})\Delta l_k.$$

Using Lagrange's theorem (Sec. 7.44), we can write this expression in the form

$$S_\Pi = \pi \sum_{k=0}^{n-1} [y(t_k) + y(t_{k+1})]\sqrt{[x'(\xi_k)]^2 + [y'(\eta_k)]^2}\Delta t_k,$$

where the points ξ_k and η_k both lie in the interval (t_k, t_{k+1}) but are in general distinct.

Besides S_Π, we also consider the expression

$$S_\Pi^* = 2\pi \sum_{k=0}^{n-1} y(t_k)\sqrt{[x'(t_k)]^2 + [y'(t_k)]^2}\Delta t_k,$$

which is clearly a Riemann sum of the continuous function

$$2\pi y(t)\sqrt{[x'(t)]^2+[y'(t)]^2}$$

(it is assumed that $x'(t)$ and $y'(t)$ are continuous), and hence approaches the right-hand side of (9) as $d(\Pi)\to 0$. Thus the proof of (9) will be complete if we manage to show that

$$\lim_{d(\Pi)\to 0}(S_\Pi-S_\Pi^*)=0. \tag{10}$$

This is done as follows. Given any $\varepsilon>0$, there is a $\delta>0$ such that $|t-t^*|<\delta$ implies

$$|x'(t)-x'(t^*)|<\varepsilon, \qquad |y'(t)-y'(t^*)|<\varepsilon, \qquad |y(t)-y(t^*)|<\varepsilon$$

(why?). Hence, in particular, $d(\Pi)<\delta$ implies

$$y(t_{k+1})=y(t_k)+\theta_1\varepsilon \qquad (|\theta_1|<1),$$
$$x'(\xi_k)=x'(t_k)+\theta_2\varepsilon \qquad (|\theta_2|<1),$$
$$y'(\eta_k)=y'(t_k)+\theta_3\varepsilon \qquad (|\theta_3|<1),$$

so that

$$[y(t_k)+y(t_{k+1})]\sqrt{[x'(\xi_k)]^2+y'(\eta_k)]^2}$$
$$=[2y(t_k)+\theta_1\varepsilon]\sqrt{[x'(t_k)+\theta_2\varepsilon]^2+[y'(t_k)+\theta_3\varepsilon]^2}$$
$$=[2y(t_k)+\theta_1\varepsilon]\sqrt{[x'(t_k)]^2+[y'(t_k)]^2+\theta_4\varepsilon},$$

where

$$|\theta_4|\leqslant 2\max_{a\leq t\leq b}\{|x'(t)|+|y'(t)|+\varepsilon\}.$$

Moreover,

$$\sqrt{a}-\sqrt{b}\leqslant\sqrt{a+b}\leqslant\sqrt{a}+\sqrt{b}$$

for arbitrary $a,b>0$,† and hence

$$\sqrt{a+b}=\sqrt{a}+\theta_5\sqrt{b} \qquad (|\theta_5|\leqslant 1).$$

It follows that

$$\sqrt{[x'(t_k)]^2+[y'(t_k)]^2+\theta_4\varepsilon}=\sqrt{[x'(t_k)]^2+[y'(t_k)]^2}+\theta_5\sqrt{\theta_4\varepsilon},$$

where $|\theta_5|\leqslant 1$. Thus, finally,

† The first inequality is obvious, and the second follows from the triangle inequality applied to the vectors ($\sqrt{a},0$) and ($0,\sqrt{b}$) in the plane.

$$S_\Pi - S_\Pi^* = \pi \sum_{k=0}^{n-1} [2y(t_k) + \theta_1 \varepsilon][\sqrt{[x'(t_k)]^2 + [y'(t_k)]^2} + \theta_5\sqrt{\theta_4 \varepsilon}]$$

$$- 2\pi \sum_{k=0}^{n-1} y(t_k)\sqrt{[x'(t_k)]^2 + [y'(t_k)]^2}$$

$$= \sum_{k=0}^{n-1} [\pi\theta_1\varepsilon\sqrt{[x'(t_k)]^2 + [y'(t_k)]^2} + 2\pi y(t_k)\theta_5\sqrt{\theta_4\varepsilon} + \pi\theta_1\theta_5\varepsilon\sqrt{\theta_4\varepsilon}],$$

where the right-hand side clearly approaches zero as $d(\Pi) \to 0$, thereby establishing (10). ∎

In the case $t = x$, formula (9) reduces to

$$S = 2\pi \int_a^b y(x)\sqrt{1 + [y'(x)]^2}\,dx. \tag{9'}$$

9.84. Examples

a. Rotating the semicircle

$$x = r \cos \theta, \qquad y = r \sin \theta \qquad (0 \leqslant \theta \leqslant \pi)$$

about the x-axis generates a sphere of radius r (see Figure 51). It follows from (9) that the area of the sphere is

$$S = 2\pi \int_0^\pi r \sin \theta \sqrt{r^2 \sin^2 \theta + r^2 \cos^2 \theta}\; d\theta$$

$$= 2\pi r^2 \int_0^\pi \sin \theta\; d\theta = -2\pi r^2 \cos \theta \Big|_0^\pi = 4\pi r^2.$$

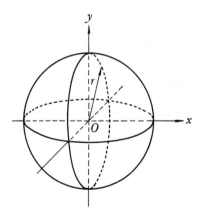

Figure   51

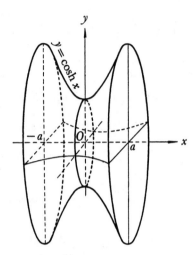

Figure   52

b. Rotating the catenary

$$y = \cosh x \qquad (-a \leqslant x \leqslant a)$$

about the x-axis, we get the surface of revolution shown in Figure 52, called a *catenoid*. Using (9'), we find that the area of the surface is

$$S = 2\pi \int_{-a}^{a} \cosh x \sqrt{1 + \sinh^2 x} \, dx = 2\pi \int_{-a}^{a} \cosh^2 x \, dx$$

$$= 2\pi \int_{-a}^{a} \frac{\cosh 2x + 1}{2} \, dx = \left(\pi \frac{\sinh 2x}{2} + x \right)\Big|_{-a}^{a} = \pi(\sinh 2a + 2a).$$

9.9. Further Applications of Integration

9.91. We begin with an almost trivial observation. Let $v = v(x)$ be a function defined on an interval $[a,b]$, and suppose we know the principal linear part (or differential)

$$dv = f(x) \, dx \qquad \qquad (1)$$

of the increment

$$\Delta v = v(x + dx) - v(x) \qquad (dx = \Delta x)$$

for all $x, x + dx \in [a,b]$. Then one value of v, say $v(\beta)$, is related to another value of v, say $v(\alpha)$, by the formula

$$v(\beta) = v(\alpha) + \int_\alpha^\beta f(x) \, dx \tag{2}$$

$(a \leqslant \alpha, \beta \leqslant b)$. To see this, we need merely recall from Sec. 7.31 that $f(x)$ is just the derivative of $v'(x)$.

In some cases where it is hard to find (1), we can establish an inequality of the form

$$dv \leqslant f(x) \, dx \tag{1'}$$

instead. Integrating (1'), we then get the estimate

$$v(\beta) \leqslant v(\alpha) + \int_\alpha^\beta f(x) \, dx. \tag{2'}$$

In finding an expression for the differential dv (as opposed to the increment Δv), things are often simplified by the fact that quantities "of a higher order of smallness than dx" can be neglected (why?).

9.92. Examples

a. How much work is needed to lift a mass m completely out of the earth's gravitational field?

Solution. Let $W = W(h)$ denote the work needed to lift the mass to a height h above the earth's surface. According to Newton's law of gravitation, the earth attracts a mass m at a height h above its surface with a force

$$F = F(h) = C\frac{m}{(R+h)^2},$$

where R is the radius of the earth (approximately 4000 miles). To determine C, we note that the gravitational force acting on a mass m at the earth's surface $(h=0)$ is just mg, where g is the "acceleration due to gravity" (approximately 32 ft/sec²), and hence $C = R^2 g$. Suppose we lift the mass from a height h above the earth's surface to a height $h + dh$. This requires an amount of work equal to

$$dW = F \, dh = \frac{R^2 mg}{(R+h)^2} \, dh, \tag{3}$$

provided F does not vary over the interval $[h, h + dh]$. Actually, of course, F does vary over $[h, h + dh]$, but only by an amount of a higher order of smallness than $F \, dh$, which can therefore be neglected in writing the differential (3). To find the amount of work needed to lift the mass m from the earth's surface to a height H, we now apply (2), obtaining

$$W = \int_0^H \frac{R^2 mg}{(R+h)^2}\, dh = \left.\frac{R^2 mg}{R+h}\right|_H^0 = Rmg - \frac{R^2 mg}{R+H}.$$

The second term on the right is negligible compared to the first for very large H, corresponding to the fact that increasing h further requires only a negligible amount of extra work against the force of gravitational attraction. Hence, finally, the amount of work needed to lift m "completely out of the earth's gravitational field" is just Rmg.

b. A funnel of height H is filled with water. Let the top of the funnel be of radius R and area S, while the bottom (i.e., the hole) is of radius r and area s. Suppose it is known that water leaves the funnel with velocity

$$v(h) = \sqrt{2gh}$$

at the moment when the height of the free surface of the water above the hole is h. Find the time T it takes for all the water to flow out of the funnel.

Solution. Let the height of the water at time t be $h = h(t)$, and let $S(h)$ be the area of the cross section at height h from the bottom (see Figure 53). During the interval $[t, t+dt]$ an amount of water of volume

$$vs\, dt = \sqrt{2gh}\, s\, dt$$

leaves the funnel. (Here we assume that v is constant during the interval $[t, t+dt]$, since taking account of the actual variation of v leads to a "correction" of a higher order of smallness than dt.) This is an amount of water equal to the volume

$$S(h)\, dh$$

of a circular cylinder of height dh and cross-sectional area $S(h)$. (Again the actual volume of water is that of a conical frustum, but this differs from the volume of the indicated cylinder by a quantity of a higher order of smallness

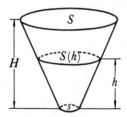

Figure   53

than dh.) It follows that

$$S(h)\ dh = \sqrt{2ghs}\ dt$$

or

$$s\ dt = \frac{S(h)}{\sqrt{2gh}}\ dh. \tag{4}$$

Substituting

$$S(h) = \pi[r + (R-r)h/H]^2$$

into (4) and integrating from 0 to T, we get

$$s\int_0^T dt = sT = \frac{\pi}{\sqrt{2g}}\int_0^H \left[\frac{r^2}{h^{1/2}} + \frac{2r(R-r)h^{1/2}}{H} + \frac{(R-r)^2 h^{3/2}}{H^2}\right] dh$$

$$= \frac{\pi}{\sqrt{2g}}\left[\frac{r^2 h^{1/2}}{\frac{1}{2}} + \frac{2r(R-r)h^{3/2}}{\frac{3}{2}H} + \frac{(R-r)^2 h^{5/2}}{\frac{5}{2}H^2}\right]\Bigg|_0^H$$

(cf. Theorem 9.53), so that finally

$$T = \sqrt{\frac{H}{2g}}\left[2 + \frac{4}{3}\left(\frac{R}{r} - 1\right) + \frac{2}{5}\left(\frac{R}{r} - 1\right)^2\right].$$

If R is very much larger than r, it is a good approximation to write

$$T = \frac{2}{5}\frac{S}{s}\sqrt{\frac{H}{2g}}.$$

9.93. The neighborhood of a plane curve. The set $V_\rho(L)$ of points belonging to all closed disks of radius ρ with centers on a given plane curve L is called a *ρ-neighborhood* of L (see Figure 54). Let L be piecewise smooth, and suppose L has no singular points in the sense of Sec. 9.74. Then it can be shown that $V_\rho(L)$ has area (cf. p. 486).

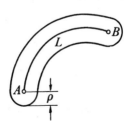

Figure 54

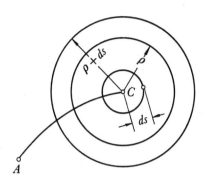

Figure   55

THEOREM. *Let $S_\rho(L)$ be the area of the set $V_\rho(L)$. Then*

$$S_\rho(L) \leqslant 2\pi\rho l + \pi\rho^2, \tag{5}$$

where l is the length of the curve L.

Proof. As in Sec. 9.74 we choose arc length along L as the parameter (s varies from 0 at the initial point A to l at the final point B). Let $S_\rho(L_\sigma)$ be the area of a ρ-neighborhood of the arc $L_\sigma \subset L$ corresponding to parameter values in the interval $0 \leqslant s \leqslant \sigma$, and let K_ρ be the disk of radius ρ centered at the point C with parameter value $s = \sigma$. The possible positions of points $C' \in L$ with parameter values in the interval $\sigma \leqslant s \leqslant \sigma + ds$ all lie in a disk of radius ds centered at C, while the possible positions of points of the disks of radius ρ centered at the points C' all lie in a disk $K_{\rho+ds}$ of radius $\rho + ds$ centered at the point C (see Figure 55). Hence the increment of $S_\rho(L_s)$ in going from the parameter value σ to the value $\sigma + ds$ cannot exceed the difference between the areas of the disks $K_{\rho+ds}$ and K_ρ, i.e., the quantity

$$\pi(\rho + ds)^2 - \pi\rho^2 = 2\pi\rho \; ds + \pi(ds)^2.$$

Thus the principal linear part of the increment satisfies the inequality

$$dS_\rho(L_s) \leqslant 2\pi\rho \; ds. \tag{6}$$

Integrating (6) with respect to s from 0 to l,† we get

$$S_\rho(L) - S_\rho(L_0) \leqslant 2\pi\rho l,$$

which implies (5), since $S_\rho(L_0) = \pi\rho^2$. ∎

9.94. Volume of a solid with horizontal cross sections of known area. It follows from the general theory of "measure" that the volume of

† As in the transition from (1′) to (2′).

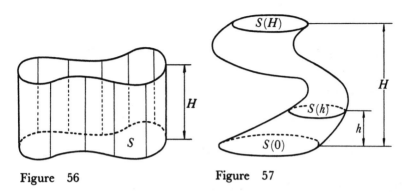

Figure   56   **Figure   57**

a right cylinder of altitude H and base of area S is just SH (see Figure 56).†
Let V be a solid whose horizontal cross section at height h above its base,
denoted by V_h, is of area $S(h)$, as shown in Figure 57.‡ Suppose the boundary
of V_h varies continuously with the height h, in the sense that given any $\varepsilon > 0$,
there is a $\delta > 0$ such that the projection of the boundary of the cross section $V_{h'}$
onto the plane of the base lies in an ε-neighborhood of the projection of the
boundary of the cross section V_h onto the same plane whenever $|h' - h| < \delta$.
But then the absolute value of the difference between the areas of the two
cross sections V_h and $V_{h'}$ does not exceed the area of the indicated ε-neigh-
borhood, and hence, by Theorem 9.93, is bounded from above by $2\pi\varepsilon L$,
where L is an upper bound for the lengths of the boundaries of all the V_h.

Now let $v(h)$ denote the volume of the part of V between the base and the
horizontal plane at height h over the base. Then the principal linear part of
the increment of $v(h)$ when the argument changes from h to $h + dh$ can be
written in the form $S(h)\, dh$, precisely as if V were cylindrical in the interval
$[h, h + dh]$. In fact, as just shown, the actual increment in volume differs from
$S(h)\, dh$ by a quantity of a higher order of smallness than dh. It follows that

$$dv = S(h)\, dh,$$

and hence, integrating from 0 to H, we find that the volume of V is just

$$v = v(H) = \int_0^H S(h)\, dh. \tag{7}$$

9.95. Examples

a. Let V be a solid cone of altitude H and upper cross-sectional area S (see

† Also see Appendix B.
‡ It is also assumed that V and the part of V between every pair of horizontal planes have
volume, and that the boundary of every cross section V_h is made up of one or more curves
with length.

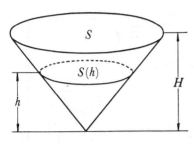   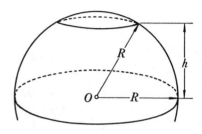

Figure   58   Figure   59

Figure 58). By similarity, the cross section at height h has area

$$S(h) = S\left(\frac{h}{H}\right)^2.$$

It follows from (7) that

$$v = \int_0^H \frac{S}{H^2} h^2 \, dh = \frac{S}{H^2} \frac{h^3}{3}\bigg|_0^H = \frac{1}{3} SH.$$

b. Let V be a solid sphere of radius R. Then, as shown in Figure 59, the area of the cross section at height h (above the center) equals $\pi(R^2 - h^2)$, and hence, by (7),

$$v = 2\int_0^R \pi(R^2 - h^2) \, dh = 2\pi\left(R^2 h - \frac{h^3}{3}\right)\bigg|_0^R = 2\pi\left(R^3 - \frac{R^3}{3}\right) = \frac{4}{3}\pi R^3.$$

c. For a solid of revolution, (7) takes the form

$$v = \pi\int_0^H r^2(h) \, dh, \tag{7'}$$

where $r(h)$ is the radius of the cross section at height h.

d. Let V be the solid cut out of a sphere of radius R by the paraboloid of revolution

$$2az = x^2 + y^2, \tag{8}$$

with vertex at the center of the sphere (see Figure 60). According to (8), the area of the cross section of the paraboloid at height h (above the center of the sphere) is just $2\pi ah$, while, on the other hand, the area of the cross section of the sphere at height h is $\pi(R^2 - h^2)$, as in the preceding example. The sphere and the paraboloid intersect in the plane $z = z_0$, where

$$2az_0 = R^2 - z_0^2.$$

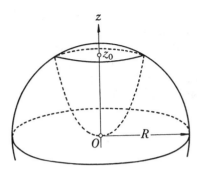

Figure   60

Hence

$$z_0 = -a + \sqrt{a^2 + R^2},$$

so that the volume of V is

$$v = \int_0^{z_0} 2\pi a h \, dh + \int_{z_0}^R \pi(R^2 - h^2) \, dh$$

(as an exercise, evaluate these simple integrals).

e. If the solid V is expanded k times along some axis, then its volume is enlarged k times. This is proved in much the same way as in the plane case (cf. Example 9.62d). In particular, we see that the volume of an ellipsoid with semiaxes a,b, and c equals $\frac{4}{3}\pi abc$, since a sphere of radius 1 has volume $\frac{4}{3}\pi$ and the ellipsoid is obtained by expanding the sphere a times along the x-axis, b times along the y-axis, and c times along the z-axis.

9.10. Integration of Sequences of Functions

9.101. Let

$$f_1(x), f_2(x), \dots, f_n(x), \dots \qquad (a \leqslant x \leqslant b)$$

be a sequence of integrable functions converging everywhere on $[a,b]$ to a limit function $f(x)$. Then two questions arise naturally:
(a) Is the function $f(x)$ also integrable?
(b) If so, does

$$\int_a^b f(x) \, dx = \lim_{n \to \infty} \int_a^b f_n(x) \, dx? \tag{1}$$

In general, the answers to both questions are negative, as shown by the

following examples:
(a) Let $f_n(x)$ equal 1 if $x \in [0,1]$ is of the form p/q with $q \leqslant n$, and let $f_n(x) = 0$ otherwise. Then every $f_n(x)$ is nonzero at only a finite number of points, and hence $f_n(x)$ is integrable with integral 0 (see Sec. 9.16d). The limit $f(x)$ of the sequence $f_n(x)$ equals 1 at every rational point of $[0,1]$ and 0 at every irrational point of $[0,1]$, i.e., the function $f(x)$ is the *Dirichlet function*, shown to be nonintegrable in Sec. 9.17.
(b) Let

$$f_n(x) = \begin{cases} n \sin nx & \text{if } 0 \leqslant x \leqslant \pi/n, \\ 0 & \text{if } \pi/n \leqslant x \leqslant \pi. \end{cases}$$

Then $f_n(x)$ approaches 0 at every point $x \in [0,\pi]$, and hence the limit function $f(x)$ is obviously integrable with integral 0. But

$$\int_0^\pi f_n(x)\ dx = \int_0^{\pi/n} n \sin nx\ dx = 2 \qquad (n = 1, 2, \ldots),$$

so that (1) fails to hold.

9.102. Thus an affirmative answer to both questions posed above can only be expected if extra conditions are imposed on the nature of the convergence of the sequence $f_n(x)$. As we will see in a moment, such an extra condition is afforded by the requirement that the convergence of $f_n(x)$ to its limit $f(x)$ be *uniform*. It will be recalled from Sec. 5.93b that a sequence of functions $f_n(x)$ is said to *converge uniformly on* $[a,b]$ to the limit function $f(x)$ if, given any $\varepsilon > 0$, there exists an integer $N > 0$ such that

$$|f(x) - f_n(x)| < \varepsilon$$

for all $n > N$ and all $x \in [a,b]$.

THEOREM. *Let $f_n(x)$ be a sequence of functions integrable on $[a,b]$ which converges uniformly on $[a,b]$ to a limit function $f(x)$. Then $f(x)$ is also integrable on $[a,b]$. Moreover, the formula*

$$\lim_{n \to \infty} \int_a^\xi f_n(x)\ dx = \int_a^\xi f(x)\ dx \qquad (2)$$

holds uniformly for all $\xi \in [a,b]$, and in particular

$$\lim_{n \to \infty} \int_a^b f_n(x)\ dx = \int_a^b f(x)\ dx.$$

Proof. For arbitrary partitions Π and Π' of the interval $[a,b]$ and every

$n = 1, 2, \ldots$, we have

$$S_\Pi(f) = S_\Pi(f_n) + S_\Pi(f - f_n), \tag{3}$$

$$S_{\Pi'} f = S_{\Pi'}(f_n) + S_{\Pi'}(f - f_n), \tag{4}$$

where $S_\Pi(f), S_\Pi(f_n), \ldots$ are Riemann sums of the functions $f, f_n, \ldots$, as in Sec. 9.12. Given any $\varepsilon > 0$, let n be such that

$$|f(x) - f_n(x)| < \frac{\varepsilon}{4(b-a)}$$

for all $x \in [a, b]$. Then, obviously,

$$|S_\Pi(f - f_n)| < \frac{\varepsilon}{4}, \qquad |S_{\Pi'}(f - f_n)| < \frac{\varepsilon}{4}.$$

Since $f_n(x)$ is integrable on $[a, b]$, there is a $\delta > 0$ such that

$$|S_\Pi(f_n) - S_{\Pi'}(f_n)| < \frac{\varepsilon}{2}$$

for any partitions Π and Π' with $d(\Pi) < \delta$, $d(\Pi') < \delta$. But then it follows from (3) and (4) with the same partitions that

$$|S_\Pi(f) - S_{\Pi'}(f)| \leqslant |S_\Pi(f_n) - S_{\Pi'}(f_n)| + |S_\Pi(f - f_n)| + |S_{\Pi'}(f - f_n)|$$

$$< \frac{\varepsilon}{2} + \frac{\varepsilon}{4} + \frac{\varepsilon}{4} = \varepsilon.$$

Therefore the Cauchy convergence criterion for the existence of a limit in the direction $d(\Pi) \to 0$ (Theorem 4.19) is satisfied for the Riemann sums of $f(x)$, and hence $f(x)$ is integrable on $[a, b]$.

To prove the second assertion, we note first that $f(x)$, like $f_n(x)$, is integrable on every interval $[a, \xi] \subset [a, b]$, by Theorem 9.15i. Since $f_n(x)$ converges uniformly to $f(x)$ on $[a, b]$, given any $\varepsilon > 0$, there is an N such that

$$|f(x) - f_n(x)| < \varepsilon$$

for all $n > N$ and all $x \in [a, b]$. But then

$$\left| \int_a^\xi f(x) \, dx - \int_a^\xi f_n(x) \, dx \right| = \left| \int_a^\xi [f(x) - f_n(x)] \, dx \right|$$

$$\leqslant \int_a^\xi |f(x) - f_n(x)| \, dx < \varepsilon(\xi - a) < \varepsilon(b - a)$$

for all $n > N$ and all $\xi \in [a, b]$, which implies (2). ∎

9.103. The above theorem has a natural analogue for a *series* of functions:

THEOREM. *Let*

$$\varphi_1(x) + \varphi_2(x) + \cdots + \varphi_n(x) + \cdots$$

be a series of functions, each integrable on $[a,b]$, which converges uniformly on $[a,b]$ to a sum function $\varphi(x)$. Then $\varphi(x)$ is also integrable on $[a,b]$. Moreover, the formula

$$\lim_{n\to\infty} \sum_{k=1}^{n} \int_a^\xi \varphi_k(x)\ dx = \int_a^\xi \varphi(x)\ dx$$

holds uniformly for all $\xi \in [a,b]$, and in particular

$$\sum_{k=1}^{\infty} \int_a^b \varphi_k(x)\ dx = \int_a^b \varphi(x)\ dx.$$

Proof. Choose

$$f_n(x) = \sum_{k=1}^{n} \varphi_k(x)$$

in Theorem 9.102. ∎

In other words, uniformly convergent series of functions can be "integrated term by term."

9.104. Examples

a. The series

$$\frac{1}{1+x} = 1 - x + x^2 - x^3 + \cdots,$$

$$\frac{1}{1+x^2} = 1 - x^2 + x^4 - x^6 + \cdots$$

both converge uniformly on every interval $[-b,b]$, $0 < b < 1$ (see Theorem 6.65a). Integrating these series term by term from 0 to ξ, we get the new series

$$\ln(1+\xi) = \xi - \frac{\xi^2}{2} + \frac{\xi^3}{3} - \frac{\xi^4}{4} + \cdots, \tag{5}$$

$$\arctan \xi = \xi - \frac{\xi^3}{3} + \frac{\xi^5}{5} - \frac{\xi^7}{7} + \cdots, \tag{6}$$

which also converge uniformly on $[-b,b]$. Moreover, since the series on the right in (5) and (6) converge for $\xi = 1$ (Theorem 6.23), while the functions on the left are continuous for $0 \leqslant \xi \leqslant 1$, we find that

$$\ln 2 = 1 - \frac{1}{2} + \frac{1}{3} - \frac{1}{4} + \cdots,$$

$$\frac{\pi}{4} = 1 - \frac{1}{3} + \frac{1}{5} - \frac{1}{7} + \cdots \tag{7}$$

(Theorem 6.68).

b. Theorem 9.103 can be used to find series expansions of nonelementary functions. For example,

$$\frac{\sin x}{x} = 1 - \frac{x^2}{3!} + \frac{x^4}{5!} - \frac{x^6}{7!} + \cdots \tag{8}$$

converges uniformly on every interval $[-T, T]$. Hence, integrating (8) term by term from 0 to ξ, we get the following expansion of the sine integral (Sec. 9.46b):

$$\text{Si } \xi = \int_0^\xi \frac{\sin x}{x} \, dx = \xi - \frac{\xi^3}{3 \cdot 3!} + \frac{\xi^5}{5 \cdot 5!} - \frac{\xi^7}{7 \cdot 7!} + \cdots. \tag{9}$$

The series (9) is also uniformly convergent on every interval $[-T, T]$.

9.105. Next we consider differentiation of sequences of functions, asking questions analogous to those posed in Sec. 9.101 for the case of integration. Let

$$f_1(x), f_2(x), \ldots, f_n(x), \ldots \qquad (a \leqslant x \leqslant b)$$

be a sequence of differentiable functions converging everywhere on $[a,b]$ to a limit function $f(x)$. Then we ask:
(a) Is the function $f(x)$ also differentiable?
(b) If so, does

$$f'(x) = \lim_{n \to \infty} f_n'(x)? \tag{10}$$

As shown by the following examples, the answers to both questions are again negative in general, even for a uniformly convergent sequence $f_n(x)$:
(a) The sequence of differentiable functions

$$f_n(x) = |x|^{1 + (1/n)}$$

converges uniformly on $[-1,1]$ to the function $f(x) = |x|$ which has no derivative at $x = 0$.
(b) The sequence of functions

$$f_n(x) = \frac{1}{n} \sin nx \qquad (0 \leqslant x \leqslant \pi)$$

converges uniformly to zero, and hence the limit function is obviously differentiable. However, (10) fails to hold, since the sequence $f_n'(x) = \cos nx$ converges nowhere except at the point $x = 0$.

9.106. The situation changes if we assume that the sequence of derivatives $f_n(x)$ converges uniformly. It then turns out that much less need be assumed about the sequence of functions $f_n(x)$ themselves:

THEOREM. *Let $f_n(x)$ be a sequence of piecewise smooth functions on $[a,b]$, which converges for at least one point $x_0 \in [a,b]$, and suppose the sequence of derivatives $f_n'(x)$ converges uniformly on $[a,b]$ to a piecewise continuous limit function $g(x)$. Then the sequence $f_n(x)$ converges uniformly on $[a,b]$ to a piecewise smooth function $f(x)$ with derivative*

$$f'(x) = \lim_{n \to \infty} f_n'(x) = g(x)$$

at every continuity point of $g(x)$.

Proof. We have

$$f_n(\xi) - f_n(x_0) = \int_{x_0}^{\xi} f_n'(x)\, dx \qquad (n = 1, 2, \ldots), \tag{11}$$

by Theorem 9.33, where the sequence (11) is uniformly convergent for all $\xi \in [a,b]$, by Theorem 9.102. But then $f_n(x)$ converges uniformly on $[a,b]$, since the numerical sequence $f_n(x_0)$ is convergent. Let $f(x)$ denote the limit of $f_n(x)$. Taking the limit as $n \to \infty$ in (11) and using Theorem 9.102 again, we get

$$f(\xi) - f(x_0) = \int_{x_0}^{\xi} g(x)\, dx.$$

But $g(x)$ is piecewise continuous, by hypothesis, and hence, by Theorem 9.31, $f(x)$ is differentiable everywhere except at the discontinuity points of $g(x)$, with derivative

$$f'(x) = g(x) = \lim_{n \to \infty} f_n'(x). \quad \blacksquare$$

9.107. As in the case of integration, the above theorem has a natural analogue for a *series* of functions:

THEOREM. *Let*

$$\varphi_1(x) + \varphi_2(x) + \cdots + \varphi_n(x) + \cdots \tag{12}$$

be a series of functions, each piecewise smooth on $[a,b]$, which converges for at least one point $x_0 \in [a,b]$, and suppose the series of derivatives

$$\varphi_1'(x) + \varphi_2'(x) + \cdots + \varphi_n'(x) + \cdots$$

converges uniformly on $[a,b]$ to a piecewise continuous sum function $f(x)$. Then the series (12) converges uniformly on $[a,b]$ to a piecewise smooth function $\varphi(x)$ with derivative

$$\varphi'(x) = g(x)$$

at every continuity point of $g(x)$.

Proof. Choose

$$f_n(x) = \sum_{k=1}^{n} \varphi_k(x)$$

in Theorem 9.106. ∎

9.11. Parameter-Dependent Integrals

9.111. THEOREM. *Let $f(x,t)$ be a real function of two variables $x \in \Delta = [a,b]$, $t \in M$, where M is a metric space with distance ρ, and suppose $f(x,t)$ is uniformly continuous in both x and t, i.e., on the product metric space $Q = \Delta \times M$.† Then the function*

$$\Phi(t) = \int_a^b f(x,t)\, dx \qquad (t \in M)$$

is uniformly continuous on M.

Proof. Let

$$\omega_f(\delta) = \sup_{\substack{|x'-x''| \leq \delta \\ \rho(t',t'') \leq \delta}} |f(x',t') - f(x'',t'')|$$

be the modulus of continuity of the function $f(x,t)$ on the space Q (see Sec. 5.17c). Clearly,

$$|\Phi(t') - \Phi(t'')| = \left| \int_a^b [f(x,t') - f(x,t'')]\, dx \right|$$
$$\leq \int_a^b |f(x,t') - f(x,t'')|\, dx \leq \omega_f(\delta)(b-a),$$

and hence

$$\omega_\Phi(\delta) \equiv \sup_{\rho(t',t'') \leq \delta} |\Phi(t') - \Phi(t'')| \leq \omega_f(\delta)(b-a). \tag{1}$$

† See Secs. 2.82, 3.16, 5.17a, and 5.18.

But given any $\varepsilon > 0$, there is a $\delta_0 > 0$ such that $\delta < \delta_0$ implies $\omega_f(\delta) < \varepsilon$, by the uniform continuity of $f(x,t)$ on Q. It follows from (1) that $\delta < \delta_0$ also implies $\omega_\Phi(\delta) < \varepsilon(b-a)$, so that $\Phi(t)$ is uniformly continuous on M. ∎

9.112. THEOREM. *If $f(x,t)$ is a continuous real function on the rectangle $a \leqslant x \leqslant b$, $\alpha \leqslant t \leqslant \beta$, then*

$$\int_a^b \left\{ \int_\alpha^\beta f(x,t) \, dt \right\} dx = \int_\alpha^\beta \left\{ \int_a^b f(x,t) \, dx \right\} dt. \tag{2}$$

Proof. The functions

$$F(x) = \int_a^b f(x,t) \, dt \qquad (a \leqslant x \leqslant b),$$

$$\Phi(t) = \int_\alpha^\beta f(x,t) \, dx \qquad (\alpha \leqslant t \leqslant \beta)$$

are both (uniformly) continuous, by the preceding theorem. Let $G(t)$ be any function continuous on $[\alpha, \beta]$, and let

$$\sum_{k=0}^{n-1} G(\tau_k) \Delta t_k \qquad (\Delta t_k = t_{k+1} - t_k, \ t_k \leqslant \tau_k \leqslant t_{k+1})$$

be a Riemann sum of $G(t)$ corresponding to a partition Π of $[\alpha, \beta]$ with $d(\Pi) < \delta$. Then, by Theorem 9.15k,

$$\left| \int_\alpha^\beta G(t) \, dt - \sum_{k=0}^{n-1} G(\tau_k) \right| \leqslant \omega_G(\delta)(\beta - \alpha), \tag{3}$$

where $\omega_G(\delta)$ is the modulus of continuity of $G(t)$ on $[\alpha, \beta]$. Applying this estimate to the function $f(x,t)$ with x fixed and $\tau_k = t_k$, we get

$$\left| \int_\alpha^\beta f(x,t) \, dt - \sum_{k=0}^{n-1} f(x,t_k) \Delta t_k \right| \leqslant \omega_f(\delta)(\beta - \alpha),$$

which gives

$$\left| \int_a^b \left\{ \int_\alpha^\beta f(x,t) \, dt \right\} dx - \int_a^b \left\{ \sum_{k=0}^{n-1} f(x,t_k) \Delta t_k \right\} dx \right| \leqslant \omega_f(\delta)(\beta - \alpha)(b-a) \tag{4}$$

after integration with respect to x. On the other hand, applying (3) to the function $\Phi(t)$ with $\tau_k = t_k$, we get

$$\left| \int_\alpha^\beta \Phi(t) \, dt - \sum_{k=0}^{n-1} \Phi(t_k) \Delta t_k \right|$$

$$= \left| \int_\alpha^\beta \left\{ \int_a^b f(x,t) \, dx \right\} dt - \sum_{k=0}^{n-1} \left\{ \int_a^b f(x,t_k) \, dx \right\} \Delta t_k \right|$$

$$\leqslant \omega_\Phi(\delta)(\beta - \alpha) \leqslant \omega_f(\delta)(b-a)(\beta - \alpha), \tag{5}$$

by (1). Since obviously

$$\int_a^b \left\{ \sum_{k=0}^{n-1} f(x,t_k) \Delta t_k \right\} dx = \sum_{k=0}^{n-1} \left\{ \int_a^b f(x,t_k) \, dx \right\} \Delta t_k,$$

it follows from (4) and (5) that

$$\left| \int_a^b \left\{ \int_\alpha^\beta f(x,t) \, dt \right\} dx - \int_\alpha^\beta \left\{ \int_a^b f(x,t) \, dx \right\} dt \right| \leqslant 2\omega_f(b-a)(\beta-\alpha),$$

which implies (2), since the right-hand side can be made arbitrarily small. ∎

9.113. Example. If

$$f(x,t) = x^t \qquad (0 \leqslant x \leqslant 1, \, 0 \leqslant \alpha \leqslant t \leqslant \beta),$$

then (2) becomes

$$\int_0^1 \left\{ \int_\alpha^\beta x^t \, dt \right\} dx = \int_\alpha^\beta \left\{ \int_0^1 x^t \, dx \right\} dt.$$

Evaluating the "inner integrals," we get

$$\int_0^1 \frac{x^\beta - x^\alpha}{\ln x} \, dx = \ln \frac{1+\beta}{1+\alpha}.$$

Any attempt to evaluate this integral directly by the methods of Secs. 9.4 and 9.5 must confront a formidable obstacle, namely the fact that the indefinite integral of the function

$$\frac{x^\beta - x^\alpha}{\ln x}$$

cannot be expressed in terms of elementary functions.

9.114. THEOREM. *Let $f(x,t)$ and its derivative with respect to t, namely the function*†

$$f_t(x,t) = \lim_{h \to 0} \frac{f(x,t+h) - f(x,t)}{h} \qquad (x \text{ fixed}), \tag{6}$$

be continuous on the rectangle

$$a \leqslant x \leqslant b, \qquad t_0 - \delta \leqslant t \leqslant t_0 + \delta \qquad (\delta > 0).$$

Then the function

$$\Phi(t) = \int_a^b f(x,t) \, dx \qquad (|t - t_0| \leqslant \delta)$$

---

† A derivative like (6), involving a function of n variables with $n-1$ variables held fixed, is called a *partial derivative*. Here, of course, we have the simplest case, where $n = 2$.

is differentiable at $t = t_0$, *with derivative*

$$\Phi'(t_0) = \int_a^b f_t(x,t_0) \, dx. \tag{7}$$

Proof. Applying Theorem 9.112 to the function $f_t(x,t)$, we get

$$\int_{t_0}^t \left\{ \int_a^b f_t(x,\tau) \, dx \right\} d\tau = \int_a^b \left\{ \int_{t_0}^t f_t(x,\tau) \, d\tau \right\} dx$$

$$= \int_a^b f(x,t) \, dx - \int_a^b f(x,t_0) \, dx \qquad (|t-t_0| \leqslant \delta), \tag{8}$$

with the help of Theorem 9.33. The left-hand side of (8) has a derivative with respect to t equal to the integrand, by Theorems 9.31b and 9.111, and hence the right-hand side has the same derivative. But the second term in the right-hand side is independent of t, so that its derivative with respect to t vanishes. Equating derivatives of both sides of (8) at the point $t = t_0$, we get (7). ∎

9.115. Example. Evaluate the integral

$$\Phi(t) = \int_0^{\pi/2} \ln (t^2 - \sin^2 x) \, dx \qquad (t > 1).$$

Solution. It follows from (7) that

$$\Phi'(t) = \int_0^{\pi/2} \frac{2t \, dx}{t^2 - \sin^2 x}$$

$$= \frac{2}{\sqrt{t^2-1}} \arctan \left(\frac{\sqrt{t^2-1}}{t} \tan x \right) \Bigg|_0^{\pi/2} = \frac{\pi}{\sqrt{t^2-1}} \tag{9}$$

(verify that the conditions of the theorem are satisfied).† Integrating (9) with respect to t, we get

$$\Phi(t) = \int \frac{\pi \, dt}{\sqrt{t^2-1}} = \pi \ln (t + \sqrt{t^2-1}) + C.$$

To find the value of the constant C, we take the limit as $t \to \infty$, or equivalently as $\tau = 1/t \to 0$, in the formula

---

† To evaluate the indefinite integral in (9), make the substitutions

$$u = \tan x, \qquad \sin^2 x = \frac{u^2}{1+u^2}, \qquad dx = \frac{du}{1+u^2},$$

as in Sec. 9.43b. Note that (9) entails making a slight generalization of Theorem 9.33 (which?).

$$C = \int_0^{\pi/2} \ln (t^2 - \sin^2 x) \, dx - \pi \ln (t + \sqrt{t^2 - 1})$$

$$= \int_0^{\pi/2} \ln \left(1 - \frac{\sin^2 x}{t^2}\right) dx - \pi \ln \frac{t + \sqrt{t^2 - 1}}{t}, \tag{10}$$

valid for all $t > 1$. First we note that since the function $\ln (1 - \tau^2 \sin^2 x)$ is continuous on the rectangle $0 \leqslant x \leqslant \pi/2$, $0 \leqslant \tau \leqslant \tau_0 < 1$, it follows from Theorem 9.111 that the function

$$\Psi(\tau) = \int_0^{\pi/2} \ln (1 - \tau^2 \sin^2 x) \, dx$$

is continuous for $0 \leqslant \tau \leqslant \tau_0$, and hence that

$$0 = \Psi(0) = \lim_{\tau \to 0} \Psi(\tau) = \lim_{\tau \to 0} \int_0^{\pi/2} \ln (1 - \tau^2 \sin^2 x) \, dx,$$

so that the limit of the first term in the right-hand side of (10) as $\tau \to 0$ vanishes. Therefore

$$C = -\pi \lim_{t \to \infty} \frac{t + \sqrt{t^2 - 1}}{t} = -\pi \lim_{\tau \to 0} \ln (1 + \sqrt{1 - \tau^2}) = -\pi \ln 2,$$

and hence finally

$$\Phi(t) = \pi \ln (t + \sqrt{t^2 - 1}) - \pi \ln 2 = \pi \ln \frac{t + \sqrt{t^2 - 1}}{2}.$$

9.116. Next we consider a somewhat more complicated case, where the limits of integration as well as the integrand depend on the parameter t:

THEOREM. *Let $f(x,t)$ and $f_t(x,t)$ be the same as in Theorem 9.114, and let $\varphi(t)$ be a differentiable function on the interval $t_0 - \delta \leqslant t \leqslant t_0 + \delta$, taking its values in the interval $a \leqslant x \leqslant b$. Then the function*

$$\Phi(t) = \int_a^{\varphi(t)} f(x,t) \, dx \qquad (|t - t_0| \leqslant \delta)$$

is differentiable at $t = t_0$, with derivative

$$\Phi'(t_0) = \int_a^{\varphi(t_0)} f_t(x,t_0) \, dx + f(\varphi(t_0),t_0) \varphi'(t_0). \tag{11}$$

Proof. Let

$$\Phi(t) = \Phi_1(t) + \Phi_2(t), \tag{12}$$

where

$$\Phi_1(t) = \int_a^{\varphi(t_0)} f(x,t) \, dx, \qquad \Phi_2(t) = \int_{\varphi(t_0)}^{\varphi(t)} f(x,t) \, dx.$$

By Theorem 9.114, $\Phi_1(t)$ is differentiable at $t=t_0$, with derivative

$$\Phi_1(t_0) = \int_a^{\varphi(t_0)} f_t(x,t_0) \, dx. \tag{13}$$

The derivative of $\Phi_2(t)$ at $t=t_0$ can be calculated directly. In fact,

$$\frac{\Phi_2(t_0+h) - \Phi_2(t_0)}{h} = \frac{1}{h} \int_{\varphi(t_0)}^{\varphi(t_0+h)} f(x,t_0+h) \, dx$$

$$= \frac{\varphi(t_0+h) - \varphi(t_0)}{h} f(\theta,t_0+h) \qquad (|h| < \delta),$$

by Theorem 9.15f, where θ is a number between $\varphi(t_0)$ and $\varphi(t_0+h)$. Taking the limit as $h \to 0$ and using the properties of the functions $f(x,t)$ and $\varphi(t)$, we get

$$\Phi_2'(t) = \varphi'(t_0) f(\varphi(t_0),t_0). \tag{14}$$

Comparison of (12)–(14) then gives (11). ∎

In particular, if $\varphi(t) = t$, then (11) reduces to

$$\left[\int_a^t f(x,t) \, dx \right]_{t=t_0}' = \int_a^{t_0} f_t(x,t_0) \, dx + f(t_0,t_0). \tag{11'}$$

9.117. Example. Differentiate the function

$$\Phi_n(t) = \frac{1}{(n-1)!} \int_a^t (t-x)^{n-1} f(x) \, dx \qquad (n=1,2,\dots), \tag{15}$$

where $f(x)$ is continuous on the interval $a \leqslant x \leqslant b$.

Solution. It follows from (11') that

$$\Phi_n'(t) = \frac{1}{(n-1)!} \int_a^t (n-1)(t-x)^{n-2} f(x) \, dx + \frac{(t-t)^{n-1}}{(n-1)!} f(t) = \Phi_{n-1}(t).$$

But obviously $\Phi_n(a) = 0$, and hence

$$\Phi_n(t) = \int_a^t \Phi_n'(x) \, dx = \int_a^t \Phi_{n-1}(x) \, dx.$$

Noting that

$$\Phi_1(t) = \int_a^t f(x)\, dx,$$

we see that $\Phi_n(t)$ is the result of n consecutive integrations of the function $f(x)$ from a to t, i.e., the operation leading to $\Phi_n(t)$ is the inverse of n-fold differentiation. Formula (15) represents this operation in the form of a single integral involving a parameter.

9.12. Line Integrals

9.121. Definitions and basic properties. As in Sec. 9.71, let L be a curve in n-dimensional space with parametric equations

$$x_1 = x_1(t), \qquad x_2 = x_2(t), \ldots, \qquad x_n = x_n(t) \qquad (a \leqslant t \leqslant b), \tag{1}$$

where the functions $x_j(t)$ are continuous with continuous derivatives $x_j'(t)$ satisfying the condition

$$\sum_{j=1}^n [x_j'(t)]^2 > 0 \qquad (a \leqslant t \leqslant b), \tag{2}$$

so that L is smooth and has no singular points (Sec. 9.74). We will think of L as equipped with the "direction of increasing t," namely, the direction of motion of the variable point $P(t)$ with coordinates (1) as t varies from a to b. Such a curve with an assigned direction will simply be called a (smooth) *path*.† As in Sec. 9.74, let $s(t)$ be the length of the arc of L joining the initial point of L (with parameter value a) to the variable point $P(t)$. Then

$$s(t) = \int_a^t \sqrt{\sum_{j=1}^n [x_j'(\tau)]^2}\, d\tau,$$

and we can choose s as our parameter instead of the original parameter t. The equations (1) then become

$$x_1 = x_1(s), \qquad x_2 = x_2(s), \ldots, \qquad x_n = x_n(s) \qquad (0 \leqslant s \leqslant l), \tag{1'}$$

where $x_j(s) \equiv x_j(t(s))$‡ and l is just the length of L. By the same token, the variable point with coordinates (1') is denoted by $P(s)$. The function $s = s(t)$

---

† It will also be tacitly assumed that a path has no singular points, i.e., that the condition (2) holds (or the analogous condition (2') below).
‡ There is a slight abuse of notation here, since the function $x_j(s)$ is not the same as the function $x_j(t)$, with t changed to s, but the context precludes any confusion. The same observation applies to writing $f(t) \equiv f(P(t))$, $g(t) \equiv g(P(t))$ below.

has a continuously differentiable inverse $t = t(s)$, with derivative

$$t'(s) = \frac{1}{s'(t)}.$$

Now let

$$\Pi_s = \{0 = s_0 \leqslant s_1 \leqslant \cdots \leqslant s_p = l\}$$

be a partition of the interval $0 \leqslant s \leqslant l$, and let

$$\Pi_t = \{a = t_0 \leqslant t_1 \leqslant \cdots \leqslant t_p = b\}$$

be the corresponding partition of the interval $a \leqslant t \leqslant b$, where

$$s_k = s(t_k), \qquad t_k = t(s_k).$$

As usual, let

$$d(\Pi_s) = \max \{\Delta s_0, \Delta s_1, \ldots, \Delta s_{p-1}\},$$
$$d(\Pi_t) = \max \{\Delta t_0, \Delta t_1, \ldots, \Delta t_{p-1}\},$$

where

$$\Delta s_j = s_{j+1} - s_j, \qquad \Delta t_j = t_{j+1} - t_j.$$

Moreover, let $f(P)$ and $g(P)$ be two real functions defined on L, i.e., at every point $P \in L$, and form the (Riemann-Stieltjes) sum

$$\sum_{j=0}^{p-1} f(P_j^*)[g(P_{j+1}) - g(P_j)], \tag{3}$$

where

$$P_j = P(s_j), \qquad P_j^* = P(s_j^*) \qquad (s_j \leqslant s_j^* \leqslant s_{j+1})$$

(so that P_j^* is an arbitrary point of the arc $\widehat{P_j P_{j+1}}$). Then the limit of the sum (3) as $d(\Pi_s) \to 0$, provided it exists, is called the (*Stieltjes*) *line integral of the function* $f(P)$ *with respect to the function* $g(P)$ *along the path* $\widehat{AB}$ and is denoted by

$$\int_{\widehat{AB}} f(P) \, dg(P). \tag{4}$$

Alternatively, we can write

$$f(t) \equiv f(P(t)), \qquad g(t) \equiv g(P(t))$$

and form the sum

$$\sum_{j=0}^{p-1} f(t_j^*)[g(t_{j+1}) - g(t_j)], \tag{3'}$$

where $t_j \leqslant t_j^* \leqslant t_{j+1}$. But $d(\Pi_s) \to 0$ implies $d(\Pi_t) \to 0$ and conversely, since

$$\Delta t_j = t'(\sigma_j) \Delta s_j \qquad (s_j < \sigma_j < s_{j+1}),$$
$$\Delta s_j = s'(\tau_j) \Delta t_j \qquad (t_j < \tau_j < t_{j+1}).$$

Therefore if either of the sums (3) or (3′) approaches a limit under arbitrary refinement of the partition, then so does the other, and investigation of the line integral (4) is equivalent to investigation of the limit of (3′) as $d(\Pi_t) \to 0$, i.e., of the ordinary Stieltjes integral

$$\int_a^b f(t) \, dg(t)$$

(Sec. 9.55a). This integral exists and equals

$$\int_a^b f(t) g'(t) \, dt \tag{5}$$

if $f(t)$ is continuous and $g(t)$ has a continuous derivative (Theorem 9.55b).

Next we establish a few properties of line integrals. Let $\widehat{AB}$ be a path with initial point A and final point B. Then obviously

$$\int_{\widehat{AB}} dg(P) = g(B) - g(A)$$

for every function $g(P)$, while

$$\int_{\widehat{AB}} f(P) \, dg(P) = 0$$

for every function $f(P)$ if $g(P)$ is constant along $\widehat{AB}$. Moreover, it is easy to see that if either of the integrals (4) or

$$\int_{\widehat{BA}} f(P) \, dg(P)$$

exists, then so does the other and

$$\int_{\widehat{AB}} f(P) \, dg(P) = - \int_{\widehat{BA}} f(P) \, dg(P).$$

Furthermore, we have the following analogues of the formulas of Sec. 9.55c (provided the integrals on the right exist):

$$\int_{\widehat{AB}} [\alpha_1 f_1(P) + \alpha_2 f_2(P)] \, dg(P) = \alpha_1 \int_{\widehat{AB}} f_1(P) \, dg(P) + \alpha_2 \int_{\widehat{AB}} f_2(P) \, dg(P),$$
$$\tag{6}$$

$$\int_{\widehat{AB}} f(P)\, d[\beta_1 g_1(P) + \beta_2 g_2(P)] = \beta_1 \int_{\widehat{AB}} f(P)\, dg_1(P) + \beta_2 \int_{\widehat{AB}} f(P)\, dg_2(P),$$

(7)

$$\int_{\widehat{AB}} f(P)\, dg(P) = \int_{\widehat{AC}} f(P)\, dg(P) + \int_{\widehat{CB}} f(P)\, dg(P) \qquad (C \in \widehat{AB}),$$

(8)

$$\int_{\widehat{AB}} f(P)\, dg(P) = f(P)\, g(P)\Big|_A^B - \int_{\widehat{AB}} g(P)\, df(P)$$

$(\alpha_1, \alpha_2, \beta_1, \beta_2 \text{ real}).$

A sum

$$\int_{\widehat{AB}} f_1(P)\, dg_1(P) + \cdots + \int_{\widehat{AB}} f_m(P)\, dg_m(P)$$

of integrals over the same path $\widehat{AB}$ is written more concisely as

$$\int_{\widehat{AB}} f_1(P)\, dg_1(P) + \cdots + f_m(P)\, dg_m(P),$$

(9)

or even just as

$$\int_{\widehat{AB}} f_1\, dg_1 + \cdots + f_m\, dg_m,$$

(9′)

where we retain only one integral sign. If the functions $f_j(t)$ are all continuous and the functions $g_j(t)$ all have continuous derivatives, the integral (9) exists and equals

$$\int_a^b [f_1(t)g_1'(t) + \cdots + f_m(t)g_m'(t)]\, dt.$$

The above theory is easily extended to the case where the functions $x_j(t)$ figuring in (1) are continuous but their derivatives $x_j'(t)$ are only piecewise continuous and satisfy the conditions

$$\sum_{j=1}^n [x_j'(t+0)]^2 > 0, \qquad \sum_{j=1}^n [x_j'(t-0)]^2 > 0 \qquad (a \leqslant t \leqslant b) \tag{2′}$$

instead of (2), so that L is piecewise smooth instead of smooth. In fact, L is then the union of a finite number of smooth arcs

$$L_1 = \widehat{AC_1},\ L_2 = \widehat{C_1 C_2}, \ldots, L_m = \widehat{C_{m-1} B},$$

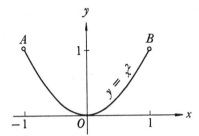

Figure 61

and we need only write

$$\int_L f(P)\, dg(P) = \int_{L_1} f(P)\, dg(P) + \int_{L_2} f(P)\, dg(P) + \cdots + \int_{L_m} f(P)\, dg(P),$$

by definition. This is consistent with the definition of the left-hand side as the limit of the sum (3) as $d(\Pi_s) \to 0$ in the case where $s(t)$ is only piecewise smooth instead of smooth.

9.122. Example. Evaluate the line integrals

$$I_1 = \int_{\widehat{AB}} (x^2 - y^2)\, dx, \qquad I_2 = \int_{\widehat{AB}} (x^2 - y^2)\, dy,$$

where $\widehat{AB}$ is the arc of the parabola $y = x^2$ joining the points $A = (-1,1)$ and $B = (1,1)$, as shown in Figure 61.

Solution. In the first case, choosing x as the parameter, we get

$$I_1 = \int_{-1}^{1} (x^2 - x^4)\, dx = \left(\frac{x^3}{3} - \frac{x^5}{5} \right)\Bigg|_{-1}^{1} = \frac{2}{3} - \frac{2}{5} = \frac{4}{15}.$$

In the second case, we cannot choose y as the parameter, since two points of the arc correspond to each value of y (except $y = 0$). Therefore we again choose x as the parameter, obtaining

$$I_2 = \int_{-1}^{1} (x^2 - x^4) 2x\, dx = \left(\frac{2x^4}{4} - \frac{2x^6}{6} \right)\Bigg|_{-1}^{1} = 0,$$

with the help of (5).

9.123. Line integrals along a closed contour. By a *closed path* we mean a path $\widehat{AB}$ whose end points A and B coincide. A closed path is often called a (*closed*) *contour*, and the line integral along a contour L is often denoted by

$$\oint_L f \, dg. \tag{10}$$

The quantity (10) does not depend on the choice of the initial point A, since if A' is any other point, then, by (8),

$$\int_{\overparen{AA'A}} f \, dg = \int_{\overparen{AA'}} f \, dg + \int_{\overparen{A'A}} f \, dg = \int_{\overparen{A'A}} f \, dg + \int_{\overparen{AA'}} f \, dg = \int_{\overparen{A'AA'}} f \, dg.$$

However, (10) does depend on the direction in which the contour is traversed, since changing this direction changes the sign of all the differences $g(P_{j+1}) - g(P_j)$ and hence the sign of the integral (10) as well. Thus, using arrowheads to distinguish directions along the contour L, we have

$$\oint_L f \, dg = -\oint_L f \, dg.$$

Note that this notation is useful only in the plane ($n=2$).

9.124. Area in terms of line integrals. Consider the plane figure shown in Figure 62 bounded on top by a curve L_1 with equation $y = \varphi_1(x)$ and on the bottom by a curve L_2 with equation $y = \varphi_2(x)$. The area of this figure is just

$$S = \int_a^b \varphi_1(x) \, dx - \int_a^b \varphi_2(x) \, dx. \tag{11}$$

By the very definition of a line integral, we can write (11) as a difference of line integrals

$$S = \int_{L_1} y \, dx - \int_{L_2} y \, dx,$$

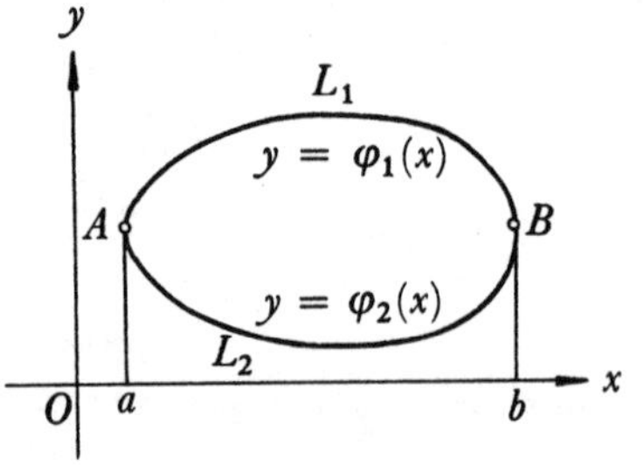

Figure   62

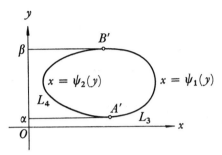

Figure 63

or equivalently, in the form

$$S= \int_{L_1} y \, dx + \int_{-L_2} y \, dx,$$

where $-L_2$ denotes the path L_2 traversed from B to A instead of from A to B, afterwards adding two more (vanishing) integrals along the vertical line segments (if any) forming the sides of the figure.† Combining all these integrals, we can express the area of the figure as the following line integral along the closed contour L forming the boundary of the figure:

$$S= \oint_L y \, dx = - \oint_L y \, dx. \tag{12}$$

To derive a more symmetric formula for S, we interchange the roles of x and y (see Figure 63). This gives

$$S= \int_\alpha^\beta \psi_1(y) \, dy - \int_\alpha^\beta \psi_2(y) \, dy = \int_{L_3} x \, dy - \int_{L_4} x \, dy$$

$$= \int_{L_3} x \, dy + \int_{-L_4} x \, dy = \oint_L x \, dy. \tag{12'}$$

Adding (12) and (12′) and multiplying by $\frac{1}{2}$, we get

$$S= \frac{1}{2} \oint_L x \, dy - y \, dx. \tag{13}$$

Formula (13) often leads to particularly simple calculations. For example, it follows from (13) that the area enclosed by the ellipse

$$x=a \cos t, \qquad y=b \sin t \qquad (0 \leqslant t \leqslant 2\pi)$$

† In Figure 62 these segments reduce to the points A and B.

is just

$$S = \frac{1}{2} \oint_L x \, dy - y \, dx = \frac{1}{2} \int_0^{2\pi} ab(\cos^2 t + \sin^2 t) \, dt = \pi ab,$$

in keeping with Example 9.62e. By contrast with (13), it is interesting to note that the formula

$$\int_{\widehat{AB}} x \, dy + y \, dx = \int_{\widehat{AB}} d(xy) = xy \Big|_A^B = x(B)y(B) - x(A)y(A)$$

implies

$$\oint_L x \, dy + y \, dx = 0$$

for any closed contour L.†

9.125. We now prove two theorems about line integrals of the special type

$$\int_L f_1(P) \, dx_1 + \cdots + f_n(P) \, dx_n. \tag{14}$$

a. THEOREM. *Let L be a path in n-dimensional space, with parametric equations* (1), *where the functions $x_j(t)$ are continuous with continuous derivatives $x_j'(t)$ satisfying the condition* (2), *and suppose the integral* (14) *exists. Then*

$$\left| \int_L f_1(P) \, dx_1 + \cdots + f_n(P) \, dx_n \right| \leqslant l \sup_{P \in L} \sqrt{\sum_{j=1}^n f_j^2(P)}, \tag{15}$$

where l is the length of L.

Proof. The expression

$$\sum_{k=0}^{p-1} \sum_{j=1}^n f_j(P_k)[x_j(t_{k+1}) - x_j(t_k)],$$

where $P_k = P(t_k)$, is a Riemann-Stieltjes sum for the integral in the left-hand side of (15) corresponding to a partition

$$\Pi = \{a = t_0 \leqslant t_1 \leqslant \cdots \leqslant t_p = b\}.$$

† Note that
$$\oint_L f(x) \, dx = \oint_L g(y) \, dy = 0$$
for arbitrary continuous functions $f(x)$, $g(y)$ and any closed contour L (why?).

Applying Cauchy's inequality (Sec. 3.14b), we get

$$\left| \sum_{k=0}^{p-1} \sum_{j=1}^{n} f_j(P_k) \Delta x_{jk} \right| \leqslant \sum_{k=0}^{p-1} \sqrt{\sum_{j=1}^{n} f_j^2(P_k)} \sqrt{\sum_{j=1}^{n} [x_j(t_{k+1}) - x_j(t_k)]^2}$$

$$\leqslant \sup_{P \in L} \sqrt{\sum_{j=1}^{n} f_j^2(P)}\, s(L_\Pi), \tag{16}$$

where $s(L_\Pi)$ is the length of the polygonal line $L_\Pi = P_0 P_1 \ldots P_p$ inscribed in L. But $s(L_\Pi) \to l$ as $d(\Pi) \to 0$, where l is the length of L. Hence, passing to the limit as $d(\Pi) \to 0$ in (16), we get (15). ∎

b. THEOREM. *Let L be the same as in the preceding theorem, and suppose the functions $f_1(P), \ldots, f_n(P)$ are continuous on L itself and on some neighborhood of L containing all polygonal lines L_Π with sufficiently small segments inscribed in L. Then*

$$\lim_{d(\Pi) \to 0} \int_{L_\Pi} f_1(P)\, dx_1 + \cdots + f_n(P)\, dx_n$$

$$= \int_{L} f_1(P)\, dx_1 + \cdots + f_n(P)\, dx_n, \tag{17}$$

so that the line integral along L can be deduced from a knowledge of the line integrals along all polygonal lines L_Π (with sufficiently small segments) inscribed in L.

Proof. It is enough to prove (17) for one term, i.e., to show that

$$\lim_{d(\Pi) \to 0} \int_{L_\Pi} f_j(P)\, dx_j = \int_{L} f_j(P)\, dx_j,$$

or simply

$$\lim_{d(\Pi) \to 0} \int_{L_\Pi} f(P)\, dx = \int_{L} f(P)\, dx, \tag{18}$$

where for brevity we drop the index j. Given any $\varepsilon > 0$, choose $\delta > 0$ such that $d(\Pi) < \delta$ implies

$$\left| \int_{L} f(P)\, dx - \sum_{k=0}^{p-1} f(P_k) \Delta x_k \right| < \varepsilon, \tag{19}$$

while $\rho(P, P') < \delta$ implies $|f(P) - f(P')| < \varepsilon$ (justify the latter implication in detail). † Let $P_k P_{k+1}$ be any segment of L_Π. Then, by (15), $d(\Pi) < \delta$

† By $\rho(P, P')$ we mean the length of the segment PP' with the usual Euclidean metric (Sec. 3.14a).

implies

$$\left|\int_{P_kP_{k+1}} f(P)\ dx - f(P_k)\Delta x_k\right| = \left|\int_{P_kP_{k+1}} f(P)\ dx - \int_{P_kP_{k+1}} f(P_k)\ dx\right|$$

$$= \left|\int_{P_kP_{k+1}} [f(P) - f(P_k)]\ dx\right| \leqslant \varepsilon l_k,\qquad (20)$$

where l_k is the length of the segment P_kP_{k+1}. Summing over all the segments of L_Π, we get

$$\left|\int_{L_\Pi} f(P)\ dx - \sum_{k=0}^{p-1} f(P_k)\Delta x_k\right| \leqslant \varepsilon s(L_\Pi).$$

But $s(L_\Pi) \leqslant l$, where l is the length of L (see Sec. 9.73). Hence (19) and (20) together imply first (18) and then (17).   ∎

Both of the above theorems can be extended to the case of a piecewise smooth path L by the usual device of dividing L into a finite number of smooth arcs.

Problems

1. Prove that if $f(x)$ is periodic with period T on $(-\infty,\infty)$, then

$$F(x) = \int_a^x f(\xi)\ d\xi$$

is the sum of a periodic function and a linear function.

2. Prove that if $f(x)$ and $\varphi(x) \geqslant 0$ are integrable on $[a,b]$, then there exists a point $\xi \in [a,b]$ such that

$$\int_a^b f(x)\varphi(x)\ dx = \varphi(a+0) \int_a^\xi f(x)\ dx$$

if $\varphi(x)$ is nonincreasing on $[a,b]$ and a point $\eta \in [a,b]$ such that

$$\int_a^b f(x)\varphi(x)\ dx = \varphi(b-0) \int_\eta^b f(x)\ dx$$

if $\varphi(x)$ is nondecreasing on $[a,b]$.

3. Prove that if $f(x)$ and $\varphi(x)$ are integrable on $[a,b]$ and if $\varphi(x)$ is monotonic on $[a,b]$, then there exists a point $\xi \in [a,b]$ such that

$$\int_a^b f(x)\varphi(x)\ dx = \varphi(a+0) \int_a^\xi f(x)\ dx + \varphi(b-0) \int_\xi^b f(x)\ dx.$$

Comment. This result is often called the *second mean value theorem for integrals,* as opposed to the (first) mean value theorem for integrals (Theorem 9.15f).

4. Prove the following formula for repeated integration by parts (valid for "sufficiently smooth" u and v):

$$\int_a^b uv^{(n+1)}\,dx = [uv^{(n)} - u'v^{(n-1)} + \cdots + (-1)^n u^{(n)}v]\Big|_a^b + (-1)^{n+1}\int_a^b u^{(n+1)}v\,dx.$$

5 (*Riemann's criterion*). Let $\Omega_f(\alpha,\beta)$ be the *oscillation of the function* $f(x)$ *on the interval* $[\alpha,\beta]$, i.e., the quantity

$$\Omega_f(\alpha,\beta) = \sup_{x'x''\in[\alpha,\beta]} |f(x') - f(x'')|.$$

Prove that a (bounded) function $f(x)$ is integrable on $[a,b]$ if and only if

$$\lim_{d(\Pi)\to 0} \sum_{k=0}^{n-1} \Omega_f(x_k, x_{k+1})\Delta x_k = 0,$$

where Π, $d(\Pi)$, and Δx_k have the usual meaning (Sec. 9.11).

6. Prove that every bounded monotonic function is integrable.

7. Prove that if $f(x)$ is integrable, then so is $|f(x)|$.

8 (*Du Bois-Reymond's criterion*). Let $\Omega_f(c)$ be the *oscillation of the function* $f(x)$ *at the point* c, i.e., the quantity

$$\Omega_f(c) = \inf_{\alpha < c < \beta} \Omega_f(\alpha,\beta),$$

where $\Omega_f(\alpha,\beta)$ is the same as in Problem 5. Prove that a (bounded) function is integrable on $[a,b]$ if and only if given any $\varepsilon > 0$ and $\delta > 0$, the set of all points $c \in [a,b]$ at which $\Omega_f(c) \geqslant \varepsilon$ can be covered (Sec. 3.93c) by a finite number of intervals the sum of whose lengths does not exceed δ.

9 (*Lebesgue's criterion*). Prove that a (bounded) function $f(x)$ is integrable on $[a,b]$ if and only if, given any $\delta > 0$, the set of all discontinuity points of $f(x)$ can be covered by a finite or countable number of intervals the sum of whose lengths does not exceed δ.

10. Prove that if $f(x)$ is integrable, then so is the function $1/f(x)$ provided it is bounded.

11. By integrating the inequality

$$\sin^{2n+1}x \leqslant \sin^{2n}x \leqslant \sin^{2n-1}x$$

from 0 to $\pi/2$ and then using formulas (1) and (2), p. 309, prove *Wallis' product*

$$\frac{\pi}{2} = \lim_{n\to\infty} \frac{2\cdot 2\cdot 4\cdot 4\cdots 2n\cdot 2n}{1\cdot 3\cdot 3\cdot 5\cdots(2n-1)(2n+1)} = \prod_{n=1}^{\infty} \frac{1}{1-(1/4n^2)}.$$

12. Let

$$\Pi = \{a = x_0, x_1, \ldots, x_n = b\}$$

be any set of points of an interval $[A,B] \supset [a,b]$, arranged in any order, and let

$$d(\Pi) = \max \{|x_1 - x_0|, |x_2 - x_1|, \ldots, |x_n - x_{n-1}|\}.$$

Consider the "generalized Riemann sum"

$$S_\Pi(f) = \sum_{k=0}^{n-1} f(\xi_k)(x_{k+1} - x_k),$$

where $f(x)$ is defined on $[A,B]$ and ξ_k is any point between x_k and x_{k+1} (including the points x_k, x_{k+1} themselves). Prove that

$$\lim_{d(\Pi) \to 0} S_\Pi(f) = \int_a^b f(x)\, dx$$

for every function $f(x)$ continuous on $[A,B]$, provided that

$$\sum_{k=0}^{n-1} |x_{k+1} - x_k| < C \tag{1}$$

for all $n = 1, 2, \ldots$

13. Prove that the conclusion of the preceding problem breaks down if the condition (1) is dropped or if $f(x)$ is allowed to be piecewise continuous.

14 (*Axiomatic definition of the integral*). Suppose that with every piecewise continuous function $f(x)$ on the interval $[a,b]$ and with every subinterval $[\alpha,\beta] \subset [a,b]$ there is associated a number $I_\alpha^\beta(f)$ satisfying the following conditions:

(a) $I_\alpha^\beta(kf) = kI_\alpha^\beta(f)$ for all real k;
(b) $I_\alpha^\beta(f) + I_\alpha^\beta(g) = I_\alpha^\beta(f+g)$ for all piecewise continuous $f(x)$ and $g(x)$;
(c) $I_\alpha^\beta(1) = \beta - \alpha$;
(d) $I_\alpha^\beta(f) = I_\alpha^\gamma(f) + I_\gamma^\beta(f)$ for all $\gamma \in (\alpha,\beta)$;
(e) $|I_\alpha^\beta(f)| \leqslant C \sup_{a \leq x \leq b} |f(x)|$ for some fixed constant C.

Prove that

$$I_\alpha^\beta(f) = \int_\alpha^\beta f(x)\, dx \tag{2}$$

for all piecewise continuous $f(x)$.

15. The series (7), p. 353 converges "too slowly" to be useful in calculating the number π. A more practical way of calculating π is to use the "rapidly converging" series

$$\frac{\pi}{6} = \frac{1}{\sqrt{3}}\left(1 - \frac{1}{3 \cdot 3} + \frac{1}{3^2 \cdot 5} - \frac{1}{3^3 \cdot 7} + \cdots\right). \tag{3}$$

Prove (3).

10 Analytic Functions

10.1 Basic Concepts

10.11. Differentiation in the complex domain

a. Let C_z be the "plane of the complex variable $z = x + iy$," i.e., the set of all complex numbers of the form $z = x + iy$ where x and y are arbitrary real numbers, and let C_w be the "plane of the complex variable $w = u + iv$," i.e., the set of all complex numbers of the form $w = u + iv$ where u and v are arbitrary real numbers.† Then a function $w = f(z)$ defined on a set $E \subset C_z$ and taking values in C_w is called a (complex) *function of a complex variable*. Let $z_0 \in E$ be a nonisolated point of E (Sec 3.75c), so that every neighborhood of z_0 contains a point of E other than z_0 itself. Then a complex number A is said to be the *derivative of the function* $w = f(z)$ *at the point* $z = z_0$ *relative to the set* E, denoted by $f_E'(z_0)$, if, given any $\varepsilon > 0$, there exists a $\delta > 0$ such that $0 < |z - z_0| < \delta$, $z \in E$ implies

$$\left| A - \frac{f(z) - f(z_0)}{z - z_0} \right| < \varepsilon.$$

In this case, we say that $w = f(z)$ is *differentiable at* $z = z_0$, with derivative $A = f_E'(z_0)$, where A is clearly the limit of the "difference quotient"

$$\frac{f(z) - f(z_0)}{z - z_0} \tag{1}$$

in the direction $z \underset{E}{\to} z_0$ determined by the set of intersections of E with all "deleted neighborhoods" $0 < |z - z_0| < \delta$ (cf. Sec. 4.15a).

b. The above definition of the derivative "in the complex domain" closely resembles that of the derivative "in the real domain," to which it reduces if E is an open interval on the real line containing the point $z_0 = x_0$ (see Sec. 7.11). Nevertheless, there are important differences between the implications of the two definitions, as will subsequently become apparent.

c. Let F be a subset of E, and let z_0 be a nonisolated point of F (and hence automatically a nonisolated point of E). Then the existence of $f_E'(z_0)$ obviously implies that of $f_F'(z_0)$ and

$$f_F'(z_0) = f_E'(z_0),$$

i.e., "differentiability relative to a larger set implies differentiability relative to a smaller set." The converse is false, as shown, say, by Example 10.12b.

d. If the function $f(z)$ is differentiable at the point $z = z_0$ relative to the set

† More concisely, C_z can be called the "z-plane" and C_w the "w-plane."

E, then, as $z \underset{E}{\to} z_0$, the quotient (1) is bounded by some constant C. Therefore

$$|f(z) - f(z_0)| \leqslant C|z - z_0|$$

for all sufficiently small $|z - z_0|$ ($z \in E$), and hence

$$\lim_{\substack{z \to z_0 \\ E}} f(z) = f(z_0).$$

Thus $f(z)$ is "continuous at $z = z_0$ relative to the set E" (cf. Theorem 7.12).

e. A set G in the plane C_z is said to be *open* if every point $z_0 \in G$ is an *interior point* of G, i.e., if whenever G contains z_0, G also contains some disk $|z - z_0| < r$ centered at z_0 (the radius of the disk depends in general on z_0). An open set $G \subset C_z$ is called a *domain* if it is *(arcwise) connected*, i.e., if every pair of points $z_0, z_1 \in G$ can be joined by a piecewise smooth curve entirely contained in G.†

f. A function $f(z)$ is said to be *analytic* (synonymously, *holomorphic* or *regular*) *on an open set* $G \subset C_z$ if $f(z)$ is differentiable on G.‡ We then denote the derivative of $f(z)$ simply by $f'(z)$, without explicit reference to the underlying set G. Obviously, if $f(z)$ is analytic on G and if $z_0 \in G$, then $f(z)$ is differentiable at $z = z_0$ relative to every set E with z_0 as a nonisolated point and

$$f'_E(z_0) = f'(z_0).$$

A function $f(z)$ is said to be *analytic at a point* $z = z_0$ (or *on a set E*) if $f(z)$ is analytic on some open set G containing z_0 (or E).

10.12. Examples

a. The function $f(z) = z$ is analytic on the whole plane C_z, since

$$\frac{f(z) - f(z_0)}{z - z_0} = \frac{z - z_0}{z - z_0} = 1,$$

so that $f'(z_0) = 1$ for every z.

b. The function $f(z) = \bar{z} = x - iy$ is differentiable at every point $z = z_0$ relative to any ray E drawn from z_0, since

$$\frac{f(z) - f(z_0)}{z - z_0} = \frac{\overline{z - z_0}}{z - z_0} = -2 \arg (z - z_0),$$

---

† Here we use the term "domain" in a precise technical sense, as opposed to its loose meaning in the phrases "in the real domain" and "in the complex domain."

‡ As in the real case, a function $f(z)$ is said to be differentiable *on a set E* if it is differentiable at every point of E.

and hence

$$f_E'(z_0) = -2 \arg (z - z_0) \qquad (z \in E, \ z \neq z_0). \tag{2}$$

However, (2) shows that $f(z) = \bar{z}$ fails to be differentiable relative to any set containing two distinct rays drawn from z_0 and hence relative to any domain containing z_0. Therefore $f(z) = \bar{z}$ fails to be analytic at every point $z = z_0 \in C_z$.

10.13. Basic rules for calculating derivatives

a. THEOREM. *Suppose $f(z)$ and $g(z)$ are differentiable at the point $z = z_0$ relative to a set E, and let k be an arbitrary complex number. Then the functions $f(z) + g(z)$, $kf(z)$, $f(z)g(z)$ and $f(z)/g(z)$ are also differentiable at $z = z_0$ relative to E (provided $g(z_0) \neq 0$ in the last case), with derivatives*

$$[kf(z)]_E' = kf_E'(z), \tag{3}$$

$$[f(z) + g(z)]_E' = f_E'(z) + g_E'(z), \tag{4}$$

$$[f(z)g(z)]_E' = f_E'(z)g(z) + f(z)g_E'(z), \tag{5}$$

$$\left[\frac{f(z)}{g(z)}\right]_E' = \frac{g(z)f_E'(z) - g_E'(z)f(z)}{g_E^2(z)}, \tag{6}$$

all evaluated at $z = z_0$.

Proof. The complex analogue of the proof of Theorem 7.13. ∎

b. Example. Given any polynomial

$$P(z) = a_0 + a_1 z + \cdots + a_n z^n$$

and any rational function

$$R(z) = \frac{P(z)}{Q(z)} = \frac{a_0 + a_1 z + \cdots + a_n z^n}{b_0 + b_1 z + \cdots + b_m z^m},$$

it follows from the above theorem that $P(z)$ is analytic everywhere, with derivative

$$P'(z) = a_1 + 2a_2 z + \cdots + na_n z^{n-1},$$

while $R(z)$ is analytic at every point $z = z_0$ such that $Q(z_0) \neq 0$.

c. THEOREM (**Differentiation of a composite function**). *Suppose $w = f(z)$ is differentiable at $z = z_0$ relative to a set E, while $\zeta = g(w)$ is defined on a set F containing all the points $w = f(z)$ with $z \in E$ sufficiently close to z_0 and is differentiable* *at $w = w_0 = f(z_0)$ relative to F. Then the composite function*

$$\zeta = h(z) \equiv g(f(z))$$

is differentiable at $z = z_0$ relative to E, with derivative

$$h'_E(z_0) = g'_F(w_0)f'_E(z_0). \tag{7}$$

Proof. The complex analogue of the proof of Theorem 7.15a. ∎

10.14. Higher derivatives. As in the real case, we define the higher derivatives

$$f_E^{(n+1)}(z) = [f_E^{(n)}(z)]'_E \qquad (n = 1, 2, \ldots)$$

of a complex function $f(z)$ inductively, assuming that the derivatives in question exist.

THEOREM. *If $f(z)$ and $g(z)$ have derivatives of order n relative to a set E, then*

$$[kf(z)]_E^{(n)} = kf_E^{(n)}(z),$$
$$[f(z) + g(z)]_E^{(n)} = f_E^{(n)}(z) + g_E^{(n)}(z)$$

for arbitrary complex k.

Proof. The complex analogues of the proofs of Theorems 8.12a and 8.12b. ∎

10.15. Differentials

a. Just as in the real case (Sec. 7.31), the increment of a complex function $w = f(z)$ differentiable at a point $z = z_0$ relative to a set E can be written in the form

$$\Delta w = f(z_0 + h) - f(z_0) = f'_E(z_0)h + \varepsilon(h)h \qquad (h = z - z_0),$$

where $\varepsilon(h) \to 0$ as $h \to 0$. Suppose we denote the quantity $f'_E(z_0)h$, called the *principal linear part* of the increment Δw, by $dw(z_0)$, or more briefly by dw or df, while writing the increment h of the independent variable z as $d_E z$ or simply dz. Then the quantities dw, df, dz, etc. are called *differentials*, and the derivative $f'_E(z_0)$ can be written as a ratio of differentials:

$$f'_E(z_0) = \frac{dw}{dz} = \frac{df}{dz}$$

(cf. Sec. 7.32).

b. The rules for calculating derivatives summarized in Theorem 10.13 lead to corresponding rules for calculating differentials. In fact, multiplying both sides of formulas (3)–(6) by dz and dropping the subscript E, we get

$$d(kf) = k \, df,\tag{3'}$$

$$d(f+g) = df + dg,\tag{4'}$$

$$d(fg) = f \, dg + g \, df,\tag{5'}$$

$$d\left(\frac{f}{g}\right) = \frac{g \, df - f \, dg}{g^2(z_0)} \qquad (g(z_0) \neq 0).\tag{6'}$$

c. Multiplying both sides of formula (7) by dz, we get

$$d\zeta = h_E'(z_0) \, dz = g_F'(w_0) f_E'(z_0) \, dz = g_F'(w_0) \, dw.$$

Thus *the differential of a function does not depend on whether its argument is the independent variable or a function of some other independent variable* (cf. Sec. 7.34).

10.16. The Cauchy-Riemann equations

a. First we introduce the concept of a "partial derivative," anticipated in the footnote on p. 357. Let $F(x, y)$ be a real function of two real variables x and y, defined on some open set G in the xy-plane, and let (x_0, y_0) be a point of G. Then the limit

$$\lim_{h \to 0} \frac{F(x_0 + h, y_0) - F(x_0, y_0)}{h}$$

(if it exists) is called the *partial derivative of $F(x, y)$ with respect to x* at the point (x_0, y_0), denoted by

$$\frac{\partial F(x_0, y_0)}{\partial x}$$

or $F_x(x_0, y_0)$, while the limit

$$\lim_{h \to 0} \frac{F(x_0, y_0 + h) - F(x_0, y_0)}{h}$$

(if it exists) is called the *partial derivative of $F(x, y)$ with respect to y* at the point (x_0, y_0), denoted by

$$\frac{\partial F(x_0, y_0)}{\partial y}$$

or $F_y(x_0, y_0)$.

b. THEOREM. *Suppose the function†*

$$w = f(z) = u(z) + iv(z) = u(x, y) + iv(x, y)$$

† To be perfectly explicit, $u(x, y)$ denotes the real function of two real variables x and y such that $u(x, y) \equiv u(x + iy)$ for all $x + iy \in E$, and similarly for $v(x, y)$.

is differentiable at the point $z = z_0 = x_0 + iy_0$ relative to a set E containing a pair of line segments through z_0 parallel to the real and imaginary axes. Then the partial derivatives of $u(x, y)$ and $v(x, y)$ at the point (x_0, y_0) satisfy the pair of equations

$$\frac{\partial u(x_0, y_0)}{\partial x} = \frac{\partial v(x_0, y_0)}{\partial y},$$

$$\frac{\partial v(x_0, y_0)}{\partial x} = -\frac{\partial u(x_0, y_0)}{\partial y}, \tag{8}$$

*known as the **Cauchy-Riemann equations**.*

Proof. First let $z = z_0 + h \in E$, where h is real. Then

$$f_E'(z_0) = \lim_{h \to 0} \frac{f(z) - f(z_0)}{z - z_0} = \lim_{h \to 0} \frac{u(z_0 + h) - u(z_0)}{h} + i \lim_{h \to 0} \frac{v(z_0 + h) - v(z_0)}{h}$$

$$= \lim_{h \to 0} \frac{u(x_0 + h, y_0) - u(x_0, y_0)}{h} + i \lim_{h \to 0} \frac{v(x_0 + h, y_0) - v(x_0, y_0)}{h}$$

$$= \frac{\partial u(x_0, y_0)}{\partial x} + i \frac{\partial v(x_0, y_0)}{\partial x}, \tag{9}$$

since $f_E'(z_0)$ exists. Next let $z = z_0 + ih \in E$, where h is again real. Then

$$f_E'(z_0) = \lim_{h \to 0} \frac{f(z) - f(z_0)}{z - z_0} = \lim_{h \to 0} \frac{u(z_0 + ih) - u(z_0)}{ih} + i \lim_{h \to 0} \frac{v(z_0 + ih) - v(z_0)}{ih}$$

$$= \lim_{h \to 0} \frac{u(x_0, y_0 + ih) - u(x_0, y_0)}{ih} + \lim_{h \to 0} \frac{v(x_0, y_0 + ih) - v(x_0, y_0)}{h}$$

$$= -i \frac{\partial u(x_0, y_0)}{\partial y} + \frac{\partial v(x_0, y_0)}{\partial y}. \tag{10}$$

Equating real and imaginary parts of (9) and (10), we get (8). ∎

The nub of the proof is that, in keeping with Sec. 10.11c, we must get the same value of $f_E'(z_0)$ whether z approaches z_0 along a line parallel to the real axis, as in (9), or along a line parallel to the imaginary axis, as in (10).

10.17. In particular, if $f(z)$ is analytic at a point $z = z_0$, then the Cauchy-Riemann equations

$$\frac{\partial u}{\partial x} = \frac{\partial v}{\partial y}, \qquad \frac{\partial v}{\partial x} = -\frac{\partial u}{\partial y}$$

hold at every point of some neighborhood of z_0 (cf. Sec. 10.11f). The converse is also true, provided we impose an extra condition on the partial derivatives of u and v:

THEOREM. *Suppose the Cauchy-Riemann equations hold in some neighborhood U of a point z_0, and suppose the partial derivatives $\partial u/\partial x$, $\partial u/\partial y$, $\partial v/\partial x$, $\partial v/\partial y$ are continuous on U. Then $f(z)$ is analytic at the point $z = z_0$.*

Proof. Let $h = p + iq$ be such that $z_0 + h \in U$, and consider the increment

$$
\begin{aligned}
f(z_0 + h) - f(z_0) &= [u(x_0 + p, y_0 + q) + iv(x_0 + p, y_0 + q)] \\
&\quad - [u(x_0, y_0,) + iv(x_0, y_0)] \\
&= [u(x_0 + p, y_0 + q) - u(x_0, y_0)] \\
&\quad + i[v(x_0 + p, y_0 + q) - v(x_0, y_0)].
\end{aligned}
\tag{11}
$$

We can write the first term in the right-hand side of (11) as

$$
\begin{aligned}
u(x_0 + p, y_0 + q) - u(x_0, y_0) &= [u(x_0 + p, y_0 + q) - u(x_0, y_0 + q)] \\
&\quad + [u(x_0, y_0 + q) - u(x_0, y_0)],
\end{aligned}
\tag{12}
$$

and similarly for the second term. Applying Lagrange's theorem (Sec. 7.44) to the first term in the right-hand side of (12) regarded as a function of x with $y = y_0 + q$ fixed, we get

$$
u(x_0 + p, y_0 + q) - u(x_0, y_0 + q) = u_x(x_0 + \theta_1 p, y_0 + q)p \qquad (0 < \theta_1 < 1), \tag{13}
$$

where u_x denotes the partial derivative of u with respect to x. Similarly, applying Lagrange's theorem to the second term in the right-hand side of (12) regarded as a function of y with $x = x_0$ fixed, we get

$$
u(x_0, y_0 + q) - u(x_0, y_0) = u_y(x_0, y_0 + \theta_2 q)q \qquad (0 < \theta_2 < 1), \tag{14}
$$

where u_y denotes the partial derivative of u with respect to y. Substituting (13) and (14) into (12) and treating the second term in the right-hand side of (11) in the same way, we find that

$$
\begin{aligned}
f(z_0 + h) - f(z_0) &= [u_x(x_0 + \theta_1 p, y_0 + q)p + u_y(x_0, y_0 + \theta_2 q)q] \\
&\quad + i[v_x(x_0 + \theta_3 p, y_0 + q)p + v_y(x_0, y_0 + \theta_4 q)q],
\end{aligned}
\tag{15}
$$

where the numbers $\theta_1, \theta_2, \theta_3, \theta_4$ all lie between 0 and 1.

We now use the Cauchy-Riemann equations to write (15) in the form

$$
\begin{aligned}
f(z_0 + h) - f(z_0) &= [u_x(x_0 + \theta_1 p, y_0 + q)p + u_y(x_0, y_0 + \theta_2 q)q] \\
&\quad + i[u_x(x_0, y_0 + \theta_4 q)q - u_y(x_0 + \theta_3 p, y_0 + q)p].
\end{aligned}
$$

By the assumed continuity of u_x and u_y, this can be transformed further into

$$f(z_0+h) - f(z_0) = u_x(x_0,y_0)(p+iq) - iu_y(x_0,y_0)(p+iq)$$
$$+ \varepsilon_1(p,q)p + \varepsilon_2(p,q)q,$$

where $\varepsilon_1(p,q)$ and $\varepsilon_2(p,q)$ both approach 0 as $p \to 0$, $q \to 0$ (i.e., as $\sqrt{p^2+q^2} \to 0$). It follows that

$$\frac{f(z_0+h) - f(z_0)}{h} = u_x(x_0,y_0) - iu_y(x_0,y_0) + \frac{\varepsilon_1(p,q)p + \varepsilon_2(p,q)q}{p+iq}.$$

But

$$\left|\frac{p}{p+iq}\right| \leqslant 1, \qquad \left|\frac{q}{p+iq}\right| \leqslant 1,$$

and hence

$$\left|\frac{\varepsilon_1(p,q)p + \varepsilon_2(p,q)q}{p+iq}\right| \leqslant |\varepsilon_1(p,q)| + |\varepsilon_2(p,q)|,$$

where the expression on the right approaches 0 as $p \to 0$, $q \to 0$. Therefore

$$\lim_{h \to 0} \frac{f(z_0+h) - f(z_0)}{h} = u_x(z_0) - iu_y(z_0) = u_x(z_0) + iv_x(z_0), \tag{16}$$

i.e., $f(z)$ is differentiable at $z = z_0$, with derivative (16). The same argument shows that $f(z)$ has a derivative at every point sufficiently near $z = z_0$. Hence $f(z)$ is analytic at $z = z_0$. ∎

By introducing the concept of a differentiable function of two variables, we can establish necessary and sufficient conditions (involving the Cauchy-Riemann equations) for a complex function to be differentiable at a point (see Problem 6).

10.18. Harmonic functions. It will be shown below (see Theorem 10.34) that every analytic function $f(z) = u + iv$ has a second derivative (and in fact derivatives of all orders). Arguing as in the proof of the Cauchy-Riemann equations themselves (Theorem 10.16b), we find that

$$f''(z) = [f'(z)]' = \frac{\partial}{\partial x}\left(\frac{\partial u}{\partial x} + i\frac{\partial v}{\partial x}\right) = \frac{\partial^2 u}{\partial x^2} + i\frac{\partial^2 v}{\partial x^2},$$

while, on the other hand,

$$f''(z) = \frac{1}{i}\frac{\partial}{\partial y}\left(\frac{1}{i}\frac{\partial u}{\partial y} + \frac{\partial v}{\partial y}\right) = -\frac{\partial^2 u}{\partial y^2} - i\frac{\partial^2 v}{\partial y^2},$$

where $\partial^2 u/\partial x^2$ denotes the second partial derivative of the function $u(x,y)$ with respect to x (i.e., the second derivative of $u(x,y)$ regarded as a function of x with y held fixed), $\partial^2 u/\partial y^2$ denotes the second partial derivative of $u(x,y)$

with respect to y, and similarly for $\partial^2 v/\partial x^2$ and $\partial^2 v/\partial y^2$. Equating real and imaginary parts of (16) and (17), we get

$$\frac{\partial^2 u}{\partial x^2} + \frac{\partial^2 u}{\partial y^2} = 0, \qquad \frac{\partial^2 v}{\partial x^2} + \frac{\partial^2 v}{\partial y^2} = 0.$$

In general, any continuous real function $\varphi = \varphi(x,y)$ satisfying the equation

$$\frac{\partial^2 \varphi}{\partial x^2} + \frac{\partial^2 \varphi}{\partial y^2} = 0$$

(known as *Laplace's equation*) at every point of a domain G is said to be *harmonic on G*. Thus we see that the real and imaginary parts of a function analytic on a domain G are harmonic on G.

Conversely, it can be shown† that every function $u(x,y)$ harmonic on a domain G is the real part of a function $f(z)$ analytic on G, or the imaginary part of the analytic function $if(z)$. Two harmonic functions $u(x,y)$ and $v(x,y)$ are called *conjugate harmonic functions* if they are the real and imaginary parts of a single analytic function $f(z)$, i.e., if they are related by the Cauchy-Riemann equations.

10.2. Line Integrals of Complex Functions

10.21. Line integrals of the form

$$\int_L f(P) \, dg(P),\tag{1}$$

where L is a piecewise smooth path with no singular points, have already been encountered in Sec. 9.121 for the case of real $f(P)$ and $g(P)$.‡ We now generalize the definition of (1) to include the case of complex functions

$$f(P) = f_1(P) + if_2(P), \qquad g(P) = g_1(P) + ig_2(P),$$

by the simple expedient of setting

$$\int_L f(P) \, dg(P) = \int_L f_1(P) \, dg_1(P) - f_2(P) \, dg_2(P)$$
$$+ i \int_L f_1(P) \, dg_2(P) + f_2(P) \, dg_1(P),\tag{2}$$

† See, e.g., R. A. Silverman, *Introductory Complex Analysis,* Dover Publications, Inc., N.Y. (1972), Theorem 13.2, p. 275.

‡ Let $P = P(t)$, $a \leqslant t \leqslant b$ be a variable point of L. Then, as on p. 363, (1) reduces to the ordinary integral

$$\int_a^b f(t) g'(t) \, dt$$

if $f(t) \equiv f(P(t))$ is continuous and $g(t) \equiv g(P(t))$ has a continuous derivative.

which is in keeping, formally at least, with formulas (6) and (7), pp. 363–364. Note that (2) reduces to

$$\int_L f(P) \, dg(P) = \int_L f_1(P) \, dg(P) + i \int_L f_2(P) \, dG(P)$$

if $f(P)$ is complex and $g(P)$ real, while (2) reduces to

$$\int_L f(P) \, dg(P) = \int_L f(P) \, dg_1(P) + i \int_L f(P) \, dg_2(P)$$

if $f(P)$ is real and $g(P)$ complex.

Now let L be a plane curve, and let

$$f(z) = f(x,y) = u(x,y) + iv(x,y), \qquad g(z) = z = x + iy.$$

Then (2) becomes

$$\int_L f(z) \, dz = \int_L u \, dx - v \, dy + i \int_L u \, dy + v \, dx \tag{3}$$

(with $P = z$). We can also give a direct definition of the integral (3). To this end, let Π be a partition of L with points of subdivision $z_k = x_k + iy_k$. Then

$$\int_L f(z) \, dz = \lim_{d(\Pi) \to 0} \sum_{k=0}^{n-1} f(z_k) \Delta z_k, \tag{4}$$

where $\Delta z_k = z_{k+1} - z_k$ and

$$d(\Pi) = \max \{|\Delta z_0|, |\Delta z_1|, \ldots, |\Delta z_{n-1}|\}.$$

In fact, the right-hand side of (4) is just

$$\sum_{k=0}^{n-1} u(x_k, y_k) \Delta x_k - \sum_{k=0}^{n-1} v(x_k, y_k) \Delta y_k + i \sum_{k=0}^{n-1} u(x_k, y_k) \Delta y_k + i \sum_{k=0}^{n-1} v(x_k, y_k) \Delta x_k,$$

which clearly approaches the right-hand side of (3) as $d(\Pi) \to 0$ (cf. Theorem 9.73 and Sec. 9.121).

The following properties of "complex line integrals" of the form

$$\int_L f(z) \, dz$$

are all immediate consequences of the definition (4):

(a) If α_1, α_2 are arbitrary complex numbers and $f_1(z), f_2(z)$ arbitrary integrable functions, then

$$\int_L [\alpha_1 f_1(z) + \alpha_2 f_2(z)] \, dz = \alpha_1 \int_L f_1(z) \, dz + \alpha_2 \int_L f_2(z) \, dz.$$

(b) If L is made up of two arcs L_1 and L_2 "joined end to end," then

$$\int_L f(z)\ dz = \int_{L_1} f(z)\ dz + \int_{L_2} f(z)\ dz.$$

(c) If the path L is of length l, then

$$\left| \int_L f(z)\ dz \right| \leqslant \int_L |f(z)|\ dz \leqslant l \sup_{z\in L} |f(z)|. \tag{5}$$

10.22. a. THEOREM. *Let $f_n(z)$ be a sequence of functions, each continuous on a piecewise smooth path L, and suppose $f_n(z)$ converges uniformly on L to a limit function $f(z)$. Then*

$$\lim_{n\to\infty} \int_L f_n(z)\ dz = \int_L f(z)\ dz. \tag{6}$$

Proof. The integrability of $f(z)$ follows from its continuity (cf. Theorem 9.55b), which in turn follows from the fact that $f(z)$ is the "uniform limit" of a sequence of continuous functions (see Corollary 5.95b). To prove (6), we deduce from (5) that

$$\left| \int_L [f(z) - f_n(z)]\ dz \right| \leqslant l \sup_{z\in L} |f(z) - f_n(z)|, \tag{7}$$

where l is the length of L. But the right-hand side approaches 0 as $n\to\infty$, by the assumed uniform convergence of $f_n(z)$. ∎

b. THEOREM. *Let $f_n(z,\lambda)$ be a sequence of functions, each continuous on a piecewise smooth path L and dependent on a parameter λ varying over some set Λ, and suppose $f_n(z,\lambda)$ converges uniformly on the direct product $L\times\Lambda$ to a limit function $f(z,\lambda)$. Then the sequence of functions*

$$F_n(\lambda) = \int_L f_n(z,\lambda)\ dz$$

converges uniformly on Λ to the function

$$F(\lambda) = \int_L f(z,\lambda)\ dz.$$

Proof. This time we have

$$\left| \int_L [f(z,\lambda) - f_n(z,\lambda)]\ dz \right| \leqslant l \sup_{\substack{z\in L \\ \lambda\in\Lambda}} |f(z,\lambda) - f_n(z,\lambda)| \tag{7'}$$

instead of (7). But the right-hand side approaches 0 as $n\to\infty$, by the assumed uniform convergence of $f_n(z,\lambda)$. ∎

10.23. Next we prove a result closely related to the mean value theorem for ordinary integrals (Theorem 9.15f):

THEOREM. *If $f(z)$ is continuous on a piecewise smooth path*

$$L = \{z: z = z(t), \, a \leqslant t \leqslant b\}$$

and if $z'(t_0) \neq 0$, then

$$\lim_{z \to z_0} \frac{1}{z - z_0} \int_{L(z_0, z)} f(z) \, dz = f(z_0), \tag{8}$$

where $L(z_0, z)$ is the arc of L joining z_0 to z.

Proof. If $f(z) \equiv f(z_0) = $ const, then (8) is an immediate consequence of (4). Hence in the general case we can assume that $f(z_0) = 0$, since otherwise we need only replace $f(z)$ by $f(z) - f(z_0)$. Given any $\varepsilon > 0$, choose $\delta > 0$ such that $|z - z_0| < \delta$, $z \in L$ implies $|f(z)| < \varepsilon$. Then, by (5),

$$\left| \int_{L(z_0, z)} f(z) \, dz \right| \leqslant \varepsilon \, s(z_0, z),$$

where $s(z_0, z)$ is the length of the arc $L(z_0, z)$, and hence

$$\left| \frac{1}{z - z_0} \int_{L(z_0, z)} f(z) \, dz \right| \leqslant \varepsilon \frac{s(z_0, z)}{|z - z_0|}. \tag{9}$$

But

$$\lim_{z \to z_0} \frac{s(z_0, z)}{|z - z_0|} = 1,$$

by Theorem 9.73, and hence for sufficiently small $|z - z_0|$ we can replace (9) by the estimate

$$\left| \frac{1}{z - z_0} \int_{L(z_0, z)} f(z) \, dz \right| < 2\varepsilon,$$

which immediately implies (8). ∎

10.24. THEOREM. *Let $f(z, \zeta)$ be a function of two complex variables z and ζ, where $z = x + iy$ varies over a piecewise smooth path L in the z-plane and $\zeta = \xi + i\eta$ varies over a piecewise smooth path Λ in the ζ-plane. Suppose $f(z, \zeta)$ is continuous in both arguments z and ζ.† Then each of the integrals*

---

† In other words, suppose $f(z, \zeta)$ is continuous on $L \times \Lambda$ (see Sec. 5.18). It then follows from Theorem 5.17b that $f(z, \zeta)$ is *uniformly* continuous on $L \times \Lambda$.

$$I(\zeta) = \int_L f(z,\zeta)\, dz, \qquad J(z) = \int_\Lambda f(z,\zeta)\, d\zeta$$

is continuous, the first on Λ and the second on L, and

$$\int_L \left\{ \int_\Lambda f(z,\zeta)\, d\zeta \right\} dz = \int_\Lambda \left\{ \int_L f(z,\zeta)\, dz \right\} d\zeta.$$

Proof. An immediate consequence of Theorems 9.111 and 9.112, since complex line integrals can be expressed as "linear combinations" of ordinary real integrals. ∎

10.25. a. THEOREM. *If $f(z)$ is differentiable on a set E containing a piecewise smooth path L, with initial point z_0 and final point z_1, then*

$$\int_L f_L'(z)\, dz = f(z_1) - f(z_0).$$

Proof. Let

$$z = z(t) \qquad (t_0 \leqslant t \leqslant t_1)$$

be the parametric representation of L, where $z(t)$ is piecewise smooth. Then

$$\int_L f_L'(z)\, dz = \int_{t_0}^{t_1} f'(z(t)) z'(t)\, dt = \int_{t_0}^{t_1} [f(z(t))]'\, dt$$
$$= f(z(t_1)) - f(z(t_0)) = f(z_1) - f(z_0)$$

(justify the various steps). ∎

b. THEOREM. *If $f(z)$ is analytic on a domain G, then*

$$\int_L f'(z)\, dz = f(z_1) - f(z_0) \tag{10}$$

for every piecewise smooth path $L \subset G$ joining z_0 to z_1.

Proof. An immediate consequence of the preceding theorem. ∎

c. THEOREM. *If $f(z)$ is analytic on a domain G, then*

$$\oint_L f'(z)\, dz = 0$$

for every piecewise smooth closed path $L \subset G$.

Proof. Let $z_1 = z_0$ in (10). ∎

d. Example. Given any polynomial

$$P(z) = a_0 + a_1 z + \cdots + a_n z^n$$

and any closed path $L \subset C_z$, we have

$$\oint_L P(z)\, dz = 0,$$

since $P(z)$ is the derivative of the polynomial

$$P_1(z) = a_0 z + a_1 \frac{z^2}{2} + \cdots + a_n \frac{z^{n+1}}{n+1},$$

which is clearly analytic on the whole plane C_z.

10.26. Next we prove the converse of Theorem 10.25c:

THEOREM. *Let $\varphi(z)$ be continuous on a domain G, and suppose the integral*

$$\int_L \varphi(z)\, dz$$

of $\varphi(z)$ along every piecewise smooth closed path $L \subset G$ vanishes. Then there exists a function $f(z)$ analytic on G such that $f'(z) = \varphi(z)$.

Proof. Let $L_z \subset G$ be any piecewise smooth path joining a fixed point $z_0 \in G$ to a variable point $z \in G$,† and let

$$f(z) = \int_{L_z} \varphi(\zeta)\, d\zeta. \tag{11}$$

Then $f(z)$ is independent of the choice of L_z. In fact, if $\widehat{z_0 p z}$ and $\widehat{z_0 q z}$ are two paths joining z_0 to z, as shown in Figure 64, then

$$\int_{\widehat{z_0 p z}} \varphi(\zeta)\, d\zeta + \int_{\widehat{z q z_0}} \varphi(\zeta)\, d\zeta = \int_{\widehat{z_0 p z q z_0}} \varphi(\zeta)\, d\zeta = 0,$$

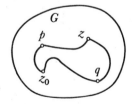

Figure 64

---

† The fact that z_0 can be joined to z by a path L entirely contained in G follows from the connectedness of G (see Sec. 10.11e).

by hypothesis, since the path $\widehat{z_0 p z q z_0}$ is closed, and hence

$$\int_{\widehat{z_0pz}} \varphi(\zeta)\, d\zeta = -\int_{\widehat{zqz_0}} \varphi(\zeta)\, d\zeta = \int_{\widehat{z_0qz}} \varphi(\zeta)\, d\zeta.$$

Thus it makes sense to write just

$$f(z) = \int_{z_0}^{z} \varphi(\zeta)\, d\zeta, \tag{11'}$$

instead of (11).

To prove that $f(z)$ is analytic on G, with derivative $f'(z) = \varphi(z)$, we note that

$$
\begin{aligned}
f(z+h) - f(z) &= \int_{z_0}^{z+h} \varphi(\zeta)\, d\zeta - \int_{z_0}^{z} \varphi(\zeta)\, d\zeta = \int_{z}^{z+h} \varphi(\zeta)\, d\zeta \\
&= \int_{z}^{z+h} \varphi(z)\, d\zeta + \int_{z}^{z+h} [\varphi(\zeta) - \varphi(z)]\, d\zeta \\
&= h\varphi(z) + \int_{h}^{z+h} [\varphi(\zeta) - \varphi(z)]\, d\zeta,
\end{aligned}
$$

where the last integral can be evaluated along any piecewise smooth path in G joining z to $z+h$. If $|h|$ is small enough, this path can be chosen to be the line segment joining z to $z+h$. But then, by (5),

$$
\begin{aligned}
\left| \frac{f(z+h) - f(z)}{h} - \varphi(z) \right| &= \left| \frac{1}{h} \int_{z}^{z+h} [\varphi(\zeta) - \varphi(z)]\, d\zeta \right| \\
&\leqslant \frac{1}{|h|} |h| \sup_{|\zeta - z| \leqslant |h|} |\varphi(\zeta) - \varphi(z)|,
\end{aligned}
$$

where the expression on the right approaches 0 as $h \to 0$, by the continuity of $\varphi(z)$. It follows that

$$f'(z) = \lim_{h \to 0} \frac{f(z+h) - f(z)}{h} = \varphi(z). \quad \blacksquare$$

10.27. THEOREM. *Let $f(z, \zeta)$ be a function of two complex variables z and ζ, where $z = x + iy$ varies over a domain G in the z-plane and $\zeta = \xi + i\eta$ varies over a piecewise smooth path Λ in the ζ-plane. Suppose $f(z, \zeta)$ is analytic on G for every fixed $\zeta \in \Lambda$, while both $f(z, \zeta)$ and its partial derivative $f_z(z, \zeta)$ with respect to z are continuous on $G \times \Lambda$. Then the function*

$$\Phi(z) = \int_{\Lambda} f(z, \zeta)\, d\zeta$$

is analytic on G, with derivative

$$\varphi(z) = \int_{\Lambda} f_z(z, \zeta)\, d\zeta.$$

Proof. It is easy to see that

$$\oint_L \varphi(z) \ dz = 0$$

for every piecewise smooth closed path $L \subset G$. In fact, by Theorems 10.24 and 10.25c,

$$\oint_L \varphi(z) \ dz = \oint_L \left\{ \int_\Lambda f_z(z,\zeta) \ d\zeta \right\} dz = \int_\Lambda \left\{ \oint_L f_z(z,\zeta) \ dz \right\} d\zeta = 0.$$

Hence, by Theorem 10.26, there exists a function $\Phi_0(z)$ analytic on G, with derivative $\Phi_0'(z) = \varphi(z)$, where

$$\Phi_0(z) = \int_{z_0}^z \varphi(\tau) \ d\tau = \int_{z_0}^z \left\{ \int_\Lambda f_z(\tau,\zeta) \ d\zeta \right\} d\tau = \int_\Lambda \left\{ \int_{z_0}^z f_z(\tau,\zeta) \ d\tau \right\} d\zeta$$

$$= \int_\Lambda f(z,\zeta) \ d\zeta - \int_\Lambda f(z_0,\zeta) \ d\zeta,$$

again by Theorem 10.24. Therefore $\Phi_0(z)$ differs from $\Phi(z)$ by only a constant, so that $\Phi(z)$ is itself analytic on G. ∎

10.28. By an *integral of the Cauchy type* we mean an expression of the form

$$F(z) = \int_\Lambda \frac{f(\zeta)}{\zeta - z} \ d\zeta \qquad (z \notin \Lambda), \tag{12}$$

where $f(\zeta)$ is analytic on a piecewise smooth path $\Lambda \subset C_z$. Clearly $F(z)$ is defined at every point of the open set $G = C_z - \Lambda$ (G may not be connected and hence may not be a domain).

THEOREM. *The function $F(z)$ defined by* (12) *is analytic on G, with derivatives of all orders on G (themselves analytic), given by*

$$F^{(n)}(z) = n! \int_\Lambda \frac{f(\zeta)}{(\zeta - z)^{n+1}} \ d\zeta \qquad (n = 1,2,\ldots).$$

Proof. For every fixed $\zeta \in \Lambda$, the function

$$f(z,\zeta) = \frac{f(\zeta)}{\zeta - z}$$

and its partial derivatives with respect to z

$$f_z^{(n)}(z,\zeta) = n! \frac{f(\zeta)}{(\zeta - z)^{n+1}} \qquad (n = 1,2,\ldots)$$

are analytic on G. Moreover, the function $f(z,\zeta)$ and all its derivatives

$f_z^{(n)}(z,\zeta)$ are continuous on $G \times \Lambda$. Now apply Theorem 10.27. ∎

10.3. Cauchy's Theorem and Its Consequences

We now present the central results of the theory of analytic functions, due mainly to Cauchy.

10.31. Let L be a curve with parametric equations

$$x = x(t), \qquad y = y(t) \qquad (a \leqslant t \leqslant b),$$

where $x(t)$ and $y(t)$ are continuous. If the same point $(x, y) \in L$ corresponds to more than one parameter value in the interval $a \leqslant t < b$, we say that (x, y) is a *multiple point* of L. A curve with no multiple points is said to be *simple*.† It is to allow for the possibility of simple *closed* curves, i.e., simple curves whose end points coincide, that we consider parameter values in the half-open interval $a \leqslant t < b$ rather than in the closed interval $a \leqslant t \leqslant b$. Every simple closed curve L partitions the plane into a bounded domain (called the *interior* of L) and an unbounded domain (called the *exterior* of L). This entirely plausible result, known as the *Jordan curve theorem*, is proved in elementary topology.

A domain G is said to be *simply connected* if whenever G contains a simple closed curve L, G also contains the interior of L. Thus an open disk or, more generally, any convex domain is simply connected,‡ but not the *annulus* (i.e., ring-shaped domain) $r_1 < |z| < r_2$ shown in Figure 65.

10.32. THEOREM. (**Cauchy's theorem**). *If $f(z)$ is analytic on a simply con-*

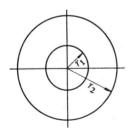

Figure 65

† Note that this is consistent with the definition of a simple closed polygonal line, given in the footnote on p. 287.
‡ A plane set E is said to be *convex* if whenever E contains two points z_1 and z_2, E also contains the line segment joining z_1 and z_2.

nected domain G, then

$$\oint_L f(z)\,dz = 0$$

for every piecewise smooth closed path $L \subset G$.

Proof. According to Theorem 9.125b, the integral along L is a limit of integrals along polygonal lines L_Π inscribed in L. Hence we need only prove the theorem for closed polygonal lines,† in fact for a closed polygonal line with no "self-intersections" (why?).

Thus let L be a simple closed polygonal line. Since G is simply connected, we can partition L and its interior into a finite number of triangles $T_1,\ldots,$ T_m, each contained in G. Then

$$\oint_L f(z)\,dz = \int_{T_1} f(z)\,dz + \cdots + \oint_{T_m} f(z)\,dz$$

(where for simplicity we use the symbols $T_1,\ldots,T_m$ both for the triangles and their boundaries), since the integrals over the sides of the triangles inside L cancel each other out, each such side being traversed twice in opposite directions (see Figure 66). Hence we need only prove Cauchy's theorem for a triangular contour.

To this end, let T be a triangle contained in G, and let

$$I = \oint_T f(z)\,dz$$

(as before, T denotes both the triangle and its boundary). Suppose we partition the triangle T into four equal triangles T_1, T_2, T_3, T_4 by joining

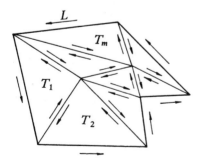

Figure   66

---

† There is a missing detail here, which the reader should prove as an exercise, namely that L_Π (like L itself) is contained in G if the segments of L_Π are sufficiently small.

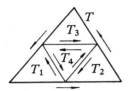

Figure 67

the midpoints of consecutive sides of T by line segments, as shown in Figure 67. Then

$$\oint_T f(z)\, dz = \oint_{T_1} f(z)\, dz + \oint_{T_2} f(z)\, dz + \oint_{T_3} f(z)\, dz + \oint_{T_4} f(z)\, dz,$$

and hence at least one of the smaller triangles, say T_i, must satisfy the inequality

$$\left|\oint_{T_i} f(z)\, dz\right| \geq \frac{|I|}{4}. \tag{1}$$

Dividing the triangle T_i in turn into four equal triangles $T_{i1},\dots,T_{i4}$, we then find a new triangle, say T_{ij}, such that

$$\left|\oint_{T_{ij}} f(z)\, dz\right| \geq \frac{1}{4}\left|\oint_{T_i} f(z)\, dz\right| \geq \frac{|I|}{4^2}.$$

Continuing this process indefinitely, we get a system of "nested triangles" $T_1^* = T_i, T_2^* = T_{ij},\dots$, each containing the next, such that

$$\left|\oint_{T_n^*} f(z)\, dz\right| \geq \frac{|I|}{4^n}.$$

Each triangle T_n^* is a compact set in the complex plane, and hence there is a point z_0 contained in all the triangles T_n^* (see Sec. 5.14e and Theorem 3.98). The function $f(z)$, being differentiable on G, is certainly differentiable at $z = z_0$. Therefore, given any $\varepsilon > 0$, there is a $\delta > 0$ such that $|z - z_0| < \delta$ implies

$$f(z) = f(z_0) + f'(z_0)(z - z_0) + \alpha(z, z_0), \tag{2}$$

where

$$|\alpha(z, z_0)| < \varepsilon |z - z_0|$$

(cf. Sec. 10.15a) and all the terms in (2) are obviously continuous (and hence integrable). Let T_n^* be such that T_n^* is contained in the disk

$|z - z_0| < \delta$ (this is the case for all sufficiently large n). Then

$$\oint_{T_n^*} f(z)\, dz = \oint_{T_n^*} f(z_0)\, dz + f'(z_0) \oint_{T_n^*} (z - z_0)\, dz + \oint_{T_n^*} \alpha(z, z_0)\, dz,$$

where the first two terms on the right vanish, by Example 10.25d. Since the perimeter of T_n^* equals $l/2^n$, where l is the perimeter of the original triangle T, it follows that

$$\left| \oint_{T_n^*} f(z)\, dz \right| = \left| \oint_{T_n^*} \alpha(z, z_0)\, dz \right| < \varepsilon \frac{l}{2^n} \frac{l}{2^n},$$

where we use the fact that the distance between any two points of a triangle must be less than the perimeter of the triangle. Comparing this inequality with (1), we find that

$$\frac{|I|}{4^n} < \varepsilon \frac{l^2}{4^n}$$

or

$$|I| < \varepsilon l^2.$$

But then $I = 0$, since ε can be made arbitrarily small. ∎

10.33. Cauchy's formula

a. The deleted neighborhood (or "punctured disk") $0 < |z - a| < R$ is an example of a *multiply connected* domain, i.e., of a domain which is not simply connected. The function

$$\frac{1}{(z - a)^n} \qquad (n = 1, 2, \ldots)$$

is analytic on this domain, but not on the whole disk $|z - a| < R$. To evaluate the integral

$$\oint_L \frac{dz}{(z - a)^n},$$

where L is a circle of radius $r < R$ centered at the point a (and traversed in the counterclockwise direction), we choose the polar angle θ as parameter, so that

$$z - a = re^{i\theta}, \qquad (z - a)^n = r^n e^{in\theta}, \qquad dz = ire^{i\theta}\, d\theta.$$

We then have

$$\oint_L \frac{dz}{(z - a)^n} = ir^{1-n} \int_0^{2\pi} e^{i(1-n)\theta}\, d\theta = \begin{cases} ir^{1-n} \dfrac{e^{i(1-n)\theta}}{i(1-n)} \Big|_0^{2\pi} = 0 & \text{if } n \neq 1, \\[2mm] i\theta \big|_0^{2\pi} = 2\pi i & \text{if } n = 1. \end{cases}$$

The fact that this integral equals $2\pi i \neq 0$ if $n = 1$ shows that the requirement that the domain G be simply connected *cannot be dropped* in Cauchy's theorem.

b. Next we establish an important general property of integrals along closed paths in the same domain. Let $f(z)$ be analytic on a domain G (which is in general multiply connected), and let L_1, L_2 be two piecewise smooth closed paths in G such that L_1 surrounds L_2.† Suppose L_1 and L_2, together with a piecewise smooth arc Γ joining them, form the boundary of a simply connected domain $G^* \subset G$, and let L be the boundary G^*, made up of the paths L_1, L_2 and the arc Γ traversed twice in opposite directions (see Figure 68). Let L be assigned the direction shown in the figure, corresponding to successively traversing L_1 in the counterclockwise direction, Γ from right to left, L_2 in the clockwise direction, and finally Γ from left to right (this arc is denoted by $-\Gamma$). Then, by Cauchy's theorem,

$$0 = \oint_L f(z)\, dz = \oint_{L_1} f(z)\, dz + \int_{\Gamma} f(z)\, dz + \oint_{L_2} f(z)\, dz + \int_{-\Gamma} f(z)\, dz$$

(concerning the meaning of the integrals with arrowheads, see Sec. 9.123). But reversing the direction of integration changes the sign of a line integral (cf. p. 363). Hence the integrals along Γ and $-\Gamma$ cancel each other out, and we are left with

$$\oint_{L_1} f(z)\, dz = -\oint_{L_2} f(z)\, dz = \oint_{L_2} f(z)\, dz. \tag{3}$$

c. THEOREM. *Let $f(z)$ be analytic on a simply connected domain G containing a point*

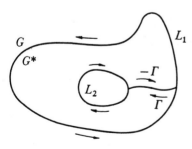

Figure 68

---

† A closed path L is said to "surround" a set E (or a point z_0) if L has no multiple points and if E (or z_0) belongs to the interior of L.

z_0 *and a piecewise smooth closed path L surrounding* z_0. *Then*

$$f(z_0) = \frac{1}{2\pi i} \oint_L \frac{f(z)}{z - z_0}\, dz, \qquad (4)$$

a result known as **Cauchy's formula.**

Proof. Let

$$I = \oint_L \frac{f(z)}{z - z_0}\, dz.$$

Then the function

$$\frac{f(z)}{z - z_0} \qquad (5)$$

in general fails to be differentiable at $z = z_0$, so that I is in general nonvanishing. But the function (5), like $f(z)$ itself, is analytic at every point of G except z_0. Hence, according to Sec. 10.33b, the integral I does not change if the contour L is replaced by a circle γ_ε of radius ε centered at the point z_0. Now let

$$f(z) = f(z_0) + \varphi(z).$$

Then the function

$$\frac{\varphi(z)}{z - z_0}, \qquad (6)$$

like $f(z)$ itself, is continuous (and hence integrable) for $z \neq z_0$, and approaches the limit $f'(z_0)$ as $z \to z_0$. Hence (6) is bounded in a neighborhood of z_0, so that

$$\left| \frac{\varphi(z)}{z - z_0} \right| \leqslant M,$$

say. It follows that

$$I = \oint_{\gamma_\varepsilon} \frac{f(z)}{z - z_0}\, dz$$

and hence

$$I = f(z_0) \oint_{\gamma_\varepsilon} \frac{dz}{z - z_0} + \oint_{\gamma_\varepsilon} \frac{\varphi(z)}{z - z_0}\, dz. \qquad (7)$$

The first integral on the right equals $2\pi i f(z_0)$, by Sec. 10.33a, while the second integral satisfies the inequality

$$\left| \oint_{\gamma_\varepsilon} \frac{\varphi(z)}{z-z_0}\, dz \right| \leqslant 2\pi M\varepsilon,$$

by formula (5), p. 383, and hence approaches zero as $\varepsilon \to 0$. But the other two terms in (7) are constant, and hence

$$I = 2\pi i f(z_0)$$

or

$$f(z_0) = \frac{I}{2\pi i},$$

which is equivalent to (4). ∎

Cauchy's formula (4), expressing the value of an analytic function $f(z)$ inside a contour L in terms of the values of $f(z)$ on L (via a line integral along L), is one of the most important results of the theory of analytic functions.

10.34. THEOREM. *If $f(z)$ is analytic on a domain G, then $f(z)$ has derivatives of all orders on G (themselves analytic), given by*

$$f^{(n)}(z) = \frac{n!}{2\pi i} \oint_L \frac{f(\zeta)}{(\zeta-z)^{n+1}}\, d\zeta \qquad (n=1,2,\dots),$$

where the integral is along any piecewise smooth closed path $L \subset G$ surrounding the point z.

Proof. An immediate consequence of Theorem 10.28 and Cauchy's formula, which shows that $f(z)$ can be represented on G as an integral of the Cauchy type. ∎

10.35. a. THEOREM. *Every function $f(z)$ analytic on a disk $K = \{z: |z-z_0| < r\}$ can be expanded in a convergent power series*

$$f(z) = \sum_{n=0}^{\infty} a_n (z-z_0)^n \tag{8}$$

on K.

Proof. Starting from Cauchy's formula

$$f(z) = \frac{1}{2\pi i} \oint_L \frac{f(\zeta)}{\zeta-z}\, d\zeta$$

valid for any circle $L = \{\zeta: |\zeta-z_0| = r-\delta\}$ and any point z inside L, we write

$$\frac{1}{\zeta-z} = \frac{1}{\zeta-z_0} \frac{1}{1-\dfrac{z-z_0}{\zeta-z_0}} = \frac{1}{\zeta-z_0} \sum_{n=0}^{\infty} \left(\frac{z-z_0}{\zeta-z_0}\right)^n.$$

This series converges uniformly in ζ on L, because of the estimate

$$\left|\frac{z-z_0}{\zeta-z_0}\right| = \frac{|z-z_0|}{r-\delta} < 1$$

and Weierstrass' test (Theorem 6.53). It follows from Theorem 10.22a that

$$f(z) = \frac{1}{2\pi i} \oint_L \frac{f(\zeta)}{\zeta-z_0} \sum_{n=0}^{\infty} \left(\frac{z-z_0}{\zeta-z_0}\right)^n d\zeta$$

$$= \sum_{n=0}^{\infty} (z-z_0)^n \frac{1}{2\pi i} \oint_L \frac{f(\zeta)}{(\zeta-z_0)^{n+1}} d\zeta,$$

which is of the form (8) with

$$a_n = \frac{1}{2\pi i} \oint_L \frac{f(\zeta)}{(\zeta-z_0)^{n+1}} d\zeta. \quad \blacksquare \tag{9}$$

b. THEOREM. *Every function $f(z)$ analytic on a disk $K = \{z : |z-z_0| < r\}$ can be expanded in a convergent power series of the form*

$$f(z) = \sum_{n=0}^{\infty} \frac{f^{(n)}(z_0)}{n!} (z-z_0)^n \tag{10}$$

on K, known as the **Taylor series** *of $f(z)$ on K (cf. Sec. 8.5).*

Proof. Because of Theorem 10.34, the "Taylor coefficients" (9) can be written as

$$a_n = \frac{f^{(n)}(z_0)}{n!}. \quad \blacksquare \tag{9'}$$

c. COROLLARY. *If $f(z)$ is analytic on a disk $K = \{z : |z-z_0| < r\}$ and if $f^{(n)}(z_0) = 0$ for all $n = 0,1,2,\ldots$, then $f(z) \equiv 0$ on K.*

Proof. An immediate consequence of (10). $\quad \blacksquare$

10.36. a. A sequence (or series) of functions defined on a domain G is said to *converge uniformly inside G* if the sequence (or series) converges uniformly on every compact set Q contained in G.†

---

† For example, it follows from Theorem 10.22b or from a familiar property of power series (Theorem 6.65a) that the series (8) converges uniformly inside the disk K.

THEOREM. *If Q is a compact set contained in a domain G, then G contains a piecewise smooth closed path L surrounding Q.*

Proof. Every point $z_0 \in Q$ is the center of an open disk $K = \{z : |z - z_0| < r\}$ such that G contains K and its boundary, and the union of all such disks clearly contains Q. By the finite covering theorem (Sec. 3.97), there is a finite number of such disks $K_1, \dots, K_m$ such that $Q \subset K_1 \cup \dots \cup K_m$. But the boundary of the domain $K_1 \cup \dots \cup K_m \subset G - Q$ is a piecewise smooth closed path L (made up of a finite number of circular arcs) surrounding Q. ∎

b. THEOREM (**Weierstrass' theorem**). *If $f_n(z)$ is a sequence of analytic functions converging uniformly inside a domain G to a limit function $f(z)$, then $f(z)$ is analytic on G. Moreover, the sequence of derivatives $f_n^{(m)}(z)$ of any order m converges uniformly inside G to the corresponding derivative $f^{(m)}(z)$ of $f(z)$.*

Proof. By Cauchy's formula,

$$f_n(z) = \frac{1}{2\pi i} \oint_L \frac{f_n(\zeta)}{\zeta - z}\, d\zeta$$

for any piecewise smooth closed path $L \subset G$ surrounding a given point $z \in G$. But

$$\sup_{\zeta \in L} \left| \frac{1}{\zeta - z} \right| \leqslant \frac{1}{\delta},$$

where

$$\delta = \inf_{\zeta \in L} |\zeta - z|,$$

while the sequence $f_n(\zeta)$ converges uniformly to $f(\zeta)$ on L (as $n \to \infty$), since L is a compact set. It follows that

$$\frac{f_n(\zeta)}{\zeta - z}$$

converges uniformly on L to

$$\frac{f(\zeta)}{\zeta - z},$$

and hence

$$f(z) = \frac{1}{2\pi i} \oint_L \frac{f(\zeta)}{\zeta - z}\, d\zeta,$$

by Theorem 10.22a, i.e., $f(z)$ can also be represented on G as an integral of the Cauchy type. Therefore $f(z)$ is analytic on G, by Theorem 10.28.

To prove the second part of the theorem, we note that, by Theorem 10.28 again,

$$f^{(m)}(z) = \frac{m!}{2\pi i} \oint_L \frac{f(\zeta)}{(\zeta - z)^{m+1}} d\zeta,$$

$$f_n^{(m)}(z) = \frac{m!}{2\pi i} \oint_L \frac{f_n(\zeta)}{(\zeta - z)^{m+1}} d\zeta,$$

where the integrals are along a piecewise smooth closed path $L \subset G$ surrounding any given compact set $Q \subset G$ (the existence of L follows from the preceding theorem). But this time

$$\sup_{\substack{\zeta \in L \\ z \in Q}} \left| \frac{1}{(\zeta - z)^{m+1}} \right| \leqslant \frac{1}{\delta^{m+1}},$$

where

$$\delta = \inf_{\substack{\zeta \in L \\ z \in Q}} |\zeta - z|,$$

while the sequence

$$\frac{f_n(\zeta)}{(\zeta - z)^{m+1}}$$

converges uniformly on $L \times Q$ to

$$\frac{f(\zeta)}{(\zeta - z)^{m+1}}.$$

Hence, by Theorem 10.22b,

$$f^{(m)}(z) = \lim_{n \to \infty} f_n^{(m)}(z),$$

where the convergence is uniform on every compact set $Q \subset G$. ∎

10.37. a. Next we prove the converse of Theorem 10.35a:

THEOREM. *Let r be the radius of convergence of the power series*

$$f(z) = \sum_{n=0}^{\infty} a_n(z - z_0)^n. \tag{11}$$

Then $f(z)$ is analytic on the disk $K = \{z: |z - z_0| < r\}$, with derivative

$$f'(z) = \sum_{n=1}^{\infty} na_n(z-z_0)^{n-1} \tag{12}$$

on K.

Proof. The power series (11) converges uniformly on any disk

$$K_1 = \{z: |z-z_0| < r_1\}$$

with $r_1 < r$ (see Theorem 6.65a). But every compact set $Q \subset K$ is contained in such a disk, since otherwise we could find a sequence of points $z_n \in Q$ such that $|z_n - z_0| \to r$, which is impossible, since the sequence z_n would then have a limit point $z^* \in Q$,† where $|z^* - z_0| = r$, contrary to the condition $Q \subset K$. Hence the sequence of partial sums

$$f_n(z) = \sum_{k=0}^{n} a_k(z-z_0)^k$$

of the series (11) converges uniformly inside K to the function (11). But the functions $f_n(z)$ are obviously analytic on K, and hence so is $f(z)$ itself, by Weierstrass' theorem. The validity of (12) is an immediate consequence of the second part of Weierstrass' theorem. ∎

b. In particular, the above theorem implies the following differentiation formulas, valid for all complex z:

$$(e^z)' = \left(1 + z + \frac{z^2}{2!} + \cdots\right)' = 1 + z + \frac{z^2}{2!} + \cdots = e^z, \tag{13}$$

$$(\sin z)' = \left(z - \frac{z^3}{3!} + \frac{z^5}{5!} + \cdots\right)' = 1 - \frac{z^2}{2!} + \frac{z^4}{4!} - \cdots = \cos z, \tag{14}$$

$$(\cos z)' = \left(1 - \frac{z^2}{2!} + \frac{z^4}{4!} - \cdots\right)' = -z + \frac{z^3}{3!} - \frac{z^5}{5!} + \cdots = -\sin z, \tag{15}$$

$$(\sinh z)' = \left(z + \frac{z^3}{3!} + \frac{z^5}{5!} + \cdots\right)' = 1 + \frac{z^2}{2!} + \frac{z^4}{4!} + \cdots = \cosh z, \tag{16}$$

$$(\cosh z)' = \left(1 + \frac{z^2}{2!} + \frac{z^4}{4!} + \cdots\right)' = z + \frac{z^3}{3!} + \frac{z^5}{5!} + \cdots = \sinh z. \tag{17}$$

We could just as well have deduced (14)–(17) from (13), by using the known expressions for the trigonometric and hyperbolic functions in terms of the exponential function (Secs. 8.63 and 8.72).

c. There is a natural connection between the radius of convergence r of the

† Recall the definition of a compact set (Sec. 3.91a).

series (11) and the "analyticity" of $f(z)$. In fact, r is just the radius of the largest disk centered at z_0 on which $f(z)$ is analytic, since, by Theorem 10.35b, $f(z)$ is the sum of its own Taylor series on every such disk. Thus, for example, the radius of convergence of the series

$$f(z) = \frac{1}{1+z^2} = 1 - z^2 + z^4 - z^6 + \cdots \tag{18}$$

equals 1, since the distance from the point $z_0 = 0$ to the nearest point at which $f(z)$ fails to be analytic, i.e., the point $z = i$ or $z = -i$, equals 1. It would be hard to relate the divergence of the series (18) at the points $z = \pm 1$ to the behavior of the function $f(z)$ for real $z = x$ only.

Points at which a function $f(z)$ fails to be analytic (like the points $z = \pm i$ in the above example) are called *singular points* or *singularities* of $f(z)$.

d. A point z_0 is called a *zero* (or *root*) of a function $f(z)$ analytic on a domain G containing z_0 if $f(z_0) = 0$. Moreover, we call z_0 a zero of *order* (or *multiplicity*) k if

$$f(z_0) = f'(z_0) = \cdots = f^{(k-1)}(z_0) = 0, f^{(k)}(z_0) \neq 0.$$

Thus an analytic function $f(z)$ with a zero of order k at the point $z = z_0$ has a Taylor series expansion of the form

$$f(z) = \frac{f^{(k)}(z_0)}{k!}(z - z_0)^k + \frac{f^{(k+1)}(z_0)}{(k+1)!}(z - z_0)^{k+1} + \cdots$$

in a neighborhood of z_0. This can also be written as

$$f(z) = (z - z_0)^k g(z),$$

where the function

$$g(z) = \frac{f^{(k)}(z_0)}{k!} + \frac{f^{(k+1)}(z_0)}{(k+1)!}(z - z_0) + \cdots$$

is analytic in a neighborhood of z_0, being the sum of a convergent power series, and hence is analytic on the whole domain G, being the quotient of two analytic functions $f(z)$ and $(z - z_0)^k$. Note that $g(z_0)$ is just

$$g(z_0) = \frac{1}{k!}f^{(k)}(z_0) \neq 0.$$

10.38. A function $f(z)$ is said to be *entire* if it is analytic on the whole complex plane C. According to Theorem 10.35, an entire function $f(z)$ has a Taylor series expansion

$$f(z) = \sum_{n=0}^{\infty} a_n z^n \qquad (z_0 = 0) \tag{19}$$

which converges for all $z \in C$.

THEOREM (**Liouville's theorem**). *If $f(z)$ is an entire function and if $|f(z)| \leqslant M$ for all $z \in C$, then $f(z) \equiv$ const, i.e., every bounded entire function is a constant.*

Proof. According to Theorem 10.35b, the coefficients of the series (19) are given by

$$a_n = \frac{1}{2\pi i} \int_L \frac{f(\zeta)}{\zeta^{n+1}} \, d\zeta, \tag{20}$$

where we choose L to be a circle of radius R centered at the origin. Using the estimate (5), p. 383 and the fact that $|f(z)| \leqslant M$ for all $z \in C$, we have

$$|a_n| \leqslant \frac{1}{2\pi} \frac{M}{R^{n+1}} 2\pi R = \frac{M}{R^n}.$$

But R can be made arbitrarily large, since $f(z)$ is entire, and hence

$$a_n = 0 \qquad (n = 1, 2, \ldots),$$

so that

$$f(z) \equiv a_0 = \text{const.} \quad \blacksquare$$

10.39. a. THEOREM (**Uniqueness theorem for analytic functions**). *If $f(z)$ is analytic on a domain G and vanishes on a sequence of points $z_0, \ldots, z_n, \ldots$ converging to a point $z_0 \in G$, then $f(z) \equiv 0$ on G.*

Proof. By Theorem 10.35a, $f(z)$ has a convergent power series expansion

$$f(z) = a_0 + a_1(z - z_0) + a_2(z - z_0)^2 + \cdots \tag{21}$$

on every disk $K = \{z : |z - z_0| < r\} \subset G$. First we show that $f(z) \equiv 0$ on K. To this end, we introduce further power series

$$f_1(z) = a_1 + a_2(z - z_0) + a_3(z - z_0)^2 + \cdots,$$
$$f_2(z) = a_2 + a_3(z - z_0) + a_4(z - z_0)^2 + \cdots, \tag{22}$$
$$\cdots$$

where each new series is obtained from the preceding one by dropping the constant term and dividing by $z - z_0$. The series (22) all converge on the same disk as the series (21) itself, and the functions $f_m(z)$ are all continuous at $z = z_0$, like $f(z)$ itself (see Corollary 6.65b). It follows by induction that

$$f(z_n) = 0, \qquad\qquad\qquad a_0 = f(z_0) = \lim_{n \to \infty} f(z_n) = 0,$$

$$f_1(z_n) = \frac{f(z_n) - f(z_0)}{z_n - z_0} = 0, \qquad a_1 = f_1(z_0) = \lim_{n \to \infty} f_1(z_n) = 0,$$

$$\ldots \qquad\qquad\qquad\qquad \ldots$$

$$f_m(z_n) = \frac{f_{m-1}(z_n) - f_{m-1}(z_0)}{z_n - z_0} = 0, \qquad a_m = f_m(z_0) = \lim_{n \to \infty} f_m(z_n) = 0,$$

$$\ldots \qquad\qquad\qquad\qquad \ldots$$

Hence the coefficients a_m all vanish. But then $f(z) \equiv 0$ on K, as asserted.

Having just shown that $f(z) \equiv 0$ on any disk $K \subset G$ centered at z_0, we now show that $f(z) \equiv 0$ on the whole domain G. Let z^* be any point of G other than z_0. Then, since G is connected, there is a piecewise smooth path $L \subset G$ joining z_0 to z^*. Let

$$\rho = \inf_{z \in L} r(z),$$

where $r(z)$ is the radius of the largest disk contained in G and centered at z. Then $\rho > 0$, by the compactness of L (why?). As just shown, $f(z) \equiv 0$ on the disk K_0 of radius ρ centered at z_0. Suppose we shift K_0 along L by sliding its center along L from z_0 to z^*, thereby generating a "chain of shifted disks," the first centered at z_0 and the last centered at z^*. Making sure that the center of each shifted disk lies inside the "preceding disk" (draw a figure), we find that $f(z) \equiv 0$ on each shifted disk, by the same argument used to show that $f(z) \equiv 0$ on K_0. It follows that $f(z^*) = 0$ and hence that $f(z) \equiv 0$ on G, since the point $z^* \in G$ is arbitrary. $\blacksquare$

Another version of the uniqueness theorem for analytic functions is given by the following

b. THEOREM. *If two functions $f_1(z)$ and $f_2(z)$ are analytic on a domain G and coincide on a sequence of points $z_1, \ldots, z_n, \ldots$ converging to a point $z_0 \in G$, then $f_1(z) \equiv f_2(z)$ on G.*

Proof. If $f(z) = f_1(z) - f_2(z)$, then $f(z)$ is analytic on G and vanishes on the sequence $z_1, \ldots, z_n, \ldots$ Hence, by the preceding theorem, $f(z) \equiv 0$ on G, i.e., $f_1(z) \equiv f_2(z)$ on G. $\blacksquare$

c. The above considerations lead at once to the idea of "analytic continuation," anticipated in Sec. 8.55. Suppose $f_1(z)$ is analytic on a domain G_1, while $f_2(z)$ is analytic on a domain G_2, and suppose the intersection of G_1 and G_2 is a (nonempty) domain G_0. Suppose further that $f_1(z)$ and $f_2(z)$ coincide on G_0 (this will be the case, for example, if $f_1(z_n) = f_2(z_n)$ on a sequence of points $z_1,...,z_n,...$ in G_0 converging to a point $z_0 \in G_0$). Then the function

$$f(z) = \begin{cases} f_1(z) \text{ if } z \in G_1, \\ f_2(z) \text{ if } z \in G_2, \end{cases}$$

called the *analytic continuation of* $f_1(z)$ *from* G_1 *into* G_2 is "single-valued" (see Sec. 2.84) and analytic on the whole domain $G = G_1 \cup G_2$. Note that the analytic continuation of $f_1(z)$ from G_1 into G is necessarily unique (why?).

d. Example. The function

$$f_1(z) = \sum_{n=0}^{\infty} z^n \qquad (|z| < 1)$$

is analytic, being the sum of a convergent power series. The function

$$f(z) = \frac{1}{1-z} \qquad (z \neq 1),$$

equal to $f_1(z)$ if $|z| < 1$, is the analytic continuation of $f_1(z)$ from the disk $|z| < 1$ into the domain $C' = C - \{1\}$ (the complex plane minus the point $z = 1$).

e. THEOREM. *If $f(z)$ is analytic on G and if $f^{(n)}(z_0) = 0$ for all $n = 0,1,...,$ then $f(z) \equiv 0$ on G.*

Proof. Let $K = \{z : |z - z_0| < r\}$ be a disk contained in G and centered at z_0. Then $f(z) \equiv 0$ on K, by Corollary 10.35c. Now apply Theorem 10.39a. ∎

f. Real analytic functions. A function $f(z)$ analytic on a domain G containing points of the real axis is called a *real analytic function* if all its values on the real axis are themselves real. Since we can calculate the values of the derivative of an analytic function $f(z)$ on the real axis using only the values of $f(z)$ on the real axis, it is clear that the derivative $f'(z)$ of a real analytic function is itself a real analytic function. By the same token, the higher-order derivatives $f''(z), f'''(z),...$ are all real analytic functions. Given any point $z_0 \in G$ lying on the real axis, let

$$f(z) = \sum_{n=0}^{\infty} a_n (z - z_0)^n$$

be the Taylor series expansion of $f(z)$ at z_0, converging on some disk K centered at z_0. Then the coefficients

$$a_n = \frac{f_n(z_0)}{n!}$$

are real, and hence

$$\overline{f(\overline{z})} = \sum_{n=0}^{\infty} \overline{a_n(\overline{z}-z_0)^n} = \sum_{n=0}^{\infty} a_n(z-z_0)^n = f(z) \tag{23}$$

for all $z \in K$, so that $\overline{f(\overline{z})} = f(z)$ on or near the real axis. More generally, we have the following

THEOREM. *Let $f(z)$ be a real analytic function on a domain G. Then $f(z)$ satisfies the condition*

$$\overline{f(\overline{z})} = f(z)$$

at every point $z \in G$ such that $\overline{z}$ also belongs to G.

Proof. By retaining only those points $z \in G$ such that $\overline{z}$ also belongs to G, we can assume that G is symmetric with respect to the real axis. Let G^+ denote the set of all points $z \in G$ with Im $z > 0$ and G^- the set of all points $z \in G$ with Im $z < 0$. Suppose we introduce a new function

$$f_1(z) = \overline{f(\overline{z})} \qquad (z \in G^+),$$

whose definition involves the values of $f(z)$ for $z \in G^-$. Let $z_0 \in G^+$, so that $\overline{z}_0 \in G^-$. Being analytic on G^-, $f(z)$ has a Taylor series expansion

$$f(z) = \sum_{n=0}^{\infty} a_n(z - \overline{z}_0)^n,$$

valid in a neighborhood of every point $\overline{z}_0 \in G^-$. Hence

$$f_1(z) = \overline{f(\overline{z})} = \sum_{n=0}^{\infty} \overline{a_n(\overline{z} - \overline{z}_0)^n} = \sum_{n=0}^{\infty} \overline{a}_n(z - z_0)^n$$

in a neighborhood of every point $z_0 \in G$, so that $f_1(z)$ is analytic on G^+. But $f_1(z) = f(z)$ on or near the real axis, by (23). It follows from Theorem 10.39b that $f_1(z) = \overline{f(\overline{z})} = f(z)$ on the "upper domain" G^+. Reversing the roles of G^+ and G^-, we find in the same way that $\overline{f(\overline{z})} = f(z)$ on the "lower domain" G^- as well. Thus, finally, $\overline{f(\overline{z})} = f(z)$ on the whole domain G. ∎

10.4. Residues and Isolated Singular Points

10.41. Residues. Let L be a piecewise smooth closed path with no multiple points (Sec. 10.31), let D be the interior of L, and let $f(z)$ be analytic on $L \cup D$. Then the integral

$$\frac{1}{2\pi i}\oint_L f(z)\,dz \tag{1}$$

vanishes, by Cauchy's theorem (Sec. 10.32). On the other hand, if $f(z)$ fails to be analytic at certain points of D, called singular points (Sec. 10.37c),† then the integral (1) may well be nonzero. For example, if

$$f(z) = \frac{g(z)}{z - z_0},$$

where $g(z)$ is analytic on $L \cup D$, then, by Cauchy's formula (Theorem 10.33c),

$$\frac{1}{2\pi i}\oint_L f(z)\,dz = g(z_0).$$

More generally, let $f(z)$ be any function analytic on $L \cup D$ except at a point $z_0 \in D$. Then the value of the integral (1) is called the *residue of $f(z)$ at $z = z_0$,* denoted by

$$\operatorname*{Res}_{z = z_0} f(z).$$

Note that the value of (1) does not change if we replace L by any other piecewise smooth closed path $L' \subset D$ surrounding z_0.

10.42. Examples

a. Let

$$f(z) = \frac{g(z)}{P(z)}, \tag{2}$$

where $g(z)$ is analytic on $L \cup D$ and $P(z)$ is a polynomial with a simple zero at $z_0 \in D$ (so that $P(z_0) = 0$, $P'(z_0) \neq 0$) and no other zeros in D. Then

$$P(z) = a_1(z - z_0) + a_2(z - z_0)^2 + \cdots = (z - z_0)P_1(z)$$

(Sec. 10.37d), where

$$P_1(z) = a_1 + a_2(z - z_0) + \cdots$$

is a polynomial equal to

$$a_1 = P'(z_0) \neq 0$$

† In particular, z_0 is a singular point of $f(z)$ if $f(z)$ fails to be defined at $z = z_0$ or fails to be differentiable at $z = z_0$.

at $z = z_0$, with no zeros in D. By Cauchy's formula,

$$\frac{1}{2\pi i} \oint_L f(z)\, dz = \frac{1}{2\pi i} \oint_L \frac{g(z)}{P(z)} = \frac{1}{2\pi i} \oint_L \frac{g(z)}{(z-z_0)P_1(z)}\, dz$$
$$= \frac{g(z_0)}{P_1(z_0)} = \frac{g(z_0)}{P'(z_0)},$$

so that

$$\operatorname*{Res}_{z=z_0} f(z) = \operatorname*{Res}_{z=z_0} \frac{g(z)}{P(z)} = \frac{g(z_0)}{P'(z_0)}. \tag{3}$$

b. Again let $f(z)$ be given by (2), where $g(z)$ is analytic on $L \cup D$, but this time let $P(z)$ be a polynomial with a zero of order k at $z_0 \in D$ (so that $P(z_0) = \cdots = P^{(k-1)}(z_0) = 0$, $P^{(k)}(z_0) \neq 0$) and no other zeros in D. Then

$$P(z) = a_k(z-z_0)^k + a_{k+1}(z-z_0)^{k+1} + \cdots = (z-z_0)^k P_k(z) \qquad (a_k \neq 0),$$

where

$$P_k(z) = a_k + a_{k+1}(z-z_0) + \cdots$$

is a polynomial equal to

$$a_k = \frac{P^{(k)}(z_0)}{k!} \neq 0$$

at $z = z_0$, with no zeros in D. Expanding the function $g(z)/P_k(z)$ as a Taylor series in powers of $z - z_0$, we get

$$\frac{g(z)}{P_k(z)} = b_0 + b_1(z-z_0) + b_2(z-z_0)^2 + \cdots,$$

and hence

$$f(z) = \frac{g(z)}{P(z)} = \frac{g(z)}{(z-z_0)^k P_k(z)} = \frac{b_0}{(z-z_0)^k} + \frac{b_1}{(z-z_0)^{k-1}} + \cdots + \frac{b_{k-1}}{z-z_0}$$
$$+ b_k + b_{k+1}(z-z_0) + \cdots.$$

Integrating $f(z)$ term by term along L (why is this justified?), we see that only the term containing the coefficient b_{k-1} survives (see Sec. 10.33a), i.e.,

$$\frac{1}{2\pi i} \oint_L f(z)\, dz = \frac{1}{2\pi i} \oint_L \frac{g(z)}{P(z)} = \frac{1}{2\pi i} \oint_L \frac{b_{k-1}}{z-z_0}\, dz$$
$$= b_{k-1} = \frac{1}{(k-1)!} \left[\frac{g(z)}{P_k(z)} \right]_{z=z_0}^{(k-1)} = \frac{1}{(k-1)!} \left[\frac{g(z)(z-z_0)^k}{P(z)} \right]_{z=z_0}^{(k-1)},$$

with the help of Theorem 10.35b. It follows that

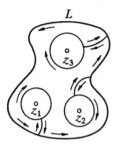

Figure 69

$$\operatorname*{Res}_{z=z_0} f(z) = \operatorname*{Res}_{z=z_0} \frac{g(z)}{P(z)} = \frac{1}{(k-1)!} \left[\frac{g(z)(z-z_0)^k}{P(z)} \right]^{(k-1)}_{z=z_0} \tag{3'}$$

10.43. a. To evaluate the integral (1) in the case where the contour L contains several, say m, singular points of the function $f(z)$, we use the fact that

$$\frac{1}{2\pi i} \oint_L f(z)\, dz = \frac{1}{2\pi i} \oint_{L_1} f(z)\, dz + \cdots + \frac{1}{2\pi i} \oint_{L_m} f(z)\, dz$$

$$= \sum_{j=1}^{m} \operatorname*{Res}_{z=z_j} f(z), \tag{4}$$

where each $L_j \subset D$ is a piecewise smooth closed path surrounding z_j but no other singular points. The "residue theorem" (4) is an immediate consequence of Cauchy's theorem and the construction given in Figure 69,† which shows the original path L, the paths $L_1, \ldots, L_m$ (for the case $m=3$) and appropriate arcs joining these paths, where each arc is traversed twice in opposite directions.

b. Example. Evaluate the integral

$$I = \oint_{|z|=1} \frac{dz}{z^2 - 2pz + 1} \qquad (p > 1).$$

Solution. The denominator has two simple zeros, whose product equals 1. One zero

$$z_1 = p - \sqrt{p^2 - 1}$$

lies inside the unit circle $|z| = 1$, while the other zero

$$z_2 = p + \sqrt{p^2 - 1}$$

† The argument leading to (4) is the natural generalization of that given in Sec. 10.33b in connection with Figure 68.

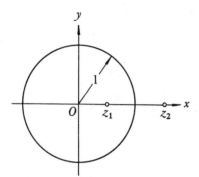

Figure   70

lies outside the unit circle (see Figure 70). It follows from formula (3) that

$$I = 2\pi i \left[\frac{1}{2(z-p)} \right]_{z=z_1} = \frac{\pi i}{\sqrt{p^2-1}}.$$

10.44. Logarithmic residues. Consider the integral

$$N = \frac{1}{2\pi i} \oint_L \frac{f'(z)}{f(z)}\, dz, \tag{5}$$

where L is a piecewise smooth closed path with no multiple points and D is the interior of L, while $f(z)$ is analytic on $L \cup D$ and does not vanish on L. If $f(z)$ is also nonvanishing on D, then (5) vanishes by Cauchy's theorem. If $f(z)$ has zeros inside L (i.e., in D), then by the uniqueness theorem for analytic functions (Theorem 10.39a), there are only a finite number of such zeros (why?). Let these zeros be denoted by $z_1, \ldots, z_m$ and their orders by $k_1, \ldots, k_m$. Then we have the following

THEOREM. *The integral N, called the* **logarithmic residue** *of $f(z)$ with respect to the contour L,† is a positive integer equal to the sum of the orders of all the zeros of the function $f(z)$ inside L.*

Proof. As in Sec. 10.43,

$$N = \sum_{j=1}^{m} N_j, \qquad N_j = \frac{1}{2\pi i} \oint \frac{f'(z)}{f(z)} dz,$$

where L_j surrounds z_j but no other singular points. For brevity, let $z_j = \zeta$,

---

† As shown in Sec. 10.37, the integrand of (5) is just the derivative of the composite function $\ln f(z)$.

$k_j = k$. Then

$$f(z) = a_k(z-\zeta)^k + a_{k+1}(z-\zeta)^{k+1} + \cdots \qquad (a_k \neq 0),$$
$$f'(z) = ka_k(z-\zeta)^{k-1} + (k+1)a_{k+1}(z-\zeta)^k + \cdots,$$
$$\frac{f'(z)}{f(z)} = \frac{1}{z-\zeta} \frac{ka_k + (k+1)a_{k+1}(z-\zeta) + \cdots}{a_k + a_{k+1}(z-\zeta) + \cdots} = \frac{\varphi(z)}{z-\zeta}$$

in a neighborhood of the point ζ. Clearly the function $\varphi(z)$ is analytic at the point $z = \zeta$, where it takes the value $\varphi(\zeta) = k$, and hence

$$\frac{1}{2\pi i} \oint_{L_j} \frac{f'(z)}{f(z)} \, dz = \frac{1}{2\pi i} \oint_{L_j} \frac{\varphi(z)}{z-\zeta} \, dz = \varphi(\zeta) = k.$$

It follows that

$$N = \sum_{j=1}^{m} N_j = \sum_{j=1}^{m} k_j. \quad \blacksquare$$

10.45. Laurent series. We now turn to the "qualitative analysis" of the behavior of a function $f(z)$ near an isolated singular point z_0 (i.e., a singular point z_0 with a neighborhood containing no other singular points), using as our basic tool an expansion of $f(z)$ in a series involving both positive and negative powers of $z - z_0$:

a. THEOREM. *Every function $f(z)$ analytic on an annulus*

$$K = \{z : r_0 < |z - z_0| < R_0\}$$

can be expanded in a convergent "two-sided series" of the form†

$$f(z) = \sum_{n=-\infty}^{\infty} a_n(z-z_0)^n \qquad (6)$$

*on K, known as the **Laurent series** of $f(z)$ on K.*

Proof. The annulus or ring-shaped domain K is shown in Figure 71. Let r and R be numbers such that $r_0 < r < R < R_0$. Then

$$f(z) = \frac{1}{2\pi i} \oint_{|\zeta - z_0| = R} \frac{f(\zeta)}{\zeta - z} \, d\zeta - \frac{1}{2\pi i} \oint_{|\zeta - z_0| = r} \frac{f(\zeta)}{\zeta - z} \, d\zeta,$$

by Cauchy's theorem applied to the simply connected domain obtained by cutting the annulus $r < |\zeta - z_0| < R$ along a radius which does not go through

---

† It will be recalled that (6) is just the limit of the sum

$$\sum_{n=-p}^{q} a_n(z-z_0)^n$$

as p and q approach infinity *independently* (see Sec. 6.48a). We often call (6) the *Laurent (series) expansion* of $f(z)$ at $z = z_0$.

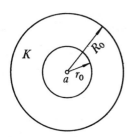

Figure 71

the point z. Making the expansion

$$\frac{1}{\zeta-z}=\frac{1}{(\zeta-z_0)-(z-z_0)}=\frac{1}{\zeta-z_0}\frac{1}{1-\frac{z-z_0}{\zeta-z_0}}=\frac{1}{\zeta-z_0}\sum_{n=0}^{\infty}\left(\frac{z-z_0}{\zeta-z_0}\right)^n$$

in the first integral and the expansion

$$\frac{1}{\zeta-z}=\frac{1}{(\zeta-z_0)-(z-z_0)}=-\frac{1}{z-z_0}\frac{1}{1-\frac{\zeta-z_0}{z-z_0}}=-\sum_{n=0}^{\infty}\frac{(\zeta-z_0)^n}{(z-z_0)^{n+1}}$$

in the second integral, where, by Weierstrass' test (Theorem 6.53), both expansions are uniformly convergent in ζ on the corresponding contours of integration. It follows from Theorem 10.22a that

$$f(z)=\sum_{n=0}^{\infty}a_n(z-z_0)^n+\sum_{n=-\infty}^{-1}a_n(z-z_0)^n, \tag{7}$$

where the "Laurent coefficients" a_n are given by

$$a_n=\begin{cases}\dfrac{1}{2\pi i}\displaystyle\oint_{|\zeta-z_0|=R}\frac{f(\zeta)}{(\zeta-z_0)^{n+1}}\,d\zeta & \text{if } n\geqslant 0,\\[3mm]\dfrac{1}{2\pi i}\displaystyle\oint_{|\zeta-z_0|=r}\frac{f(\zeta)}{(\zeta-z_0)^{n+1}}\,d\zeta & \text{if } n<0.\end{cases} \tag{8}$$

Note that since the integrand

$$\frac{f(\zeta)}{(\zeta-z_0)^{n+1}}$$

is analytic on the annulus K, we can evaluate these integrals along any circle $|z-a|=\rho$, $r_0<\rho<R_0$, or, for that matter, along any piecewise smooth closed path $L\subset K$ with no multiple points. In particular, if $r_0=0$, the coefficient a_{-1} is just the residue of $f(z)$ at $z=z_0$. Finally, we can write the ex-

pansion (7) in the form (6), using the fact that each of the "one-sided series" in (7) converges separately (see Theorem 6.48b). ∎

b. According to Theorem 10.22b, the Laurent series (6) converges uniformly inside the annulus K. In particular, the series (6) converges uniformly on every closed annulus $r \leqslant |z - z_0| \leqslant R$, where $r_0 < r < R < R_0$.

The Laurent series (6) is uniquely determined by the function $f(z)$, as we see from (8) or equivalently from the easily verified formula

$$\oint_{|z-z_0|=\rho} \frac{f(z)}{(z-z_0)^m} \, dz = \sum_{n=-\infty}^{\infty} a_n \oint_{|z-z_0|=\rho} (z-z_0)^{n-m} \, dz = 2\pi i a_{m-1},$$

where $r_0 < \rho < R$, $m = 0, \pm 1, \pm 2, \ldots$

c. The first series on the right in (7), i.e., the series

$$f^+(z) = \sum_{n=0}^{\infty} a_n (z-z_0)^n,$$

is called the *regular part* of the Laurent series (6), while the second series

$$f^-(z) = \sum_{n=-\infty}^{-1} a_n (z-z_0)^n$$

is called the *principal part* of (6). The regular part of the Laurent series is a series in positive powers of $z - z_0$, convergent for $|z - z_0| = R$ and hence convergent on the whole disk $|z - z_0| \leqslant R$ (see Theorem 6.65a) including the whole "inner region" $|z - z_0| \leqslant r_0$ excluded from the annulus K. On the other hand, the principal part of the Laurent series is a series in negative powers of $z - z_0$, convergent for $|z - z_0| = r$. Setting

$$z - z_0 = \frac{1}{\zeta},$$

we get a series in positive powers of ζ converging for $|\zeta| = 1/r$ and hence converging for all $|\zeta| \leqslant 1/r$. Thus the principal part of the Laurent series converges for all $|z - z_0| \geqslant r$ including the whole "outer region" $|z - z_0| \geqslant R_0$ excluded from the annulus K.

10.46. Classification of isolated singular points. Let $w = f(z)$ be a function analytic at every point of a disk $|z - z_0| < R_0$ except at the center z_0. Then $f(z)$ can be expanded in a Laurent series

$$f(z) = \sum_{n=-\infty}^{\infty} a_n (z-z_0)^n \tag{9}$$

on the "punctured disk" $0 < |z - z_0| < R_0$. There are now essentially three kinds of structure of the series (9), leading to three designations of the isolated singular point z_0:

a. If all the Laurent coefficients a_n with $n < 0$ vanish, we call z_0 a *removable singular point* of $f(z)$. In this case, the function $f(z)$ has a power series expansion of the form

$$f(z) = \sum_{n=0}^{\infty} a_n (z - z_0)^n$$

on the domain $0 < |z - z_0| < R_0$, whose sum is a function analytic on the whole disk $|z - z_0| < R_0$ including the point z_0 at which it takes the value a_0. Hence, if we merely set $f(z_0) = a_0$, the original function $f(z)$ becomes analytic on the whole disk $|z - z_0| < R_0$, with no singular point at z_0. This explains the term "removable singular point."

b. If all the Laurent coefficients a_n with $n < -m$ $(m > 0)$ vanish and if a_{-m} is nonzero, the point z_0 is called a *pole of order m* of $f(z)$.

c. If there are nonzero Laurent coefficients with negative indices of arbitrarily large absolute value, we call z_0 an *essential singular point* of $f(z)$.

10.47. The point at infinity

a. In the interest of both generality and simplicity, we now "extend" the complex z-plane by adding an extra "point" $z = \infty$, called the *point at infinity*. By $z \to \infty$ we mean the direction in the z-plane consisting of the sets $A_r = \{z: |z| > r\}$, where r varies over all positive numbers, or at least over all numbers larger than some number $r_0 > 0$ (cf. Sec. 4.73f). The complex w-plane is extended in the same way by introducing the "point" $w = \infty$ and the direction $w \to \infty$. The points of the original z- and w-planes (as opposed to the points $z = \infty$ and $w = \infty$) are now said to be *finite*, and the set of all finite numbers together with the point at infinity (the "number" ∞) is called the *extended complex plane*. Given any direction S in the z-plane and any function $w = f(z)$ defined on the sets of S, we say that $f(z)$ *approaches infinity in the direction S* and write

$$\lim_S f(z) = \infty$$

if, given any $r > 0$, there exists a set $A_r \in S$ such that

$$|f(z)| > r$$

for all $z \in A_r$ (cf. Theorem 4.34a). This, together with the definition of the direction $z \to \infty$, explains the meaning of the expressions

$$\lim_{z \to z_0} f(z) = \infty, \qquad \lim_{z \to \infty} f(z) = a, \qquad \lim_{z \to \infty} f(z) = \infty$$

(where the points z_0 and a are finite). Thus, for example,

(a) $\displaystyle\lim_{z \to z_0} \frac{1}{(z - z_0)^m} = \infty \qquad (m = 1,2,\ldots)$;

(b) $\displaystyle\lim_{z \to z_0} \alpha(z)f(z) = \infty$ if $\displaystyle\lim_{z \to z_0} \alpha(z) = a \neq 0$, $\displaystyle\lim_{z \to z_0} f(z) = \infty$;

(c) $\displaystyle\lim_{z \to z_0} f(z) = \infty$ if and only if $\displaystyle\lim_{z \to z_0} \frac{1}{f(z)} = 0$

(cf. Theorems 4.36b and 4.37a).

b. The point $w = \infty$ is said to be a *limit point of* $f(z)$ *in the direction S* if, given any $r > 0$ and any $A_r \in S$, there exists a point $z_r \in A_r$ such that

$$|f(z_r)| > r$$

(cf. Sec. 4.45). If $f(z)$ is unbounded in the direction S (cf. Sec. 4.32), then obviously $f(z)$ has $w = \infty$ as a limit point in the direction S, and conversely.

c. Suppose $f(z)$ is analytic on the domain $G = \{z : |z| > r\}$. Then, by Theorem 10.45a, $f(z)$ can be represented on G as a Laurent series

$$f(z) = \sum_{n=-\infty}^{\infty} a_n z^n. \tag{10}$$

The substitution $z = 1/\zeta$ carries G into the domain $H = \{\zeta : 0 < |\zeta| < 1/R\}$ and $f(z)$ into the function

$$\varphi(\zeta) = \sum_{n=-\infty}^{\infty} a_n \zeta^{-n}. \tag{10'}$$

We say that the original function $f(z)$ has a *removable singular point, pole,* or *essential singular point at* $z = \infty$ if the new function $\varphi(\zeta)$ has the same kind of singular point at $\zeta = 0$. In other words, the point $z = \infty$ is a *removable singular point* of $f(z)$ if all the Laurent coefficients a_n figuring in (10) with $n > 0$ vanish, a *pole of order m* if the coefficients $a_{m+1}, a_{m+2}, \ldots$ vanish but $a_m \neq 0$, and an *essential singular point* if there are nonzero Laurent coefficients with arbitrarily large positive indices.

10.48. Behavior of a function at a pole. Suppose $f(z)$ has a pole of order m at the point $z = z_0$, so that the Laurent expansion of $f(z)$ at $z = z_0$ is of the form

$$f(z) = \sum_{n=-m}^{\infty} a_n (z - z_0)^n,$$

where $a_n = 0$ if $n < -m$ $(m > 0)$ and $a_{-m} \neq 0$. Writing

$$\varphi(z) = (z - z_0)^m f(z) = a_{-m} + a_{-m+1}(z - z_0) + \cdots,$$

we see that $\varphi(z)$ has a removable singular point at $z = z_0$ and hence becomes analytic at $z = z_0$ if we set $\varphi(z_0) = a_{-m}$. It follows that

$$f(z) = \frac{\varphi(z)}{(z - z_0)^m},$$

where

$$\lim_{z \to z_0} f(z) = \infty$$

(Sec. 10.47a) since $\varphi(z_0) \neq 0$. This is the kind of behavior exhibited by the integrands in Examples 10.22a and 10.22b.

10.49. Behavior of a function at an essential singular point

a. THEOREM. *If the function $f(z)$ has an essential singular point at $z = z_0$, then $f(z)$ is unbounded as $z \to z_0$, i.e., $f(z)$ has the number $A = \infty$ as a limit point as $z \to z_0$.*

Proof. If the function $f(z)$ were bounded as $z \to z_0$, then its principal part $f^-(z)$ would also be bounded as $z \to z_0$, since its regular part $f^+(z)$ is obviously bounded as $z \to z_0$. The series

$$f^-(z) = \sum_{n=-\infty}^{-1} a_n (z - z_0)^n$$

converges for all $z \neq z_0$ (see Sec. 10.45c). Hence, replacing $z - z_0$ by $1/\zeta$, we get a series

$$\varphi(\zeta) = \sum_{n=1}^{\infty} a_{-n} \zeta^n$$

which converges in the whole ζ-plane. If $f^-(z)$ were bounded in some deleted neighborhood $0 < |z - z_0| < \varepsilon$, then $\varphi(\zeta)$ would be bounded for $|\zeta| > 1/\varepsilon$. Since $\varphi(\zeta)$ is certainly bounded for $|\zeta| \leqslant 1/\varepsilon$, by its continuity (Theorem 5.16b), $\varphi(\zeta)$ would be bounded on the whole ζ-plane. But then $\varphi(\zeta) \equiv$ const, by Liouville's theorem (Theorem 10.38) and hence $\varphi(\zeta) \equiv 0$ since $\varphi(0) = 0$. This implies $a_n = 0$ for all $n < 0$, contrary to hypothesis, thereby proving the theorem by contradiction. ∎

b. THEOREM. *If $f(z)$ has an essential singular point at $z = z_0$, then $f(z)$ has every number A in the extended complex plane as a limit point as $z \to z_0$.*†

Proof. We have just seen that $A = \infty$ is a limit point of $f(z)$ as $z \to z_0$. Given any finite complex number A, suppose $f(z)$ does not have A as a limit point as $z \to z_0$. Then $f(z)$ does not take the value A in some neighborhood of z_0, and hence the function

$$\varphi(z) = \frac{1}{f(z) - A}$$

has no singular points in this neighborhood except at the point z_0 itself. If z_0 is a pole or essential singular point of $\varphi(z)$, then, as shown above, there exists a sequence of points $z_n \to z_0$ such that $\varphi(z_n) \to \infty$ and hence $f(z_n) \to A$, so that A is a limit point of $f(z)$ as $z \to z_0$, contrary to hypothesis. If z_0 is a removable singular point of $\varphi(z)$, then the limit

$$\lim_{z \to z_0} \varphi(z) = a$$

exists and is finite (cf. Sec. 10.46a). But there exists a sequence $z_n \to z_0$ such that $f(z_n) \to \infty$, and hence

$$a = \lim_{z \to z_0} \varphi(z) = \lim_{n \to \infty} \frac{1}{f(z_n) - A} = 0.$$

It follows that

$$\varphi(z) = (z - z_0)^m \psi(z)$$

for some $m \geqslant 1$, where $\psi(z_0) \neq 0$. The function $1/\psi(z)$ is analytic at $z = z_0$ and hence has a Taylor series expansion

$$\frac{1}{\psi(z)} = b_0 + b_1 (z - z_0) + \cdots$$

at $z = z_0$, where $b_0 = 1/\psi(z_0) \neq 0$. But then

$$f(z) = A + \frac{1}{\varphi(z)} = A + \frac{1}{(z - z_0)^m \psi(z)}$$
$$= A + (z - z_0)^{-m}[b_0 + b_1(z - z_0) + \cdots],$$

so that the Laurent expansion of $f(z)$ has only a finite number of terms in its principal part, contrary to the assumption that z_0 is an essential singular

† There is a more exact theorem due to Picard, which asserts that $f(z)$ *actually takes* every finite value (with one possible exception) infinitely many times in every neighborhood of an essential singular point. See e.g., A. I. Markushevich, *Theory of Functions of a Complex Variable*, Vol. III (translated by R. A. Silverman), Chelsea Publishing Co., N.Y. (1977), Theorem 9.16, p. 343.

point of $f(z)$. Thus, in any event, the assumption that $f(z)$ fails to have any given finite complex number A as a limit point as $z \to z_0$ leads to a contradiction. ∎

c. If $f(z)$ has an isolated singular point at $z = z_0$, then obviously there are just three possibilities:
(1) $f(z)$ is bounded in every neighborhood of z_0;
(2) $f(z)$ approaches the limit ∞ as $z \to z_0$;
(3) $f(z)$ has at least two limit points as $z \to z_0$, one of which equals ∞. On the other hand, the following three possibilities are also mutually exclusive and exhaustive:
(1') z_0 is a removable singular point of $f(z)$;
(2') z_0 is a pole of $f(z)$;
(3') z_0 is an essential singular point of $f(z)$.
It has already been shown that 1' implies 1, 2' implies 2, and 3' implies 3. But then the converse assertions, i.e., that 1 implies 1', 2 implies 2', and 3 implies 3', are also valid, since 1, say, must imply one of the three possibilities 1', 2', and 3', but certainly not 2' or 3'! It should be emphasized that these assertions are nontrivial. For example, the assertion that 1 implies 1' means that a function which is analytic and bounded in a neighborhood of a point z_0 actually approaches a finite limit as $z \to z_0$. This can only be deduced by examining the behavior of a function near a pole (Sec. 10.48) and near an essential singular point (Sec. 10.49).

d. Using Sec. 10.47c, we can transcribe all these results to the case of an isolated singular point at infinity. In particular, if $f(z)$ is analytic for all finite z and if $f(z)$ has a unique limit point $A = \infty$ as $z \to \infty$, then $f(z)$ is a (nonconstant) polynomial. Put somewhat differently, if an entire function $f(z)$ is not a polynomial,† then $f(z)$ has every number in the extended complex plane as a limit point as $z \to \infty$.

10.5. Mappings and Elementary Functions

10.51. Conformal mapping. Consider the *linear function*

$$w = f(z) = a(z - z_0) + w_0,$$
(1)

where z_0, w_0, and $a \neq 0$ are given complex numbers. Obviously $f(z_0) = w_0$ and $f'(z_0) = 0$. Moreover $w - w_0 = a(z - z_0)$, so that

$$|w - w_0| = |a||z - z_0|$$
(2)

† By Liouville's theorem (Theorem 10.38), the polynomial reduces to a constant if $f(z)$ is bounded.

and

$$\arg (w - w_0) = \arg a + \arg (z - z_0) \tag{3}$$

(to within an integral multiple of 2π). According to (2) and (3), the mapping $w = f(z)$ has the following geometrical meaning: *Every disk $|z - z_0| \leqslant r$ is expanded $|a|$ times in all directions, then rotated about its center z_0 through the angle $\arg a$, and finally shifted until its center lies at the point w_0.*

Now let $w = f(z)$ be any function analytic on the disk $|z - z_0| \leqslant r$ such that

$$f(z_0) = w_0, \qquad f'(z_0) = a \neq 0.$$

Then, by the very definition of the derivative,

$$w = w_0 + a(z - z_0) + \varepsilon(z, z_0)(z - z_0), \tag{4}$$

where $\varepsilon(z, z_0) \to 0$ as $z \to z_0$. According to (4), the mapping $w = f(z)$ is described by formula (1) to within "quantities of the second order of smallness," and hence, to this accuracy, the mapping is the result of $|a|$-fold expansion, rotation through the angle $\arg a$, and shifting to a new center at the point w_0. In particular, *the mapping carries two curves through the point z_0 making the angle α with each other into two curves through the point w_0 making the same angle α with each other.*† A mapping of this kind is said to be *conformal* (i.e., "angle-preserving") at $z = z_0$.

10.52. a. Let $w = f(z)$ be a function defined on a set E, and let F be the set of all "images" of the points of E, i.e., the set

$$F = \{w : w = f(z), \ z \in E\}.$$

Then, given any point $w \in F$, every point $z \in E$ such that $f(z) = w$ is called an *inverse image* of w.

b. THEOREM. *If the function $w = f(z)$ is analytic at $z = z_0$ and if*

$$f(z_0) = w_0, \qquad f'(z_0) = a \neq 0,$$

then z_0 has a neighborhood U (in the z-plane) and w_0 has a neighborhood V (in the w-plane) such that every point $w \in V$ has a unique inverse image $z \in U$.

Proof. The theorem is obvious in the case of the function (1), where we can solve the equation directly for z, but a proof is needed in the general case. It follows from Theorem 10.39a that the zeros of the analytic function $f(z) - w_0$ must be isolated (why?). Hence there is a closed disk $|z - z_0| \leqslant \varepsilon$

---

† The *angle between two intersecting curves* (with tangents) is defined as the angle between their tangents at the point of intersection.

on which $f(z) - w_0$ vanishes only once, namely at the point $z = z_0$. Then, by Theorem 10.44,

$$\frac{1}{2\pi i} \oint_{|z|=\varepsilon} \frac{f'(z)}{f(z)-w_0}\, dz = 1,$$

since the only zero of $f(z) - w_0$ inside or on the circle $|z| = \varepsilon$ is of order 1. The function

$$N(w) = \frac{1}{2\pi i} \oint_{|z|=\varepsilon} \frac{f'(z)}{f(z)-w}\, dz$$

is defined for $w = w_0$ (where it equals 1) and for neighboring values of w, say for $|w - w_0| < \delta$, since the denominator is nonvanishing on the circle $|z| = \varepsilon$ not only for $w = w_0$ but also for all values of w sufficiently close to w_0. Moreover, $N(w)$ is analytic and hence certainly continuous for $|w - w_0| < \delta$, by Theorem 10.27. But $N(w)$ takes only integral values (Theorem 10.44) and hence $N(w)$ must equal 1 for $|w - w_0| < \delta$. By Theorem 10.44 again, this implies that the function $f(z) - w$ vanishes only once inside the circle $|z| = \varepsilon$, i.e., for any w in the neighborhood $V = \{w : |w - w_0| < \delta\}$ there is one and only one point z in the neighborhood $U = \{z : |z - z_0| < \varepsilon\}$ such that $f(z) = w$. ∎

c. THEOREM. *Let $f(z)$, U, and V be the same as in the preceding theorem. Then the function $f(z)$ is one-to-one on some neighborhood of z_0.*

Proof. Some points of the set $f(U) = \{w : w = f(z), z \in U\}$ may fail to lie in V. However, $f(z)$ is continuous at $z = z_0$, and hence there is another neighborhood $U^* \subset U$ of the point z_0 such that the set $f(U^*) = \{w : w = f(z), z \in U^*\}$ is contained in V. But, by the preceding theorem, every $w \in f(U^*)$ has a unique inverse image in U, and this inverse image can only be the point $z \in U^*$ such that $f(z) = w$. ∎

10.53. THEOREM. *Let $f(z)$, U^*, and V be the same as in the preceding theorem, and let $z = \varphi(w)$ be the inverse of the function $w = f(z)$ on U^*, i.e., the function assigning to each point $w \in f(U^*)$ the unique point $z \in U^*$ such that $f(z) = w$. Then $\varphi(w)$ is differentiable at $w = w_0 = f(z_0)$, with derivative*

$$\varphi'(w_0) = \frac{1}{f'(z_0)}.$$

Proof. The function $\varphi(w)$ is continuous at $w = w_0$. In fact, given *any* sufficiently small neighborhood $U = \{z : |z - z_0| < \varepsilon\}$ of the point z_0, the same argument as in the proof of Theorem 10.52b shows that there is a corresponding neighborhood $V = \{w : |w - w_0| < \delta\}$ of the point w_0 such that the inverse image of every point $w \in V$ belongs to U. But this is precisely what is meant by saying that $\varphi(w)$ is continuous at $w = w_0$.

Now consider the "difference quotient"

$$\frac{\varphi(w) - \varphi(w_0)}{w - w_0}. \tag{5}$$

As just shown, $z = \varphi(w) \to z_0 = \varphi(w_0)$, and hence (cf. Theorem 7.16) the limit of (5) as $w \to w_0$ exists and equals

$$\varphi'(w_0) = \lim_{w \to w_0} \frac{\varphi(w) - \varphi(w_0)}{w - w_0} = \lim_{z \to z_0} \frac{1}{\dfrac{w - w_0}{z - z_0}} = \frac{1}{\displaystyle\lim_{z \to z_0} \frac{f(z) - f(z_0)}{z - z_0}}$$

$$= \frac{1}{f'(z_0)}. \quad \blacksquare$$

10.54. a. A function $w = f(z)$ is said to be *univalent* on a domain G if it is one-to-one and analytic on G. According to Theorem 10.52c, if $f(z)$ is analytic at $z = z_0$ and if $f'(z_0) \neq 0$, then $f(z)$ is univalent on some neighborhood of z_0. Moreover, as we will see later (Sec. 10.59b), if $f(z)$ is univalent on a domain G, then $f'(z) \neq 0$ for all $z \in G$.

b. The linear function (1) is obviously univalent on the whole z-plane, while the function

$$w = \frac{1}{z} \tag{6}$$

is univalent on the whole z-plane minus the point $z = 0$.†

10.55. By a *fractional linear function* we mean any mapping of the form

$$w = \frac{az + b}{cz + d} \qquad (bc \neq ad, \ c \neq 0). \tag{7}$$

Writing (7) as

$$w = \frac{a}{c} + \frac{bc - ad}{c(cz + d)}, \tag{7'}$$

we see that (7) is the result of applying first a linear function of the form (1), then the function (6), and finally another linear function. But (1) is a one-to-one mapping of the extended z-plane onto the extended w-plane, carrying the point $z = \infty$ into the point $w = \infty$, while (6) is a one-to-one mapping of the extended z-plane onto the extended w-plane, carrying the point $z = 0$

† Note that the mapping (6) is conformal at every finite point $z \neq 0$ (see Sec. 10.51).

into the point $w = \infty$ and the point $z = \infty$ into the point $w = 0$.[†] Hence the fractional linear function (7) is itself a one-to-one mapping of the extended z-plane into the extended w-plane, carrying the point $z = -d/c$ into the point $w = \infty$ and the point $z = \infty$ into the point $w = a/c$.

10.56. Let

$$w = z^n, \tag{8}$$

where n is an integer greater than 1. Then

$$|w| = |z|^n, \qquad \arg w = n \arg z,$$

so that the function (8) is univalent on the "wedge" or "angular domain"

$$-\alpha < \arg z < \alpha \qquad (\alpha \leqslant \pi/n) \tag{9}$$

which it maps in a one-to-one fashion onto the wedge

$$-n\alpha < \arg w < n\alpha.$$

Clearly the largest wedge of the form (9) on which the function (8) remains univalent is the wedge

$$-\pi/n < \arg z < \pi/n, \tag{9'}$$

which is mapped by (8) onto the domain W, consisting of all points of the w-plane except those on the negative real axis.

The inverse of the function (8) is denoted by

$$z = \sqrt[n]{w} \tag{10}$$

and called the *nth root* of w as in the case of positive real w, for which it reduces to the familiar function of elementary analysis (Sec. 1.63). The function (10) is univalent on the domain W, which it maps onto the wedge (9').

According to Sec. 10.51, the mapping (8) is conformal at every point $z \neq 0$, since the derivative $w' = nz^{n-1}$ is nonzero if $z \neq 0$. However, (8) fails to be conformal at the point $z = 0$, where w' vanishes. In fact, the mapping obviously produces an n-fold enlargement of angles between rays intersecting at $z = 0$.

10.57. Next consider the *exponential*

$$w = e^z. \tag{11}$$

Writing $w = u + iv$, $z = x + iy$, we have

[†] More exactly, if w is the linear function (1), then $w \to \infty$ as $z \to \infty$, while if w is the function (6), then $w \to \infty$ as $z \to 0$ and $w \to 0$ as $z \to \infty$ (see Sec. 10.47a).

$$u + iv = e^{x+iy} = e^x(\cos y + i \sin y)$$

(Sec. 8.63), so that in particular

$$|e^z| = e^x, \qquad \arg e^z = y + 2k\pi \qquad (k = 0, \pm 1, \pm 2, \ldots). \tag{12}$$

The function (11) maps the horizontal line $y = y_0$ in the z-plane into the curve

$$u = e^x \cos y_0, \qquad v = e^x \sin y_0,$$

i.e., the ray drawn from the origin of the w-plane making the angle y_0 with the positive u-axis. By the same token, the strip

$$-\alpha < y < \alpha \qquad (\alpha \leqslant \pi) \tag{13}$$

in the z-plane is mapped by (11) in a one-to-one fashion onto the wedge

$$-\alpha < \arg w < \alpha$$

in the w-plane. The largest strip of the form (13) on which the function (11) remains univalent is the strip

$$-\pi < y < \pi, \tag{13'}$$

which is mapped by (11) onto the same domain W as in Sec. 10.56.

The inverse of the function (11) is denoted by

$$z = \ln w \tag{14}$$

and called the *logarithm* as in the case of positive real w, for which it reduces to the familiar function of elementary analysis (Sec. 5.4). The function (14) is univalent on the domain W, which it maps onto the strip (13'). Solving (12) for x and y, we get

$$x = \ln |e^z|, \qquad y = \arg e^z,$$

and hence

$$\ln w = z = x + iy = \ln |e^z| + i \arg e^z = \ln |w| + i \arg w \qquad (-\pi < \arg w < \pi). \tag{15}$$

Formula (15) allows us to take real and imaginary parts of $\ln w$. Moreover, "inverting" the formula

$$e^{z_1 + z_2} = e^{z_1} e^{z_2}$$

(Theorem 8.64), we get the corresponding property of complex logarithms:

$$\ln w_1 w_2 = \ln w_1 + \ln w_2.$$

To differentiate the logarithm, we use Theorem 10.53 and the formula

$$(e^z)' = e^z$$

proved in Sec. 10.37b, obtaining

$$(\ln w)' = \frac{1}{(e^z)'} = \frac{1}{e^z} = \frac{1}{w}.$$

It follows from Theorem 10.25b that

$$\ln w = \ln w_0 + \int_{w_0}^{w} \frac{d\omega}{\omega}$$

for every $w \in W$, where the path of integration is any piecewise smooth curve $L \subset W$ joining w_0 to w. The simplest choice is $w_0 = 1$. Then $\ln w_0 = 0$ and we have

$$\ln w = \int_1^w \frac{d\omega}{\omega}.$$

Thus the logarithm has now been defined for all complex w with the exception of negative real w. Later we will enlarge the domain of definition of $\ln w$ to include negative values of w as well (see Example 10.59d).

10.58. We now use the formula

$$z^\lambda = e^{\lambda \ln z} \qquad (\lambda > 0)$$

to define an arbitrary positive real power of z, assuming that z is not a negative real number. Consider the wedge

$$-\alpha < \arg z < \alpha \qquad (\alpha \leqslant \min \{\pi, \pi/\lambda\}) \tag{16}$$

or equivalently

$$-\alpha < \operatorname{Im} \ln z < \alpha. \tag{17}$$

Multiplication of (17) by λ gives

$$-\lambda\alpha < \operatorname{Im} (\lambda \ln z) < \lambda\alpha, \tag{18}$$

which corresponds to

$$-\lambda\alpha < \arg w < \lambda\alpha.$$

where

$$w = z^\lambda. \tag{19}$$

The largest wedge of the form (16) on which the function (19) remains univalent is the wedge

$$-\pi/\lambda < \arg z < \pi/\lambda, \tag{16'}$$

which is mapped by (19) onto the same domain W as in Secs. 10.56 and 10.57. Using the usual rules, we find that the derivative of (19) is just

$$(e^{\lambda \ln z})' = e^{\lambda \ln z}(\lambda \ln z)' = \frac{\lambda}{z}e^{\lambda \ln z} = \lambda z^{\lambda-1}.$$

10.59. Riemann surfaces

a. The function

$$w = (z - z_0)^n \qquad (n > 1) \tag{20}$$

is univalent on the wedge

$$-\pi/n < \arg(z - z_0) < \pi/n$$

(cf. Sec. 10.56) and analytic on the disk $K = \{z: |z - z_0| < r\}$, but not univalent on K. In fact, (20) maps the circle $L = \{z: |z - z_0| = \rho < r\}$ onto the circle $\Lambda = \{w: w = \rho^n\}$ in the w-plane, but as the point z makes one circuit around L, the corresponding point w makes n circuits around Λ.

b. The situation is similar near a zero z_0 of order n of any analytic function

$$w = f(z) = a_n(z - z_0)^n + a_{n+1}(z - z_0)^{n+1} + \cdots \qquad (a_n \neq 0, n > 1).$$

In fact, let U be a neighborhood of z_0 containing no other zeros of $f(z)$ and $f'(z)$, and let $L \subset U$ be a piecewise smooth closed path surrounding z_0. By Theorem 10.44, the logarithmic residue of $f(z)$ with respect to L is just

$$\frac{1}{2\pi i}\oint_L \frac{f'(z)}{f(z)}\,dz = n.$$

But then

$$\frac{1}{2\pi i}\oint_L \frac{f'(z)}{f(z) - w}\,dz = n$$

for all $w \neq 0$ with sufficiently small absolute value, by the same argument as in the proof of Theorem 10.52b. It follows, again by Theorem 10.44, that $f(z)$ takes any such value w precisely n times in U, with due regard for order. But $f'(z)$ does not vanish in U except at the point $z = z_0$ itself, and hence $f(z)$ takes the value w at n *distinct* points of U.

In cases like that just considered, the mapping can be made one-to-one by introducing the notion of a *Riemann surface*. We will not give a general definition of a Riemann surface here, confining ourselves instead to a discussion of two typical examples and some subsequent remarks of a qualitative nature.†

c. Example. To construct the Riemann surface for the function $w = z^n$ ($n > 1$) we start with n "samples" $D_0, D_1, \ldots, D_{n-1}$ of the ordinary w-plane, all "cut along the negative real axis" (i.e., with the points of the negative real axis deleted), regarding the point $w = 0$ as the same for all n "samples" or "sheets." The sheets are then "pasted together" (i.e., the corresponding points of the sheets are then identified) in the following fashion: The lower edge of the cut on the first sheet D_0 is pasted to the upper edge of the cut on the second sheet D_1, the lower edge of the cut on the second sheet D_1 is pasted to the upper edge of the cut on the third sheet D_2, and so on, until finally the lower edge of the cut on the nth sheet D_{n-1} is pasted to the upper edge of the cut on the first sheet D_0, as shown in Figure 72 (for the case $n = 4$).‡ At the same time, each sheet is regarded as equipped with its original real and imaginary axes.

The function $w = z^n$ can now be regarded as a one-to-one mapping of the whole z-plane onto the Riemann surface just constructed. In fact, suppose the point z traverses the circle $|z| = \rho$ in the counterclockwise direction,

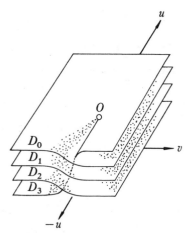

Figure   72

---

† Readers interested in pursuing the study of Riemann surfaces are referred to the abundant literature on the subject. See e.g., Markushevich *Theory of Functions of a Complex Variable*, Vol. III, Chapter 7 and G. Springer, *Introduction to Riemann Surfaces*, Addison-Wesley Publishing Co., Inc., Reading, Mass. (1957).
‡ The fact that this "pasting together" cannot actually be accomplished in three-dimensional space without further "self-intersections" does not matter.

starting from the initial position $z = \rho > 0$. Then the corresponding point w traverses the circle $|w| = \rho^n$ on the first sheet of the Riemann surface until the point z reaches the boundary of the "domain of univalence" $-\pi/n <$ arg $z < \pi/n$ (cf. Sec. 10.56), corresponding to the value arg $z = \pi/n$. Then arg $w = \pi$, so that w lies on the cut along which the first sheet is pasted to the second sheet. As z traverses the circle $|z| = \rho$ further in the same direction, the point w moves onto the second sheet, then onto the third sheet, and so on, finally returning to its original position on the first sheet after z has made one complete circuit around the circle $|z| = \rho$. These considerations show that the function $w = z^n$ is in fact a one-to-one mapping of the z-plane onto the Riemann surface shown in Figure 72. The inverse function, which we continue to denote by $z = \sqrt[n]{w}$, is defined on the same Riemann surface, which it maps back onto the whole z-plane in a one-to-one fashion.

Any n distinct points of the Riemann surface with the same real and imaginary parts but on different sheets of the surface correspond to n distinct points $z_0, z_1, \ldots, z_{n-1}$ of the z-plane, namely the points with absolute value $\sqrt[n]{|w|}$ and arguments†

$$\arg z_k = \frac{1}{n}(\arg w + 2k\pi) \qquad (k = 0, 1, \ldots, n-1).$$

d. Example. Next we construct the Riemann surface for the function $w = e^z$, starting from a (countably) infinite number of samples D_k ($k = 0$, $\pm 1, \pm 2, \ldots$) of the w-plane, all cut along the negative real axis. The sheets are pasted together in the same way as in the preceding example, i.e., they are all joined together at the point $w = 0$ and for each k the lower edge of the cut on the sheet D_k is pasted to the upper edge of the cut on the sheet D_{k+1} (see Figure 73).

The function $w = e^z$ can now be regarded as a one-to-one mapping of the whole z-plane onto this Riemann surface. In fact, suppose the point z moves upward along the vertical line $x = x_0$, starting from the initial position $z = x_0 > 0$. Then the corresponding point w traces out the circle $|w| = e^{x_0}$ on the first sheet of the Riemann surface until z reaches the boundary of the "domain of univalence" $-\pi < y < \pi$ (cf. Sec. 10.57), corresponding to the value $y = \pi$. Then arg $w = \pi$, so that w lies on the cut along which the first sheet (D_0 in the figure) is pasted to the second sheet. As z moves further up the line $x = x_0$, the point w moves onto the second sheet (D_1 in the figure), then onto the third sheet (D_2), and so on indefinitely. (As an exercise, describe what happens as the point x moves *downward* along the line $x = x_0$.)

The inverse function, which we continue to denote by $z = \ln w$, is defined

† Here we restrict arg z_k and arg w to their *principal values*, i.e., to their unique values in the half-open interval $(-\pi, \pi]$.

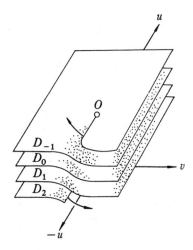

Figure   73

on the same Riemann surface, which it maps onto the whole z-plane in a one-to-one fashion, with the value of w on the nth sheet corresponding to the value of z in the strip

$$(2n-1)\pi < y < (2n+1)\pi.$$

On the other hand, the inverse of the function $w = e^z$, regarded as a function defined on the whole w-plane (rather than on the Riemann surface), is a "multiple-valued" function (Sec. 2.84), in fact an "infinite-valued" function, which we denote by Ln w. The values of Ln w at the point w are given by the formula

$$\text{Ln } w = \ln |w| + i \text{ arg } w,$$

where arg w varies over all its possible values (differing by integral multiples of 2π). In particular, for a negative real number $w = -p$ ($p > 0$), we have

$$\text{Ln } (-p) = \ln |p| + (2k+1)\pi i \qquad (k = 0, \pm 1, \pm 2, \dots),$$

where there is no natural reason to prefer one of these infinitely many values of the logarithm to any other.

e. Riemann surfaces can be constructed for other analytic functions in much the same way. To construct a Riemann surface for a given function $w = f(z)$, we first prepare a suitable number of identical sheets, equipped with cuts making them into "images" of domains of univalence of $f(z)$. We then paste the edges of the cuts on the different sheets together (just as was done in the above examples), thereby guaranteeing the one-to-one character of the resulting mapping of the whole z-plane onto the whole Riemann surface for w.

Suppose z_0 is a point of the z-plane with the property that every circuit around a circle of arbitrarily small radius centered at z_0 causes the function $w = f(z)$ to take a new value. Then z_0 is called a *branch point* of $f(z)$.† Branch points play an important role in determining the structure of a Riemann surface. Given any point z_0 which is not a branch point of $f(z)$, there is a neighborhood of z_0 in which $f(z)$ reduces to a family of single-valued analytic functions. No such neighborhood exists for a branch point.

Problems

1. Given a series

$$f(z) = \sum_{n=0}^{\infty} a_n z^n \qquad (1)$$

with radius of convergence 1, suppose the coefficients a_n are all positive. Prove that $z = 1$ is a singular point of $f(z)$.

2. Given a series (1) with radius of convergence 1, suppose the only singular point of $f(z)$ on the circle $|z| = 1$ is a pole z_0 of order m. Prove that $|a_n| \leqslant An^{m-1}$, where A is a constant.

3 (*Maximum modulus principle*). Let $f(z)$ be analytic on a domain G bounded by a contour L, and let

$$M = \sup_{z \in L} |f(z)|.$$

Prove that $|f(z)| \leqslant M$ for all $z \in G$. Prove that if $|f(z_0)| = M$ at some (interior) point $z_0 \in G$, then $f(z)$ is constant on G.

4 (*Schwarz's lemma*). Let $f(z)$ be analytic on the disk $K = \{z : |z| < R\}$, where $f(0) = 0$ and $|f(z)| \leqslant M$ for all $z \in K$. Prove that

$$|f(z)| \leqslant \frac{M}{R}|z|$$

for all $z \in K$.

5. Let $f(z)$ be analytic on the disk $K = \{z : |z| < R\}$, and let

$$M(r) = \max_{|z| = r} |f(z)| \qquad (r < R).$$

Prove that $\ln M(r)$ is a convex function of $\ln r$. (Hadamard)

---

† As an exercise, show that the points $z = 0$ and $z = \infty$ are branch points (and the only ones) of the functions $w = \sqrt[n]{z}$ and $w = \operatorname{Ln} z$.

6. A function $u(x, y)$ of two real variables x and y is said to be *differentiable* at the point $(x, y) = (x_0, y_0)$ if there exist numbers A_u and B_u such that

$$u(x_0 + h, y_0 + k) = u(x_0, y_0) + A_u h + B_u k + \varepsilon(h, k)$$

in a neighborhood of (x_0, y_0), where

$$\frac{\varepsilon(h, k)}{\sqrt{h^2 + k^2}} \to 0$$

as $h \to 0$, $k \to 0$. Prove that the complex function $w = u(x, y) + iv(x, y)$ is differentiable at $z = z_0$ if and only if $u(x, y)$ and $v(x, y)$ are differentiable at the point $(x, y) = (x_0, y_0)$ and $A_u = B_v$, $A_v = -B_u$.

7. Prove that the fractional linear function

$$w = \frac{az + b}{cz + d}$$

carries the family of all circles (including straight lines, regarded as circles of infinite radius) into itself.

8. The "growth" of the function

$$f(z) = \sum_{n=0}^{\infty} a_n z^n$$

can be estimated by using the functions

$$M(r) = \max_{|z| = r} |f(z)|$$

and

$$M_1(r) = \sum_{n=0}^{\infty} |a_n| r^n,$$

where obviously $M(r) \leqslant M_1(r)$. Prove that

$$M_1(r) \leqslant \frac{r + \delta}{\delta} M(r + \delta)$$

for every $\delta > 0$, so that "$M_1(r)$ does not grow too rapidly compared to $M(r)$."

9 (*Argument principle*). Let $f(z)$ be analytic on a piecewise smooth simple closed contour L and its interior G, except possibly for poles in G, and let $f(z)$ be nonzero on L. Prove that the number of circuits around the point $w = 0$ made by the point $w = f(z)$ as z traverses L once in the counterclockwise direction equals the number of zeros of $f(z)$ in G minus the number of poles.

10. Let $f_n(z)$ be a sequence of analytic functions which converges uniformly inside a domain G (see Sec. 10.36a) to a function $f(z) \not\equiv 0$, and let Z be the set of all zeros of all the functions $f_n(z)$ in G. Prove that the set of all zeros of the function $f(z)$ in G (with a zero of order m counted as m zeros) coincides with the set of all limit points of Z in G.

11. Prove that if the series

$$\sum_{n=0}^{\infty} g^{(n)}(z)$$

converges at even one point of analyticity of the function $g(z)$, then $g(z)$ is entire and the series converges uniformly on every disk. (Pólya)

12. Construct the Riemann surface, find the branch points, and describe the domains of univalence of the function

$$w = \frac{1}{2}\left(z + \frac{1}{z}\right) \tag{2}$$

and its inverse

$$z = w \pm \sqrt{w^2 - 1}. \tag{2'}$$

Do the same for the function

$$w = \cos z \tag{3}$$

and its inverse

$$z = \text{Arc cos } w. \tag{3'}$$

13. Prove that $i \text{ Arc cos } w = \text{Ln}(w \pm \sqrt{w^2 - 1})$.

14. Prove that if the functions $f_1(z), \ldots, f_n(z)$ are all analytic on a domain G, then $|f_1(z)| + \cdots + |f_n(z)|$ cannot have a maximum in G.

15. Suppose $f(z)$ is analytic on the whole z-plane which it maps into a subset of the upper half-plane. Prove that $f(z)$ is a constant.

16. Derive the fundamental theorem of algebra (Theorem 5.85c) from Liouville's theorem (Sec. 10.38).

17. Derive the fundamental theorem of algebra from Theorem 10.44.

18. Derive the fundamental theorem of algebra from the maximum modulus principle (Problem 3).

19 (*Partial fraction expansions*). Let $f(z)$ be analytic on the whole z-plane except at the points $z_1, \ldots, z_n, \ldots$, where $f(z)$ has simple poles with residues $b_1, \ldots, b_n, \ldots$ Suppose that for some C the set $\{z : |f(z)| \leqslant C\}$ contains a family of circles Γ_n centered at $z = 0$ with radii $R_n \to \infty$. Prove that

$$f(z) = f(0) + \sum_{n=1}^{\infty} b_n\left(\frac{1}{z - z_n} + \frac{1}{z_n}\right). \tag{4}$$

20 (*Infinite product expansions*). Let $g(z)$ be analytic on the whole z-plane, with simple zeros at the points $z_1,...,z_n,...$, and suppose the function

$$f(z) = \frac{g'(z)}{g(z)}$$

satisfies the conditions of the preceding problem. Prove that

$$g(z) = g(0)e^{g'(0)z/g(0)} \prod_{n=1}^{\infty} \left\{ \left(1 - \frac{z}{z_n}\right)e^{z/z_n} \right\}. \tag{5}$$

21. Verify the expansions

$$\sin z = z \prod_{n=1}^{\infty} \left(1 - \frac{z^2}{n^2\pi^2}\right),$$

$$\cos z = \prod_{n=1}^{\infty} \left(1 - \frac{z^2}{(n+\frac{1}{2})^2\pi^2}\right).$$

22 (*Phragmén-Lindelöf theorem*). Suppose $f(z)$ is a function analytic on a domain

$$G = \{z: -\alpha \leqslant \arg z \leqslant \alpha < \pi/2\}$$

such that

$$|f(z)| \leqslant Ae^{B|z|}$$

for all $z \in G$, while $|f(z)| \leqslant 1$ on the rays $\arg z = \pm\alpha$. Prove that the inequality $|f(z)| \leqslant 1$ also holds for all $z \in G$.

23. Suppose $f(z)$ is a function analytic on the upper half-plane G such that

$$|f(x)| \leqslant C \qquad (-\infty < x < \infty),$$

while, given any $\varepsilon > 0$,

$$|f(z)| \leqslant A_\varepsilon e^{\varepsilon|z|}$$

for all $z \in G$ and some A_ε. Prove that $|f(z)| \leqslant C$ for all $z \in G$.

24. Suppose $f(z)$ is an entire function such that

$$|f(x)| \leqslant C, \qquad \inf|f(x)| = 0 \qquad (-\infty < x < \infty),$$

while, given any $\varepsilon > 0$,

$$|f(z)| \leqslant A_\varepsilon e^{\varepsilon|z|}$$

for all z and some A_ε. Prove that $f(z) \equiv 0$.

11 Improper Integrals

11.1. Improper Integrals of the First Kind

11.11. Let $y = f(x)$ be a complex-valued function of a real variable x, defined on the interval $a \leqslant x \leqslant b < \infty$. Then, in keeping with Sec. 10.21, the integral of $f(x)$ over the interval $[a,b]$ is defined as

$$\int_a^b f(x) \; dx = \int_a^b u(x) \; dx + i \int_a^b v(x) \; dx,$$

where $u(x)$ is the real part and $v(x)$ the imaginary part of $f(x)$. Thus $f(x)$ is integrable on $[a,b]$ if and only if $u(x)$ and $v(x)$ are both integrable on $[a,b]$.

Suppose $f(x)$ is integrable (being piecewise continuous, say) on every finite interval $a \leqslant x \leqslant X < \infty$, where a is fixed and X variable, and consider the (complex) function

$$I(X) = \int_a^X f(x) \; dx, \tag{1}$$

defined for all $X \geqslant a$. Suppose $I(X)$ approaches a finite (complex) limit I as $X \to \infty$. Then the expression

$$\int_a^\infty f(x) \; dx, \tag{2}$$

called an *improper integral of the first kind*, is said to be *convergent*, with *value I*. In other words, we set

$$\int_a^\infty f(x) \; dx = I = \lim_{X \to \infty} I(X) = \lim_{X \to \infty} \int_a^X f(x) \; dx,$$

by definition. On the other hand, if $I(X)$ does not approach a finite limit as $X \to \infty$, we call the integral (2) *divergent* and assign it no value at all.

Examples

a. The improper integral

$$\int_1^\infty x^\alpha \; dx$$

is convergent for $\alpha < -1$, with value $-(\alpha+1)^{-1}$, and divergent for $\alpha \geqslant -1$, since

$$\int_1^X x^\alpha \; dx = \begin{cases} \dfrac{X^{\alpha+1} - 1}{\alpha + 1} & \text{if } \alpha \neq -1, \\ \ln X & \text{if } \alpha = -1. \end{cases}$$

b. The improper integrals

$$\int_0^\infty \cos x \, dx, \qquad \int_0^\infty e^{ix} \, dx$$

are divergent, since both expressions

$$\int_0^X \cos x \, dx = \sin X, \qquad \int_0^X e^{ix} \, dx = \frac{e^{iX}-1}{i}$$

fail to approach a limit as $X \to \infty$.

11.12. THEOREM. *Let $I(X)$ be given by* (1). *Then the following four assertions are equivalent*:
(a) *Given any $\varepsilon > 0$, there exists a number $X_0 > a$ such that*

$$|I - I(X)| = \left| I - \int_a^X f(x) \, dx \right| < \varepsilon$$

for all $X \geqslant X_0$ and some number I;
(b) *The sequence $I(X_n)$ has a finite limit for every sequence $X_n \to \infty$*;
(c) *The series*

$$\sum_{n=1}^\infty [I(X_{n+1}) - I(X_n)] = \sum_{n=1}^\infty \int_{X_n}^{X_{n+1}} f(x) \, dx$$

converges for every sequence $X_n \to \infty$;
(d) *Given any $\varepsilon > 0$, there exists a number $X_0 > a$ such that*

$$|I(X') - I(X'')| = \left| \int_{X'}^{X''} f(x) \, dx \right| < \varepsilon$$

*for all $X' \geqslant X_0$, $X'' \geqslant X_0$ (the **Cauchy convergence criterion for improper integrals**)*.

Proof. Assertion a is simply the definition of what is meant by the function $I(X)$ having a limit as $X \to \infty$, while Assertion d is the Cauchy convergence criterion for the existence of this limit (Theorem 4.19) and hence is equivalent to Assertion a. Assertion b is the equivalent of Assertion a in the language of sequences (see Theorem 4.65), while Assertion c merely expresses the usual connection between convergence of a series and convergence of the sequence of partial sums of the series (see Sec. 6.11a). ∎

Note that in Assertion b it is essential that we allow *any* sequence $X_n \to \infty$, since the convergence of the improper integral (2) is not implied by the convergence of the sequence $I(X_n)$ just for *some* sequence $X_n \to \infty$. For example,

$$\int_0^{2n\pi} \cos x \, dx = 0 \qquad (n=1,2,\ldots),$$

but the integral

$$\int_0^\infty \cos x \, dx$$

is divergent (Example 11.11b).

11.13. a. If the integrand $f(x)$ of the improper integral (2) is nonnegative, then the function $I(X)$ defined by (1) is nondecreasing. Then either $I(X)$ is unbounded on the interval $a \leqslant X < \infty$, in which case $I(X) \to \infty$ as $X \to \infty$ (see Theorem 4.55) and we say that (2) *diverges to* $+\infty$ (Sec. 4.61), or else $I(X)$ is bounded on $a \leqslant X < \infty$, in which case

$$\lim_{X \to \infty} I(X) = \sup_{a \leqslant X < \infty} I(X) \tag{3}$$

(see Theorem 4.53) and the integral (1) is convergent, with value (3).

b. THEOREM (**Comparison test for improper integrals**). *Let $f_1(x)$ and $f_2(x)$ be two nonnegative functions on $[a,\infty)$ integrable on every finite subinterval $[a,X] \subset [a,\infty)$, and suppose $f_1(x) \leqslant cf_2(x)$ for all $x > X_0$. Then convergence of the integral*

$$\int_a^\infty f_2(x) \, dx \tag{4}$$

implies convergence of

$$\int_a^\infty f_1(x) \, dx, \tag{5}$$

while divergence of (5) implies divergence of (4).

Proof. An immediate consequence of Sec. 11.13a and the inequality

$$I_1(X) - I_1(X_0) = \int_{X_0}^X f_1(x) \, dx \leqslant c \int_{X_0}^X f_2(x) \, dx = c[I_2(X) - I_2(X_0)],$$

valid for all $X > X_0$. ∎

c. Example. Let $P(x)$ and $Q(x)$ be polynomials with complex coefficients of degrees p and q, respectively, where $Q(x)$ has no zeros in the interval $a \leqslant x < \infty$. Then the integral

$$\int_a^\infty \frac{P(x)}{Q(x)} \, dx \tag{6}$$

of the rational function $P(x)/Q(x)$ converges if $q \geqslant p+2$ and diverges if $q \leqslant p+1$. In fact, if $q \geqslant p+2$, then

$$\left| \frac{P(x)}{Q(x)} \right| \leqslant \frac{A}{x^2}$$

for some constant A and all sufficiently large x. But then the convergence of (6) follows by using the function $1/x^2$ in the comparison test (see Example 11.11a).† On the other hand, suppose $q \leqslant p+1$. Then the rational function $xP(x)/Q(x)$ has a nonzero "polynomial part," say

$$x \frac{P(x)}{Q(x)} = \frac{P_1(x)}{Q_1(x)} + b_0 + b_1 x + \cdots + b_n x^n,$$

where $n \geqslant 0$, $b_n \neq 0$, and the degree of $P_1(x)$ is less than that of $Q_1(x)$. It follows that

$$\frac{P(x)}{Q(x)} = \frac{1}{x} \frac{P_1(x)}{Q_1(x)} + \frac{b_0}{x} + b_1 + \cdots + b_n x^{n-1},$$

and hence

$$\int_a^X \frac{P(x)}{Q(x)}\,dx = \int_a^X \frac{1}{x} \frac{P_1(x)}{Q_1(x)}\,dx + b_0 \ln \frac{X}{a} + b_1 (X-a) + \cdots + b_n (X^n - a^n).$$

As just shown, the first term on the right has a finite limit (the degree of the denominator is at least 2 greater than the degree of the numerator), but the sum of the remaining terms obviously has no limit as $X \to \infty$, since its absolute value approaches infinity as $X \to \infty$. Therefore (6) diverges if $q \leqslant p+1$, as asserted.

d. Example. Let

$$P(x) = b_0 + b_1 x + \cdots + b_n x^n \qquad (b_n > 0)$$

be a polynomial with real coefficients. Then

$$\frac{c_1}{x^{n/m}} \leqslant \frac{1}{\sqrt[m]{b_0 + b_1 x + \cdots + b_n x^n}} \leqslant \frac{c_2}{x^{n/m}}$$

for sufficiently large x, say $x \geqslant a$. Hence, by Example 11.11a, the integral

$$\int_a^\infty \frac{dx}{\sqrt[m]{P(x)}}$$

converges if $n/m > 1$ and diverges if $n/m \leqslant 1$.

† Here we anticipate the almost trivial Theorem 11.21a.

11.14. Next we use improper integrals to prove a useful test (due to Cauchy) for the convergence of numerical series:

THEOREM (**Integral test**). *Given a numerical series*

$$\sum_{n=1}^{\infty} a_n \tag{7}$$

with positive nonincreasing terms $(a_{n+1} \leqslant a_n)$, *let* $a(x)$ *be a positive nonincreasing function such that* $a(n) = a_n$. *Then convergence of the integral*

$$\int_1^{\infty} a(x) \, dx \tag{8}$$

implies convergence of the series (7), *while divergence of* (8) *implies divergence of* (7).

Proof. An immediate consequence of the fact that

$$a_2 + \cdots + a_n \leqslant \int_1^n a(x) \, dx \leqslant a_1 + \cdots + a_{n-1},$$

which is in turn clear from Figure 74. ∎

Examples

a. According to Example 11.11a, the integral

$$\int_1^{\infty} \frac{dx}{x^{\alpha}} \qquad (\alpha \geqslant 0)$$

converges if $\alpha > 1$ and diverges if $0 \leqslant \alpha \leqslant 1$. Hence, by the integral test, the corresponding series

$$\sum_{n=1}^{\infty} \frac{1}{n^{\alpha}}$$

converges if $\alpha > 1$ and diverges if $0 \leqslant \alpha \leqslant 1$ (see Example 6.15a).

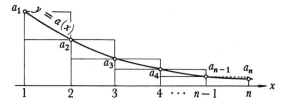

Figure   74

b. The formula

$$\int_a^\infty \frac{dx}{x(\ln x)^\alpha} = \int_{\ln a}^\infty \frac{du}{u^\alpha} \qquad (u = \ln x, \, a > 1, \, \alpha \geqslant 0)$$

shows that the improper integral on the left converges if $\alpha > 1$ and diverges if $0 \leqslant \alpha \leqslant 1$. Hence the corresponding series

$$\sum_{n=2}^\infty \frac{1}{n(\ln n)^\alpha}$$

converges if $\alpha > 1$ and diverges if $0 \leqslant \alpha \leqslant 1$ (see Example 6.15c).

c. Invoking geometrical considerations like those used to prove the integral test itself, we can deduce various estimates for the remainder of converging series and for partial sums of both convergent and divergent series. For example, Figure 75 shows at once that

$$\sum_{k=n}^\infty \frac{1}{k^\lambda} \leqslant \int_{n-1}^\infty \frac{dx}{x^\lambda} = \frac{1}{\lambda-1} \frac{1}{(n-1)^{\lambda-1}} \qquad (\lambda > 1),$$

while choosing $a(x) = 1/x$ in Figure 74, we get

$$\sum_{k=2}^n \frac{1}{k} \leqslant \int_1^n \frac{dx}{x} = \ln n \leqslant \sum_{k=1}^{n-1} \frac{1}{k}.$$

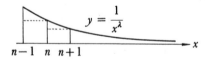

Figure   75

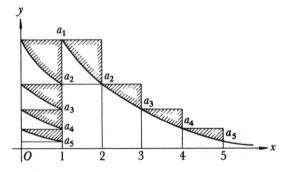

Figure   76

Similarly, it follows from Figure 76 that

$$\sum_{k=1}^{n} \left\{ a_k - \int_k^{k+1} a(x)\, dx \right\} \leqslant a_1$$

where the quantity on the left increases with n, approaching a limit C as $n \to \infty$, so that

$$\sum_{k=1}^{n} a_k = \int_1^{n+1} a(x)\, dx + C - \gamma_n, \tag{9}$$

where $\gamma_n \searrow 0$ as $n \to \infty$. The value of the constant C for $a_k = 1/k$, known as *Euler's constant*, is often encountered in analysis. This constant, first calculated by Euler in 1734, equals $C = 0.5772\ldots$ Thus, choosing $a_k = 1/k$ in (9), we find that

$$1 + \frac{1}{2} + \frac{1}{3} + \cdots + \frac{1}{n} = \ln(n+1) + 0.5772\ldots - \gamma_n \qquad (\gamma_n \searrow 0).$$

11.15. Next we establish the analogues for improper integrals of the formulas for integration by parts and integration by substitution.

a. Let $u(x)$ and $v(x)$ be two functions defined on $[a, \infty)$, both piecewise smooth on every finite subinterval $[a, X] \subset [a, \infty)$. Then integration by parts gives

$$\int_a^X u\, dv = uv \Big|_a^X - \int_a^X v\, du, \tag{10}$$

as in Sec. 9.51a. Taking the limit as $X \to \infty$ in (10), we get

$$\int_a^\infty u\, dv = uv \Big|_a^\infty - \int_a^\infty v\, du, \tag{10'}$$

provided that at least two of the limits in (10′) exist (in which case the other exists automatically). Here, of course, the "integrated term" in (10′) is just

$$uv \Big|_a^\infty = \lim_{X \to \infty} uv \Big|_a^X = \lim_{X \to \infty} [u(X)v(X) - u(a)v(a)].$$

b. Similarly, suppose $u(t)$ has a continuous derivative on every finite subinterval $[\alpha, \xi] \subset [\alpha, \infty)$, and suppose $f(u)$ is continuous on the interval $\{u : u = u(t),\ \alpha \leqslant t < \infty\}$. Then

$$\int_a^X f(u)\, du = \int_\alpha^\xi f(u(t))u'(t)\, dt, \tag{11}$$

as in Sec. 9.53, where $u(\alpha) = a$, $u(\xi) = X$. Suppose $\xi \to \infty$ implies $X \to b$

$(b \leqslant \infty)$. Then, taking the limit as $\xi \to \infty$ in (11), we get

$$\int_a^b f(u) \ du = \int_\alpha^\infty f(n(t))u'(t) \ dt, \tag{11'}$$

provided that at least one of the integrals in (11) exists (in which case the other exists automatically).

11.2. Convergence of Improper Integrals

11.21. Consider the improper integral

$$\int_a^\infty f(x) \ dx, \tag{1}$$

where $f(x)$ is a complex function which is piecewise continuous on every finite subinterval $[a,X] \subset [a,\infty)$.

a. THEOREM. *If the integral*

$$\int_a^\infty |f(x)| \ dx \tag{2}$$

converges, then so does the integral (1).

Proof. An immediate consequence of the Cauchy convergence criterion for improper integrals (see Theorem 11.12) and the inequality

$$\left| \int_{X'}^{X''} f(x) \ dx \right| \leqslant \int_{X'}^{X''} |f(x)| \ dx,$$

valid for all $X' \geqslant a$, $X'' \geqslant a$. ∎

b. Definition. An improper integral (1) is said to be *absolutely convergent* if the integral (2) converges.† It may turn out that the integral (1) converges while the integral (2) diverges (an example will be given below). We then say that (1) is *conditionally convergent*, just as in the analogous situation for series (see Sec. 6.22).

11.22. Next we consider convergence tests for nonabsolutely convergent improper integrals.

THEOREM (**Leibniz's test for improper integrals**). *Suppose $f(x)$ is real and has infinitely many zeros $a_1, a_2, \ldots, a_n, \ldots$ in the interval $[a,\infty)$, where $a_1 < a_2 <$*

† In this case, the function $f(x)$ is said to be *absolutely integrable* on $[a,\infty)$.

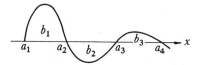

Figure   77

$\cdots < a_n < \cdots$ and $a_n \to \infty$ as $n \to \infty$. Suppose $f(x) > 0$ if $a_{2n-1} < x < a_{2n}$, while $f(x) < 0$ if $a_{2n} < x < a_{2n+1}$ (see Figure 77), and let

$$b_n = \int_{a_n}^{a_{n+1}} f(x) \, dx.$$

Suppose further that

$$|b_n| \geqslant |b_{n+1}| \qquad (n = N, \, N+1, \ldots)$$

and $|b_n| \to 0$ as $n \to \infty$. Then the integral (1) converges.

Proof. Given any $X > a$, let n be such that $a_n \leqslant X < a_{n+1}$. Then

$$\int_a^X f(x) \, dx = \int_a^{a_1} f(x) \, dx + [b_1 + \cdots + b_{n-1}] + \int_{a_n}^X f(x) \, dx. \qquad (3)$$

The first term on the right is constant, while the term in brackets approaches a (finite) limit as $n \to \infty$, by Leibniz's test for series (Theorem 6.23). The last term does not exceed

$$\int_{a_n}^{a_{n+1}} |f(x)| \, dx = |b_n|$$

in absolute value, and hence approaches zero as $n \to \infty$. It follows that the left-hand side of (3) approaches a finite limit as $X \to \infty$. ∎

Examples

a. Consider the improper integral

$$\int_a^\infty g(x) \sin x \, dx, \qquad (4)$$

where $g(x) > 0$ and $g(x) \searrow 0$ as $x \to \infty$. The consecutive zeros of the function $g(x)\sin x$ are at the points $x = n\pi$, $(n+1)\pi, \ldots$ (where $n\pi \geqslant a$). Since, obviously,

$$g(x)|\sin x| \geqslant g(x+\pi)|\sin (x+\pi)|,$$

we have

$$|b_n| = \int_{n\pi}^{(n+1)\pi} g(x)|\sin x| \, dx \geqslant \int_{n\pi}^{(n+1)\pi} g(x+\pi)|\sin (x+\pi)| \, dx$$
$$= \int_{(n+1)\pi}^{(n+2)\pi} g(x)|\sin x| \, dx = |b_{n+1}|.$$

Moreover,

$$|b_n| = \int_{n\pi}^{(n+1)\pi} g(x)|\sin x| \, dx \leqslant \pi g(n\pi) \to 0$$

as $n \to \infty$. Thus the conditions of Leibniz's test are satisfied, and the integral (4) converges.

If

$$\int_a^\infty g(x) \, dx < \infty,$$

the integral (4) is absolutely convergent. On the other hand, if

$$\int_a^\infty g(x) \, dx = \infty,$$

then the integral

$$\int_a^\infty g(x)|\sin x| \, dx \tag{5}$$

diverges. In fact, the inequalities

$$|\sin x| \geqslant \frac{1}{2}, \qquad g(x) \geqslant g(n\pi)$$

hold in the interval

$$(n-1)\pi + \frac{\pi}{6} \leqslant x \leqslant n\pi - \frac{\pi}{6},$$

and hence

$$\int_{(n-1)\pi}^{n\pi} g(x)|\sin x| \, dx \geqslant \int_{(n-1)\pi+(\pi/6)}^{n\pi-(\pi/6)} g(x)|\sin x| \, dx \geqslant g(n\pi)\frac{1}{2}\frac{2\pi}{3}.$$

But the series with general term $g(n\pi)$ diverges, by the integral test (Theorem 11.14), since

$$\int_a^\infty g(\pi x) \, dx = \frac{1}{\pi} \int_{a\pi}^\infty g(u) \, du \qquad (u = \pi x),$$

where the integral on the right diverges. Therefore the integral (5) also diverges, by Theorem 11.12.

b. The substitution $x^\gamma = u$ transforms the integral

$$\int_a^\infty \sin(x^\gamma)\, dx \qquad (\gamma > 0) \tag{6}$$

into the integral

$$\frac{1}{\gamma} \int_{a^\gamma}^\infty u^{(1/\gamma)-1} \sin u\, du,$$

which, as just shown, converges if $(1/\gamma) - 1 < 0$, i.e., if $\gamma > 1$.† Note that the integrand $\sin(x^\gamma)$ of the convergent integral (6) does not approach zero as $x \to \infty$, as might be expected on the basis of a formal analogy with the case of series. In fact, the analogue of the assertion "the general term of a convergent numerical series approaches zero" is not the assertion "the integrand of a convergent improper integral approaches zero as $x \to \infty$," which, as we see, is false, but rather the assertion "the integral of the integrand of a convergent improper integral taken over an interval of fixed length approaches zero as the interval moves off to infinity." In other words,

$$\lim_{x \to \infty} \int_x^{x+h} f(u)\, du = 0$$

for any fixed $h > 0$, as follows at once from the Cauchy convergence criterion for improper integrals (see Theorem 11.12).

11.23. The Abel-Dirichlet test for improper integrals. The Abel-Dirichlet test gives conditions under which an improper integral of the form

$$\int_a^\infty f(x)g(x)\, dx \tag{7}$$

converges. First we consider the general case where the functions $f(x)$ and $g(x)$ are both complex:

a. THEOREM. *Let $f(x)$ be piecewise smooth on every finite subinterval $[a,X] \subset [a,\infty)$, with derivative $f'(x)$, and suppose $f(x) \to 0$ as $x \to \infty$ while $f'(x)$ is absolutely integrable on $[a,\infty)$. Moreover, let $g(x)$ be piecewise continuous on every finite subinterval $[a,X] \subset [a,\infty)$, and suppose*

$$|G(x)| \leqslant C \qquad (a \leqslant x < \infty),$$

where

$$G(x) = \int_a^x g(\xi)\, d\xi \qquad (a \leqslant x < \infty)$$

† We could also apply Leibniz's test directly to the integral (6).

and C is independent of $x \in [a, \infty)$. *Then the integral* (7) *converges*.

Proof. Integrating by parts between finite limits p and q, we get

$$\int_p^q f(x)g(x)\,dx = f(x)G(x)\Big|_p^q - \int_p^q G(x)f'(x)\,dx \qquad (a < p < q). \tag{8}$$

The first term on the right satisfies the inequality

$$f(x)G(x)\Big|_p^q = |f(q)G(q) - f(p)G(p)| \leqslant 2C \max\{|f(p)|, |f(q)|\},$$

and hence approaches zero as $p, q \to \infty$ since $f(x) \to 0$ as $x \to \infty$, while the second term satisfies the inequality

$$\left| \int_p^q G(x)f'(x)\,dx \right| \leqslant C \int_p^q |f'(x)|\,dx,$$

and hence also approaches zero as $p, q \to \infty$, by the absolute integrability of $f'(x)$ on $[a, \infty)$. Therefore the left-hand side of (8) approaches zero as $p, q \to \infty$, i.e., the integral (7) satisfies the Cauchy convergence criterion and hence converges. ∎

b. If $f(x)$ is real, we can modify the conditions on $f(x)$ somewhat, obtaining the following version of the Abel-Dirichlet test:

THEOREM.† *Let the real function* $f(x)$ *be piecewise smooth on every finite subinterval* $[a, X] \subset [a, \infty)$, *and suppose* $f(x) \searrow 0$ (*or* $\nearrow 0$) *as* $x \to \infty$. *Moreover, let the* (*complex*) *function* $g(x)$ *be the same as in the preceding theorem. Then the integral* (7) *converges*.

Proof. Under these conditions, $f'(x)$ does not change sign, and hence

$$\int_a^X |f'(x)|\,dx = \left| \int_a^X f'(x)\,dx \right| = |f(X) - f(a)|$$

has a finite limit as $X \to \infty$, i.e., $f'(x)$ is absolutely integrable on $[a, \infty)$, as before. ∎

c. Example. Let $P(x)$ and $Q(x)$ be two polynomials in x with complex coefficients, and let α be any nonzero real number. Consider the integral

$$\int_a^\infty \frac{P(x)}{Q(x)} e^{i\alpha x}\,dx, \tag{9}$$

assuming that $Q(x)$ has no zeros in the interval $a \leqslant x < \infty$, and that the degree of $Q(x)$ exceeds that of $P(x)$ by at least 1. Writing

† Note the similarity between this theorem and the Abel-Dirichlet test for series (Theorem 6.47c).

$$f(x) = \frac{P(x)}{Q(x)},$$

we find that

$$f'(x) = \frac{P'(x)Q(x) - P(x)Q'(x)}{Q^2(x)}$$

is a rational function with a denominator whose degree exceeds that of its numerator by at least 2. Therefore $f'(x)$ is absolutely integrable on $[a,\infty)$, by Example 11.13c. Moreover, the function

$$g(x) = e^{i\alpha x}$$

has the antiderivative

$$G(x) = \frac{e^{i\alpha x}}{i\alpha},$$

which is bounded on $[a,\infty)$. Applying Theorem 11.23a, we see at once that (9) converges.

11.3. Improper Integrals of the Second and Third Kinds

11.31. Improper integrals of the second kind

a. Let $f(x)$ be a complex function defined on a finite interval $[a,b]$ and integrable on every subinterval $[a+\varepsilon,b] \subset [a,b]$, but possibly not integrable (e.g., unbounded) on the whole interval $[a,b]$.† Suppose the function

$$I(\varepsilon) = \int_{a+\varepsilon}^{b} f(x) \ dx$$

approaches a finite limit I as $\varepsilon \searrow 0$. Then the expression

$$\int_{a}^{b} f(x) \ dx, \tag{1}$$

called an *improper integral of the second kind*, is said to be *convergent*, with *value* I. In other words, we set

$$\int_{a}^{b} f(x) \ dx = \lim_{\varepsilon \searrow 0} I(\varepsilon) = \lim_{\varepsilon \searrow 0} \int_{a+\varepsilon}^{b} f(x) \ dx, \tag{2}$$

by definition. On the other hand, if $I(\varepsilon)$ does not approach a finite limit as $\varepsilon \searrow 0$, we call the integral (1) *divergent* and assign it no value at all.

† Here we allow $f(x)$ to become infinite at the point $x=a$, but nowhere else.

b. THEOREM. *If the integral* (1) *exists as an ordinary integral with value I, then it exists in the sense of the definition* (2) *and has the same value I.*

Proof. If (1) exists in the ordinary sense, then the real and imaginary parts of $f(x)$ are bounded (by Theorem 11.15c), and hence $f(x)$ is bounded, so that $|f(x)| \leqslant C$, say. Then, given any $\varepsilon > 0$,

$$I = \int_a^b f(x)\ dx = \int_a^{a+\varepsilon} f(x)\ dx + \int_{a+\varepsilon}^b f(x)\ dx,$$

where

$$\left| \int_a^{a+\varepsilon} f(x)\ dx \right| \leqslant C\varepsilon.$$

Therefore

$$\lim_{\varepsilon \searrow 0} \int_{a+\varepsilon}^b f(x)\ dx = I - \lim_{\varepsilon \searrow 0} \int_a^{a+\varepsilon} f(x)\ dx = I. \quad \blacksquare$$

c. The above definition of an improper integral of the second kind closely resembles the definition of an improper integral of the first kind (i.e., with an infinite upper limit), given in Sec. 11.11. In fact, an improper integral of the second kind can be immediately transformed into an improper integral of the first kind by making the substitution

$$x - a = \frac{1}{u}.$$

Hence, starting from the theory of improper integrals of the first kind, we can deduce a whole theory of improper integrals of the second kind, formulating analogues of all the theorems of Secs. 11.1 and 11.2.† Thus, at this point, we can confine ourselves to a few subsidiary comments.

d. The above considerations have obvious analogues for the case of a function $f(x)$ defined on a finite interval $[a,b]$ and integrable on every subinterval $[a, b - \varepsilon] \subset [a,b]$, but possibly not integrable on the whole interval $[a,b]$. In this case, the appropriate improper integral (of the second kind) is defined as

$$\int_a^b f(x)\ dx = \lim_{\varepsilon \searrow 0} \int_a^{b-\varepsilon} f(x)\ dx$$

(give further details).

---

† As an exercise, the reader should state (and prove) the Cauchy convergence criterion, the comparison test, the definition of absolute convergence, and the analogue of Theorem 11.21a, the Leibniz and Abel-Dirichlet tests, etc., for improper integrals of the second kind.

e. As always, to apply the comparison test we need suitable "standard integrals." In the case of improper integrals of the second kind, the most commonly used standard integral is

$$\int_a^b \frac{dx}{(x-a)^\lambda}.\tag{3}$$

Since

$$\int_{a+\varepsilon}^b \frac{dx}{(x-a)^\lambda} = \begin{cases} \dfrac{1}{\lambda-1}\left[\dfrac{1}{(b-a)^{\lambda-1}} - \dfrac{1}{\varepsilon^{\lambda-1}}\right] & \text{if } \lambda\neq 1, \\ \ln\dfrac{b-a}{\varepsilon} & \text{if } \lambda=1, \end{cases}$$

we see that (3) converges if $\lambda<1$ and diverges if $\lambda\geqslant 1$. Applying the comparison test, we find that if

$$\frac{c_1}{(x-a)^\lambda} \leqslant f(x) \leqslant \frac{c_2}{(x-a)^\lambda}$$

for all $a<x\leqslant b_0\leqslant b$, then the integral (1) converges if $\lambda<1$ and diverges if $\lambda\geqslant 1$.

f. In particular, consider the integral

$$\int_0^b |\ln x|^\gamma \ dx \qquad (\gamma>0).\tag{4}$$

For small x, we have

$$|\ln x| \leqslant cx^{-\alpha}$$

for every $\alpha>0$ (cf. Theorem 5.56). Choosing $\alpha=1/2\gamma$ and comparing (4) with the convergent integral

$$\int_0^{b_0} x^{-1/2}\, dx,$$

we find that (4) converges for all $\gamma>0$.

11.32. Improper integrals of the third kind

a. Let $f(x)$ be a complex function defined on an interval $[a,b]$, where the end points may be infinite, i.e., where the values $a=-\infty$, $b=+\infty$ are allowed.† Then by a *singular point*‡ of $f(x)$ we mean

(a) The point a if $a=-\infty$;
(b) The point b if $b=+\infty$;

† Instead of $[a,b]$ we write $(a,b]$ if $a=-\infty$ and $[a,b)$ if $b=+\infty$.
‡ Here we use the term "singular point" in a sense different from that of Sec. 10.37c.

(c) Any point $c \in (a,b)$ such that $f(x)$ fails to be integrable in the ordinary sense in every neighborhood of c;

(d) The point a if a is finite and $f(x)$ fails to be integrable in every one-sided neighborhood $a \leqslant x < a + \varepsilon$;

(e) The point b if b is finite and $f(x)$ fails to be integrable in every one-sided neighborhood $b - \varepsilon < x \leqslant b$.

b. Suppose $f(x)$ has no more than finitely many singular points in $[a,b]$, and let $f(x)$ be continuous or piecewise continuous on every set obtained by deleting neighborhoods of these points from the interval $[a,b]$. Choosing a point p_i between every pair of consecutive singular points c_i, c_{i+1} of $f(x)$, we get a set of intervals $[c_i, p_i]$, $[p_i, c_{i+1}]$, each containing only one singular point (as an end point). The integral over each of these intervals is defined as the appropriate improper integral of the first or second kind, i.e.,

$$\int_{c_i}^{p_i} f(x)\ dx = \lim_{X \searrow c_i} \int_X^{p_i} f(x)\ dx,$$

$$\int_{p_i}^{c_{i+1}} f(x)\ dx = \lim_{X \nearrow c_{i+1}} \int_{p_i}^{X} f(x)\ dx.$$

Suppose all these improper integrals converge. Then the integral

$$\int_a^b f(x)\ dx,$$

called an *improper integral of the third kind*, is said to be *convergent*, with *value*

$$\int_a^b f(x)\ dx = \sum \left\{ \int_{c_i}^{p_i} f(x)\ dx + \int_{p_i}^{c_{i+1}} f(x)\ dx \right\}, \tag{5}$$

where the sum is over all the singular points c_i of $f(x)$.

We must still verify that (5) does not depend on the choice of the points p_i. It is clearly enough to do this for one interval $[c_i, c_{i+1}]$. Let $c_i < p_i < q_i < c_{i+1}$. Then

$$\int_{c_i}^{p_i} f(x)\ dx + \int_{p_i}^{c_{i+1}} f(x)\ dx = \lim_{X \searrow c_i} \int_X^{p_i} + \lim_{Y \nearrow c_{i+1}} \int_{p_i}^{Y}$$

$$= \lim_{X \searrow c_i} \int_X^{p_i} + \lim_{Y \nearrow c_{i+1}} \left\{ \int_{p_i}^{q_i} + \int_{q_i}^{Y} \right\}$$

$$= \lim_{X \searrow c_i} \left\{ \int_X^{p_i} + \int_{p_i}^{q_i} \right\} + \lim_{Y \nearrow c_{i+1}} \int_{q_i}^{Y}$$

$$= \lim_{X \searrow c_i} \int_X^{q_i} + \lim_{Y \nearrow c_{i+1}} \int_{q_i}^{Y} = \int_{c_i}^{q_i} f(x)\ dx + \int_{q_i}^{c_{i+1}} f(x)\ dx,$$

as asserted.†

c. Example. Consider the integral

$$\int_{-\infty}^{\infty} \frac{dx}{|x - c_1|^{\alpha_1} \cdots |x - c_n|^{\alpha_n}} \qquad (-\infty < c_1 < \cdots < c_n < \infty) \tag{6}$$

with singular points $-\infty, c_1, \ldots, c_n, \infty$. The integrals

$$\int_{c_i}^{p_i}, \quad \int_{p_{i-1}}^{c_i}$$

(again we omit the integrand) converge if and only if $\alpha_i < 1$, since the factors other than

$$\frac{1}{|x - c_i|^{\alpha_i}}$$

are bounded in a neighborhood of the point c_i. Moreover, the integrals

$$\int_{-\infty}^{p_0}, \quad \int_{p_n}^{\infty}$$

converge if and only if $\alpha_1 + \cdots + \alpha_n > 1$. Note that the integral

$$\int_{-\infty}^{\infty} \frac{dx}{|x - c|^{\alpha}},$$

to which (6) reduces for $n = 1$, diverges for every α, since we cannot have both $\alpha < 1$ and $\alpha > 1$. The same is true of both integrals

$$\int_{-\infty}^{c} \frac{dx}{(c - x)^{\alpha}}, \quad \int_{c}^{\infty} \frac{dx}{(x - c)^{\alpha}}.$$

11.4. Evaluation of Improper Integrals by Residues

11.41. Integrals of rational functions

a. Consider the improper integral

$$\int_{-\infty}^{\infty} \frac{P(x)}{Q(x)} \, dx, \tag{1}$$

where the integrand is a rational function, i.e., the quotient $P(x)/Q(x)$ of two polynomials (with complex coefficients). Just as in Example 11.13c, the integral (1) converges if the denominator has no real zeros and if the degree

† For brevity, we omit the integrand $f(x)$ in the intermediate steps of the calculation.

of the denominator is at least 2 greater than the degree of the numerator. We now examine the problem of actually evaluating (1). Of course, it follows from the natural generalization of Theorem 9.33 to the case of convergent improper integrals that (1) equals the quantity

$$G(\infty) - G(-\infty) = \lim_{\substack{a \to -\infty \\ b \to \infty}} [G(b) - G(a)],$$

where $G(x)$ is the indefinite integral of $P(x)/Q(x)$, which can be found by the technique of Sec. 9.42. However, as we now show, (1) can often be evaluated much more quickly by taking advantage of the analyticity of the function $P(x)/Q(x)$, or, more exactly, of the fact that the function $P(z)/Q(z)$ of the complex variable z is analytic at every point of the z-plane except at the finitely many zeros of the denominator $Q(z)$.

Thus let L_R^+ be the (piecewise smooth) closed path in the upper half-plane made up of the interval $[-R,R]$ of the real axis and the semicircle

$$C_R^+ = \{z = Re^{i\theta}, \ 0 \leqslant \theta \leqslant \pi\}, \tag{2}$$

where L_R^+ and C_R^+ are traversed in the counterclockwise direction and R is so large that all the zeros of the denominator $Q(z)$ in the upper half-plane, say $z_1,\ldots,z_q$, lie inside L_R^+ (see Figure 78). It follows from formula (4), p. 407, that

$$\oint_{L_R^+} \frac{P(z)}{Q(z)} \, dz = \int_{-R}^{R} \frac{P(x)}{Q(x)} \, dx + \int_{C_R^+} \frac{P(z)}{Q(z)} \, dz = 2\pi i \sum_{j=1}^{q} \operatorname*{Res}_{z=z_j} \frac{P(z)}{Q(z)}. \tag{3}$$

Using the condition on the degrees of $P(z)$ and $Q(z)$, we see that the inequality

$$\left| \frac{P(z)}{Q(z)} \right| \leqslant \frac{A}{R^2}$$

holds on C_R^+ for some constant A and all sufficiently large R. Therefore

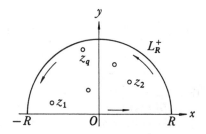

Figure 78

$$\lim_{R\to\infty} \int_{C_R^+} \frac{P(z)}{Q(z)}\, dz = 0,$$

since

$$\left| \int_{C_R^+} \frac{P(z)}{Q(z)}\, dz \right| \leqslant \frac{A}{R^2} 2\pi R = \frac{\pi A}{R},$$

and hence

$$\lim_{R\to\infty} \oint_{L_R^+} \frac{P(z)}{Q(z)}\, dz = \lim_{R\to\infty} \int_{-R}^{R} \frac{P(x)}{Q(x)}\, dx = 2\pi i \sum_{j=1}^{q} \operatorname*{Res}_{z=z_j} \frac{P(z)}{Q(z)}.$$

But since the integral (1) converges, it must coincide with

$$\lim_{R\to\infty} \int_{-R}^{R} \frac{P(x)}{Q(x)}\, dx,$$

so that, finally,

$$\int_{-\infty}^{\infty} \frac{P(x)}{Q(x)}\, dx = 2\pi i \sum_{j=1}^{q} \operatorname*{Res}_{z=z_j} \frac{P(z)}{Q(z)}. \tag{4}$$

b. If the zeros $z_1,\dots,z_q$ are all simple, then

$$\operatorname*{Res}_{z=z_j} \frac{P(z)}{Q(z)} = \frac{P(z_j)}{Q'(z_j)}, \tag{5}$$

by formula (3), p. 406, and hence (4) becomes

$$\int_{-\infty}^{\infty} \frac{P(x)}{Q(x)}\, dx = 2\pi i \sum_{j=1}^{q} \frac{P(z_j)}{Q'(z_j)}. \tag{6}$$

For example,

$$\int_{-\infty}^{\infty} \frac{dx}{x^2+1} = 2\pi i \frac{1}{2z}\bigg|_{z=i} = \pi.$$

c. If the zeros $z_1,\dots,z_q$ are multiple, with orders $k_1,\dots,k_q$, respectively, then (5) must be replaced by the more general formula (3'), p. 407,

$$\operatorname*{Res}_{z=z_j} \frac{P(z)}{Q(z)} = \frac{1}{(k_j-1)!} \left[(z-z_j)^{k_j} \frac{P(z)}{Q(z)} \right]^{(k_j-1)}_{z=z_j}, \tag{5'}$$

so that

$$\int_{-\infty}^{\infty} \frac{P(x)}{Q(x)}\, dx = 2\pi i \sum_{j=1}^{q} \frac{1}{(k_j-1)!} \left[(z-z_j)^{k_j} \frac{P(z)}{Q(z)} \right]^{(k_j-1)}_{z=z_j}. \tag{6'}$$

For example,

$$\int_{-\infty}^{\infty} \frac{dx}{(x^2+1)^2} = 2\pi i \left[(z-i)^2 \frac{1}{(z^2+1)^2} \right]'_{z=i} = 2\pi i \left[\frac{1}{(z+i)^2} \right]'_{z=i}$$

$$= \frac{-4\pi i}{(z+i)^3} = -4\pi i \frac{1}{8i^3} = \frac{\pi}{2}.$$

d. We have just shown that the improper integral (1) equals $2\pi i$ times the sum of the residues of the function $P(z)/Q(z)$ in the upper half-plane, by evaluating the integral of $P(z)/Q(z)$ along the contour L_R^+ made up of the segment $[-R,R]$ of the real axis and the semicircle (2). But we could just as well have carried out a similar calculation, using the contour L_R^- made up of the interval $[-R,R]$, this time traversed from right to left, and the semicircle

$$C_R^- = \{z = Re^{i\theta}, \ \pi \leqslant \theta \leqslant 2\pi\} \tag{2'}$$

in the *lower* half-plane (L_R^- and C_R^- are again traversed in the counterclockwise direction). This gives

$$\oint_{L_R^-} \frac{P(z)}{Q(z)} dz = \int_R^{-R} \frac{P(x)}{Q(x)} dx + \int_{C_R^-} \frac{P(z)}{Q(z)} dz = 2\pi i \sum_{k=1}^r \operatorname*{Res}_{z=z_k^*} \frac{P(z)}{Q(z)}, \tag{3'}$$

instead of (3), where $z_1^*,\ldots,z_r^*$ are now the zeros of the denominator $Q(z)$ in the lower half-plane. Taking the limit as $R \to \infty$, we get

$$\lim_{R\to\infty} \oint_{L_R^-} \frac{P(z)}{Q(z)} dz = \lim_{R\to\infty} \int_R^{-R} \frac{P(x)}{Q(x)} dx = 2\pi i \sum_{k=1}^r \operatorname*{Res}_{z=z_k^*} \frac{P(z)}{Q(z)},$$

since

$$\lim_{R\to\infty} \int_{C_R^-} \frac{P(z)}{Q(z)} dz = 0,$$

as before, and hence

$$\int_{-\infty}^{\infty} \frac{P(x)}{Q(x)} dx = -2\pi i \sum_{k=1}^r \operatorname*{Res}_{z=z_k^*} \frac{P(z)}{Q(z)}. \tag{4'}$$

Note that (4'), unlike (4), has a minus sign in the right-hand side. Since the left-hand sides of (4) and (4') must coincide, it follows as once that *the sum of the residues of the function $P(z)/Q(z)$ at all the zeros of $Q(z)$, in both the upper and lower half-planes, must vanish.*

The italicized assertion can easily be proved directly. In fact, the sum of the residues of $P(z)/Q(z)$ at all the zeros of $Q(z)$ must equal the integral

$$\oint_{|z|=R} \frac{P(z)}{Q(z)}\, dz \tag{7}$$

along the full circle $|z|=R$, provided that the radius R is large enough so that all the zeros of $Q(z)$ lie inside the circle. But (7) does not depend on R (for sufficiently large R), while at the same time (7) satisfies the estimate

$$\left| \oint_{|z|=R} \frac{P(z)}{Q(z)}\, dz \right| \leqslant \max_{|z|=R}\left| \frac{P(z)}{Q(z)} \right| 2\pi R \leqslant \frac{A}{R^2} 2\pi R \to 0$$

as $R \to \infty$. Hence the integral (7) vanishes, so that

$$\sum_{j=1}^{q} \operatorname*{Res}_{z=z_j} \frac{P(z)}{Q(z)} + \sum_{k=1}^{r} \operatorname*{Res}_{z=z_{k^*}} \frac{P(z)}{Q(z)} = 0,$$

as asserted.

11.42. Fourier integrals

a. By a *Fourier integral* we mean one of the frequently encountered integrals

$$\int_{-\infty}^{\infty} f(x)e^{i\sigma x}\, dx, \tag{8}$$

$$\int_{-\infty}^{\infty} f(x)\cos \sigma x\, dx, \tag{9}$$

$$\int_{-\infty}^{\infty} f(x)\sin \sigma x\, dx, \tag{10}$$

involving a real parameter σ. If the condition

$$\int_{-\infty}^{\infty} |f(x)|\, dx < \infty$$

is satisfied, then all three integrals (8)–(10) are absolutely convergent. If the function $f(x)$ is real and approaches zero monotonically as $|x| \to \infty$, then the integrals (9) and (10) converge for $\sigma \neq 0$ (in general nonabsolutely), by Example 11.22a. Suppose (9) and (10) converge. Then (10) vanishes if $f(x)$ is an even function, i.e., if $f(-x) \equiv f(x)$, while (9) vanishes if $f(x)$ is an odd function, i.e., if $f(-x) \equiv -f(x)$. Moreover, obviously

$$\int_{-\infty}^{\infty} f(x)e^{i\sigma x}\, dx = \int_{-\infty}^{\infty} f(x)\cos \sigma x\, dx + i \int_{-\infty}^{\infty} f(x)\sin \sigma x\, dx,$$

and hence the integrals (9) and (10) are the real and imaginary parts of (8) if $f(x)$ is real.

b. We now show how contour integration can be used to calculate Fourier integrals of rational functions. Let

$$f(x) = \frac{P(x)}{Q(x)}$$

be a rational function, i.e., a quotient of polynomials, and suppose the denominator $Q(x)$ has no real zeros and is of degree at least 1 greater than the degree of the numerator $P(x)$. Then the integrals (8)–(10) converge, by Example 11.23c. As before, let $z_1, \ldots, z_q$ be the zeros of $Q(x)$ in the upper half-plane, and let L_R^+ be the contour consisting of the interval $[-R, R]$ of the real axis and the semicircle (2). Then, by formula (4), p. 407,

$$\oint_{L_R^+} \frac{P(z)}{Q(z)} e^{i\sigma z} \, dz = \int_{-R}^{R} \frac{P(x)}{Q(x)} e^{i\sigma x} \, dx + \int_{C_R^+} \frac{P(z)}{Q(z)} e^{i\sigma z} \, dz$$

$$= 2\pi i \sum_{j=1}^{q} \operatorname*{Res}_{z=z_j} \frac{P(z) e^{i\sigma z}}{Q(z)}, \tag{11}$$

provided R is large enough so that L_R^+ surrounds the points $z_1, \ldots, z_q$. If $\sigma > 0$, then

$$\lim_{R \to \infty} \int_{C_R^+} \frac{P(z)}{Q(z)} e^{i\sigma z} \, dz = 0. \tag{12}$$

This follows by the same argument as in Sec. 11.41a if the degree of $Q(z)$ exceeds that of $P(z)$ by at least 2, since

$$|e^{i\sigma z}| = |e^{i\sigma x - \sigma y}| = e^{-\sigma y} \leqslant 1$$

if $y \geqslant 0$, $\sigma \geqslant 0$. However, the argument given there does not work if the degree of $Q(z)$ exceeds that of $P(z)$ by only 1, and we must then resort to a new argument, based on the following two propositions:

c. LEMMA. *If $\sigma > 0$, then*

$$\int_0^{\pi} e^{-R\sigma \sin \theta} \, d\theta \leqslant \frac{A}{R\sigma}, \tag{13}$$

where A is a constant.

Proof. Since $\sin(\pi - \theta) = \sin \theta$, we need only consider the integral

$$\int_0^{\pi/2} e^{-R\sigma \sin \theta} \, d\theta,$$

equal to half the integral in (13). But

$$\int_0^{\pi/2} e^{-R\sigma \sin \theta} \, d\theta = \int_0^{\pi/6} e^{-R\sigma \sin \theta} \, d\theta + \int_{\pi/6}^{\pi/2} e^{-R\sigma \sin \theta} \, d\theta$$

$$< \int_0^{\pi/6} \frac{\cos \theta}{\cos (\pi/6)} e^{-R\sigma \sin \theta} \, d\theta + \int_{\pi/6}^{\pi/2} e^{-R\sigma \sin \theta} \, d\theta$$

$$< \left. \frac{e^{-R\sigma \sin \theta}}{R\sigma \cos(\pi/6)} \right|_{\pi/6}^0 + \frac{\pi}{3} e^{-R\sigma/2}$$

$$< \frac{1}{R\sigma \cos(\pi/6)} + \frac{\pi}{3} e^{-R\sigma/2}$$

$$= \frac{1}{R\sigma} \left[\frac{1}{\cos (\pi/6)} + \frac{\pi}{3} R\sigma e^{-R\sigma/2} \right] \leqslant \frac{A}{R\sigma},$$

since the function xe^{-x} is bounded for $x > 0$. ∎

d. COROLLARY (**Jordan's lemma**). *Given a function $f(z)$ defined for* Im $z \geqslant 0$, $|z| \geqslant R_0$, *suppose*

$$\lim_{R \to \infty} \sup_{|z|=R} |f(z)| = 0. \tag{14}$$

Then

$$\lim_{R \to \infty} \int_{C_R^+} f(z) e^{i\sigma z} \, dz = 0$$

if $\sigma > 0$, where C_R^+ is the path (2).

Proof. It follows from (13) and (14) that

$$\left| \int_{C_R^+} f(z) e^{i\sigma z} \, dz \right| = \left| \int_0^{\pi} f(z) e^{i\sigma x} e^{-\sigma y} R e^{i\theta} \, d\theta \right|$$

$$\leqslant R \sup_{z \in C_R^+} |f(z)| \int_0^{\pi} e^{-R\sigma \sin \theta} \, d\theta$$

$$\leqslant \frac{A}{\sigma} \sup_{z \in C_R^+} |f(z)| \to 0$$

as $R \to \infty$. ∎

e. We can now complete the proof of formula (12) in the case where the degree of $Q(z)$ exceeds that of $P(z)$ by only 1. Here

$$f(z) = \frac{P(z)}{Q(z)},$$

and hence the inequality

$$|f(z)| \leqslant \frac{A}{R}$$

holds on C_R^+ for some constant A and all sufficiently large R. But then (14) holds, and hence so does (12). Thus, finally, to get the Fourier integral of $f(z)$ for $\sigma > 0$, we need only take the limit as $R \to \infty$ in (11), obtaining

$$\int_{-\infty}^{\infty} \frac{P(x)}{Q(x)} e^{i\sigma x} \, dx = 2\pi i \sum_{j=1}^{q} \operatorname*{Res}_{z=z_j} \frac{P(z) e^{i\sigma z}}{Q(z)}, \tag{15}$$

since clearly

$$\int_{-\infty}^{\infty} \frac{P(x)}{Q(x)} e^{i\sigma x} \, dx = \lim_{R \to \infty} \int_{-R}^{R} \frac{P(x)}{Q(x)} e^{i\sigma x} \, dx.$$

f. If $\sigma < 0$, the above argument breaks down, since $e^{i\sigma x}$ becomes arbitrarily large on the path C_R^+. In this case, we carry out an analogous construction in the lower half-plane rather than in the upper half-plane. Thus let C_R^- and L_R^- be the same as in Sec. 11.41d, and let R be large enough so that all the zeros $z_1^*, \dots, z_r^*$ of $Q(z)$ in the lower half-plane lie inside L_R^-. Then

$$\oint_{L_R^-} \frac{P(z)}{Q(z)} e^{i\sigma z} \, dz = \int_{R}^{-R} \frac{P(x)}{Q(x)} e^{i\sigma x} \, dx + \int_{C_R^-} \frac{P(z)}{Q(z)} e^{i\sigma z} \, dz$$

$$= 2\pi i \sum_{k=1}^{r} \operatorname*{Res}_{z=z_k^*} \frac{P(z) e^{i\sigma z}}{Q(z)},$$

where the integral along C_R^- satisfies the estimate

$$\left| \int_{C_R^-} \frac{P(z)}{Q(z)} e^{i\sigma z} \, dz \right| = \left| \int_{\pi}^{2\pi} \frac{P(z)}{Q(z)} e^{i\sigma x} e^{-\sigma y} R e^{i\theta} \, d\theta \right|$$

$$\leqslant R \sup_{z \in C_R^-} \left| \frac{P(z)}{Q(z)} \right| \int_{0}^{\pi} e^{R\sigma \sin \theta} \, d\theta,$$

and hence approaches 0 as $R \to \infty$, by Lemma 11.42c. It follows that

$$\int_{-\infty}^{\infty} \frac{P(x)}{Q(x)} e^{i\sigma x} \, dx = -2\pi i \sum_{k=1}^{r} \operatorname*{Res}_{z=z_k^*} \frac{P(z) e^{i\sigma z}}{Q(z)}, \tag{15'}$$

which gives the value of the Fourier integral of $P(x)/Q(x)$ for $\sigma < 0$. Note that the left-hand side of (15) or (15') fails to exist if $\sigma = 0$ and if the degree of $Q(x)$ exceeds that of $P(x)$ by only 1.

g. Example. Using (15) and (15′), we find that

$$\int_{-\infty}^{\infty} \frac{xe^{i\sigma x}}{x^2+1}\, dx = \begin{cases} 2\pi i \operatorname*{Res}_{z=i} \dfrac{ze^{i\sigma z}}{z^2+1} = i\pi e^{-\sigma} & \text{if } \sigma > 0, \\[2ex] -2\pi i \operatorname*{Res}_{z=-i} \dfrac{ze^{i\sigma z}}{z^2+1} = -i\pi e^{\sigma} & \text{if } \sigma < 0. \end{cases}$$

The power of the method of contour integration is particularly apparent in this case, since the corresponding indefinite integral

$$\int \frac{xe^{i\sigma x}}{x^2+1}\, dx$$

cannot be expressed in terms of elementary functions.

h. The Fourier integral (9) or (10) can sometimes exist even if $f(x)$ has singular points on the x-axis, provided the behavior of $f(x)$ at these points is compensated by appropriate zeros of the functions $\cos \sigma x$ or $\sin \sigma x$. For example, consider the integral

$$\int_{-\infty}^{\infty} \frac{\sin \sigma x}{x}\, dx, \tag{16}$$

of importance in mathematical physics.† To evaluate (16), let Λ_R be the contour shown in Figure 79, obtained by consecutively traversing the interval $[-R, -\varepsilon]$ of the real axis, a small semicircle

$$C_\varepsilon = \{z: z = \varepsilon e^{i\theta}, \; 0 \leqslant \theta \leqslant \pi\}$$

in the clockwise direction, the interval $[\varepsilon, R]$ of the real axis, and finally the

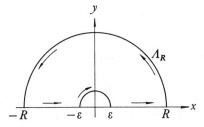

Figure   79

† Note that in this case the complex Fourier integral
$$\int_{-\infty}^{\infty} \frac{e^{i\sigma x}}{x}\, dx$$
fails to exist.

same large semicircle C_R^+ as above, given by (2). Since the function $e^{i\sigma z}/z$ has no singular points inside or on Λ_R, we have

$$\oint_{\Lambda_R} \frac{e^{i\sigma z}}{z}\,dz = \int_{-R}^{-\varepsilon} \frac{e^{i\sigma x}}{x}\,dx + \int_{C_\varepsilon} \frac{e^{i\sigma z}}{z}\,dz + \int_{\varepsilon}^{R} \frac{e^{i\sigma x}}{x}\,dx + \int_{C_R^+} \frac{e^{i\sigma z}}{z}\,dz = 0 \qquad (17)$$

(why?). Moreover

$$\lim_{R\to\infty} \int_{C_R^+} \frac{e^{i\sigma z}}{z}\,dz = 0$$

if $\sigma > 0$, by Jordan's lemma, and

$$\int_{-R}^{-\varepsilon} \frac{e^{i\sigma x}}{x}\,dx + \int_{\varepsilon}^{R} \frac{e^{i\sigma x}}{x}\,dx$$

$$= \left[\int_{-R}^{-\varepsilon} \frac{\cos \sigma x}{x}\,dx + \int_{\varepsilon}^{R} \frac{\cos \sigma x}{x}\,dx \right] + i\left[\int_{-R}^{-\varepsilon} \frac{\sin \sigma x}{x}\,dx + \int_{\varepsilon}^{R} \frac{\sin \sigma x}{x}\,dx \right]$$

$$= i\left[\int_{-R}^{-\varepsilon} \frac{\sin \sigma x}{x}\,dx + \int_{\varepsilon}^{R} \frac{\sin \sigma x}{x}\,dx \right],$$

since the integrals of the odd function

$$\frac{\cos \sigma x}{x}$$

over the intervals $[-R, -\varepsilon]$ and $[\varepsilon, R]$ cancel each other out. Furthermore

$$\int_{C_\varepsilon} \frac{e^{i\sigma z}}{z}\,dz = \int_{C_\varepsilon} \frac{dz}{z} + \int_{C_\varepsilon} \frac{e^{i\sigma z}-1}{z}\,dz = \int_{\pi}^{0} \frac{i\varepsilon e^{i\theta}}{\varepsilon e^{i\theta}}\,d\theta = -i\pi + \rho_\varepsilon,$$

where

$$|\rho_\varepsilon| = \left| \int_{C_\varepsilon} \frac{e^{i\sigma z}-1}{z}\,dz \right| \leqslant \max_{z\in C_\varepsilon}\left| \frac{e^{i\sigma z}-1}{z} \right| \varepsilon\pi \leqslant A\varepsilon$$

for some constant A, since the function

$$\frac{e^{i\sigma z}-1}{z}$$

is continuous at $z=0$ and hence bounded near $z=0$. Therefore, letting $\varepsilon\to 0$, $R\to\infty$ in (17), we get

$$\int_{-\infty}^{\infty} \frac{\sin \sigma x}{x}\,dx = \pi.$$

If $\sigma < 0$, then

$$\sin \sigma x = -\sin(-\sigma)x,$$

where $-\sigma > 0$, and hence

$$\int_{-\infty}^{\infty} \frac{\sin \sigma x}{x}\, dx = -\int_{-\infty}^{\infty} \frac{\sin (-\sigma)x}{x}\, dx = -\pi.$$

Thus, finally,

$$\int_{0}^{\infty} \frac{\sin \sigma x}{x}\, dx = \begin{cases} \pi/2 \text{ if } \sigma > 0, \\ -\pi/2 \text{ if } \sigma < 0, \end{cases}$$

as found by Euler in 1781.

11.5. Parameter-Dependent Improper Integrals

11.51. The Fourier integral is an example of an improper integral containing a parameter (σ in this case). More generally, consider an arbitrary improper integral (of the first kind, say)

$$\Phi(t) = \int_{a}^{\infty} f(x,t)\, dx \tag{1}$$

containing a real parameter t, where (1) is assumed to converge for all t in some interval $\alpha \leqslant t \leqslant \beta$. We now ask the following questions about $\Phi(t)$, regarded as a function of t, thereby generalizing the considerations of Sec. 9.11 to the case of improper integrals:
(a) Under what conditions is $\Phi(t)$ continuous (cf. Theorem 9.111)?
(b) When does the formula

$$\int_{a}^{\infty} \left\{ \int_{\alpha}^{\beta} f(x,t)\, dt \right\} dx = \int_{\alpha}^{\beta} \left\{ \int_{a}^{\infty} f(x,t)\, dx \right\} dt \tag{2}$$

hold (cf. Theorem 9.112)?
(c) When does the formula

$$\Phi'(t) = \int_{a}^{\infty} f_t(x,t)\, dx \tag{3}$$

hold?†
 To illustrate the import of these questions, consider the integral

$$\Phi(t) = \int_{a}^{\infty} \frac{\sin xt}{x}\, dx,$$

---

† As in Theorem 9.114, $\Phi'(t)$ denotes the derivative of $\Phi(t)$ and $f_t(x,t)$ the partial derivative of $f(x, t)$ with respect to t.

which converges for all real t. As shown in Sec. 11.42h,

$$\Phi(t) = \begin{cases} \pi/2 \text{ if } t > 0, \\ -\pi/2 \text{ if } t < 0, \end{cases}$$

while obviously $\Phi(0) = 0$. Thus, despite the continuity of the integrand

$$f(x,t) = \frac{\sin xt}{x} \tag{4}$$

in both variables x and t, the integral $\Phi(t)$ is discontinuous at $t = 0$. Hence continuity of the function $f(x,t)$ is certainly not enough to guarantee continuity of the function $\Phi(t)$. Moreover, differentiating (4), we get

$$f_t(x,t) = \cos xt,$$

so that (3) fails to hold, since the integral on the right is divergent. Admittedly, formula (2) holds for the function (4) if $\alpha > 0$ (see Example 11.59b), but this is due to properties of (4) other than its mere continuity. In fact, Problem 6 gives an example where (2) fails to hold although the integrand satisfies the continuity requirements of Theorem 11.54.

Thus it is clear that extra conditions must be imposed on the integral (1) if $\Phi(t)$ is to be continuous and if formulas (2) and (3) are to hold. As we will see below, such an extra condition is afforded by the requirement that the convergence of the improper integral be *uniform*. Although we will develop the theory only for improper integrals of the first kind, everything can easily be carried over to the case of improper integrals of the second and third kinds (do this as an exercise).

11.52. Definition. Suppose the "parameter-dependent" integral (1) converges for all t in some set M. Then (1) is said to *converge uniformly on M*, if, given any $\varepsilon > 0$, there exists a number $X_0 > a$ such that

$$\left| \Phi(t) - \int_a^X f(x,t) \, dx \right| = \left| \int_X^\infty f(x,t) \, dx \right| < \varepsilon$$

for all $X \geqslant X_0$ and all $t \in M$.

Given any numerical sequence $X_n \nearrow \infty$ ($X_n > a$), consider the sequence of functions

$$\Phi_n(t) = \int_a^{X_n} f(x,t) \, dx \qquad (n = 1, 2, \ldots), \tag{5}$$

and suppose the integral (1) converges uniformly on M. Then clearly the sequence (5) converges uniformly on M to its limit, given by (1).

11.53. THEOREM. *Given a metric space M, suppose $f(x,t)$ is uniformly continuous on* *every product space*

$$[a,X] \times M \qquad (a < X < \infty),$$

and suppose the integral (1) *converges uniformly on* M. *Then the function* $\Phi(t)$ *defined by* (1) *is continuous on* M.

Proof. By Theorem 9.111, each function $\Phi_n(t)$ defined by (5) is (uniformly) continuous on M. Hence $\Phi(t)$ is also continuous on M, being the limit of a uniformly convergent sequence of continuous functions (see Corollary 5.95b). ∎

11.54. THEOREM. *If* $f(x,t)$ *is continuous on every rectangle*

$$a \leqslant x \leqslant X, \qquad \alpha \leqslant t \leqslant \beta \qquad (a < X < \infty),$$

and if the integral (1) *converges uniformly on the interval* $\alpha \leqslant t \leqslant \beta$, *then* (2) *holds.*

Proof. Since the sequence of continuous functions $\Phi_n(t)$ defined by (5) converges uniformly to the function $\Phi(t)$ defined by (1), it follows from Theorem 9.102 that

$$\lim_{n \to \infty} \int_\alpha^\beta \Phi_n(t) \, dt = \int_\alpha^\beta \Phi(t) \, dt. \qquad (6)$$

On the other hand, by Theorem 9.112,

$$\int_\alpha^\beta \Phi_n(t) \, dt = \int_\alpha^\beta \left\{ \int_a^{X_n} f(x,t) \, dx \right\} dt = \int_a^{X_n} \left\{ \int_\alpha^\beta f(x,t) \, dt \right\} dx.$$

Therefore we can write (6) in the form

$$\lim_{n \to \infty} \int_a^{X_n} \left\{ \int_\alpha^\beta f(x,t) \, dt \right\} dx = \int_\alpha^\beta \Phi(t) \, dt. \qquad (7)$$

But, by Theorem 11.12, the existence of the limit on the left for arbitrary $X_n \nearrow \infty$ guarantees the existence of the improper integral

$$\int_a^\infty \left\{ \int_\alpha^\beta f(x,t) \, dt \right\} dx, \qquad (8)$$

where, of course, the limit equals (8). Hence, recalling the definition of $\Phi(t)$, we see that (7) is equivalent to (2). ∎

11.55. a. THEOREM. *Let* $f(x,t)$ *and its partial derivative* $f_t(x,t)$ *be continuous on every rectangle*

$$a \leqslant x \leqslant X, \qquad \alpha \leqslant t \leqslant \beta \qquad (a < X < \infty),$$

and suppose the integral

$$\int_a^\infty f(x,\alpha)\ dx$$

converges while the integral

$$\int_a^\infty f_t(x,t)dx$$

converges uniformly on the interval $\alpha \leqslant t \leqslant \beta$. Then the function $\Phi(t)$ defined by (1) *exists and is differentiable on $\alpha \leqslant t \leqslant \beta$, with derivative* (3).

Proof. Consider the sequence of functions $\Phi_n(t)$ defined by (5). It follows from Theorem 9.114 that

$$\Phi_n'(t) = \int_a^{X_n} f_t(x,t)\ dx \qquad (n=1,2,\ldots). \tag{9}$$

By hypothesis, the sequence $\Phi_n(t)$ converges at the point $t=\alpha$, while the sequence $\Phi_n'(t)$ converges uniformly on the interval $\alpha \leqslant t \leqslant \beta$. Hence, by Theorem 9.106, the sequence $\Phi_n(t)$ converges uniformly on $\alpha \leqslant t \leqslant \beta$, and its limit function $\Phi(t)$ is differentiable on $\alpha \leqslant t \leqslant \beta$, with derivative

$$\Phi'(t) = \lim_{n\to\infty} \Phi_n'(t).$$

This proves the existence and differentiability of the integral

$$\Phi(t) = \lim_{n\to\infty} \Phi_n(t) = \int_a^\infty f(x,t)\ dx$$

and the validity of (3). ∎

b. The above theorem takes a somewhat different form for analytic functions. Suppose the function $f(x,t)$ is defined for all $x \in [a,\infty)$ and all $t \in G$, where G is some domain in the plane of the complex variable $t=\sigma+i\tau$, and suppose the integral

$$\int_a^\infty f(x,t)\ dx \tag{10}$$

converges for all $t \in G$. Then (10) is said to *converge uniformly inside G* (cf. Sec. 10.36a) if, given any compact set $Q \subset G$ and any $\varepsilon > 0$, there exists a number $X_0 = X_0(Q,\varepsilon) > a$ such that

$$\left| \int_X^\infty f(x,t)\ dx \right| < \varepsilon$$

for all $X \geqslant X_0$ and all $t \in Q$ (cf. Sec. 11.52).

THEOREM. *Let $f(x,t)$ and its partial derivative $f_t(x,t)$ be continuous on every "cylinder"*

$$a \leqslant x \leqslant X, \qquad |t - t_0| \leqslant r \qquad (a < X < \infty),$$

where the disk $|t - t_0| \leqslant r$ is contained in a domain G, and suppose $f(x,t)$ is analytic in t for every $x \in [a, \infty)$ while the integral (10) converges uniformly inside G. Then the function $\Phi(t)$ defined by (1) is analytic on G, with derivative (3).

Proof. By Theorem 10.27, every function (5) is analytic on G, with derivative (9). Moreover, $\Phi_n(t)$ converges uniformly to $\Phi(t)$ inside G, and hence $\Phi(t)$ is analytic on G, by Weierstrass' theorem (Theorem 10.36b). But $\Phi_n'(t)$ converges uniformly to $\Phi'(t)$ inside G, by the same theorem, and hence

$$\Phi'(t) = \lim_{n \to \infty} \int_a^{X_n} f_t(x,t) \, dx = \int_a^\infty f_t(x,t) \, dx. \quad \blacksquare$$

11.56. Next we prove a test for uniform convergence of improper integrals similar to the *Cauchy convergence criterion* (see Theorem 11.12).

THEOREM. *The improper integral*

$$\int_a^\infty f(x,t) \, dx \qquad (t \in M) \tag{11}$$

converges uniformly on the set M if and only if the following condition is satisfied: Given any $\varepsilon > 0$, there exists a number $X_0 > a$ such that

$$\left| \int_{X'}^{X''} f(x,t) \, dx \right| < \varepsilon \tag{12}$$

for all $X', X'' \geqslant X_0$ and all $t \in M$.

Proof. Suppose (11) is uniformly convergent on M. Then, given any $\varepsilon > 0$, there is an $X_0 > a$ such that

$$\left| \int_{X'}^\infty f(x,t) \, dt \right| < \frac{\varepsilon}{2}$$

for all $X' \geqslant X_0$ and all $t \in M$. Choosing any other $X'' \geqslant X_0$, we also have

$$\left| \int_{X''}^\infty f(x,t) \, dt \right| < \frac{\varepsilon}{2}$$

for all $t \in M$. Therefore (12) holds for all $X', X'' \geqslant X_0$ and all $t \in M$, since obviously

$$\left|\int_{X'}^{X''} f(x,t)\ dx\right| = \left|\int_{X'}^{\infty} f(x,t)\ dx - \int_{X''}^{\infty} f(x,t)\ dx\right|$$
$$\leqslant \left|\int_{X'}^{\infty} f(x,t)\ dx\right| + \left|\int_{X''}^{\infty} f(x,t)\ dx\right|.$$

Conversely, suppose (2) holds for all X', $X'' \geqslant X_0$ and all $t \in M$. Then the integral (11) converges (for all $t \in M$) by Theorem 11.12. Taking the limit as $X'' \to \infty$ in (12), we get

$$\left|\int_{X'}^{\infty} f(x,t)\ dx\right| \leqslant \varepsilon$$

for all $X' \geqslant X_0$ and all $t \in M$, so that (11) is uniformly convergent on M. ∎

11.57. a. Let $\varphi(x)$ be a nonnegative function such that the integral

$$\int_a^{\infty} \varphi(x)\ dx$$

converges. Then $\varphi(x)$ is called an *integrable majorant* of every (possibly complex) function $f(x)$ such that $|f(x)| \leqslant \varphi(x)$.

THEOREM. *Suppose $\varphi(x)$ is an integrable majorant of the function $f(x,t)$ for all $t \in M$. Then the integral*

$$\int_a^{\infty} f(x,t)\ dx$$

is uniformly convergent on M.

Proof. An immediate consequence of Theorem 11.56 and the estimate

$$\left|\int_{X'}^{X''} f(x,t)\ dx\right| \leqslant \int_{X'}^{X''} |f(x,t)|\ dx \leqslant \int_{X'}^{X''} \varphi(x)\ dx$$

(see also Theorem 11.12).

b. Example. The improper integral

$$\int_0^{\infty} \frac{\cos \alpha x - \cos \beta x}{x^2}\ dx \qquad (0 \leqslant \alpha \leqslant 1, 0 \leqslant \beta \leqslant 1) \tag{13}$$

is uniformly convergent on the interval $0 \leqslant \alpha \leqslant 1$ for any fixed β. To see this, we first prove the formula

$$\left|\frac{\cos \alpha x - \cos \beta x}{x^2}\right| \leqslant \begin{cases} 2/x^2 & \text{if } x \geqslant 1, \\ 2 & \text{if } 0 \leqslant x \leqslant 1, \end{cases}$$

valid for all α and β in the unit interval $[0,1]$. The first inequality is obvious.

To prove the second, we use the formula

$$\cos \gamma x = 1 - \frac{\gamma^2 x^2}{2} + x^4 g(\gamma, x)$$

(see formula (7), p. 254), where

$$|g(\gamma, x)| = \left| \frac{\gamma^4}{4!} - x^2 \frac{\gamma^6}{6!} + \cdots \right| \leqslant \frac{1}{4!} + \frac{1}{6!} + \cdots < \frac{1}{2}$$

if $|\gamma| \leqslant 1$, $|x| \leqslant 1$, which in turn implies

$$\left| \frac{\cos \alpha x - \cos \beta x}{x^2} \right| = \left| \frac{\beta^2 - \alpha^2}{2} + x^2 [g(\alpha, x) - g(\beta, x)] \right| < \frac{1}{2} + \frac{1}{2} + \frac{1}{2} < 2$$

if $\alpha, \beta, x \in [0,1]$. Thus the function

$$\varphi(x) = \begin{cases} 2/x^2 & \text{if } x \geqslant 1, \\ 2 & \text{if } 0 \leqslant x \leqslant 1 \end{cases}$$

is an integrable majorant for the function

$$\frac{\cos \alpha x - \cos \beta x}{x^2}$$

for all $\alpha, \beta \in [0,1]$. Therefore the integral (13) is uniformly convergent on the interval $0 \leqslant \alpha \leqslant 1$ for any fixed α (or, for that matter, on the square $0 \leqslant \alpha \leqslant 1, 0 \leqslant \beta \leqslant 1$). Hence, by Theorem 11.53, (13) represents a continuous function of α on the interval $0 \leqslant \alpha \leqslant 1$. An explicit expression for this function will be given in Sec. 11.59b.

11.58. Convolutions

a. By the *convolution* of two complex functions $f(t)$ and $g(t)$, defined for all real t, we mean the integral

$$h(t) = \int_{-\infty}^{\infty} f(x) g(t-x) \, dx \qquad (14)$$

(an improper integral of the third kind). The function $h(t)$ does not always exist. Conditions for the existence of $h(t)$ are given by the following theorem, which at the same time establishes the properties of $h(t)$:

THEOREM. *If $f(t)$ and $g(t)$ are bounded, continuous, and absolutely integrable on the real line $-\infty < t < \infty$, then $h(t)$ exists for all t. Moreover $h(t)$ is itself bounded, continuous, and absolutely integrable on the real line, and satisfies the relation*

$$\int_{-\infty}^{\infty} h(t) \, dt = \int_{-\infty}^{\infty} f(t) \, dt \int_{-\infty}^{\infty} g(t) \, dt. \qquad (15)$$

Proof. Let $|g(t)| \leqslant C$. Then the integrand of (14) satisfies the inequality

$$|f(x) g(t-x)| \leqslant C |f(x)|,$$

so that the integral (14) is convergent, in fact uniformly convergent on the whole line $-\infty < t < \infty$. Moreover, $h(t)$ is bounded, since

$$|h(t)| \leqslant \int_{-\infty}^{\infty} |f(x)||g(t-x)|\, dx \leqslant C \int_{-\infty}^{\infty} |f(x)|\, dx.$$

Next, with a view to proving the continuity of $h(t)$, we show that the integrand of (14) is continuous in both arguments on every finite rectangle

$$a \leqslant x \leqslant b, \qquad \alpha \leqslant t \leqslant \beta. \tag{16}$$

Given any $\varepsilon > 0$, there exists a $\delta > 0$ such that $x',x'' \in [a,b]$, $|x'-x''| < \delta$ implies $|f(x')-f(x'')| < \varepsilon$, while $t',t'' \in [\alpha-b,\beta-a]$, $|t'-t''| < 2\delta$ implies $|g(t')-g(t'')| < \varepsilon$.† But $x',x'' \in [a,b]$, $|x'-x''| < \delta$, $t',t'' \in [\alpha,\beta]$, $|t'-t''| < \delta$ implies $t'-x'$, $t''-x'' \in [\alpha-b,\beta-a]$, $|(t'-x')-(t''-x'')| < 2\delta$, and hence

$$|f(x')g(t'-x')-f(x'')g(t''-x'')|$$
$$\leqslant |f(x')-f(x'')||g(t'-x')| + |f(x'')||g(t'-x')-g(t''-x'')|$$
$$\leqslant C\varepsilon + C\varepsilon = 2C\varepsilon$$

if $|f(x)| \leqslant C$, $|g(x)| \leqslant C$. Thus $f(x)g(t-x)$ is continuous in both x and t on every rectangle (16). It follows from Theorem 11.53 that $h(t)$ is a continuous function of t.

We now verify the absolute integrability of $h(t)$. First we note that the functions $|f(t)|$ and $|g(t)|$ have the same properties as $f(t)$ and $g(t)$ themselves, so that the integral

$$\int_{-\infty}^{\infty} |f(x)||g(t-x)|\, dx$$

is also uniformly convergent on the whole line $-\infty < t < \infty$, with an integrand which is continuous in both x and t on every rectangle (16). Applying Theorem 11.54, we get

$$\int_{-\tau}^{\tau} |h(t)|\, dt \leqslant \int_{-\tau}^{\tau} \left\{ \int_{-\infty}^{\infty} |f(x)||g(t-x)|\, dx \right\} dt$$
$$= \int_{-\infty}^{\infty} |f(x)| \left\{ \int_{-\tau}^{\tau} |g(t-x)|\, dt \right\} dx$$
$$= \int_{-\infty}^{\infty} |f(x)| \left\{ \int_{-\tau-x}^{\tau-x} |g(t)|\, dt \right\} dx$$
$$\leqslant \int_{-\infty}^{\infty} |f(x)| \left\{ \int_{-\infty}^{\infty} |g(t)|\, dt \right\} dx$$
$$= \int_{-\infty}^{\infty} |g(t)|\, dt \int_{-\infty}^{\infty} |f(x)|\, dx,$$

† Here we use the fact that the continuous functions $f(x)$ and $g(x)$ are uniformly continuous on every closed interval (see Theorem 5.17b).

from which the existence of the integral

$$\int_{-\infty}^{\infty} |h(t)|\, dt$$

follows at once.

Finally, to prove (15), we again use Theorem 11.54, this time obtaining

$$\int_{-\tau}^{\tau} h(t)\, dt = \int_{-\tau}^{\tau} \left\{ \int_{-\infty}^{\infty} f(x)g(t-x)\, dx \right\} dt$$

$$= \int_{-\infty}^{\infty} f(x) \left\{ \int_{-\tau}^{\tau} g(t-x)\, dt \right\} dx$$

$$= \int_{-\infty}^{\infty} f(x) \left\{ \int_{-\tau-x}^{\tau-x} g(t)\, dt \right\} dx$$

$$= \int_{-\infty}^{\infty} f(x) \left\{ \int_{-\infty}^{\infty} g(t)\, dt - \int_{|t+x|\geqslant\tau} g(t)\, dt \right\} dx$$

$$= \int_{-\infty}^{\infty} g(t)\, dt \int_{-\infty}^{\infty} f(x)\, dx - \int_{-\infty}^{\infty} f(x) \left\{ \int_{|t+x|\geqslant\tau} g(t)\, dt \right\} dx. \quad (17)$$

Given any $\varepsilon > 0$, we first choose ρ large enough so that

$$\int_{|x|\geqslant\rho} |f(x)|\, dx < \varepsilon$$

and then τ large enough so that

$$\int_{|t|\geqslant\tau+\rho} |g(t)|\, dt < \varepsilon.$$

We then have

$$\left| \int_{-\infty}^{\infty} f(x) \left\{ \int_{|t+x|\geqslant\tau} g(t)\, dt \right\} dx \right|$$

$$\leqslant \left| \int_{|x|\geqslant\rho} f(x) \left\{ \int_{|t+x|\geqslant\tau} g(t)\, dt \right\} dx \right| + \left| \int_{|x|\leqslant\rho} f(x) \left\{ \int_{|t+x|\geqslant\tau} g(t)\, dt \right\} dx \right|$$

$$\leqslant \int_{|x|\geqslant\rho} |f(x)| \left\{ \int_{-\infty}^{\infty} |g(t)|\, dt \right\} dx + \int_{|x|\leqslant\rho} |f(x)| \left\{ \int_{|t|\geqslant\tau+\rho} |g(t)|\, dt \right\} dx$$

$$< \varepsilon \left\{ \int_{-\infty}^{\infty} |g(t)|\, dt + \int_{-\infty}^{\infty} |f(x)|\, dx. \right.$$

Hence the last integral in (17) approaches 0 as $\tau \to \infty$, and the proof of (15) is complete. ∎

b. Denoting the convolution of the functions $f(t)$ and $g(t)$ by $f(t) * g(t)$, we can write (15) in the form

$$\int_{-\infty}^{\infty} [f(t) * g(t)] \, dt = \int_{-\infty}^{\infty} f(t) \, dt \int_{-\infty}^{\infty} g(t) \, dt. \tag{15'}$$

Making the substitution $t - x = \xi$, $x = t - \xi$ in the "proper" integral

$$h_\tau(t) = \int_{-\tau}^{\tau} f(x) g(t - x) \, dx,$$

we get

$$h_\tau(t) = \int_{t-\tau}^{t+\tau} f(t - \xi) g(\xi) \, d\xi.$$

Then, taking the limit as $\tau \to \infty$, we find that

$$f(t) * g(t) = \int_{-\infty}^{\infty} f(x) g(t - x) \, dx = \int_{-\infty}^{\infty} f(t - \xi) g(\xi) \, d\xi = g(t) * f(t),$$

i.e., the convolution of the functions $f(t)$ and $g(t)$ does not change if we reverse the order of the functions.

11.59. a. The following test for uniform convergence of improper integrals is the natural generalization of the *Abel-Dirichlet test* (Theorem 11.23a) and works in some cases where the "majorant test" (Theorem 11.57a) is not applicable:

THEOREM. *Let $f(x)$ be piecewise smooth on every finite subinterval $[a,X] \subset [a,\infty)$, with derivative $f'(x)$, and suppose $f(x) \to 0$ as $x \to \infty$ while $f'(x)$ is absolutely integrable on $[a,\infty)$. Moreover, let $g(x,t)$ be piecewise smooth on every finite subinterval $[a,X] \subset [a,\infty)$ for every $t \in M$, and suppose*

$$|G(x,t)| \leqslant C \qquad (a \leqslant x < \infty, \, t \in M),$$

where

$$G(x,t) = \int_a^x g(\xi,t) \, d\xi \qquad (a \leqslant x < \infty)$$

and C is independent of $x \in [a,\infty)$ and $t \in M$. Then the integral

$$\int_a^\infty f(x) g(x,t) \, dx$$

is uniformly convergent on M.

Proof. The exact analogue of the proof of Theorem 11.23a, with the use of the Cauchy convergence criterion for uniform convergence of improper integrals (Theorem 11.56). ∎

b. Example. The integral

$$\int_0^\infty \frac{\sin xt}{x}\, dx = \frac{\pi}{2} \qquad (t>0)$$

(calculated in Sec. 11.42h) is uniformly convergent on any interval $0 < \alpha \leqslant t \leqslant \beta$. This follows from the Abel-Dirichlet test, since the function

$$f(x) = \frac{1}{x}$$

approaches 0 as $x \to \infty$ and has an (absolutely) integrable derivative

$$f'(x) = -\frac{1}{x^2},$$

while

$$\left| \int_a^x g(\xi,t)\, d\xi \right| = \left| \int_a^x \sin \xi t\, d\xi \right| = \frac{\cos xt - \cos at}{t} \leqslant \frac{2}{t} \leqslant \frac{2}{\alpha}.$$

Using Theorem 11.54 to integrate with respect to t from α to β, we get the integral considered in Example 11.57b:

$$\int_0^\infty \frac{\cos \alpha x - \cos \beta x}{x^2}\, dx = \int_0^\infty \left\{ \int_\alpha^\beta \frac{\sin xt}{x}\, dx \right\} dt$$
$$= \int_\alpha^\beta \left\{ \int_0^\infty \frac{\sin xt}{x}\, dx \right\} dt = \frac{\pi}{2}(\beta - \alpha). \qquad (18)$$

11.6. The Gamma and Beta Functions

11.61. By the *gamma function* we mean the function defined by the integral

$$\Gamma(\tau) = \int_0^\infty t^{\tau-1} e^{-t}\, dt. \qquad (1)$$

This improper integral of the third kind is the sum of an improper integral

$$\int_1^\infty t^{\tau-1} e^{-t}\, dt$$

of the first kind and an improper integral

$$\int_0^1 t^{\tau-1} e^{-t}\, dt$$

of the second kind, where the first integral converges for all real τ and the second converges for all $\tau > 0$. Therefore (1) defines $\Gamma(\tau)$ for all $\tau > 0$.

Suppose $0 < \alpha \leqslant \tau \leqslant \beta$. Then the integrand $t^{\tau-1}e^{-t}$ has the integrable majorant

$$\varphi(t) = \begin{cases} t^{\alpha-1}e^{-t} & \text{if } 0 \leqslant t \leqslant 1, \\ t^{\beta-1}e^{-t} & \text{if } t \geqslant 1. \end{cases}$$

It follows that the integral (1) converges uniformly on every interval $[\alpha, \beta]$, and hence, by Theorem 11.53, represents a continuous function of τ on $(0, \infty)$, since $t^{\tau-1}e^{-t}$ is clearly continuous on every rectangle $0 < a \leqslant t \leqslant T$, $0 < \alpha \leqslant \tau \leqslant \beta$ (why?). By the same token, the derivative $t^{\tau-1}e^{-t}\ln t$ of the integrand with respect to τ has the integrable majorant

$$\psi(t) = \begin{cases} t^{a-1}e^{-1}\,|\ln t| & \text{if } 0 \leqslant t \leqslant 1, \\ t^{\beta}e^{-t} & \text{if } t \geqslant 1, \end{cases}$$

so that the integral

$$\int_0^\infty t^{\tau-1}e^{-t}\ln t\ dt$$

also converges uniformly on every interval $[\alpha, \beta]$. Hence, by Theorem 11.55a, $\Gamma(\tau)$ has the derivative

$$\Gamma'(\tau) = \int_0^\infty t^{\tau-1}e^{-t}\ln t\ dt \qquad (\tau > 0),$$

since $t^{\tau-1}e^{-t}\ln t$ and its partial derivative with respect to τ are both continuous on every rectangle $0 < a \leqslant t \leqslant T$, $0 < \alpha \leqslant \tau \leqslant \beta$. Moreover, $\Gamma'(\tau)$ is continuous for all $\tau > 0$ for the same reasons as the function $\Gamma(\tau)$ itself. We can repeatedly differentiate $\Gamma(\tau)$, each time applying Theorem 11.55a. Thus $\Gamma(\tau)$ has derivatives of all orders.

11.62. Integrating by parts in the definition (1) of $\Gamma(\tau)$, we find that

$$\Gamma(\tau) = \int_0^\infty t^{\tau-1}e^{-t}\ dt = \frac{t^\tau}{\tau}e^{-t}\bigg|_0^\infty + \frac{1}{\tau}\int_0^\infty t^\tau e^{-t}\ dt = \frac{1}{\tau}\int_0^\infty t^\tau e^{-t}\ dt,$$

so that

$$\int_0^\infty t^\tau e^{-t}\ dt = \tau\Gamma(\tau)$$

or

$$\Gamma(\tau+1) = \tau\Gamma(\tau). \tag{2}$$

Formula (2) is the basic *functional equation for the gamma function.* Applying (2) repeatedly, we get

$$\Gamma(\tau+n) = (\tau+n-1)(\tau+n-2)\cdots(\tau+1)\tau\Gamma(\tau).$$

Thus from a knowledge of the values of the function $\Gamma(\tau)$ on any interval of length 1, we can find its values on the rest of the half-line $\tau > 0$. Moreover, since

$$\Gamma(1) = \int_0^\infty e^{-t}\, dt = 1,$$

we see that

$$\Gamma(n+1) = n(n-1)\cdots 1 = n!.$$

It follows that the gamma function is a continuous extension to all positive real numbers of the function $n!$ (n factorial) defined for all positive integers. Note that (2) implies

$$\lim_{\tau\to 0} \Gamma(\tau) = \lim_{\tau\to 0} \frac{\Gamma(\tau+1)}{\tau} = \infty,$$

since $\Gamma(\tau)$ is continuous at $\tau = 1$.

11.63. The beta function and its relation to the gamma function.
By the *beta function* we mean the function of two parameters p and q defined by the integral

$$B(p,q) = \int_0^1 x^{p-1}(1-x)^{q-1}dx. \tag{3}$$

This integral exists for all positive values of p and q, as an ordinary "proper" integral if $p,q > 0$ and as a convergent improper integral otherwise. The substitution

$$x = \frac{\theta}{1+\theta}$$

transforms (3) into the form

$$B(p,q) = \int_0^\infty \frac{\theta^{p-1}}{(1+\theta)^{p+q}}\, d\theta. \tag{4}$$

Next we show how the beta function can be expressed in terms of the gamma function. Making the substitution $t = \theta y$ in the expression (1) for the gamma function, we get

$$\frac{\Gamma(\tau)}{\theta^\tau} = \int_0^\infty y^{\tau-1} e^{-\theta y} \, dy,$$

which becomes

$$\frac{\Gamma(p+q)}{(1+\theta)^{p+q}} = \int_0^\infty y^{p+q-1} e^{-y} e^{-\theta y} \, dy \tag{5}$$

after replacing τ by $p+q$ and θ by $1+\theta$. Then, multiplying both sides of (5) by θ^{p-1} and integrating with respect to θ from 0 to n, we find that

$$\Gamma(p+q) \int_0^n \frac{\theta^{p-1}}{(1+\theta)^{p+q}} \, d\theta = \int_0^n \left\{ \int_0^\infty \theta^{p-1} y^{p+q-1} e^{-y} e^{-\theta y} \, dy \right\} d\theta, \tag{6}$$

where the left-hand side approaches $\Gamma(p+q)\mathrm{B}(p,q)$ as $n \to \infty$, because of (4).

As for the right-hand side of (6), the integrand has the integrable majorant $n^{p-1} y^{p+q-1} e^{-y}$. Hence, by Theorem 11.54, we can reverse the order of integration, obtaining

$$\int_0^n \left\{ \int_0^\infty \theta^{p-1} y^{p+q-1} e^{-y} e^{-\theta y} \, dy \right\} d\theta$$

$$= \int_0^\infty y^{p+q-1} e^{-y} \left\{ \int_0^n \theta^{p-1} e^{-\theta y} \, d\theta \right\} dy$$

$$= \int_0^\infty y^{p+q-1} e^{-y} \left\{ \frac{1}{y^p} \int_0^{ny} t^{p-1} e^{-t} \, dt \right\} dy$$

$$= \int_0^\infty y^{q-1} e^{-y} F_n(y) \, dy,$$

where the function

$$F_n(y) = \int_0^{ny} t^{p-1} e^{-t} \, dt$$

approaches the function

$$F(y) = \begin{cases} 0 & \text{if } y = 0, \\ \Gamma(p) & \text{if } y > 0 \end{cases}$$

as $n \to \infty$. This convergence is nonuniform on the interval $0 \leqslant y \leqslant b$ (as can be seen from the discontinuity of the limit function), but the convergence is indeed uniform on every interval $0 < h \leqslant y \leqslant b$, since

$$0 < \Gamma(p) - \int_0^{ny} t^{p-1} e^{-t} \, dt \leqslant \Gamma(p) - \int_0^{nh} t^{p-1} e^{-t} \, dt.$$

Moreover, since $F_n(y)$ is an increasing nonnegative function converging to $\Gamma(p)$, the set of functions

$$y^{q-1}e^{-y}F_n(y) \qquad (n=1,2,\dots).$$

has the integrable majorant $y^{q-1}e^{-y}\Gamma(p)$, and hence, by a suitable "sequence analogue" of Theorem 11.53 (give the details),

$$\lim_{n\to\infty}\int_h^\infty y^{q-1}e^{-y}F_n(y)\,dy = \Gamma(p)\int_h^\infty y^{q-1}e^{-y}\,dy.$$

On the other hand,

$$0\leqslant\int_0^h y^{q-1}e^{-y}F_n(y)\,dy \leqslant \Gamma(p)\int_0^h y^{q-1}\,dy = \Gamma(p)\frac{h^q}{q} \tag{7}$$

for every $h>0$. Given any $\varepsilon>0$, first let $h>0$ be such that

$$\Gamma(p)\frac{h^q}{q} < \frac{\varepsilon}{3}, \tag{8}$$

and then choose N such that

$$0<\Gamma(p)\int_h^\infty y^{q-1}e^{-y}\,dy - \int_h^\infty y^{q-1}e^{-y}F_n(y)\,dy < \frac{\varepsilon}{3} \tag{9}$$

for all $n>N$, at the same time noting that

$$0<\Gamma(p)\Gamma(q) - \Gamma(p)\int_h^\infty y^{q-1}e^{-y}\,dy$$
$$= \Gamma(p)\int_0^h y^{q-1}e^{-y}\,dy < \Gamma(p)\frac{h^q}{q} < \frac{\varepsilon}{3} \tag{10}$$

for our choice of h. It follows from (7)–(10) that

$$0<\Gamma(p)\Gamma(q) - \int_0^\infty y^{q-1}e^{-y}F_n(y)\,dy < \varepsilon,$$

and hence that

$$\lim_{n\to\infty}\int_0^\infty y^{q-1}e^{-y}F_n(y)\,dy = \Gamma(p)\Gamma(q).$$

Thus, finally, taking the limit as $n\to\infty$ in (6), we get the desired expression for the beta function in terms of the gamma function:

$$\mathrm{B}(p,q) = \frac{\Gamma(p)\Gamma(q)}{\Gamma(p+q)}. \tag{11}$$

11.64. Many trigonometric integrals can in turn be expressed in terms of the beta function. For example, making the substitution $x = \sin^2 \theta$ in the integral

$$I = \int_0^{\pi/2} \sin^{p-1} \theta \, \cos^{q-1} \theta \; d\theta,$$

we get

$$I = \int_0^1 x^{(p-1)/2}(1-x)^{(q-1)/2} \frac{dx}{2\sqrt{x(1-x)}}$$

$$= \int_0^1 x^{(p/2)-1}(1-x)^{(q/2)-1} \, dx = \frac{1}{2} \mathrm{B}\left(\frac{p}{2}, \frac{q}{2}\right).$$

11.65. Setting $q = 1 - p$ in (11) and recalling (4), we get

$$\Gamma(p)\Gamma(1-p) = \mathrm{B}(p, 1-p) = \int_0^\infty \frac{\theta^{p-1}}{1+\theta} \, d\theta. \tag{12}$$

The integral on the right can easily be calculated by using contour integration. To this end, consider the analytic function

$$f(z) = \frac{z^{p-1}}{1+z}, \tag{13}$$

defined on the complex z-plane cut along the positive real axis. The function (13) is uniquely determined if we set

$$f(x+i0) = \frac{x^{p-1}}{1+x} \tag{14}$$

on the upper edge of the cut (x^{p-1} being defined as in Sec. 5.53). Making one circuit around the origin in the counterclockwise direction, we get $z = x - i0 = xe^{2\pi i}$, and hence we set

$$f(x-i0) = \frac{x^{p-1}}{1+x} e^{2\pi i(p-1)} \tag{15}$$

on the lower edge of the cut.

Now let L_R be the contour shown in Figure 80, consisting of the interval $[0,R]$, $R > 1$ of the real axis along the upper edge of the cut, the circle C_R of radius R centered at the origin (traversed in the counterclockwise direction) and the interval $[0,R]$ of the real axis along the lower edge of the cut (traversed from right to left). The only singular point of $f(z)$ inside L is at the point $z = -1$. Therefore, by formula (4), p. 407,

$$\oint_{L_R} f(z) \, dz = 2\pi i \operatorname*{Res}_{z=-1} f(z) = 2\pi i e^{\pi i(p-1)}, \tag{16}$$

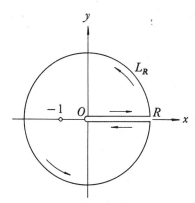

Figure 80

while on the other hand,

$$\oint_{L_R} f(z)\,dz = \int_0^R f(x+i0)\,dx + \int_{C_R} f(z)\,dz + \int_R^0 f(x-i0)\,dx. \tag{17}$$

On C_R we have

$$|f(z)| \leqslant \frac{R^{p-1}}{R-1} \leqslant AR^{p-2}$$

for some constant A and all sufficiently large R, and hence

$$\left| \int_{C_R} f(z)\,dz \right| \leqslant AR^{p-2} 2\pi R = 2\pi AR^{p-1} \to 0$$

as $R \to \infty$. Thus, taking the limit as $R \to \infty$ in (17) and using (14)–(16), we get

$$[1 - e^{2\pi i(p-1)}] \int_0^\infty \frac{x^{p-1}}{1+x}\,dx = 2\pi i e^{\pi i(p-1)},$$

and hence

$$\int_0^\infty \frac{x^{p-1}}{1+x}\,dx = 2\pi i \frac{e^{\pi i(p-1)}}{1 - e^{2\pi i(p-1)}} = \frac{2\pi i}{e^{-\pi i(p-1)} - e^{\pi i(p-1)}}$$

$$= -\frac{\pi}{\sin \pi(p-1)} = \frac{\pi}{\sin \pi p}. \tag{18}$$

Finally, comparing (12) and (18), we obtain the following "complementation formula" for the gamma function:

$$\Gamma(p)\Gamma(1-p) = B(p,1-p) = \frac{\pi}{\sin \pi p}.$$ (19)

In particular, setting $p = \frac{1}{2}$ in (19) gives

$$\Gamma\left(\frac{1}{2}\right) = \sqrt{\pi}$$

which together with the basic functional equation (2) implies

$$\Gamma\left(\frac{3}{2}\right) = \frac{1}{2}\Gamma\left(\frac{1}{2}\right) = \frac{1}{2}\sqrt{\pi},$$

$$\Gamma\left(\frac{5}{2}\right) = \frac{3}{2}\frac{1}{2}\sqrt{\pi},\ldots,$$

and more generally,

$$\Gamma\left(\frac{2n+1}{2}\right) = \frac{(2n-1)(2n-3)\cdots 3 \cdot 1}{2^n}\sqrt{\pi}.$$

The graph of the gamma function $\Gamma(x)$ for $0 < x \leqslant 5$ is shown in Figure 81.

11.66. Using the gamma function, we can easily evaluate the following important integral encountered in probability theory:

$$I_m = \int_0^\infty x^m e^{-ax^2}\, dx \qquad (a > 0,\ m > -1).$$

In fact, the substitution

$$ax^2 = t, \qquad x = \sqrt{\frac{t}{a}}, \qquad dx = \frac{1}{2}\frac{dt}{\sqrt{at}}$$

reduces the integral I_m to the form

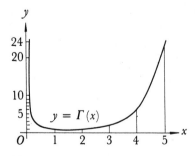

Figure   81

$$I_m = \frac{1}{2a^{(m+1)/2}} \int_0^\infty t^{(m-1)/2} e^{-t} \, dt = \frac{1}{2a^{(m+1)/2}} \Gamma\left(\frac{m+1}{2}\right).$$

In particular, we have

$$I_0 = \int_0^\infty e^{-x^2} \, dx = \frac{1}{2}\Gamma\left(\frac{1}{2}\right) = \frac{1}{2}\sqrt{\pi}.$$

11.67. Asymptotic representation of the gamma function. We now find an asymptotic representation of the gamma function

$$\Gamma(\tau) = \int_0^\infty t^{\tau-1} e^{-1} \, dt$$

for large values of τ.

a. LEMMA. *Let $f(x)$ be a nonnegative function defined for $x > 0$, such that*
(a) $f(1) = 0$;
(b) $f(x)$ *is decreasing for $0 < x < 1$ and increasing for $x > 1$;*
(c) $f(x)$ *has the representation*

$$f(x) = a(x-1)^2 + (x-1)^{2+2\delta}\psi(x),$$

where $a > 0$, $\delta > 0$, $|\psi(x)| \leqslant M$ in a neighborhood of the point $x = 1$;
(d) $f(x)$ *satisfies the inequality*

$$|f(x)| \geqslant bx \qquad (b > 0) \tag{20}$$

for $x \geqslant c > 1$.
    Then the function

$$I(s) = \int_0^\infty e^{-sf(x)} \, dx \tag{21}$$

satisfies the limiting relation

$$\lim_{s \to \infty} \frac{I(s)}{\sqrt{\pi/sa}} = 0. \tag{22}$$

Proof. The convergence of the integral (21) for $s > 0$ follows from the estimate (20). For $0 < \varepsilon < 1$ we have

$$I(s) = \left\{ \int_0^{1-\varepsilon} + \int_{1-\varepsilon}^{1+\varepsilon} + \int_{1+\varepsilon}^c + \int_c^\infty \right\} e^{-sf(x)} \, dx. \tag{23}$$

By hypothesis,

$$\int_0^{1-\varepsilon} e^{-sf(x)} \, dx < e^{-sf(1-\varepsilon)} < e^{-s(a/2)\varepsilon^2}, \tag{24}$$

$$\int_{1+\varepsilon}^{c} e^{-sf(x)}\, dx < (c-1)e^{-sf(1+\varepsilon)} < (c-1)e^{-s(a/2)\varepsilon^2} \tag{25}$$

for sufficiently small ε, while

$$\int_{0}^{\infty} e^{-sf(x)}\, dx < \int_{c}^{\infty} e^{-sbx}\, dx = \frac{e^{-sbc}}{sb}. \tag{26}$$

With a view to calculating the integral from $1-\varepsilon$ to $1+\varepsilon$, we note that

$$|e^{-s(x-1)^2+2\delta}\psi(x) - 1| = |s(x-1)^{2+2\delta}\psi(x) + \tfrac{1}{2}s^2(x-1)^{2(2+2\delta)}\psi^2(x) + \cdots|$$
$$\leqslant Ms\varepsilon^{2+2\delta} + \tfrac{1}{2}M^2s^2\varepsilon^{2(2+2\delta)} + \cdots \leqslant 2Ms\varepsilon^{2+2\delta}$$

on the interval $[1-\varepsilon, 1+\varepsilon]$, provided only that

$$Ms\varepsilon^{2+2\delta} \leqslant \tfrac{1}{2}. \tag{27}$$

So far, the number ε is arbitrary. Now let

$$\varepsilon = s^{-1/(2+\delta)}$$

for large s. Then

$$s\varepsilon^{2+2\delta} = s^{1-(2+2\delta)/(2+\delta)} = s^{-\delta/(2+\delta)} \to 0$$

as $s \to \infty$, and the condition (27) is indeed satisfied. On the other hand, for this choice of ε,

$$s\varepsilon^2 = s^{1-2/(2+\delta)} = s^{\delta/(2+\delta)} \to \infty$$

as $s \to \infty$. Hence

$$|e^{-s(x-1)^2+2\delta}\psi(x) - 1| \leqslant 2Ms\varepsilon^{2+2\delta} \to 0$$

for any $x \in [1-\varepsilon, 1+\varepsilon]$ as $s \to \infty$, so that

$$e^{-s(x-1)^2+2\delta}\psi(x) = 1 + o(1),$$

where $o(1) \to 0$ as $s \to \infty$ (cf. Sec. 4.32). Therefore

$$\int_{1-\varepsilon}^{1+\varepsilon} e^{-sf(x)}\, dx = \int_{1-\varepsilon}^{1+\varepsilon} e^{-sa(x-1)^2}[1+o(1)]\, dx$$
$$= [1+o(1)] \int_{1-\varepsilon}^{1+\varepsilon} e^{-sa(x-1)^2}dx$$
$$= [1+o(1)]\frac{1}{\sqrt{sa}} \int_{-\varepsilon\sqrt{sa}}^{\varepsilon\sqrt{sa}} e^{-u^2}du.$$

But $\varepsilon\sqrt{s} \to \infty$ as $s \to \infty$, and hence the last integral approaches the limit

$$\int_{-\infty}^{\infty} e^{-u^2} \, du = \sqrt{\pi}$$

as $s \to \infty$ (see Sec. 11.66), so that

$$\int_{-\varepsilon\sqrt{sa}}^{\varepsilon\sqrt{sa}} e^{-u^2} \, du = [1 + o(1)]\sqrt{\pi},$$

where $o(1) \to 0$ as $s \to \infty$. It follows that

$$\int_{1-\varepsilon}^{1+\varepsilon} e^{-sf(x)} \, dx = [1 + o(1)]\sqrt{\frac{\pi}{sa}}.$$

The other terms in (23), namely (24), (25), and (26), all approach zero exponentially as $s \to \infty$. Thus, finally,

$$I(s) = \int_0^\infty e^{-sf(x)} \, dx = [1 + o(1)]\sqrt{\frac{\pi}{sa}} + o\left(\frac{1}{\sqrt{s}}\right) = \sqrt{\frac{\pi}{sa}}\,[1 + o(1)],$$

which is equivalent to (22). ∎

b. We now transform the expression for the gamma function into a form suitable for application of the lemma. Making the substitution $t = sx$, we get

$$\Gamma(s+1) = \int_0^\infty t^s e^{-t} \, dt = s^{s+1} \int_0^\infty x^s e^{-sx} \, dx = s^{s+1} \int_0^\infty e^{-sx + s\ln x} \, dx$$

$$= s^{s+1} e^{-s} \int_0^\infty e^{-s(x - \ln x - 1)} \, dx = s^{s+1} e^{-s} \int_0^\infty e^{-sf(x)} \, dx,$$

where

$$f(x) = x - \ln x - 1.$$

It is easy to see that $f(x)$ satisfies all the conditions of the lemma. Clearly

$$f(1) = 0, \qquad f'(1) = 0, \qquad f''(1) = 1, \qquad f'''(1) = -2,$$

and hence

$$f(x) = \frac{1}{2}(x-1)^2 - \frac{E}{3}(x-1)^3, \qquad \lim_{x \to 1} E = 1$$

in a neighborhood of the point $x = 1$, so that the constant a equals $\frac{1}{2}$. Therefore, applying the lemma, we get the desired asymptotic representation of the gamma function:†

$$\Gamma(s+1) = s^{s+1} e^{-s} \sqrt{\frac{2\pi}{s}}\,[1 + o(1)] \approx \sqrt{2\pi s}\, s^s e^{-s}. \qquad (28)$$

† The symbol ≈ denotes approximate equality (for large values of the argument).

In particular, if $s = n$ is a positive integer n, then $\Gamma(n+1) = n!$ and (28) reduces to *Stirling's formula*

$$n! \approx \sqrt{2\pi n}\, n^n e^{-n} \tag{28'}$$

(dating from 1730).

11.68. The gamma function in the complex domain. The formula

$$\Gamma(z) = \int_0^\infty t^{z-1} e^{-t}\, dt = \int_0^\infty e^{(z-1)\ln t} e^{-t}\, dt \tag{29}$$

defining the gamma function is applicable not only for positive real z, but also for certain complex values of z. In fact, the integral (29) is also convergent if $z = x + iy$, $x > 0$, since the integrand

$$e^{(x+iy-1)\ln t} e^{-t} = e^{(x-1)\ln t} e^{-t} e^{iy\ln t} = t^{x-1} e^{-t} e^{iy\ln t} \tag{30}$$

differs from the function $t^{x-1} e^{-t}$ only by a factor of absolute value 1. Thus (30) defines $\Gamma(z)$ directly for all $z = x + iy$ with $x = \mathrm{Re}\ z > 0$, i.e., for all z in the whole (open) right half-plane G. Moreover, the integral (29) converges uniformly inside G, since the quantity

$$d = \inf_{z \in Q} \mathrm{Re}\ z$$

is positive for every compact set $Q \subset G$, so that (30) has the integrable majorant $t^{d-1} e^{-t}$ for all $z \in Q$. It follows from Theorem 11.55b that $\Gamma(z)$ is analytic on G.

Next we examine the possibility of continuing the function $\Gamma(z)$ analytically into the left half-plane (cf. Sec. 10.39c). To this end, we use the formula

$$\Gamma(z+n) = z(z+1)\cdots(z+n-1)\Gamma(z), \tag{31}$$

proved for positive real z in Sec. 11.62. Since the two sides of (31) are obviously analytic on G and since they coincide on the positive real axis, it follows from the uniqueness theorem for analytic functions (Theorem 10.39b) that they coincide on the whole half-plane G. Writing (31) in the form

$$\Gamma(z) = \frac{\Gamma(z+n)}{z(z+1)\cdots(z+n-1)}, \tag{32}$$

we observe that the right-hand side is defined and analytic on the domain $G_n = \{z : \mathrm{Re}\ z > -n\}$. The formally distinct definitions obtained for different values of n all give the same value of $\Gamma(z)$ for any given z, by the uniqueness of the analytic continuation. Since we can make n arbitrarily large, this defines $\Gamma(z)$ everywhere in the z-plane except for isolated singular points $z=0,\ -1,...,\ -n+1,...$ It is clear from (32) that these singular points are all poles of the first order. To calculate the residue of $\Gamma(z)$ at the point $z=-n+1$, we use formula (3), p. 406, obtaining

$$\operatorname*{Res}_{z=-n+1}\ \Gamma(z)=\frac{\Gamma(z+n)}{z(z+1)\cdots(z+n-2)}\Bigg|_{z=-n+1}$$

$$=\frac{\Gamma(1)}{(-1)(-2)\cdots(-n+1)}=\frac{(-1)^{n+1}}{(n-1)!}.$$

Problems

1. Prove *Dirichlet's formula*

$$\int_0^1\left\{\int_0^{1-x}x^{\lambda-1}y^{\mu-1}(1-x-y)^{\nu-1}f(x,y)\ dy\right\}\ dx$$

$$=\int_0^1\left\{\int_0^{1-y}x^{\lambda-1}y^{\mu-1}(1-x-y)^{\nu-1}f(x,y)\ dx\right\}\ dy,$$

where $0<\lambda\leqslant1,\ 0<\mu\leqslant1,\ 0<\nu\leqslant1$ and $f(x,y)$ is continuous for $0\leqslant x\leqslant1$, $0\leqslant y\leqslant1$.

2. Prove *Frullani's formula*

$$\int_0^\infty\frac{f(ax)-f(bx)}{x}\ dx=f(0)\ \ln\frac{b}{a}\qquad(a>0,\ b>0),$$

where $f(x)$ is continuous for $0\leqslant x<\infty$ and the integral

$$\int_1^\infty\frac{f(x)}{x}\ dx$$

converges.

3. Use integration in the complex plane to evaluate

$$I(p)=\int_0^\infty e^{-px^2}\ dx,$$

where $\operatorname{Re} p\geqslant0$.

4. Evaluate the *Fresnel integrals*

$$F_1=\int_0^\infty\sin(x^2)\ dx,\qquad F_2=\int_0^\infty\cos(x^2)\ dx.$$

5. Evaluate the singular Fourier integral

$$I = \int_{-\infty}^{\infty} \frac{P(x)}{Q(x)} \frac{\sin x}{x} \, dx,$$

where $P(x)$ and $Q(x)$ are polynomials such that $Q(x)$ has no real zeros and the degree of $P(x)$ does not exceed that of $Q(x)$.

6. Prove that

$$\int_0^{\infty} \left\{ \int_0^1 \frac{t-x}{(x+t)^3} \, dt \right\} dx = -1,$$

$$\int_0^1 \left\{ \int_0^{\infty} \frac{t-x}{(x+t)^3} \, dx \right\} dt = 0.$$

Why doesn't this contradict Theorem 11.54?

7. Starting from the integral (18), p. 467, prove that

$$\int_0^{\infty} \frac{1-\cos \beta x}{x^2} \, dx = \frac{\pi \beta}{2}$$

and hence that

$$\int_0^{\infty} \frac{\sin^2 x}{x^2} \, dx = \frac{\pi}{2}.$$

8. Prove that

$$\int_a^b \frac{dx}{\sqrt{(x-a)(b-x)}} = \pi \qquad (a<b).$$

9. Prove that

$$\int_0^{\pi/2} \ln \sin x \, dx = -\frac{\pi}{2} \ln 2.$$

10. Prove that

$$\int_0^{\infty} \left(\frac{x}{e^x - e^{-x}} - \frac{1}{2} \right) = -\frac{1}{2} \ln 2.$$

11. Prove that if the function $f(x)$ is continuous and has a bounded integral

$$\Phi(x) = \int_a^x f(x) \, dx$$

for all $x \geqslant a$, then the integral

$$\int_a^{\infty} \frac{f(x)}{x^{\alpha}} \, dx \qquad (a, \alpha > 0)$$

converges.

12. Use residues to evaluate the integral

$$\int_{-\infty}^{\infty} \frac{x^{2m}}{1+x^{2n}}\, dx \qquad (m<n),$$

where m and n are positive integers.

13. Let L be the path in the plane of the complex variable s consisting of the interval $[\varepsilon,\infty)$ of the real axis traversed from right to left, the circle $|s|=\varepsilon$ traversed once in the counterclockwise direction, and the interval $[\varepsilon,\infty)$ traversed once again, this time from left to right. Prove that the formula

$$\Gamma(z) = \frac{1}{e^{2\pi i z}-1}\int_{L} e^{-s}s^{z-1}\, ds$$

represents the gamma function for all complex z except the poles $z=0,\ -1,\ -2,\ldots$

Appendix A
Elementary Symbolic Logic

A.1. Logicians have long made use of an economical notation for writing out mathematical demonstrations. We now give a brief sketch of the simplest and most useful notation of this kind.

Suppose we are interested not so much in the concrete nature of a given proposition as in its relation to other propositions. Then we can designate the proposition by a single letter, such as α or β. We now introduce the following notations, each with the meanings given in parentheses:

(1) $\alpha \Rightarrow \beta$ ("the proposition α implies the proposition β");
(2) $\alpha \Leftrightarrow \beta$ ("each of the propositions α and β implies the other, i.e., α and β are equivalent propositions");
(3) $\forall x \in E: \alpha$ ("for all $x \in E$ the proposition α is true");
(4) $\exists x \in E: \alpha$ ("there exists an element $x \in E$ such that the proposition α is true").

For example, saying that "the number ξ is the least upper bound of the set E" means that the following two conditions hold (see Sec. 1.24):

(a) $\forall x \in E: x \leqslant \xi$ ("the inequality $x \leqslant \xi$ holds for all $x \in E$");
(b) $\forall a \geqslant E: a \geqslant \xi$ ("every a greater than or equal to every element of E is greater than or equal to ξ").

A.2. Let $\bar{\alpha}$ denote "not α," i.e., the negation of the proposition α. Then, clearly,

$$\bar{\bar{\alpha}} \Leftrightarrow \alpha,$$
$$(\alpha \Rightarrow \beta) \Leftrightarrow (\bar{\beta} \Rightarrow \bar{\alpha}),$$
$$(\alpha \Leftrightarrow \beta) \Leftrightarrow (\bar{\alpha} \Leftrightarrow \bar{\beta}).$$

We now construct the negation of the proposition $\forall x \in E: \alpha$ ("every $x \in E$ has the property α"). If the proposition in question is false, then the property α does not hold for every $x \in E$, i.e., there exists an element $x \in E$ which does not have the property α. Therefore

$$\overline{\forall x \in E: \alpha} \Leftrightarrow \exists x \in E: \bar{\alpha}.$$

The negation of the proposition $\exists x \in E: \alpha$ ("there exists an $x \in E$ with the property α") can be found similarly: If the proposition is false, then there is no such $x \in E$, i.e., every $x \in E$ fails to have the property α. Therefore

$$\overline{\exists x \in E: \alpha} \Leftrightarrow \forall x \in E: \bar{\alpha}.$$

Thus putting an expression under the overbar has the effect of changing $\forall$ to $\exists$ or $\exists$ to $\forall$, and then replacing the property appearing after the colon by its negative. For example, the negation of Condition b above is just

($\overline{\text{b}}$) $\overline{\forall a \geqslant E: a \geqslant \xi} \Leftrightarrow \exists a \geqslant E: \overline{a \geqslant \xi} \Leftrightarrow \exists a \geqslant E: a < \xi$

("there exists an a greater than or equal to every $x \in E$ and less than ξ"). Writing out $a \geqslant E$ in more detail, we get

($\overline{\text{b}}$) $\exists a < \xi, \forall x \in E: x \leqslant a.$

A.3. Consider the proposition

(c) $\forall \varepsilon > 0: \exists x \in E, x > \xi - \varepsilon$

("for every $\varepsilon > 0$ there exists an $x \in E$ greater than $\xi - \varepsilon$"). The negation of this proposition is given by

($\overline{\text{c}}$) $\exists \varepsilon > 0: \forall x \in E, \overline{x > \xi - \varepsilon} \Leftrightarrow \exists \varepsilon > 0: \forall x \in E, x \leqslant \xi - \varepsilon.$

Replacing $\xi - \varepsilon$ by a, we get

($\overline{\text{c}}$) $\exists a < \xi: \forall x \in E, x \leqslant a,$

which is identical with Condition $\overline{\text{b}}$. It follows that Conditions c and b coincide. Thus, in defining the least upper bound, we can replace Condition b by Condition c.

A.4. We have just seen how operations involving logical symbols can have interesting implications. Naturally, the arguments given above could have been carried out without using logical symbols at all. However, "symbolic logic" of the type just described is often very useful. Although we have avoided using it in the text (where maximum economy of style was not an objective), we recommend that the reader make use of it in his own work.

Appendix B
Measure and Integration on a Compact Metric Space

In this appendix, we sketch the main features of a general theory of measure and integration on a compact metric space, stating a number of theorems without proof.

B.1. We begin by defining the Riemann integral on a compactum, i.e., on a compact metric space (Sec. 3.91). A compactum K is said to be *weighted* if K has a family of subsets, called *cells*, with the following properties:
(1) K itself and the empty set are cells;
(2) The intersection of any pair of cells sharing interior points is a cell;
(3) If a cell Q is contained in a cell P, then P can be represented as a union of nonintersecting cells with Q as one of the cells;
(4) Given any $\delta > 0$, K is the union of a finite number of cells sharing no interior points (Sec. 3.21), all of diameter less than δ (Sec. 3.12a);
(5) Every cell Q is assigned a nonnegative number $m(Q)$, called the *measure* of Q;
(6) If Q is a union of cells $Q_1,\dots,Q_n$ sharing no interior points, then

$$m(Q_1) = m(Q_1) + \cdots + m(Q_n),$$

a formula called the (*finite*) *additivity condition*.

The (Riemann) integral over a weighted compactum K is defined by analogy with the definition of the integral over a closed interval $[a,b]$, as described in Sec. 9.1. Let $f(x)$ be a function defined on K, and let Π be a partition of K into a finite number of cells $Q_1,\dots,Q_n$ sharing no interior points. Moreover, let $d(\Pi)$ be the maximum diameter of the cells $Q_1,\dots,Q_n$. Choosing an arbitrary point ξ_k in each cell Q_k, we form the Riemann sum

$$S_\Pi(f) = \sum_{k=1}^{n} f(\xi_k) m(Q_k). \tag{1}$$

Then a finite number I is called the (*Riemann*) *integral of the function $f(x)$ over the weighted compactum K* if, given any $\varepsilon > 0$, there exists a $\delta > 0$ such that

$$|I - S_\Pi(f)| < \varepsilon$$

for every partition Π with $d(\Pi) < \delta$ (cf. Sec. 9.13). In other words, the integral of $f(x)$ over K is the limit of the Riemann sum (1) under arbitrary refinement of the partition Π. If this integral exists, we say that $f(x)$ is *integrable on K*. Just as in Sec. 9.13, this definition is comprised in the general scheme of a limit in a direction T. Moreover, we have the following analogue of Theorem 9.14e, proved in virtually the same way: *If $f(x)$ is continuous on K, then $f(x)$ is integrable on K.*

B.2. Examples

a. Let K be the interval $[a,b]$ equipped with the usual metric, and let the cells be all possible intervals $\alpha \leqslant x \leqslant \beta$, $\alpha < x \leqslant \beta$, $\alpha \leqslant x < \beta$, $\alpha < x < \beta$, where $\alpha \leqslant \beta$, with the length of each interval being chosen as its measure. Then the integral over K reduces to the ordinary integral over $[a,b]$, as defined in Sec. 9.13.

b. Let K be the "closed n-dimensional block" specified by the inequalities

$$a_1 \leqslant x_1 \leqslant b_1, \ldots, a_n \leqslant x_n \leqslant b_n$$

and equipped with the metric of n-dimensional Euclidean space (see Sec. 3.14a). Let the cells be all possible "subblocks" $Q \subset K$ specified by inequalities of the form

$$\alpha_1 \leqslant x_1 \leqslant \beta_1, \ldots, \alpha_n \leqslant x_n \leqslant \beta_n$$

and also by the inequalities obtained by replacing some or all of the signs $\leqslant$ by $<$. As the measure of Q, we choose its "n-dimensional volume," i.e., the number

$$m(Q) = \prod_{k=1}^{n} (\beta_k - \alpha_k).$$

Then it is easy to see that this system of cells and measures satisfies Properties 1–5 above. In this case, the corresponding integral is called the *integral over the n-dimensional block K*.

B.3. A set $Z \subset K$ is said to be a *set of (Jordan) measure zero* if, given any $\varepsilon > 0$, there is a finite collection of cells $Q_1, \ldots, Q_n$ such that every point $x \in Z$ is an interior point of

$$\bigcup_{k=1}^{n} Q_k$$

and

$$\sum_{k=1}^{n} m(Q_k) < \varepsilon.$$

There is a theorem, generalizing the theorem on integration of a piecewise-continuous function (Theorem 9.16c), which asserts that *if $f(x)$ is bounded on a weighted compactum K and continuous outside of a set of measure zero, then $f(x)$ is integrable on K*.

B.4. Next, with every set $E \subset K$ we associate a function

$$\chi_E(x) = \begin{cases} 1 \text{ if } x \in E, \\ 0 \text{ if } x \notin E, \end{cases}$$

called the *characteristic function* of E. A set E is said to have volume (or Jordan measure) if the function $\chi_E(x)$ is integrable, and we then call the number

$$m(E) = I(\chi_E),$$

where $I(\chi_E)$ is the integral of χ_E, the *volume* (or *Jordan measure*) of E. Every set of measure zero has zero volume (why?). It follows from the basic properties of the integral that $m(E_1) \leqslant m(E_2)$ if $E_1 \subset E_2$ (provided that $m(E_1)$ and $m(E_2)$ exist) and that if

$$E = E_1 \cup \cdots \cup E_n, \tag{2}$$

where the sets $E, E_1, \ldots, E_n$ all have volume and the sets $E_1, \ldots, E_n$ share no interior points, then

$$m(E) = m(E_1) + \cdots + m(E_n) \tag{3}$$

(volume is additive).

Which sets $E \subset K$ have volume? This question can be answered by examining the boundary of E. By the *boundary* of E we mean the set of all points $x \in K$ which are limit points of both E and its complement (i.e., of all points $x \in K$ such that every neighborhood of x contains both points of E and points of $K - E$). It turns out that *a set $E \subset K$ has volume if and only if the Jordan measure of its boundary is zero*. Sets with volume will henceforth be called *Jordan sets*. In particular, every cell Q is a Jordan set whose volume is its originally assigned measure $m(Q)$. Clearly (2) implies (3) if $E, E_1, \ldots, E_n$ are all Jordan sets and if $E_1, \ldots, E_n$ share no interior points.

B.5. A simple but "sufficiently rich" class of Jordan sets can easily be found in the case where K is an n-dimensional block (see Example B.2b). In fact, it turns out that a bounded set $E \subset K$ is a Jordan set if its boundary is the union of a finite number of "surfaces" with equations of the form

$$x_k = \varphi_k(x_1, \ldots, x_{k-1}, x_{k+1}, \ldots, x_n),$$

where φ_k is a continuous function of the point $(x_1, \ldots, x_{k-1}, x_{k+1}, \ldots, x_n)$ defined on some domain of $(n-1)$-dimensional space. In particular, every "polyhedron" (i.e., every set bounded by a finite number of planes) is a Jordan set.

B.6. Two sets F and G in the n-dimensional Euclidean space R_n are said to be *congruent* if there exists an isometric mapping of R_n into itself carrying F into G (Sec. 3.15b). It can be shown that *congruent Jordan sets have the same volume*. This is true in particular if G is obtained from F by a shift or by reflection in some plane (Sec. 3.15a), or by a rotation (Sec. 5.75b). Moreover, if a Jordan set G is obtained from a Jordan set F by λ-fold expansion

along some axis (Sec. 2.67), then

$$m(G) = \lambda m(F),$$

while if G is obtained from F by λ_1-fold expansion along the x_1-axis, λ_2-fold expansion along the x_2-axis, and so on, then

$$m(G) = \lambda_1 \lambda_2 \cdots \lambda_n m(F).$$

In particular, if G is obtained from F by λ-fold expansion along all axes, i.e., if G is similar to F with ratio of similitude λ, then

$$m(G) = \lambda^n m(E).$$

B.7. In the plane ($n=2$), the curvilinear trapezoid Φ considered in Sec. 9.21, bounded by the x-axis and the curves $x=a$, $x=b$, $y=f(x) \geqslant 0$ is a Jordan set if the function $f(x)$ is continuous.† Let Φ_1 be an elementary figure made up of circumscribed rectangles, as in Figure 22, p. 288, while Φ_2 is an elementary figure made up of inscribed rectangles, as in Figure 23. Then $\Phi_1 \subset \Phi \subset \Phi_2$, and hence

$$m(\Phi_1) \leqslant m(\Phi) \leqslant m(\Phi_2),$$

as in Sec. B.4. The proof that

$$m(\Phi) = \int_a^b f(x)\, dx \tag{4}$$

now follows by the argument given in Sec. 9.21. Moreover, in keeping with (2) and (3), the area of a figure made up of a finite number of curvilinear trapezoids (or figures congruent to such trapezoids), is just a sum of integrals like (4). Such a figure, made up of four curvilinear trapezoids, is shown in . Figure 82.

Thus the theory of area in the plane reduces largely to the evaluation of

Figure 82

† In this case, the curve $y=f(x)$, $a \leqslant x \leqslant b$ automatically has Jordan measure zero (see Sec. B.5).

Riemann integrals. However, it should be noted that this theory is not capable of answering all the questions that naturally arise. In particular, if E is a union of sets with volume, then the present theory guarantees the existence of $m(E)$ only in the case where E is a union of a *finite* number of sets. The case of a countable union of sets is allowed in the more general (but more complicated) theory of *Lebesgue measure*.†

† See e.g., G. E. Shilov and B. L. Gurevich, *Integral, Measure and Derivative: A Unified Approach*, Part 3.

Selected Hints and Answers

Chapter 1

2. *Hint.* (a) It follows from $|x| = |x - y + y| \leqslant |x - y| + |y|$ that $|x| - |y| \leqslant |x - y|$. Now interchange x and y.

4. *Hint.* First prove that the square of an integer is divisible by 3 if and only if the integer itself is divisible by 3.

5. *Ans.* $\sqrt{2} + \sqrt{6}$.

6. *Hint.* Obviously $(a - 1)^2 \geqslant 0$, and hence $a^2 - 2a + 1 \geqslant 0$ or equivalently $a^2 + 1 \geqslant 2a$.

7. *Hint.* Use induction and Problem 6.

8. *Hint.* Apply Problem 7.

9. *Hint.* Apply Problem 8.

10. *Ans.* (a) $x = -1, 3$; (b) $x = 2$; (c) $x = -2$; (d) $x = -2, \frac{2}{3}$.

11. *Ans.* The interval $(-\frac{3}{2}, 15)$.

12. *Ans.* The union of the two intervals $[-4, -1)$ and $(1, 8]$.

13. *Ans.* If $a = 0$, max $A = \min A = \sup A = \inf A = 0$. If $a = 1$, max $A = \min A = \sup A = \inf A = 1$. If $0 < a < 1$, max $A = \sup A = a$ and $\inf A = 0$, but min A does not exist. If $a > 1$, min $A = \inf A = a$ and $\sup A = \infty$, but max A does not exist. If $a = -1$, max $A = \sup A = 1$, min $A = \inf A = -1$. If a is negative and $0 < |a| < 1$, then min $A = \inf A = a$ and max $A = \sup A = a^2$. If a is negative and $|a| > 1$, then $\sup A = \infty$, $\inf A = -\infty$, but max A and min A do not exist.

14. *Hint.* Let $x = -\frac{7}{2}$, say.

15. *Hint.* If

$$\beta = \frac{2}{3} \frac{4}{5} \frac{6}{7} \cdots \frac{100}{101},$$

then $\alpha < \beta$ and hence $\alpha^2 < \alpha\beta = \frac{1}{101}$.

17. *Ans.* (a) $\frac{23}{25}$; (b) $\frac{139}{333}$; (c) $\frac{2329}{999}$; (d) $-\frac{811}{99}$.

18. *Hint.* Verify that the number $\sup A + \sup B$ has the defining properties of the least upper bound of the set $A + B$.

19. *Hint.* Verify that the number $\sup A \cdot \sup B$ has the defining properties of the least upper bound of the set AB.

20. *Hint.* Do the same thing as in Problems 18 and 19.

21. *Hint.* The set A is bounded from above and $\gamma = \sup A$.

Chapter 2

1. *Hint.* Use Theorems 2.34 and 2.35.

2. *Hint.* (a) A rational point can be chosen in each interval; (b) A point with rational coordinates can be chosen in each of the two loops of the figure eight; (c) Only a finite number of points of M can lie outside every interval $[0,\varepsilon]$.

3. *Hint.* Use a construction like that in the proof of Theorem 2.33.

4. *Hint.* There are countably many brothers N.

5. *Hint.* Let C be a countable subset of A, and let $D = A - C$. Then $A = D + C$, $A + B = D + C + B$. Now use the equivalence of the sets C and $C + B$.

6. *Hint.* Use Problem 5, with $A = I$ and B the set of all rational numbers. Treat the case of the set T similarly.

7. *Hint.* The set A is the union of two sets, the set of all sequences containing only a finite number of zeros and the set of all sequences containing infinitely many zeros. The first set is countable, while the second set has the power of the continuum, being in one-to-one correspondence with the set of points of the unit interval written in binary notation (see Sec. 1.78). Now use Problem 5.

8. *Hint.* Every sequence $n_1, n_2, \ldots$ can be associated with a sequence consisting only of zeros and ones, with ones in the places with numbers $n_1, n_2, \ldots$. Now use Problem 7.

9. *Hint.* Every sequence $n_1, n_2, \ldots$ can be associated with an increasing sequence

$$k_1 = n_1, \ k_2 = n_1 + n_2, \ldots, \ k_m = n_1 + \cdots + n_m, \ldots.$$

10. *Hint.* With every sequence $\xi_1, \xi_2, \ldots$ we can associate an array

$$n_{11}, n_{12}, \ldots, n_{1m}, \ldots$$
$$n_{21}, n_{22}, \ldots, n_{2m}, \ldots$$
$$\ldots$$

whose elements are natural numbers (first write ξ_1 as a decimal, say, then ξ_2, and so on indefinitely). This array can be written as a single sequence, as in the proof of Theorem 2.33. Now use Problem 9.

11. *Hint.* Make an appropriate simplification of the solution of Problem 10.

12. *Hint.* Suppose we associate a function $f_t(x) \in E$ with every point $t \in [0,1]$. Then the set $W \subset E$ of all such functions $f_t(x)$ does not exhaust the whole set E, since W does not contain the function $\varphi(x)$ whose value at every point x differs from $f_x(x)$.

13. *Hint.* Generalize the solution of Problem 12, replacing $[0,1]$ by A and specifying every subset $B \subset A$ by a function equal to 1 on B and 0 outside B.

Chapter 3

1. *Hint.* If $A = \{1, \frac{1}{2}, \frac{1}{3}, \ldots\}$, then $A' = \{0\}$ while A'' is empty. Continue the construction recursively.

2. *Hint.* Every limit point of A' is a limit point of A.

3. *Hint.* It is enough to consider the case $n=1$. A point of the set A which is not a limit point of A can be covered by an open interval with rational end points containing no other points of A. Now use Chapter 2, Problem 1a.

4. *Hint.* A point which is not a condensation point of the set A can be covered by an open interval with rational end points containing at most countably many points of A.

5. *Hint.* "Mark" all intervals with rational end points contained in the intervals of $\mathscr{B}$, and then keep an interval of $\mathscr{B}$ containing each such "marked" interval.

8. *Hint.* Let

$$G_1 = \bigcup_{x \in F_1} \{ y \in M : \rho(x,y) < \tfrac{1}{2}\,\rho(x,F_2)\},$$

and similarly for G_2.

10. *Hint.* Use a pair of neighborhoods to make the construction if $n=3$. The space M consisting of four points x_1, x_2, x_3, x_4 with distances

$$\rho(x_1,x_2) = \rho(x_2,x_3) = \rho(x_3,x_4) = \rho(x_4,x_1) = 1,$$
$$\rho(x_1,x_3) = \rho(x_2,x_4) = 2$$

already fails to be isometric to any subset of the Euclidean space R_3.

11. *Hint.* Consider the sphere $S_n \subset R_{n+1}$ of radius 2 centered at the point $(0,0,\ldots,0,1)$. Map the space $R_n \subset R_{n+1}$ of all points $(\xi_1,\xi_2,\ldots,\xi_n,0)$ onto the sphere S_n, using "straight lines" going through the point $(0,0,\ldots,0,2)$, and make this latter point correspond to the point $\infty \in \bar{R}_n$. Then for the metric r choose the usual metric for the corresponding points of S_n in the space R_{n+1} ("stereographic projection").

12. *Hint.* Given any three points $a,x,y \in M$ (with a fixed), find a triple of points A,X,Y in the Euclidean plane R_2 the same distances apart (cf. Problem 10). The new metric $r(x,y)$ is then defined by stereographic projection of R_2 onto the sphere S_2 tangent to R_2 at the point A.

13. *Hint.* Two elements on every line through the origin of coordinates.

Chapter 4

1. *Hint.* See Chapter 1, Problem 1. Consider the sequence

$$x_n = (-1)^n \qquad (n=1,2,\ldots).$$

2. *Hint.* Note that $\underline{\lim}$ and $\overline{\lim}$ are limits of certain sequences.

3. *Ans.* Yes.

5. *Hint.* If $a_{n_k} \to \underline{\lim} \, a_n < \overline{\lim} \, a_n$, let

$$b_n = \begin{cases} 0 & \text{for } n = n_k, \ k = 1, 2, \ldots, \\ \underline{\lim} \, a_n - \overline{\lim} \, a_n & \text{for } n \neq n_k, \end{cases}$$

say.

6. *Ans.* $\frac{1}{3}a + \frac{2}{3}b$.

7. *Hint.* If

$$\varepsilon_n = \frac{x_n}{\sqrt{a}} - 1,$$

then

$$\varepsilon_{n+1} = \frac{\varepsilon_n^2}{2(1 + \varepsilon_n)} < \frac{\varepsilon_n}{2}.$$

for sufficiently large n.

8. *Hint.* Note that $y_n < y_{n+1} < x_{n+1} < x_n$.

9. *Hint.* Suppose $p_1 = \cdots = p_r \geqslant p_{r+1} \geqslant \cdots \geqslant p_n$. Then

$$\lim_{n \to \infty} \left(\frac{p_k}{p_1} \right)^n = 0$$

if $k > r$.

12. *Ans.* No, in both cases. For example, let

$$X = \{1, 2, \ldots, n, \ldots\} \qquad S = \{n \to \infty\},$$

$$Y = \left\{ 0, 1, \frac{1}{2}, \ldots, \frac{1}{n}, \ldots \right\} \qquad Z = \{0, 1\},$$

and let

$$y_1(2n) = 1/2n, \qquad y_1(2n + 1) = 0,$$
$$y_2(n) = 0 \qquad (n = 1, 2, \ldots),$$
$$z(0) = 1, \qquad z(y) = 0 \text{ if } y \neq 0.$$

Then $z(y_1(n))$ has no limit as $n \to \infty$, while $z(y_2(n))$ has the limit

$$1 \neq \lim_{y \to 0} z(y).$$

13. *Hint.* Parts a and b can be verified directly, while Part c follows from Theorem 4.16c.

Chapter 5

1. *Ans.* No.

2. *Ans.* No.

3. *Hint.* Use Chapter 1, Problem 1.

4. *Ans.* Both functions are discontinuous for integral values of x and continuous for all other values of x.

7. *Hint.* Note that

$$\min \{\xi_1,\ldots,\xi_n\} \leqslant \frac{1}{n}\sum_{k=1}^{n}\xi_k \leqslant \max \{\xi_1,\ldots,\xi_n\}.$$

8. *Hint.* Use the existence of $f(a+0)$ and $f(b-0)$.

9. *Hint.* Set $x = \log_c t$ and apply Theorem 5.41.

10. *Ans.* $[-\infty, -1-\delta]$, $[-1+\delta, 1-\delta]$, $[1+\delta, \infty]$ for any $\delta > 0$.

11. *Hint.* Consider the pairs of functions $\sin ax$, $\cos ax$ and $a^x \sin x$, $a^x \cos x$.

12. *Hint.* Consider the function

$$y = \begin{cases} \sin\dfrac{1}{x} & \text{if } x \neq 0, \\ 0 & \text{if } x = 0. \end{cases}$$

13. *Hint.* Use formulas (1)–(3), p. 157.

14. *Hint.* Every point c with the property that

$$\left|\lim_{x\to c}f(x) - f(c)\right| \geqslant q > 0$$

has a neighborhood in which there is no other point with the same property.

15. *Ans.* The function $x(t)$ is continuous everywhere except at points such that all the digits $t_{n_1}, t_{n_2},\ldots$ are nines, beginning with some index n_k.

16. *Hint.* If an arc $\Delta_0 \subset \Gamma$ contains no points of M, then the arcs $\Delta_1 = \Delta - 1$, $\Delta_2 = \Delta - 2,\ldots$, obtained by shifting Δ_0 back 1 unit, 2 units,... along Γ, also contain no points of M. But the arcs $\Delta_0, \Delta_1, \Delta_2,\ldots \subset \Gamma$ are all of equal length, and hence some pairs of arcs must intersect. If Δ_k intersects Δ_{k+m}, say, then the union of the arcs $\Delta_k, \Delta_{k+m}, \Delta_{k+2m},\ldots$ covers the whole circle Γ.

17. *Hint.* Let q be any point of the set $M - P$, and let p_n be a sequence of points of P converging to q. Use the uniform continuity of $f(x)$ on P to show that the sequence $f(p_n)$ has a limit which is independent of the choice of the sequence $p_n \to q$. Setting $f(q) = \lim_{n\to\infty} f(p_n)$, verify the continuity of $f(q)$ on the whole set M.

18. *Hint.* The intervals $(f(x-0), f(x+0))$, constructed for all discontinuity points of $f(x)$, are *pairwise* nonintersecting (i.e., no two of the intervals intersect).

Chapter 6

1. *Hint.* Use Raabe's test (Theorem 6.17a).

2. *Hint.* Use Problem 1 and the tests of Sec. 6.1.

3. *Hint.* For any n and $m < n$,

$$(n-m)a_n \leqslant a_{m+1} + \cdots + a_n \leqslant r_m = \sum_{n=m+1}^{\infty} a_n,$$

and hence

$$na_n \leqslant \frac{n}{n-m} r_m.$$

4. *Ans.* One solution is to find $N = N(n)$ for each $n = 1, 2, \ldots$ such that

$$\sum_{m=N}^{\infty} a_m < \frac{1}{n^3},$$

and then set $b_m = n$ for $N(n) \leqslant m < N(n+1)$.

5. *Hint.* Use the inequality $k(n-k) \leqslant n^2$ $(k = 1, 2, \ldots, n-1)$.

6. *Hint.* $s_{2n} < s < s_{2n+1} = s_{2n} + a_{2n+1}$.

7. *Hint.* Use D'Alembert's test (Theorem 6.14a).

8. *Hint.* Clearly

$$L = \lim_{t \to 1} f(t) \leqslant \sum_{n=0}^{\infty} a_n,$$

where the existence of L follows from the fact that $f(t)$ is nondecreasing and bounded. On the other hand, given any $\varepsilon > 0$, we can choose N such that

$$\sum_{n=0}^{N} a_n > \sum_{n=0}^{\infty} a_n - \varepsilon,$$

afterwards choosing t such that

$$\sum_{n=0}^{N} a_n t^n > \sum_{n=0}^{\infty} a_n - \varepsilon.$$

9. *Hint.* The idea is the same as in Problem 8.

10. *Hint.* $z_n = P_{n+1}/P_n$.

11. *Hint.* $\log_b P_n = \sum_{k=1}^{n} \log_b x_k$.

12. *Hint.* Use Theorem 5.58c.

13. *Hint.* Calculate $(1-x)P_n$.

14. *Hint.* Expand

$$\frac{1}{1 - (1/p_k^x)}$$

in a geometric series.

15. *Hint.* Euler's formula continues to hold for $x=1$, in which case both sides become infinite.

16. *Hint.* There are $9^m - 9^{m-1}$ integers between $10^{m-1}-1$ and $10^m - 1$ containing no nines at all. Hence the sum in question is less than

$$\frac{9-1}{1} + \frac{9^2-9}{10} + \frac{9^3-9^2}{100} + \cdots = 80.$$

17. *Hint.* The length of a vector is no less than the length of any of its projections and does not exceed the sum of the lengths of all its projections on the coordinate axes.

18. *Hint.* Use the compactness of a sphere in R_m.

19. *Hint.* Use Theorem 6.31 (cf. Sec. 6.45).

20. *Hint.* Let $V_1 \supset \cdots \supset V_n \supset \cdots$ be a sequence of solid angles containing the vector q such that

$$\sup_{\substack{x \in V_n \\ |x|=1}} \sin \langle x, e \rangle = \sigma_n \searrow 0$$

and

$$\sum_{n=1}^{\infty} \sigma_n < \infty.$$

Now consecutively choose terms of the series (2) contained in $V_1, \ldots, V_n, \ldots$ the sum of whose lengths lies between 1 and 2.

21. *Hint.* The proof is by induction. For $n=1$ the assertion reduces to Riemann's theorem (Sec. 6.37). Let f be any vector in R_m, and let R_{m-1} be the orthogonal complement of f, i.e., the set of all $x \in R_m$ such that $(x, f) = 0$. By the induction hypothesis, there exists a rearrangement of the series (2) the sum of whose projections onto R_{m-1} equals a given vector in R_{m-1}. Using Problem 20, we choose a part of the series (2) for which the components along f form a series converging to $+\infty$ while the components orthogonal to f form an absolutely convergent series. The components of the complementary part of the series along f then have the sum $-\infty$. By rearranging these parts we can obtain any desired component along f. But, by Problem 19, this has no effect on the sum of the projections onto R_{m-1}.

22. *Hint.* Apply the result of Problem 21 to the orthogonal complement of the subspace A, i.e., to the set of all $x \in R_m$ such that $(x, y) = 0$ for all $y \in A$.

Chapter 7

1. *Hint.* $f'(x) \equiv 0$.

2. *Hint.* Use Lagrange's theorem (Sec. 7.44).

3. *Hint.* Use Lagrange's theorem.

4. *Hint.* $y'(0) = 0$, but $y'(x)$ does not approach 0 as $x \to 0$.

5. *Hint.* It suffices to consider the case $f'(a) < 0 < f'(b)$, proving the existence of a point $c \in (a,b)$ such that $f'(c) = 0$. This is just the point where $f(x)$ achieves its minimum.

6. *Hint.* Give a proof by contradiction, using Lagrange's theorem.

7. *Hint.* If the inequality (1) holds for two points $\lambda = (\lambda_1, ..., \lambda_n)$ and $\mu = (\mu_1, ..., \mu_n)$ in n-dimensional real space R_n (Sec. 2.6), then it holds for the entire segment in R_n connecting λ and μ. The inequality is obvious for the points $e_1 = (1,0,...,0)$, $e_2 = (0,1,...,0),...,$ $e_n = (0,0,...,1)$. From this deduce its validity for an arbitrary point $\lambda = (\lambda_1, ..., \lambda_n)$.

8. *Hint.* The inequality

$$\frac{f(x) - f(\alpha)}{x - \alpha} \leqslant \frac{f(\beta) - f(\alpha)}{\beta - \alpha} \qquad (\alpha < x < \beta)$$

is equivalent to the inequality (1).

9. *Hint.* Taking the limit as $\beta \searrow x$ in the inequality

$$\frac{f(x) - f(\alpha)}{x - \alpha} \leqslant \frac{f(\beta) - f(\alpha)}{\beta - \alpha} \leqslant \frac{f(\beta) - f(x)}{\beta - x} \qquad (\alpha < x < \beta)$$

proves the existence of $f'_r(x)$ and the formula $f(x + 0) = f(x)$, while taking the limit as $\alpha \nearrow x$ proves the existence of $f'_l(x)$ and the formula $f(x - 0) = f(x)$.

10. *Hint.* Use the same inequality as above.

11. *Hint.* The intervals $(f'_l(x), f'_r(x))$, constructed for all points at which $f'(x)$ fails to exist, are pairwise nonintersecting.

12. *Hint.* It follows from (2) that the only points which the curve $y = f(x)$ shares with any of its chords are the end points of the chord. Hence $y = f(x)$ must either lie above the chord or below the chord.

13. *Hint.* If there is a point $(x_1, f(x_1))$ on the curve $y = f(x)$, $x > x_0$ going below the right-hand tangent, then the curve must go below the chord joining the points with abscissas x_0 and x_1. Now recall the inequality (5), p. 234.

14. *Hint.* Apply L'Hospital's rules to the limits given in Chapter 4, Problem 10.

15. *Hint.* Apply Lagrange's theorem to the ratio

$$\frac{f(a + h) - f(a)}{h}.$$

16. *Hint.* The slope of the tangent at the point $x_n = b/n$ cannot exceed the slope of the chord going through the points x_n and x_{n+1}.

17. *Hint.* Fixing m for given x_0, let the increment be $h = \pm 1/4^m$. Then the increments of all the functions $\varphi_n(x)$ vanish, starting from the mth. The function $\varphi_{m-1}(x)$ has intervals of length $2/4^m$ without "corners." The interval containing x_0 also contains one of the intervals

$$\left[x_0, x_0 + \frac{1}{4^m}\right], \qquad \left[x_0 - \frac{1}{4^m}, x_0\right].$$

None of the preceding functions $\varphi_n(x)$ with $n < m-1$ has "corners" in this interval, and the increments of $\varphi_n(x)$ equals the increment of x in absolute value. It follows that

$$\frac{\Delta f(x_0)}{h} = \frac{f(x_0 + h) - f(x_0)}{h} = \sum_{m=0}^{n-1} \frac{\Delta \varphi_n(x_0)}{h} = \begin{cases} \text{an even number if } m \text{ is even,} \\ \text{an odd number if } m \text{ is odd.} \end{cases}$$

Therefore $\Delta f(x_0)/h$ approaches no limit as $h \to 0$.

Chapter 8

1. *Hint.* The derivatives of e^{-1/x^2} all vanish at $x = 0$ (cf. Chapter 2, Problem 15).

2. *Hint.* Use Leibniz's rule (Theorem 8.12b).

3. *Hint.* Use Rolle's theorem (Sec. 7.41) repeatedly.

4. *Hint.* Let x_0 be such that $|f'(x_0)| > M_1 - \varepsilon$, and use the fact that

$$f(x) - f(x_0) \geqslant f'(x_0)(x - x_0) - \tfrac{1}{2}M_2(x - x_0)^2.$$

5. *Ans.*

$$f^{(n)}(x) = \lim_{h \to 0} \frac{f(x) - nf(x+h) + \dfrac{n(n-1)}{2!}f(x+2h) - \cdots + (-1)^n f(x+nh)}{h^n}.$$

6. *Hint.* If $f''(x_0) < 0$, then the curve $y = f(x)$ lies below its tangent at $x = x_0$ (see Theorem 8.31). Now use Chapter 7, Problem 13.

7. *Hint.* The presence of a chord lying even partially below the curve leads to the existence of a point $P = (x_0, f(x_0))$ in a neighborhood of which the curve lies below its tangent at P. But this is incompatible with the condition $f''(x) > 0$.

8. *Hint.* Use formula (7), p. 269 to get a lower bound for $|\sin z|$, bearing in mind that $\cosh y > \sinh y$.

9. *Hint.* For $|y| > \varepsilon$, use the estimate found in solving Problem 8. For $|y| < \varepsilon$, bear in mind that $|\sin x| > 1 - \delta$.

10. *Ans.* $\omega < 10^{-7}$, $h \leqslant 1.03$, $n \geqslant 10$. In particular, the error made in replacing the number e by the sum

$$1 + \frac{1}{2!} + \cdots + \frac{1}{10!}$$

does not exceed 10^{-7}.

11. *Ans.*

	$f^{(n+1)}(x_0) > 0$	$f^{(n+1)}(x_0) < 0$
n even		
n odd		

12. *Ans.* No. Use Problem 2 to construct a counterexample.

13. *Hint.* Use the inequality

$$\frac{1}{(n+k)!} < \frac{1}{n!n^k} \qquad (k = 1,2,\ldots).$$

14. *Hint.* If e is rational, then $n!e$ is an integer for some n.

15. *Hint.* Use formulas (2′) and (5′), p. 258, together with the result of Chapter 6, Problem 11.

16. *Hint.* The inequality (3) certainly holds for all $n = 1,2,\ldots$ in some deleted neighborhood of $x = 0$ (cf. Chapter 6, Problem 6). If the inequality fails to hold for all $x \neq 0$, there is a smallest $x_0 > 0$ at which it leads to an equality. Now apply Rolle's theorem (Sec. 7.41) to the interval $[0, x_0]$.

17. *Hint.* Let $n = 1$, $x = \pi/2$.

18. *Ans.* $f(x + dx) = \sum_{k=0}^{n} \frac{1}{k!} d^k f(x) + \frac{1}{(n+1)!} d^{n+1} f(c)$.

Chapter 9

1. *Hint.* The slope of the linear component is

$$\frac{1}{T} \int_0^T f(\xi)\, d\xi.$$

2. *Hint.* Regard the right-hand side as a continuous function of ξ, assuming first that $f(x)$ does not change sign.

3. *Hint.* If $\varphi(x)$ is nondecreasing, say, then the nonnegative function $\varphi(x) - \varphi(a+0)$ is also nondecreasing. Now use the result of Problem 2.

4. *Hint.* Use mathematical induction. Alternatively, replace b by x, differentiate with respect to x, and verify that this gives an identity. Then return to the original formula by integration, verifying that the two sides coincide for $x = a$.

5. *Hint.* Suitably generalize the proof of Theorem 9.14e.

6. *Hint.* Use Riemann's criterion (Problem 5).

7. *Hint.* Use the inequality $||f(x')| - |f(x'')|| \leqslant |f(x') - f(x'')|$ and Riemann's criterion.

8. *Hint.* Use Riemann's criterion.

9. *Hint.* Let $\varepsilon = 1, \frac{1}{2}, \frac{1}{3}, \ldots$ in Du Bois-Reymond's criterion (Problem 8), and correspondingly write $\delta/2, \delta/4, \delta/8, \ldots$ instead of δ.

10. *Hint.* Use Lebesgue's criterion (Problem 9).

11. *Hint.* The ratio of the left and right-hand sides of the inequality

$$\left[\frac{2n(2n-2)\cdots 4\cdot 2}{(2n-1)(2n-3)\cdots 3\cdot 1}\right]^2 \frac{1}{2n+1} \leqslant \frac{\pi}{2} \leqslant \left[\frac{2n(2n-2)\cdots 4\cdot 2}{(2n-1)(2n-3)\cdots 3\cdot 1}\right]^2 \frac{1}{2n}$$

approaches 1 as $n \to \infty$.

12. *Hint.* Given any $\varepsilon > 0$, use the uniform continuity of $f(x)$ on $[A, B]$ to find a $\delta > 0$ such that $|x' - x''| < \delta$ implies $|f(x') - f(x'')| < \varepsilon$. Then $d(\Pi) < \delta$ implies

$$f(\xi_k)(x_{k+1} - x_k) = \int_{x_k}^{x_{k+1}} f(x)\,dx + \varepsilon_k |x_{k+1} - x_k|,$$

where $|\varepsilon_k| < \varepsilon$.

13. *Hint.* If

$$\sum_{k=0}^{n-1} |x_{k+1} - x_k|$$

fails to be bounded for all $n = 1, 2, \ldots$, then the sum of the errors made in replacing $f(\xi_k)(x_{k+1} - x_k)$ by the integrals

$$\int_{x_k}^{x_{k+1}} f(x)\,dx \qquad (k = 0, 1, \ldots, n-1)$$

can become arbitrarily large. If $f(x)$ has a discontinuity point c, we can choose arbitrarily many pairs of consecutive points x_k, x_{k+1} on opposite sides of c.

14. *Hint.* To verify (2) for an arbitrary piecewise constant function $f(x)$, use

Conditions a–d. To verify (2) for an arbitrary piecewise continuous function $f(x)$, approximate $f(x)$ by a piecewise constant function, using Condition e.

15. *Hint.* Set $\xi = 1/\sqrt{3}$ in formula (6), p. 352.

Chapter 10

1. *Hint.* Suppose $f(z)$ is analytic at $z = 1$. Then

$$f(z) = \sum_{n=0}^{\infty} \frac{f^{(n)}(1)}{n!} (z-1)^n \qquad (|z-1| < \rho),$$

where

$$f(1) = \lim_{z \to 1} f(z) = \sum_{k=0}^{\infty} a_k,$$

...

$$f^{(n)}(1) = \sum_{k=0}^{\infty} k(k-1)\cdots(k-n+1)a_k,$$

...

by Chapter 6, Problem 8, while

$$f(e^{i\theta}) = \sum_{k=0}^{\infty} a_k e^{ik\theta},$$

...

$$f^{(n)}(e^{i\theta}) = \sum_{k=0}^{\infty} k(k-1)\cdots(k-n+1)a_k e^{ik\theta},$$

...

by Abel's theorem (Sec. 6.68) and Chapter 7, Problem 15. It follows that the series

$$\sum_{n=0}^{\infty} \frac{f^{(n)}(e^{i\theta})}{n!} (z - e^{i\theta})^n$$

is convergent for $|z - e^{i\theta}| < \rho$.

2. *Hint.* Start from the formula

$$\sum_{n=0}^{\infty} a_n z^n - \frac{c}{(z-z_0)^m} = \sum_{n=0}^{\infty} b_n z^n,$$

where $\varlimsup \sqrt[n]{|a_n|} < 1$.

3. *Hint.* Examine the mapping $w = f(z)$ in a neighborhood of the point z_0.
4. *Hint.* Apply the maximum modulus principle to the function $f(z)/z$.

5. *Hint.* To prove the required inequality

$$\ln M(r_2) \leqslant \frac{\ln r_2 - \ln r_1}{\ln r_3 - \ln r_1} \ln M(r_3) + \frac{\ln r_3 - \ln r_2}{\ln r_3 - \ln r_1} \ln M(r_1)$$

(cf. Chapter 7, Problem 7), consider the function $z^{\alpha}f(z)$, where α is such that $r_1^{\alpha} M(r_1) = r_3^{\alpha} M(r_3)$.

6. *Hint.* See Secs. 10.16 and 10.17.

7. *Hint.* Verify the assertion separately for the functions az, $az+b$, and $1/z$, and then use Sec. 10.55.

8. *Hint.* Use formula (20), p. 401 to estimate $|a_n|$.

9. *Hint.* Calculate the increment of the argument in traversing a small circle about each pole and zero.

10. *Hint.* See Sec. 10.44.

11. *Hint.* Let

$$g(z) = \sum_{k=0}^{\infty} \frac{g^{(k)}(a)}{k!}(z-a)^k.$$

Then

$$|g^{(m)}(z) + \cdots + g^{(m+n)}(z)| = \left| \sum_{k=0}^{\infty} [g^{(m+k)}(a) + \cdots + g^{(m+n+k)}(a)] \frac{(z-a)^k}{k!} \right|.$$

12. *Ans.* Each sheet of the Riemann surface of the function (2) consists of the whole w-plane cut along any curve joining the points -1 and $+1$, e.g., along the interval $[-1,1]$ or along the part of the real axis complementary to $[-1,1]$. The Riemann surface is two-sheeted, with the upper edge of the cut on the first sheet being pasted to the lower edge of the cut on the second sheet, and vice versa. The branch points in the w-plane are at $w = \pm 1$. The domains of univalence in the z-plane are the interior and the exterior of the unit circle $|z| = 1$.

As for the function (3), each sheet of the Riemann surface consists of the w-plane cut along the real axis from $-\infty$ to -1 and from 1 to ∞. The Riemann surface is infinite-sheeted, with the upper edge of the left half of the cut on the kth sheet being pasted to the lower edge of the left half of the cut on the $(k+1)$st sheet and the upper edge of the right half of the cut on the kth sheet being pasted to the lower edge of the right half of the cut on the $(k-1)$st sheet. The branch points in the w-plane are at $w = \pm 1$, ∞. The domains of univalence in the z-plane are the strips $k\pi < x < (k+1)\pi$, $k = 0$, $\pm 1, \pm 2, \ldots$

13. *Hint.* Use the representation of $\cos z$ in terms of exponentials.

14. *Hint.* Consider the function $c_1 f_1(z) + \cdots + c_n f_n(z)$, where the numbers

c_k are such that $|c_k| = 1$ and

$$c_1 f_1(z_0) + \cdots + c_n f_n(z_0) = |f_1(z_0)| + \cdots + |f_n(z_0)|$$

at the point z_0 where $|f_1(z)| + \cdots + |f_n(z)|$ has its maximum.

15. *Hint.* See Sec. 10.49d.

16. *Hint.* Consider $1/P(z)$.

17. *Hint.* Note that

$$\frac{P'(z)}{P(z)} = \frac{n}{z} + \frac{g(z)}{z^2},$$

where $g(z)$ is bounded as $z \to \infty$. Now integrate this formula along a circle of sufficiently large radius.

18. *Hint.* If $P(z)$ does not vanish, then $1/P(z)$ achieves its maximum at a finite point.

19. *Hint.* On the one hand,

$$\frac{1}{2\pi i} \oint_{\Gamma_n} \frac{f(\zeta)}{\zeta - z} \, d\zeta = f(z) + \sum_k \frac{b_k}{z_k - z},$$

where the sum is over all poles of $f(z)$ inside Γ_n, while on the other hand,

$$\frac{1}{2\pi i} \oint_{\Gamma_n} \frac{f(\zeta)}{\zeta - z} \, d\zeta = \frac{1}{2\pi i} \oint_{\Gamma_n} \frac{f(\zeta)}{\zeta} \, d\zeta + \frac{z}{2\pi i} \oint_{\Gamma_n} \frac{f(\zeta)}{\zeta(\zeta - z)} \, d\zeta$$

$$= f(0) + \sum_k \frac{b_k}{z_k} + \frac{z}{2\pi i} \oint_{\Gamma_n} \frac{f(\zeta)}{\zeta(\zeta - z)} \, d\zeta,$$

where the last integral approaches 0 as $n \to \infty$.

20. *Hint.* Concerning finite products, see Chapter 6, Problem 10. To get (5), integrate (4) and then take exponentials.

21. *Hint.* Use Chapter 8, Problem 9. To get the first expansion, let

$$g(z) = \frac{\sin z}{z}$$

in (5).

22. *Hint.* Consider the function

$$f(z) e^{-\varepsilon z^\lambda} \qquad (\lambda < \pi/2\alpha)$$

in the sector $|\arg z| \leqslant \alpha$, $|z| \leqslant r$, and apply the maximum modulus principle.

23. *Hint.* Apply Problem 22 to the function $f(z) e^{i\gamma z}$ $(\gamma > 0)$ in each of the two quadrants forming the upper half-plane, obtaining the estimate

$$|f(z) e^{i\gamma z}| \leqslant \max \{C, M_\gamma\},$$

where

$$M_\gamma = \max_r A_\varepsilon e^{(\varepsilon-\gamma)r}.$$

If $M_\gamma > C$, then $|f(z)|$ has a maximum at a point of the imaginary axis, which is impossible if $f(z) \not\equiv \text{const}$ (see Problem 3). Therefore $M_\gamma \leqslant C$, $|f(z)| \leqslant C|e^{-i\gamma z}|$. Now let $\gamma \to 0$.

24. *Hint.* Apply Problem 23 and Liouville's theorem.

Chapter 11

1. *Hint.* Prove the formula for an "inner triangle" on which the integrand is continuous, and then pass to the limit.

2. *Hint.* Note that

$$\int_0^\infty \frac{f(ax)-f(bx)}{x}\,dx = \int_{a\delta}^\infty \frac{f(x)}{x}\,dx - \int_{b\delta}^\infty \frac{f(x)}{x}\,dx$$

$$= \int_{a\delta}^{b\delta} \frac{f(x)}{x}\,dx = f(\xi)\,\ln\frac{b}{a} \qquad (0<\delta,\ a<b,\ a\delta<\xi<b\delta),$$

where the last step involves a simple generalization of Theorem 9.15f.

3. *Ans.* Using a contour of integration L made up of a segment of the real axis, an arc of a circle centered at the origin, and a segment of the ray $\theta = \theta_0$ (so that L is the boundary of a circular sector), we get $I(p) = \frac{1}{2}\sqrt{\pi/p}$ for a suitable value of $\sqrt{p}$.

4. *Ans.* Using Problem 3, we find that $F_1 = F_2 = \frac{1}{2}\sqrt{\pi/2}$.

5. *Ans.*

$$I = 2\pi i \sum \operatorname{Res} \frac{P(x)}{Q(x)}\frac{\sin x}{x},$$

where the sum is over all the zeros of $Q(x)$ in the upper half-plane.

6. *Hint.* The integral with the infinite limit is not uniformly convergent in the parameter t.

7. *Hint.* Let α approach 0, appealing to Theorem 11.53.

8. *Hint.* Make the substitution $x = a\cos^2\theta + b\sin^2\theta$.

9. *Hint.* Denoting the integral by I and setting $x = 2t$, we get

$$I = 2\int_0^{\pi/4} \ln\sin 2t\,dt = \frac{\pi}{2}\ln 2 + 2\int_0^{\pi/4}\ln\sin t\,dt + 2\int_0^{\pi/4}\ln\cos t\,dt.$$

But the substitution $t = (\pi/2) - u$ reduces the last term on the right to

$$2\int_{\pi/4}^{\pi/2}\ln\sin u\,du = 2I.$$

10. *Hint.* First note that

$$\frac{1}{x^2}\left(\frac{x}{e^x - e^{-x}} - \frac{1}{2}\right) = -\frac{1}{2x}(e^{-x} - e^{-2x}) + \frac{1}{x}\left(\frac{1}{e^x - 1} - \frac{1}{x} + \frac{1}{2}e^{-x}\right)$$
$$-\frac{1}{x}\left(\frac{1}{e^{2x} - 1} - \frac{1}{2x} + \frac{1}{2}e^{-2x}\right),$$

where the integrals of the second and third terms on the right cancel each other out (after an obvious substitution). Now use Problem 2.

11. *Hint.* Use integration by parts (Sec. 11.15a).

12. *Ans.* $\dfrac{\pi}{n}\dfrac{1}{\sin\dfrac{2m+1}{2n}\pi}$.

13. *Hint.* Compare the values of s^{z-1} on the interval $[\varepsilon, \infty)$ traversed first in one direction and then in the other.

Index

Abel-Dirichlet test
  for improper integrals, 441–443, 466
  for series, 205–209
Abel-Liouville theorem, 305 n
Abel's theorem, 217
Abel's transformation, 205
Absolute value
  of a complex number, 166
  of a real number, 10
Addition
  associativity of, 3
  commutativity of, 3
Addition axioms, 3
  consequences of, 4–5
Additivity condition, 484
Algebraic number, 33, 34
Alternating series, 194
Analytic continuation, 262, 402–403
Analyticity
  at a point, 374
  on a set, 374
Analytic function(s), 374
  infinite differentiability of, 395
  power series expansion of, 395
  real, 403
  uniqueness theorem for, 401
  zero (root) of, 400
    order (multiplicity) of, 400
Angle between two vectors, 168
Antiderivative, 292
Arc length, 289–291, 328–335
  of a catenary, 331
  of a circle, 331–332
  of an ellipse, 332
  as parameter, 334–335
Archimedes, 1 n
  principle of, 16
    multiplicative version of, 16
Area
  behavior of, under k-fold expansion, 321–322
  of a catenoid, 342
  of a circular disk, 320
  of a circular sector, 321
  of a conical band, 338
  over a curve, 318
  under a curve, 288, 317
  enclosed by an ellipse, 322
  of a geometric figure, 287
  in terms of line integrals, 366–368
  of the neighborhood of a curve, 346
  in polar coordinates, 327
  of a sphere, 341
  of a surface of revolution, 339

Argument of a complex number, 166
  principal value of, 425 n
Argument of a function, 49, 166 n
Argument principle, 428
Arithmetic mean, 24
Arithmetic nth power of a set, 26
Arithmetic product of two sets, 25
Arithmetic sum of two sets, 25
Asymptote, 130, 248
Asymptotic unit, 116
Automorphisms
  of n-dimensional space, 41–43, 170
  of a structure, 35
Axioms, complete system of, 36

Baire's theorem, 85
Ball(s)
  closed, 55, 76 n
  open, 55, 61 n
  principle of nested, 85
  radius of, 55
Banach, S., ix
Basis, 41
Beta function, 469
  in terms of the gamma function, 471
Binary system, 21
Binomial coefficients, 249, 313
Birkhoff, G., 177 n, 178 n
Block
  closed, 134
  open, 134
Bolzano's theorem, 140
Bolzano-Weierstrass principle, 75
  for sequences, 75
Boundary of a set, 486
Bounded function, 108
Bounded set, 15
  in a metric space, 54
Bourbaki, N., ix, 35 n
Branch point, 427

Cantor, G., 21, 33
Cartan, H., x
Catenoid, 342
Cauchy, A. L., 190, 435
Cauchy convergence criterion
  for a function with values in a metric space, 106
  for improper integrals, 432, 461
  for a numerical sequence, 81, 123
  for a numerical series, 187
  for a sequence of functions, 182
  for a series of functions, 212
  for a vector function, 129

A CATALOG OF SELECTED
DOVER BOOKS
IN SCIENCE AND MATHEMATICS

Astronomy

BURNHAM'S CELESTIAL HANDBOOK, Robert Burnham, Jr. Thorough guide to the stars beyond our solar system. Exhaustive treatment. Alphabetical by constellation: Andromeda to Cetus in Vol. 1; Chamaeleon to Orion in Vol. 2; and Pavo to Vulpecula in Vol. 3. Hundreds of illustrations. Index in Vol. 3. 2,000pp. 6⅛ x 9¼.
Vol. I: 0-486-23567-X
Vol. II: 0-486-23568-8
Vol. III: 0-486-23673-0

EXPLORING THE MOON THROUGH BINOCULARS AND SMALL TELESCOPES, Ernest H. Cherrington, Jr. Informative, profusely illustrated guide to locating and identifying craters, rills, seas, mountains, other lunar features. Newly revised and updated with special section of new photos. Over 100 photos and diagrams. 240pp. 8¼ x 11. 0-486-24491-1

THE EXTRATERRESTRIAL LIFE DEBATE, 1750–1900, Michael J. Crowe. First detailed, scholarly study in English of the many ideas that developed from 1750 to 1900 regarding the existence of intelligent extraterrestrial life. Examines ideas of Kant, Herschel, Voltaire, Percival Lowell, many other scientists and thinkers. 16 illustrations. 704pp. 5⅜ x 8½. 0-486-40675-X

THEORIES OF THE WORLD FROM ANTIQUITY TO THE COPERNICAN REVOLUTION, Michael J. Crowe. Newly revised edition of an accessible, enlightening book recreates the change from an earth-centered to a sun-centered conception of the solar system. 242pp. 5⅜ x 8½. 0-486-41444-2

A HISTORY OF ASTRONOMY, A. Pannekoek. Well-balanced, carefully reasoned study covers such topics as Ptolemaic theory, work of Copernicus, Kepler, Newton, Eddington's work on stars, much more. Illustrated. References. 521pp. 5⅜ x 8½.
0-486-65994-1

A COMPLETE MANUAL OF AMATEUR ASTRONOMY: TOOLS AND TECHNIQUES FOR ASTRONOMICAL OBSERVATIONS, P. Clay Sherrod with Thomas L. Koed. Concise, highly readable book discusses: selecting, setting up and maintaining a telescope; amateur studies of the sun; lunar topography and occultations; observations of Mars, Jupiter, Saturn, the minor planets and the stars; an introduction to photoelectric photometry; more. 1981 ed. 124 figures. 25 halftones. 37 tables. 335pp. 6½ x 9¼. 0-486-40675-X

AMATEUR ASTRONOMER'S HANDBOOK, J. B. Sidgwick. Timeless, comprehensive coverage of telescopes, mirrors, lenses, mountings, telescope drives, micrometers, spectroscopes, more. 189 illustrations. 576pp. 5⅝ x 8¼. (Available in U.S. only.)
0-486-24034-7

STARS AND RELATIVITY, Ya. B. Zel'dovich and I. D. Novikov. Vol. 1 of *Relativistic Astrophysics* by famed Russian scientists. General relativity, properties of matter under astrophysical conditions, stars, and stellar systems. Deep physical insights, clear presentation. 1971 edition. References. 544pp. 5⅝ x 8¼. 0-486-69424-0

Chemistry

THE SCEPTICAL CHYMIST: THE CLASSIC 1661 TEXT, Robert Boyle. Boyle defines the term "element," asserting that all natural phenomena can be explained by the motion and organization of primary particles. 1911 ed. viii+232pp. 5⅜ x 8½.
0-486-42825-7

RADIOACTIVE SUBSTANCES, Marie Curie. Here is the celebrated scientist's doctoral thesis, the prelude to her receipt of the 1903 Nobel Prize. Curie discusses establishing atomic character of radioactivity found in compounds of uranium and thorium; extraction from pitchblende of polonium and radium; isolation of pure radium chloride; determination of atomic weight of radium; plus electric, photographic, luminous, heat, color effects of radioactivity. ii+94pp. 5⅜ x 8½.   0-486-42550-9

CHEMICAL MAGIC, Leonard A. Ford. Second Edition, Revised by E. Winston Grundmeier. Over 100 unusual stunts demonstrating cold fire, dust explosions, much more. Text explains scientific principles and stresses safety precautions. 128pp. 5⅜ x 8½.   0-486-67628-5

THE DEVELOPMENT OF MODERN CHEMISTRY, Aaron J. Ihde. Authoritative history of chemistry from ancient Greek theory to 20th-century innovation. Covers major chemists and their discoveries. 209 illustrations. 14 tables. Bibliographies. Indices. Appendices. 851pp. 5⅜ x 8½.   0-486-64235-6

CATALYSIS IN CHEMISTRY AND ENZYMOLOGY, William P. Jencks. Exceptionally clear coverage of mechanisms for catalysis, forces in aqueous solution, carbonyl- and acyl-group reactions, practical kinetics, more. 864pp. 5⅜ x 8½.
0-486-65460-5

ELEMENTS OF CHEMISTRY, Antoine Lavoisier. Monumental classic by founder of modern chemistry in remarkable reprint of rare 1790 Kerr translation. A must for every student of chemistry or the history of science. 539pp. 5⅜ x 8½. 0-486-64624-6

THE HISTORICAL BACKGROUND OF CHEMISTRY, Henry M. Leicester. Evolution of ideas, not individual biography. Concentrates on formulation of a coherent set of chemical laws. 260pp. 5⅜ x 8½.   0-486-61053-5

A SHORT HISTORY OF CHEMISTRY, J. R. Partington. Classic exposition explores origins of chemistry, alchemy, early medical chemistry, nature of atmosphere, theory of valency, laws and structure of atomic theory, much more. 428pp. 5⅜ x 8½. (Available in U.S. only.)   0-486-65977-1

GENERAL CHEMISTRY, Linus Pauling. Revised 3rd edition of classic first-year text by Nobel laureate. Atomic and molecular structure, quantum mechanics, statistical mechanics, thermodynamics correlated with descriptive chemistry. Problems. 992pp. 5⅜ x 8½.   0-486-65622-5

FROM ALCHEMY TO CHEMISTRY, John Read. Broad, humanistic treatment focuses on great figures of chemistry and ideas that revolutionized the science. 50 illustrations. 240pp. 5⅜ x 8½.   0-486-28690-8

Engineering

DE RE METALLICA, Georgius Agricola. The famous Hoover translation of greatest treatise on technological chemistry, engineering, geology, mining of early modern times (1556). All 289 original woodcuts. 638pp. 6¾ x 11.   0-486-60006-8

FUNDAMENTALS OF ASTRODYNAMICS, Roger Bate et al. Modern approach developed by U.S. Air Force Academy. Designed as a first course. Problems, exercises. Numerous illustrations. 455pp. 5⅜ x 8½.   0-486-60061-0

DYNAMICS OF FLUIDS IN POROUS MEDIA, Jacob Bear. For advanced students of ground water hydrology, soil mechanics and physics, drainage and irrigation engineering and more. 335 illustrations. Exercises, with answers. 784pp. 6⅛ x 9¼.
0-486-65675-6

THEORY OF VISCOELASTICITY (Second Edition), Richard M. Christensen. Complete consistent description of the linear theory of the viscoelastic behavior of materials. Problem-solving techniques discussed. 1982 edition. 29 figures. xiv+364pp. 6⅛ x 9¼.   0-486-42880-X

MECHANICS, J. P. Den Hartog. A classic introductory text or refresher. Hundreds of applications and design problems illuminate fundamentals of trusses, loaded beams and cables, etc. 334 answered problems. 462pp. 5⅜ x 8½.   0-486-60754-2

MECHANICAL VIBRATIONS, J. P. Den Hartog. Classic textbook offers lucid explanations and illustrative models, applying theories of vibrations to a variety of practical industrial engineering problems. Numerous figures. 233 problems, solutions. Appendix. Index. Preface. 436pp. 5⅜ x 8½.   0-486-64785-4

STRENGTH OF MATERIALS, J. P. Den Hartog. Full, clear treatment of basic material (tension, torsion, bending, etc.) plus advanced material on engineering methods, applications. 350 answered problems. 323pp. 5⅜ x 8½.   0-486-60755-0

A HISTORY OF MECHANICS, René Dugas. Monumental study of mechanical principles from antiquity to quantum mechanics. Contributions of ancient Greeks, Galileo, Leonardo, Kepler, Lagrange, many others. 671pp. 5⅜ x 8½.   0-486-65632-2

STABILITY THEORY AND ITS APPLICATIONS TO STRUCTURAL MECHANICS, Clive L. Dym. Self-contained text focuses on Koiter postbuckling analyses, with mathematical notions of stability of motion. Basing minimum energy principles for static stability upon dynamic concepts of stability of motion, it develops asymptotic buckling and postbuckling analyses from potential energy considerations, with applications to columns, plates, and arches. 1974 ed. 208pp. 5⅜ x 8½.
0-486-42541-X

METAL FATIGUE, N. E. Frost, K. J. Marsh, and L. P. Pook. Definitive, clearly written, and well-illustrated volume addresses all aspects of the subject, from the historical development of understanding metal fatigue to vital concepts of the cyclic stress that causes a crack to grow. Includes 7 appendixes. 544pp. 5⅜ x 8½. 0-486-40927-9

ROCKETS, Robert Goddard. Two of the most significant publications in the history of rocketry and jet propulsion: "A Method of Reaching Extreme Altitudes" (1919) and "Liquid Propellant Rocket Development" (1936). 128pp. 5⅜ x 8½.   0-486-42537-1

STATISTICAL MECHANICS: PRINCIPLES AND APPLICATIONS, Terrell L. Hill. Standard text covers fundamentals of statistical mechanics, applications to fluctuation theory, imperfect gases, distribution functions, more. 448pp. 5⅜ x 8½.
0-486-65390-0

ENGINEERING AND TECHNOLOGY 1650–1750: ILLUSTRATIONS AND TEXTS FROM ORIGINAL SOURCES, Martin Jensen. Highly readable text with more than 200 contemporary drawings and detailed engravings of engineering projects dealing with surveying, leveling, materials, hand tools, lifting equipment, transport and erection, piling, bailing, water supply, hydraulic engineering, and more. Among the specific projects outlined-transporting a 50-ton stone to the Louvre, erecting an obelisk, building timber locks, and dredging canals. 207pp. 8⅜ x 11¼.
0-486-42232-1

THE VARIATIONAL PRINCIPLES OF MECHANICS, Cornelius Lanczos. Graduate level coverage of calculus of variations, equations of motion, relativistic mechanics, more. First inexpensive paperbound edition of classic treatise. Index. Bibliography. 418pp. 5⅜ x 8½.   0-486-65067-7

PROTECTION OF ELECTRONIC CIRCUITS FROM OVERVOLTAGES, Ronald B. Standler. Five-part treatment presents practical rules and strategies for circuits designed to protect electronic systems from damage by transient overvoltages. 1989 ed. xxiv+434pp. 6⅛ x 9¼.   0-486-42552-5

ROTARY WING AERODYNAMICS, W. Z. Stepniewski. Clear, concise text covers aerodynamic phenomena of the rotor and offers guidelines for helicopter performance evaluation. Originally prepared for NASA. 537 figures. 640pp. 6⅛ x 9¼.
0-486-64647-5

INTRODUCTION TO SPACE DYNAMICS, William Tyrrell Thomson. Comprehensive, classic introduction to space-flight engineering for advanced undergraduate and graduate students. Includes vector algebra, kinematics, transformation of coordinates. Bibliography. Index. 352pp. 5⅜ x 8½.   0-486-65113-4

HISTORY OF STRENGTH OF MATERIALS, Stephen P. Timoshenko. Excellent historical survey of the strength of materials with many references to the theories of elasticity and structure. 245 figures. 452pp. 5⅜ x 8½.   0-486-61187-6

ANALYTICAL FRACTURE MECHANICS, David J. Unger. Self-contained text supplements standard fracture mechanics texts by focusing on analytical methods for determining crack-tip stress and strain fields. 336pp. 6⅛ x 9¼.   0-486-41737-9

STATISTICAL MECHANICS OF ELASTICITY, J. H. Weiner. Advanced, self-contained treatment illustrates general principles and elastic behavior of solids. Part 1, based on classical mechanics, studies thermoelastic behavior of crystalline and polymeric solids. Part 2, based on quantum mechanics, focuses on interatomic force laws, behavior of solids, and thermally activated processes. For students of physics and chemistry and for polymer physicists. 1983 ed. 96 figures. 496pp. 5⅜ x 8½.
0-486-42260-7

Mathematics

FUNCTIONAL ANALYSIS (Second Corrected Edition), George Bachman and Lawrence Narici. Excellent treatment of subject geared toward students with background in linear algebra, advanced calculus, physics and engineering. Text covers introduction to inner-product spaces, normed, metric spaces, and topological spaces; complete orthonormal sets, the Hahn-Banach Theorem and its consequences, and many other related subjects. 1966 ed. 544pp. 6⅛ x 9¼. 0-486-40251-7

ASYMPTOTIC EXPANSIONS OF INTEGRALS, Norman Bleistein & Richard A. Handelsman. Best introduction to important field with applications in a variety of scientific disciplines. New preface. Problems. Diagrams. Tables. Bibliography. Index. 448pp. 5⅜ x 8½. 0-486-65082-0

VECTOR AND TENSOR ANALYSIS WITH APPLICATIONS, A. I. Borisenko and I. E. Tarapov. Concise introduction. Worked-out problems, solutions, exercises. 257pp. 5⅝ x 8¼. 0-486-63833-2

AN INTRODUCTION TO ORDINARY DIFFERENTIAL EQUATIONS, Earl A. Coddington. A thorough and systematic first course in elementary differential equations for undergraduates in mathematics and science, with many exercises and problems (with answers). Index. 304pp. 5⅜ x 8½. 0-486-65942-9

FOURIER SERIES AND ORTHOGONAL FUNCTIONS, Harry F. Davis. An incisive text combining theory and practical example to introduce Fourier series, orthogonal functions and applications of the Fourier method to boundary-value problems. 570 exercises. Answers and notes. 416pp. 5⅜ x 8½. 0-486-65973-9

COMPUTABILITY AND UNSOLVABILITY, Martin Davis. Classic graduate-level introduction to theory of computability, usually referred to as theory of recurrent functions. New preface and appendix. 288pp. 5⅜ x 8½. 0-486-61471-9

ASYMPTOTIC METHODS IN ANALYSIS, N. G. de Bruijn. An inexpensive, comprehensive guide to asymptotic methods–the pioneering work that teaches by explaining worked examples in detail. Index. 224pp. 5⅜ x 8½ 0-486-64221-6

APPLIED COMPLEX VARIABLES, John W. Dettman. Step-by-step coverage of fundamentals of analytic function theory–plus lucid exposition of five important applications: Potential Theory; Ordinary Differential Equations; Fourier Transforms; Laplace Transforms; Asymptotic Expansions. 66 figures. Exercises at chapter ends. 512pp. 5⅜ x 8½. 0-486-64670-X

INTRODUCTION TO LINEAR ALGEBRA AND DIFFERENTIAL EQUATIONS, John W. Dettman. Excellent text covers complex numbers, determinants, orthonormal bases, Laplace transforms, much more. Exercises with solutions. Undergraduate level. 416pp. 5⅜ x 8½. 0-486-65191-6

RIEMANN'S ZETA FUNCTION, H. M. Edwards. Superb, high-level study of landmark 1859 publication entitled "On the Number of Primes Less Than a Given Magnitude" traces developments in mathematical theory that it inspired. xiv+315pp. 5⅜ x 8½. 0-486-41740-9

CALCULUS OF VARIATIONS WITH APPLICATIONS, George M. Ewing. Applications-oriented introduction to variational theory develops insight and promotes understanding of specialized books, research papers. Suitable for advanced undergraduate/graduate students as primary, supplementary text. 352pp. 5⅜ x 8½. 0-486-64856-7

COMPLEX VARIABLES, Francis J. Flanigan. Unusual approach, delaying complex algebra till harmonic functions have been analyzed from real variable viewpoint. Includes problems with answers. 364pp. 5⅜ x 8½. 0-486-61388-7

AN INTRODUCTION TO THE CALCULUS OF VARIATIONS, Charles Fox. Graduate-level text covers variations of an integral, isoperimetrical problems, least action, special relativity, approximations, more. References. 279pp. 5⅜ x 8½. 0-486-65499-0

COUNTEREXAMPLES IN ANALYSIS, Bernard R. Gelbaum and John M. H. Olmsted. These counterexamples deal mostly with the part of analysis known as "real variables." The first half covers the real number system, and the second half encompasses higher dimensions. 1962 edition. xxiv+198pp. 5⅜ x 8½. 0-486-42875-3

CATASTROPHE THEORY FOR SCIENTISTS AND ENGINEERS, Robert Gilmore. Advanced-level treatment describes mathematics of theory grounded in the work of Poincaré, R. Thom, other mathematicians. Also important applications to problems in mathematics, physics, chemistry and engineering. 1981 edition. References. 28 tables. 397 black-and-white illustrations. xvii + 666pp. 6⅛ x 9¼. 0-486-67539-4

INTRODUCTION TO DIFFERENCE EQUATIONS, Samuel Goldberg. Exceptionally clear exposition of important discipline with applications to sociology, psychology, economics. Many illustrative examples; over 250 problems. 260pp. 5⅜ x 8½. 0-486-65084-7

NUMERICAL METHODS FOR SCIENTISTS AND ENGINEERS, Richard Hamming. Classic text stresses frequency approach in coverage of algorithms, polynomial approximation, Fourier approximation, exponential approximation, other topics. Revised and enlarged 2nd edition. 721pp. 5⅜ x 8½. 0-486-65241-6

INTRODUCTION TO NUMERICAL ANALYSIS (2nd Edition), F. B. Hildebrand. Classic, fundamental treatment covers computation, approximation, interpolation, numerical differentiation and integration, other topics. 150 new problems. 669pp. 5⅜ x 8½. 0-486-65363-3

THREE PEARLS OF NUMBER THEORY, A. Y. Khinchin. Three compelling puzzles require proof of a basic law governing the world of numbers. Challenges concern van der Waerden's theorem, the Landau-Schnirelmann hypothesis and Mann's theorem, and a solution to Waring's problem. Solutions included. 64pp. 5⅜ x 8½. 0-486-40026-3

THE PHILOSOPHY OF MATHEMATICS: AN INTRODUCTORY ESSAY, Stephan Körner. Surveys the views of Plato, Aristotle, Leibniz & Kant concerning propositions and theories of applied and pure mathematics. Introduction. Two appendices. Index. 198pp. 5⅜ x 8½. 0-486-25048-2

CATALOG OF DOVER BOOKS

INTRODUCTORY REAL ANALYSIS, A.N. Kolmogorov, S. V. Fomin. Translated by Richard A. Silverman. Self-contained, evenly paced introduction to real and functional analysis. Some 350 problems. 403pp. 5⅜ x 8½. 0-486-61226-0

APPLIED ANALYSIS, Cornelius Lanczos. Classic work on analysis and design of finite processes for approximating solution of analytical problems. Algebraic equations, matrices, harmonic analysis, quadrature methods, much more. 559pp. 5⅜ x 8½. 0-486-65656-X

AN INTRODUCTION TO ALGEBRAIC STRUCTURES, Joseph Landin. Superb self-contained text covers "abstract algebra": sets and numbers, theory of groups, theory of rings, much more. Numerous well-chosen examples, exercises. 247pp. 5⅜ x 8½. 0-486-65940-2

QUALITATIVE THEORY OF DIFFERENTIAL EQUATIONS, V. V. Nemytskii and V.V. Stepanov. Classic graduate-level text by two prominent Soviet mathematicians covers classical differential equations as well as topological dynamics and ergodic theory. Bibliographies. 523pp. 5⅜ x 8½. 0-486-65954-2

THEORY OF MATRICES, Sam Perlis. Outstanding text covering rank, nonsingularity and inverses in connection with the development of canonical matrices under the relation of equivalence, and without the intervention of determinants. Includes exercises. 237pp. 5⅜ x 8½. 0-486-66810-X

INTRODUCTION TO ANALYSIS, Maxwell Rosenlicht. Unusually clear, accessible coverage of set theory, real number system, metric spaces, continuous functions, Riemann integration, multiple integrals, more. Wide range of problems. Undergraduate level. Bibliography. 254pp. 5⅜ x 8½. 0-486-65038-3

MODERN NONLINEAR EQUATIONS, Thomas L. Saaty. Emphasizes practical solution of problems; covers seven types of equations. ". . . a welcome contribution to the existing literature...."–*Math Reviews.* 490pp. 5⅜ x 8½. 0-486-64232-1

MATRICES AND LINEAR ALGEBRA, Hans Schneider and George Phillip Barker. Basic textbook covers theory of matrices and its applications to systems of linear equations and related topics such as determinants, eigenvalues and differential equations. Numerous exercises. 432pp. 5⅜ x 8½. 0-486-66014-1

LINEAR ALGEBRA, Georgi E. Shilov. Determinants, linear spaces, matrix algebras, similar topics. For advanced undergraduates, graduates. Silverman translation. 387pp. 5⅜ x 8½. 0-486-63518-X

ELEMENTS OF REAL ANALYSIS, David A. Sprecher. Classic text covers fundamental concepts, real number system, point sets, functions of a real variable, Fourier series, much more. Over 500 exercises. 352pp. 5⅜ x 8½. 0-486-65385-4

SET THEORY AND LOGIC, Robert R. Stoll. Lucid introduction to unified theory of mathematical concepts. Set theory and logic seen as tools for conceptual understanding of real number system. 496pp. 5⅜ x 8¼. 0-486-63829-4

TENSOR CALCULUS, J.L. Synge and A. Schild. Widely used introductory text covers spaces and tensors, basic operations in Riemannian space, non-Riemannian spaces, etc. 324pp. 5⅜ x 8¼. 0-486-63612-7

ORDINARY DIFFERENTIAL EQUATIONS, Morris Tenenbaum and Harry Pollard. Exhaustive survey of ordinary differential equations for undergraduates in mathematics, engineering, science. Thorough analysis of theorems. Diagrams. Bibliography. Index. 818pp. 5⅜ x 8½. 0-486-64940-7

INTEGRAL EQUATIONS, F. G. Tricomi. Authoritative, well-written treatment of extremely useful mathematical tool with wide applications. Volterra Equations, Fredholm Equations, much more. Advanced undergraduate to graduate level. Exercises. Bibliography. 238pp. 5⅜ x 8½. 0-486-64828-1

FOURIER SERIES, Georgi P. Tolstov. Translated by Richard A. Silverman. A valuable addition to the literature on the subject, moving clearly from subject to subject and theorem to theorem. 107 problems, answers. 336pp. 5⅜ x 8½. 0-486-63317-9

INTRODUCTION TO MATHEMATICAL THINKING, Friedrich Waismann. Examinations of arithmetic, geometry, and theory of integers; rational and natural numbers; complete induction; limit and point of accumulation; remarkable curves; complex and hypercomplex numbers, more. 1959 ed. 27 figures. xii+260pp. 5⅜ x 8½. 0-486-63317-9

POPULAR LECTURES ON MATHEMATICAL LOGIC, Hao Wang. Noted logician's lucid treatment of historical developments, set theory, model theory, recursion theory and constructivism, proof theory, more. 3 appendixes. Bibliography. 1981 edition. ix + 283pp. 5⅜ x 8½. 0-486-67632-3

CALCULUS OF VARIATIONS, Robert Weinstock. Basic introduction covering isoperimetric problems, theory of elasticity, quantum mechanics, electrostatics, etc. Exercises throughout. 326pp. 5⅜ x 8½. 0-486-63069-2

THE CONTINUUM: A CRITICAL EXAMINATION OF THE FOUNDATION OF ANALYSIS, Hermann Weyl. Classic of 20th-century foundational research deals with the conceptual problem posed by the continuum. 156pp. 5⅜ x 8½. 0-486-67982-9

CHALLENGING MATHEMATICAL PROBLEMS WITH ELEMENTARY SOLUTIONS, A. M. Yaglom and I. M. Yaglom. Over 170 challenging problems on probability theory, combinatorial analysis, points and lines, topology, convex polygons, many other topics. Solutions. Total of 445pp. 5⅜ x 8½. Two-vol. set. Vol. I: 0-486-65536-9   Vol. II: 0-486-65537-7

INTRODUCTION TO PARTIAL DIFFERENTIAL EQUATIONS WITH APPLICATIONS, E. C. Zachmanoglou and Dale W. Thoe. Essentials of partial differential equations applied to common problems in engineering and the physical sciences. Problems and answers. 416pp. 5⅜ x 8½. 0-486-65251-3

THE THEORY OF GROUPS, Hans J. Zassenhaus. Well-written graduate-level text acquaints reader with group-theoretic methods and demonstrates their usefulness in mathematics. Axioms, the calculus of complexes, homomorphic mapping, p-group theory, more. 276pp. 5⅜ x 8½. 0-486-40922-8

Math–Decision Theory, Statistics, Probability

ELEMENTARY DECISION THEORY, Herman Chernoff and Lincoln E. Moses. Clear introduction to statistics and statistical theory covers data processing, probability and random variables, testing hypotheses, much more. Exercises. 364pp. 5⅜ x 8½. 0-486-65218-1

STATISTICS MANUAL, Edwin L. Crow et al. Comprehensive, practical collection of classical and modern methods prepared by U.S. Naval Ordnance Test Station. Stress on use. Basics of statistics assumed. 288pp. 5⅜ x 8½. 0-486-60599-X

SOME THEORY OF SAMPLING, William Edwards Deming. Analysis of the problems, theory and design of sampling techniques for social scientists, industrial managers and others who find statistics important at work. 61 tables. 90 figures. xvii +602pp. 5⅜ x 8½. 0-486-64684-X

LINEAR PROGRAMMING AND ECONOMIC ANALYSIS, Robert Dorfman, Paul A. Samuelson and Robert M. Solow. First comprehensive treatment of linear programming in standard economic analysis. Game theory, modern welfare economics, Leontief input-output, more. 525pp. 5⅜ x 8½. 0-486-65491-5

PROBABILITY: AN INTRODUCTION, Samuel Goldberg. Excellent basic text covers set theory, probability theory for finite sample spaces, binomial theorem, much more. 360 problems. Bibliographies. 322pp. 5⅜ x 8½. 0-486-65252-1

GAMES AND DECISIONS: INTRODUCTION AND CRITICAL SURVEY, R. Duncan Luce and Howard Raiffa. Superb nontechnical introduction to game theory, primarily applied to social sciences. Utility theory, zero-sum games, n-person games, decision-making, much more. Bibliography. 509pp. 5⅜ x 8½. 0-486-65943-7

INTRODUCTION TO THE THEORY OF GAMES, J. C. C. McKinsey. This comprehensive overview of the mathematical theory of games illustrates applications to situations involving conflicts of interest, including economic, social, political, and military contexts. Appropriate for advanced undergraduate and graduate courses; advanced calculus a prerequisite. 1952 ed. x+372pp. 5⅜ x 8½. 0-486-42811-7

FIFTY CHALLENGING PROBLEMS IN PROBABILITY WITH SOLUTIONS, Frederick Mosteller. Remarkable puzzlers, graded in difficulty, illustrate elementary and advanced aspects of probability. Detailed solutions. 88pp. 5⅜ x 8½. 65355-2

PROBABILITY THEORY: A CONCISE COURSE, Y. A. Rozanov. Highly readable, self-contained introduction covers combination of events, dependent events, Bernoulli trials, etc. 148pp. 5⅜ x 8¼. 0-486-63544-9

STATISTICAL METHOD FROM THE VIEWPOINT OF QUALITY CONTROL, Walter A. Shewhart. Important text explains regulation of variables, uses of statistical control to achieve quality control in industry, agriculture, other areas. 192pp. 5⅜ x 8½. 0-486-65232-7

Math–Geometry and Topology

ELEMENTARY CONCEPTS OF TOPOLOGY, Paul Alexandroff. Elegant, intuitive approach to topology from set-theoretic topology to Betti groups; how concepts of topology are useful in math and physics. 25 figures. 57pp. 5⅜ x 8½. 0-486-60747-X

COMBINATORIAL TOPOLOGY, P. S. Alexandrov. Clearly written, well-organized, three-part text begins by dealing with certain classic problems without using the formal techniques of homology theory and advances to the central concept, the Betti groups. Numerous detailed examples. 654pp. 5⅜ x 8½. 0-486-40179-0

EXPERIMENTS IN TOPOLOGY, Stephen Barr. Classic, lively explanation of one of the byways of mathematics. Klein bottles, Moebius strips, projective planes, map coloring, problem of the Koenigsberg bridges, much more, described with clarity and wit. 43 figures. 210pp. 5⅜ x 8½. 0-486-25933-1

THE GEOMETRY OF RENÉ DESCARTES, René Descartes. The great work founded analytical geometry. Original French text, Descartes's own diagrams, together with definitive Smith-Latham translation. 244pp. 5⅜ x 8½. 0-486-60068-8

EUCLIDEAN GEOMETRY AND TRANSFORMATIONS, Clayton W. Dodge. This introduction to Euclidean geometry emphasizes transformations, particularly isometries and similarities. Suitable for undergraduate courses, it includes numerous examples, many with detailed answers. 1972 ed. viii+296pp. 6⅛ x 9¼. 0-486-43476-1

PRACTICAL CONIC SECTIONS: THE GEOMETRIC PROPERTIES OF ELLIPSES, PARABOLAS AND HYPERBOLAS, J. W. Downs. This text shows how to create ellipses, parabolas, and hyperbolas. It also presents historical background on their ancient origins and describes the reflective properties and roles of curves in design applications. 1993 ed. 98 figures. xii+100pp. 6½ x 9¼. 0-486-42876-1

THE THIRTEEN BOOKS OF EUCLID'S ELEMENTS, translated with introduction and commentary by Sir Thomas L. Heath. Definitive edition. Textual and linguistic notes, mathematical analysis. 2,500 years of critical commentary. Unabridged. 1,414pp. 5⅜ x 8½. Three-vol. set.
Vol. I: 0-486-60088-2 Vol. II: 0-486-60089-0 Vol. III: 0-486-60090-4

SPACE AND GEOMETRY: IN THE LIGHT OF PHYSIOLOGICAL, PSYCHOLOGICAL AND PHYSICAL INQUIRY, Ernst Mach. Three essays by an eminent philosopher and scientist explore the nature, origin, and development of our concepts of space, with a distinctness and precision suitable for undergraduate students and other readers. 1906 ed. vi+148pp. 5⅜ x 8½. 0-486-43909-7

GEOMETRY OF COMPLEX NUMBERS, Hans Schwerdtfeger. Illuminating, widely praised book on analytic geometry of circles, the Moebius transformation, and two-dimensional non-Euclidean geometries. 200pp. 5⅝ x 8¼. 0-486-63830-8

DIFFERENTIAL GEOMETRY, Heinrich W. Guggenheimer. Local differential geometry as an application of advanced calculus and linear algebra. Curvature, transformation groups, surfaces, more. Exercises. 62 figures. 378pp. 5⅜ x 8½. 0-486-63433-7

History of Math

THE WORKS OF ARCHIMEDES, Archimedes (T. L. Heath, ed.). Topics include the famous problems of the ratio of the areas of a cylinder and an inscribed sphere; the measurement of a circle; the properties of conoids, spheroids, and spirals; and the quadrature of the parabola. Informative introduction. clxxxvi+326pp. 5⅜ x 8½.
0-486-42084-1

A SHORT ACCOUNT OF THE HISTORY OF MATHEMATICS, W. W. Rouse Ball. One of clearest, most authoritative surveys from the Egyptians and Phoenicians through 19th-century figures such as Grassman, Galois, Riemann. Fourth edition. 522pp. 5⅜ x 8½.
0-486-20630-0

THE HISTORY OF THE CALCULUS AND ITS CONCEPTUAL DEVELOP-MENT, Carl B. Boyer. Origins in antiquity, medieval contributions, work of Newton, Leibniz, rigorous formulation. Treatment is verbal. 346pp. 5⅜ x 8½.   0-486-60509-4

THE HISTORICAL ROOTS OF ELEMENTARY MATHEMATICS, Lucas N. H. Bunt, Phillip S. Jones, and Jack D. Bedient. Fundamental underpinnings of modern arithmetic, algebra, geometry and number systems derived from ancient civiliza-tions. 320pp. 5⅜ x 8½.
0-486-25563-8

A HISTORY OF MATHEMATICAL NOTATIONS, Florian Cajori. This classic study notes the first appearance of a mathematical symbol and its origin, the com-petition it encountered, its spread among writers in different countries, its rise to pop-ularity, its eventual decline or ultimate survival. Original 1929 two-volume edition presented here in one volume. xxviii+820pp. 5⅜ x 8½.
0-486-67766-4

GAMES, GODS & GAMBLING: A HISTORY OF PROBABILITY AND STATISTICAL IDEAS, F. N. David. Episodes from the lives of Galileo, Fermat, Pascal, and others illustrate this fascinating account of the roots of mathematics. Features thought-provoking references to classics, archaeology, biography, poetry. 1962 edition. 304pp. 5⅜ x 8½. (Available in U.S. only.)
0-486-40023-9

OF MEN AND NUMBERS: THE STORY OF THE GREAT MATHEMATICIANS, Jane Muir. Fascinating accounts of the lives and accom-plishments of history's greatest mathematical minds–Pythagoras, Descartes, Euler, Pascal, Cantor, many more. Anecdotal, illuminating. 30 diagrams. Bibliography. 256pp. 5⅜ x 8½.
0-486-28973-7

HISTORY OF MATHEMATICS, David E. Smith. Nontechnical survey from ancient Greece and Orient to late 19th century; evolution of arithmetic, geometry, trigonometry, calculating devices, algebra, the calculus. 362 illustrations. 1,355pp. 5⅜ x 8½. Two-vol. set.   Vol. I: 0-486-20429-4   Vol. II: 0-486-20430-8

A CONCISE HISTORY OF MATHEMATICS, Dirk J. Struik. The best brief his-tory of mathematics. Stresses origins and covers every major figure from ancient Near East to 19th century. 41 illustrations. 195pp. 5⅜ x 8½.   0-486-60255-9

Physics

OPTICAL RESONANCE AND TWO-LEVEL ATOMS, L. Allen and J. H. Eberly. Clear, comprehensive introduction to basic principles behind all quantum optical resonance phenomena. 53 illustrations. Preface. Index. 256pp. 5⅜ x 8½.　0-486-65533-4

QUANTUM THEORY, David Bohm. This advanced undergraduate-level text presents the quantum theory in terms of qualitative and imaginative concepts, followed by specific applications worked out in mathematical detail. Preface. Index. 655pp. 5⅜ x 8½.　0-486-65969-0

ATOMIC PHYSICS (8th EDITION), Max Born. Nobel laureate's lucid treatment of kinetic theory of gases, elementary particles, nuclear atom, wave-corpuscles, atomic structure and spectral lines, much more. Over 40 appendices, bibliography. 495pp. 5⅜ x 8½.　0-486-65984-4

A SOPHISTICATE'S PRIMER OF RELATIVITY, P. W. Bridgman. Geared toward readers already acquainted with special relativity, this book transcends the view of theory as a working tool to answer natural questions: What is a frame of reference? What is a "law of nature"? What is the role of the "observer"? Extensive treatment, written in terms accessible to those without a scientific background. 1983 ed. xlviii+172pp. 5⅜ x 8½.　0-486-42549-5

AN INTRODUCTION TO HAMILTONIAN OPTICS, H. A. Buchdahl. Detailed account of the Hamiltonian treatment of aberration theory in geometrical optics. Many classes of optical systems defined in terms of the symmetries they possess. Problems with detailed solutions. 1970 edition. xv + 360pp. 5⅜ x 8½. 0-486-67597-1

PRIMER OF QUANTUM MECHANICS, Marvin Chester. Introductory text examines the classical quantum bead on a track: its state and representations; operator eigenvalues; harmonic oscillator and bound bead in a symmetric force field; and bead in a spherical shell. Other topics include spin, matrices, and the structure of quantum mechanics; the simplest atom; indistinguishable particles; and stationary-state perturbation theory. 1992 ed. xiv+314pp. 6⅛ x 9¼.　0-486-42878-8

LECTURES ON QUANTUM MECHANICS, Paul A. M. Dirac. Four concise, brilliant lectures on mathematical methods in quantum mechanics from Nobel Prize-winning quantum pioneer build on idea of visualizing quantum theory through the use of classical mechanics. 96pp. 5⅜ x 8½.　0-486-41713-1

THIRTY YEARS THAT SHOOK PHYSICS: THE STORY OF QUANTUM THEORY, George Gamow. Lucid, accessible introduction to influential theory of energy and matter. Careful explanations of Dirac's anti-particles, Bohr's model of the atom, much more. 12 plates. Numerous drawings. 240pp. 5⅜ x 8½. 0-486-24895-X

ELECTRONIC STRUCTURE AND THE PROPERTIES OF SOLIDS: THE PHYSICS OF THE CHEMICAL BOND, Walter A. Harrison. Innovative text offers basic understanding of the electronic structure of covalent and ionic solids, simple metals, transition metals and their compounds. Problems. 1980 edition. 582pp. 6⅛ x 9¼.　0-486-66021-4

CATALOG OF DOVER BOOKS

HYDRODYNAMIC AND HYDROMAGNETIC STABILITY, S. Chandrasekhar. Lucid examination of the Rayleigh-Benard problem; clear coverage of the theory of instabilities causing convection. 704pp. 5⅜ x 8¼. 0-486-64071-X

INVESTIGATIONS ON THE THEORY OF THE BROWNIAN MOVEMENT, Albert Einstein. Five papers (1905–8) investigating dynamics of Brownian motion and evolving elementary theory. Notes by R. Fürth. 122pp. 5⅜ x 8½. 0-486-60304-0

THE PHYSICS OF WAVES, William C. Elmore and Mark A. Heald. Unique overview of classical wave theory. Acoustics, optics, electromagnetic radiation, more. Ideal as classroom text or for self-study. Problems. 477pp. 5⅜ x 8½. 0-486-64926-1

GRAVITY, George Gamow. Distinguished physicist and teacher takes reader-friendly look at three scientists whose work unlocked many of the mysteries behind the laws of physics: Galileo, Newton, and Einstein. Most of the book focuses on Newton's ideas, with a concluding chapter on post-Einsteinian speculations concerning the relationship between gravity and other physical phenomena. 160pp. 5⅜ x 8½. 0-486-42563-0

PHYSICAL PRINCIPLES OF THE QUANTUM THEORY, Werner Heisenberg. Nobel Laureate discusses quantum theory, uncertainty, wave mechanics, work of Dirac, Schroedinger, Compton, Wilson, Einstein, etc. 184pp. 5⅜ x 8½. 0-486-60113-7

ATOMIC SPECTRA AND ATOMIC STRUCTURE, Gerhard Herzberg. One of best introductions; especially for specialist in other fields. Treatment is physical rather than mathematical. 80 illustrations. 257pp. 5⅜ x 8½. 0-486-60115-3

AN INTRODUCTION TO STATISTICAL THERMODYNAMICS, Terrell L. Hill. Excellent basic text offers wide-ranging coverage of quantum statistical mechanics, systems of interacting molecules, quantum statistics, more. 523pp. 5⅜ x 8½. 0-486-65242-4

THEORETICAL PHYSICS, Georg Joos, with Ira M. Freeman. Classic overview covers essential math, mechanics, electromagnetic theory, thermodynamics, quantum mechanics, nuclear physics, other topics. First paperback edition. xxiii + 885pp. 5⅜ x 8½. 0-486-65227-0

PROBLEMS AND SOLUTIONS IN QUANTUM CHEMISTRY AND PHYSICS, Charles S. Johnson, Jr. and Lee G. Pedersen. Unusually varied problems, detailed solutions in coverage of quantum mechanics, wave mechanics, angular momentum, molecular spectroscopy, more. 280 problems plus 139 supplementary exercises. 430pp. 6½ x 9¼. 0-486-65236-X

THEORETICAL SOLID STATE PHYSICS, Vol. 1: Perfect Lattices in Equilibrium; Vol. II: Non-Equilibrium and Disorder, William Jones and Norman H. March. Monumental reference work covers fundamental theory of equilibrium properties of perfect crystalline solids, non-equilibrium properties, defects and disordered systems. Appendices. Problems. Preface. Diagrams. Index. Bibliography. Total of 1,301pp. 5⅜ x 8½. Two volumes. Vol. I: 0-486-65015-4 Vol. II: 0-486-65016-2

WHAT IS RELATIVITY? L. D. Landau and G. B. Rumer. Written by a Nobel Prize physicist and his distinguished colleague, this compelling book explains the special theory of relativity to readers with no scientific background, using such familiar objects as trains, rulers, and clocks. 1960 ed. vi+72pp. 5⅜ x 8½. 0-486-42806-0

CATALOG OF DOVER BOOKS

A TREATISE ON ELECTRICITY AND MAGNETISM, James Clerk Maxwell. Important foundation work of modern physics. Brings to final form Maxwell's theory of electromagnetism and rigorously derives his general equations of field theory. 1,084pp. 5⅜ x 8½. Two-vol. set. Vol. I: 0-486-60636-8 Vol. II: 0-486-60637-6

QUANTUM MECHANICS: PRINCIPLES AND FORMALISM, Roy McWeeny. Graduate student-oriented volume develops subject as fundamental discipline, opening with review of origins of Schrödinger's equations and vector spaces. Focusing on main principles of quantum mechanics and their immediate consequences, it concludes with final generalizations covering alternative "languages" or representations. 1972 ed. 15 figures. xi+155pp. 5⅜ x 8½. 0-486-42829-X

INTRODUCTION TO QUANTUM MECHANICS With Applications to Chemistry, Linus Pauling & E. Bright Wilson, Jr. Classic undergraduate text by Nobel Prize winner applies quantum mechanics to chemical and physical problems. Numerous tables and figures enhance the text. Chapter bibliographies. Appendices. Index. 468pp. 5⅜ x 8½. 0-486-64871-0

METHODS OF THERMODYNAMICS, Howard Reiss. Outstanding text focuses on physical technique of thermodynamics, typical problem areas of understanding, and significance and use of thermodynamic potential. 1965 edition. 238pp. 5⅜ x 8½. 0-486-69445-3

THE ELECTROMAGNETIC FIELD, Albert Shadowitz. Comprehensive undergraduate text covers basics of electric and magnetic fields, builds up to electromagnetic theory. Also related topics, including relativity. Over 900 problems. 768pp. 5⅝ x 8¼. 0-486-65660-8

GREAT EXPERIMENTS IN PHYSICS: FIRSTHAND ACCOUNTS FROM GALILEO TO EINSTEIN, Morris H. Shamos (ed.). 25 crucial discoveries: Newton's laws of motion, Chadwick's study of the neutron, Hertz on electromagnetic waves, more. Original accounts clearly annotated. 370pp. 5⅜ x 8½. 0-486-25346-5

EINSTEIN'S LEGACY, Julian Schwinger. A Nobel Laureate relates fascinating story of Einstein and development of relativity theory in well-illustrated, nontechnical volume. Subjects include meaning of time, paradoxes of space travel, gravity and its effect on light, non-Euclidean geometry and curving of space-time, impact of radio astronomy and space-age discoveries, and more. 189 b/w illustrations. xiv+250pp. 8⅜ x 9¼. 0-486-41974-6

STATISTICAL PHYSICS, Gregory H. Wannier. Classic text combines thermodynamics, statistical mechanics and kinetic theory in one unified presentation of thermal physics. Problems with solutions. Bibliography. 532pp. 5⅜ x 8½. 0-486-65401-X

Paperbound unless otherwise indicated. Available at your book dealer, online at **www.doverpublications.com**, or by writing to Dept. GI, Dover Publications, Inc., 31 East 2nd Street, Mineola, NY 11501. For current price information or for free catalogues (please indicate field of interest), write to Dover Publications or log on to **www.doverpublications.com** and see every Dover book in print. Dover publishes more than 500 books each year on science, elementary and advanced mathematics, biology, music, art, literary history, social sciences, and other areas.